Searching for ORDER IN the COMPLEXITY of Evolving Worlds ·

ACKNOWLEDGMENTS

The SFI Press would not exist without the support of William H. Miller and the Miller Omega Program.

[THE COMPLEX ALTERNATIVE]

Complexity Scientists on the COVID-19 Pandemic

DAVID C. KRAKAUER
GEOFFREY WEST
editors

SFI PRESS
THE SANTA FE INSTITUTE PRESS

1399 Hyde Park Road
Santa Fe, New Mexico 87501

The Complex Alternative:
Complexity Scientists on the COVID-19 Pandemic

ISBN (HARDCOVER): 978-1-947864-40-5
Library of Congress Control Number: 2021948512

The SFI Press is supported by the
Miller Omega Program.

IN NATURE we never see anything isolated, but everything in connection with something else which is before it, beside it, under it, and over it.

— JOHANN WOLFGANG VON GOETHE
Sämtliche Werke (1850)

CONTENTS

(2)

COVID-19 & COMPLEX TIME IN BIOLOGY & ECONOMICS

[Batch 2, released 6 April 2020]

(3)

ON CORONAVIRUS, CRISIS & CREATIVE OPPORTUNITY

[Batch 3, released 13 April 2020]

CONTRIBUTORS

Danielle Allen, *Harvard University*

Lauren Ancel Meyers, *University of Texas at Austin; Santa Fe Institute*

Mahzarin Banaji, *Harvard University; Santa Fe Institute*

Caroline Buckee, *T.H. Chan School of Public Health, Harvard University*

Michelle C. Carlson, *The Johns Hopkins University School of Medicine*

Carrie Cowan, *Santa Fe Institute*

Casey Cox, *Santa Fe Institute*

Stefani Crabtree, *Utah State University; Santa Fe Institute*

Simon DeDeo, *Carnegie Mellon University; Santa Fe Institute*

Andrew Dobson, *Princeton University; Santa Fe Institute*

J. Doyne Farmer, *University of Oxford; Santa Fe Institute*

Anthony Eagan, *Santa Fe Institute*

Santiago F. Elena, *Spanish National Research Council; Santa Fe Institute*

Brian Enquist, *University of Arizona; Santa Fe Institute*

Douglas H. Erwin, *Smithsonian Institution; Santa Fe Institute*

Susan Fitzpatrick, *James S. McDonnell Foundation; Santa Fe Institute*

Jessica C. Flack, *Santa Fe Institute*

Stephanie Forrest, *Arizona State University Biodesign Institute; Santa Fe Institute*

Spencer Fox, *University of Texas at Austin*

Miguel Fuentes, *Santa Fe Institute*

Mirta Galesic, *Santa Fe Institute*

Michael Garfield, *Santa Fe Institute*

Gerd Gigerenzer, *Max Planck Institute for Human Development, Berlin*

— CONTINUED ON NEXT PAGE —

CONTRIBUTORS (CONTINUED)

Amos Golan, *American University; Santa Fe Institute; Pembroke College, Oxford University*

John Harte, *University of California, Berkeley; Santa Fe Institute*

Laurent Hébert-Dufresne, *Vermont Complex Systems Center*

Ricardo Hausmann, *Center for International Development, Harvard University; Santa Fe Institute*

Luu Hoang Duc, *Max Planck Institute for Mathematics in the Sciences, Leipzig*

Michael Hochberg, *University of Montpellier, France; Santa Fe Institute*

Jürgen Jost, *Max Planck Institute for Mathematics in the Sciences, Leipzig; Santa Fe Institute*

Christopher P. Kempes, *Santa Fe Institute*

David Kinney, *Santa Fe Institute*

David C. Krakauer, *Santa Fe Institute*

John W. Krakauer, *The Johns Hopkins University School of Medicine; Santa Fe Institute*

Michael Lachmann, *Santa Fe Institute*

Manfred D. Laubichler, *Arizona State University; Santa Fe Institute*

Ramanan Laxminarayan, *Center for Disease Dynamics, Economics & Policy, Princeton University*

Simon Levin, *Princeton University; Santa Fe Institute*

Ian MacGregor-Fors, *University of Helsinki*

Jon Machta, *University of Massachusetts Amherst; Santa Fe Institute*

Eric Maskin, *Harvard University*

William H. (Bill) Miller III, *Miller Value Partners*

Melanie Mitchell, *Portland State University; Santa Fe Institute*

Cristopher Moore, *Santa Fe Institute*

Melanie Moses, *University of New Mexico; Santa Fe Institute*

Suresh Naidu, *Columbia University; Santa Fe Institute*

Henrik Olsson, *University of Warwick; Santa Fe Institute*

Kathy Powers, *University of New Mexico; Santa Fe Institute*

Sidney Redner, *Santa Fe Institute*

Dan Rockmore, *Dartmouth College; Santa Fe Institute*

Van Savage, *University of California, Los Angeles; Santa Fe Institute*

Samuel V. Scarpino, *The Rockefeller Foundation; Santa Fe Institute; Vermont Complex Systems Center; Northeastern University*

Marten Scheffer, *Wageningen University; Santa Fe Institute*

Rajiv Sethi, *Barnard College; Columbia University; Santa Fe Institute*

Lenny Smith, *London School of Economics*

Gillian Tett, *Financial Times*

William Tracy, *Santa Fe Institute*

David Tuckett, *University College London*

Geoffrey West, *Santa Fe Institute*

E. Glen Weyl, *Microsoft Office of the Chief Technology Officer; RadicalxChange Foundation*

David Wolpert, *Santa Fe Institute*

Pamela Yeh, *University of California, Los Angeles; Santa Fe Institute*

PRODUCTION ASSISTANCE

Chapter Art Curator: Caitlin L. McShea

Copyeditors: Joy Drohan and Rachel Fudge

Glossary Writer: Jennie Dusheck

Symposium Materials Wrangler: Kayla Savard

INTRODUCTION

THE COMPLEX ALTERNATIVE

DATE: *30 September 2021*

FROM: *David Krakauer, Santa Fe Institute*
Geoffrey West, Santa Fe Institute

The fact would seem to be, if in my situation one may speak of facts, not only that I shall have to speak of things of which I cannot speak, but also, which is even more interesting, but also that I, which is if possible even more interesting, that I shall have to, I forget, no matter. And at the same time I am obliged to speak. I shall never be silent. Never . . . . The thing to avoid, I don't know why, is the spirit of system. People with things, people without things, things without people, what does it matter, I flatter myself it will not take me long to scatter them, whenever I choose, to the winds.

—SAMUEL BECKETT, *The Unnamable* (1958)

COMPLEXITY ANXIETY

In this book we shall speak of facts and, where possible, what Samuel Beckett called "the spirit of systems," something that he described in his novel *The Unnamable* as "the thing to avoid." It sometimes seems as if the entire world has heeded The Unnamable's advice and gone out of its way to avoid *systems* where the implications of following the multitudinous connections are too difficult or too grave to contemplate. In contrast, there is often great comfort, reassurance, and even relief in embracing *simplicity*. But there can also be a certain naiveté in the excessive elegance that sometimes accompanies simplicity. And of greater concern is the danger in the desire for expediency—to explain every event or phenomenon as if it had a single cause and to flatten reality

---

OPPOSITE: *Giovanni Battista Piranesi, "The Round Tower," 1761 (detail)*

into a linear chain-like narrative whose future is as predictable as its past is comprehensible.

In late 2019 a new virus was unleashed upon the world. Not entirely new, since it was a coronavirus rather like hundreds of others in circulation around the world at that time. These viruses typically emerge zoonotically, that is, they jump a species barrier into humans from another species. But unlike other related viruses, this one achieved a global spread with near cataclysmic social, economic, political, and psychological impact. The contributions in this book—with articles and interviews from March of 2020 through September of 2021—seek to understand and explore how this happened, its multidimensionality, and its many lessons and consequences. Although they are grounded in scientific research and scholarship by a community of complexity scientists, the contents are nontechnical, provocative, and intended for a general audience.

In order to understand in part what went so wrong in 2020, we need to investigate the enduring allure of the *simple*.

The Santa Fe Institute has endeavored to present an alternative, complementary perspective to conventional thinking about the pandemic in simple terms. In contrast to Samuel Beckett, we do not avoid the spirit of the system; we embrace it. Simplicity wants to reduce the multidimensional complexity of the pandemic to one or two simple factors, such as: treating it as a bounded epidemic to be eradicated by simply getting the infamous R_0 below 1, or by simple behavioral and psychological denial as practiced by anti-vaxxers, or by adopting questionable remedies with no proven efficacy, or by achieving safety and prosperity through total isolation, and so on—the whole panoply of simple one-size-fits-all approaches for communities and nations. Every one of these factors or explanations—and many more—represents an interactive, interdependent component of the complex, systemic phenomenon that we call COVID-19. We ignore this essential multicomponent interdependency at our peril.

This book presents our community's research record as a real-time response to the last two years. The work disavows simple explanations and solutions in favor of a concerted effort to come to terms with the whole matrix of the pandemic and all its messy parts. It does not present a fully coherent position, and the community is far from reaching consensus about what might be done to prevent another like catastrophe in the future. But many of the elements of what should be part of a future response are present in these contributions. In order to understand in part what went so wrong in 2020, we need to investigate the enduring allure of the *simple*.

SIMPLICITY SYNDROME

The great revolutions in mathematics and science have for the most part been simplicity rebellions. The invention of complex numbers transformed the great mass of insoluble polynomial equations into those that can be solved using complex roots. The Greeks described the heart as both hearthstone and soul, and then William Harvey showed it was just a pump. Aristotle conceived of a unique law of creation—spontaneous generation—for every organism smaller than an insect until Francesco Redi demonstrated that every visible lifeform is just another reproductive generation. And the theory of plate tectonics was able to complete the complicated jigsaw made from the lithosphere by providing a convective mechanism for continental drift. Natural science has made great progress by taking conceptual jumbles and revealing them to be tangles comprising a small number of simple underlying analytical threads.

The Greeks described the heart as both hearthstone and soul, and then William Harvey showed it was just a pump.

This approach, which can be characterized as parsimonious, minimal, or economical, has proven itself to be predictive, extraordinarily powerful, and very often intuitive. There are certainly domains where parsimony is far from intuitive, as, for example, the enigmatic mathematics and interpretation of quantum mechanics, but this lack of transparency is more often than not compensated for by a significant increase in predictive capacity.

But there are areas where theory has failed to make predictive contributions comparable in scope and precision to those achieved in the sciences of physics, chemistry, and geology. These are all areas dominated by living processes, from simple cells up through to complicated assemblies of organisms in societies. There is something about life that resists—at least at first blush—the attempt to unravel its various knots into a few simple threads. The reasons for this are by now well known:

1. Whereas non-living matter is dominated by symmetries and precision, life is pervaded by broken symmetries and/or apparent accidents of history;
2. Energy and conservation laws dominate physics but information is coeval and integral in life, and it is not subject to similar constraints;
3. All life is contextual and needs to be understood in functional terms, whereas physical reality exhibits minimal or zero agency; and
4. Complex phenomena retain multiple causes or complex causality, whereas physical phenomena are dominated by a small number of causes, and in that sense, can be deemed simple.

For all these reasons and more, parsimonious theories, though oftentimes useful for a coarse-grained understanding, fail to account for a high degree of detailed microscopic behavior in the living world.

In our attempt to understand the messy world around us, a threefold response has emerged. Firstly, society is loath to forfeit simplicity and its impressive record of achievement, leading to a variety of faux parsimonies. These include superstitions, pseudoscience, and largely irrefutable but plausible-sounding narrative explanations. Secondly, society has turned to statistics, machine learning, and algorithmic science to buttress human cognitive limitations. These generate predictions based on analyses and extrapolations of large datasets that remain hard for humans to comprehend. And thirdly, science has pursued new, nondisciplinary frameworks that operate at more coarse-grained, low-resolution or average levels of explanation. These deal explicitly with living processes—the constraints and processes acting on living histories that structure adaptive information, energy, and matter at multiple interconnected scales.

This last member of the triumvirate—partly implied by the title of this volume—is what we call complexity science. It almost goes without saying that combining the second and third approaches can be very powerful, whereas the first is always a mistake.

COMPLEX SYSTEMS RHYME

The contingencies that pervade complex systems were for a long time thought to be insurmountable, implying perhaps that general theories for complex systems would not be forthcoming. After all, science is the pursuit of quantitative regularity residing in multiplicity. The proliferation of disciplines—which by and large implies focus on a given system and at a preferred scale—reflects the idea that deep insights have rarely been very general or multiscale and only become evident upon repeated measurements within a rather narrow set of observations.

After all, science is the pursuit of quantitative regularity residing in multiplicity.

There is plenty of evidence to suggest that being disciplinary and reductionistic is prudent. The theory of particle physics at a microscopic scale provides virtually no predictions into macroscopic regimes where condensed matter physics operates, and similarly, genetics has contributed rather little to the understanding of cognitive mechanisms. This list could be extended to include just about any area that is dominated by emergent regularities, meaning that features are best described, predicted, and explained using domain-specific effective theories at the appropriate scale, without having to recruit the principles that describe, predict, and explain features from another domain or significantly different scale.

Of course, there have been some very notable violations of this pattern, including the theories of gravity and electromagnetism and the theory of evolution. And at the boundaries of inquiry, theories and models tend to melt together. Single-cell physiology and circuit-based neuroscience have significant overlaps, as do developmental genetics and models of morphogenesis.

In fact, the whole area of inquiry described as applied mathematics can in principle be of value in any area in which repeated observations give rise to regular patterns

But there are frameworks and theories, and certainly techniques, that span many different scales and disciplines. There are those that cleave very closely to mathematics and methods. For instance, calculus and probability theory are used by many disciplines, reflecting the fact that most systems are dynamical systems evolving in time and most observations are partial or imperfect. In fact, the whole area of inquiry described as applied mathematics can in principle be of value in any area in which repeated observations give rise to regular patterns.

Most intriguingly, there are features of complex systems that appear to be general and that emerge from universal adaptive requirements. These include optimization under constraints of growth and repair, the use of energy to store and process information, strategic mechanisms of competition and cooperation, traits that promote robustness and evolvability, and mechanisms of communication and computation. This list applies just as profoundly to cells as to whole organisms, ecologies, cities, economies, and societies. The study of complex systems is in large part the inquiry into these unifying multiscale characteristics shared by all evolved multiscale organizations, even though they manifest in strikingly different ways.

And this is the broad sense in which our community has applied itself to the COVID-19 pandemic. We asked

- *How does competition between virus and host spill over into cooperation and competition in society as a whole?*

- *How do human mechanisms of cultural evolution that might mitigate transmission (e.g., isolation and mask use) diminish or even accelerate organic evolution of the pathogen (e.g., promote resistance mutations)?*
- *How do the economic advantages of urban life inadvertently accelerate the rate of transmission of a virus? and*
- *How do technologies that communicate disease incidence also, and often more effectively, communicate falsehoods and fake remedies?*

These are some of the questions that arise naturally when considering the pandemic from a complexity perspective—questions that respect the idea that space and timescales are dynamically intertwined, that any intervention involves the consideration of trade-offs, and that every optimization is a constrained optimization. The more we can make complex-systems science, with its inherent holistic, integrated perspective, part of a societal response to a crisis, the less vulnerable we shall be to unintended consequences and future simplistic ideas and policies that in the long run threaten the wellbeing and ultimately the long-term sustainability of our planet.

TRANSMITTING THE COVID-19 CRISIS

This book is structured into three sections, each of which is chronologically ordered.

In March of 2020, we began our Transmission series, which sought to introduce ideas from complexity science—including why systems collapse—the nature of an evolving virus and its ecology, how networks spread disease and economic instability, the mathematics of modeling outbreaks, the way decision-making modifies disease spread, and many other ideas that touch on the disease. The series was organized into batches, each of which was then capped with a podcast interview between David Krakauer and SFI's *Complexity* podcast host, Michael Garfield. In 2021 we asked the authors of the Transmissions to comment on their earlier articles, concentrating on what they either successfully had understood or, perhaps more importantly, had failed to anticipate in their thinking. These contributions constitute Part One.

In Part Two, we include faculty-contributed essays and opinion pieces covering the pandemic, plus podcast interviews with members of our community who were working on the front lines of the pandemic in the spring of 2020. These are longer-form contributions that cover a topic in more depth than were permitted by the constraints of the Transmission pieces.

And in Part Three, we feature the proceedings of SFI's November 2020 Applied Complexity Network symposium on "The Complexity of Crisis." This part includes talks and a panel discussion among mem-

bers of SFI's faculty, followed by questions and answers.

This book is designed to be a real-time archive of the events of the COVID-19 pandemic as they unfolded throughout the world and within our community. We all felt compelled to contribute useful science to the global debate and, when possible, to keep as many ideas in play before going down policy paths that might disregard the needs of diverse demographics and communities or pursue short-sighted objectives in their desire for rhetorical and executive simplicity. Ultimately, as scientists in society, our job is to promote understanding, improve standards of living, and present sensible options in difficult circumstances.

All of these are at best only as effective as our ability to relay them. This is a record of SFI's efforts to communicate imperfectly through the chaos.

THE STRUCTURE OF THIS BOOK

Introduction

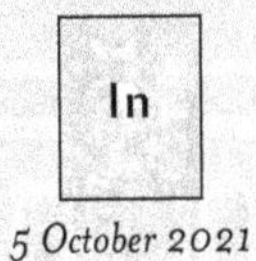

5 October 2021

Part 1: Transmissions

Transmissions ∘ Reflections ∘ Podcast

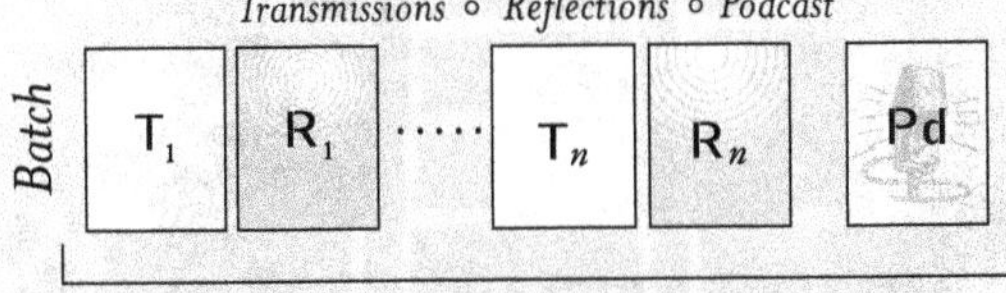

20 March 2020–14 December 2020

Part 2: Interviews & Essay

Interviews ∘ Essays

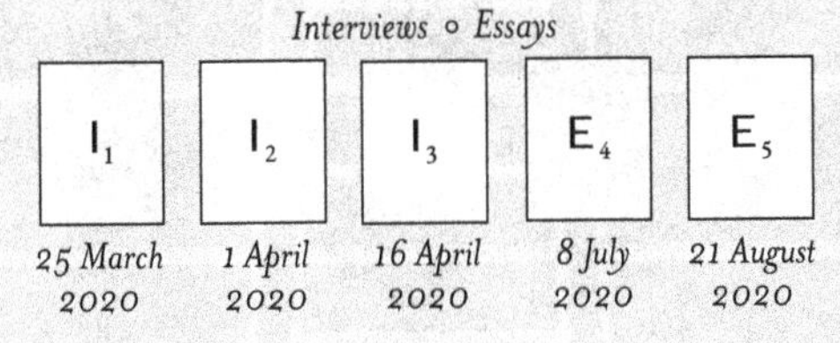

Part 3: Crisis

Talks ∘ Panel Discussion

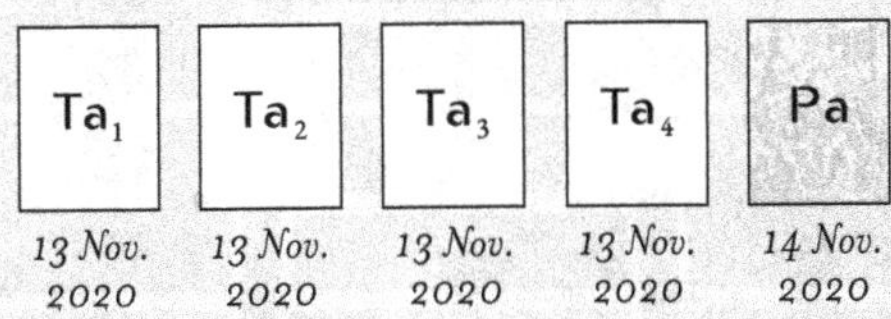

13 Nov. 2020 · *13 Nov. 2020* · *13 Nov. 2020* · *13 Nov. 2020* · *14 Nov. 2020*

Part One: Transmissions

1

[BATCH 1, RELEASED 30 MARCH 2020]

We use our understanding

of transmission to our

to **MOBILIZE** *the*

transmission network the

TECHNOLOGIES

enable the collective action

transmission of the

of the common factor advantage: continue largest information-world has ever seen—our of communication—to needed to **ELIMINATE VIRUS.**

—DAVID KRAKAUER, Transmission 000

TRANSMISSION NO. 000

CITIZEN-BASED MEDICINE

DATE: *30 March 2020*

FROM: *David Krakauer, Santa Fe Institute*

STRATEGIC INSIGHT: *By using transmission to our advantage, we can eliminate coronavirus through citizen-based medicine.*

Imagine a world where one had the opportunity to prevent cancer. And that this involved no medication and could be developed by every one of us without any special training, entirely from home. And that by preventing the disease for several months we would provide researchers the window of time required to develop a treatment, and doctors, nurses, and hospitals, the relief to effectively deploy it across the global population.

I suspect that we would all stay at home, strive as best as we could to remain productive, and thereby become an active part in one of the greatest prophylactic achievements in the history of public health. The Nobel Prize for medicine would be justly awarded to all the citizens of the world.

This is an impossible scenario for cancer. It is an impossible scenario for effectively all of the top twenty causes of deadly disease in the world. For heart disease, cancer, stroke, and Alzheimer's we have at best a rather patchy understanding of their origins, how they cause illness, and how we might treat them.

Each of these diseases is correlated to different degrees with our genetics, behavioral habits, social systems, economies, and ecosystems. For example, heart disease has a strong genetic component; it is highly dependent on our diets and addictions and our behavior—particularly how active we are. The same factors have an impact on cancer but with

---

OPPOSITE: *Edward Hopper, "Office in a Small City," 1953*

a stronger influence from genetics and environmental factors and conditions. And these factors feed back on one another to make isolating a single optimum point of intervention nearly impossible. This is what we describe as Complex Causality. And it makes prevention and treatment of disease very hard.

But the world that we can only dream of for cancer, and for most of the top causes of mortality, is a reality for COVID-19—we will have a treatment within a couple of years or less and it will work. So why the huge system shock with COVID-19? It has to do with a rather amplified property of causality that reaches out beyond the disease to touch the complex systems of the world.

Unlike with these other diseases, in the case of COVID-19 there is a rather unique convergence of causes that reestablishes a kind of simple causality, and these causes are transmission networks. The virus is transmitted initially from animal hosts to humans, typically through diet. Humans then transmit the virus to other humans by contact. These contacts are then transmitted through our transport systems and professional and social lives. This is perhaps the principal reason why the markets and society have been so volatile; the shared factor of transmission is so integral to modern life, it is to such a great extent the foundation of the modern world, that it touches nearly every factor of production. Just as monocultures can generate lethal simplicity in agriculture, transmission has generated lethal simplicity across the globe.

Strategic isolation is our antiviral flash-anti-mob.

But there is a flip side to this entanglement of complex systems: transmission, unlike the complexity of genetics, and social systems, economies, and ecosystems, can be relatively easily understood, and, by extension, controlled. By following the few simple behavioral rules that we all have come to know well—quarantine, maintaining social distance in public, practicing appropriate hygiene, and developing new habits for home-work when possible—every citizen plays a meaningful and significant part in eliminating this scourge.

We use our understanding of the common factor of transmission to our advantage: continue to mobilize the largest information-transmission network the world has ever seen—our technologies of communication—to enable the collective action needed to eliminate the transmission of the virus. Strategic isolation is our antiviral flash-anti-mob.

And we recognize the extraordinary economic sacrifices that are being made to make citizen-based medicine a reality—position economic relief as fair sharing in the reward for the unprecedented scale of teamwork required to rid the world of a terrible disease. If we can transmit insight at the speed of light, then we should do the same for compassion and support. If the economy is going to rebound anywhere near as fast as it declined, we need to understand the complex nature of transmission, in aligning emotion, reason, science, policy, and economies toward recovery as effectively as these alignments produced collapse.

By using transmission to our advantage, we can control coronavirus through citizen-based medicine.

REFLECTION

PREVENTATIVE CITIZEN-BASED MEDICINE

DATE: *5 October 2021*

FROM: *David Krakauer, Santa Fe Institute*

In my Transmission, I proposed what in retrospect proved to be an overly optimistic approach to the pandemic. The premise was that the alignment of transmission mechanisms promoting viral outbreak, spread, and growth should be turned against the virus. That human collective action could work as a countertransmission mechanism. A large suite of relatively cheap and rational actions, including isolation, masking, and remote work, would reduce the threat of the pandemic. Such inexpensive and readily available behavioral remedies would do nothing in our efforts to reduce the incidence of diseases like cancer or Alzheimer's but could accomplish a great deal in moderating a pandemic. And, unlike these diseases against which individual citizens often feel powerless and passive, in the case of COVID-19, everyone on the planet would become an active mechanism of defense against further infection.

We have witnessed that many citizens of the world understood they were a part of such a solution. However, trends quickly emerged that made truly effective collective-action measures a challenge. Men proved to be more resistant to using masks in indoor spaces than women (but no comparable difference was found for outdoor mask use). And while time spent watching COVID-19 news was associated with increased social distancing, several studies found contrary correlations between political affiliation, mask use, and a willingness to isolate.[1]

It is very difficult to come to terms with the unanticipated ways in which such a simple form of viral life morphed into chimerical ideology when encountering the complex economic and political doctrines of our time. It is true that history has shown us how even unequivocally positive steps have been met with resistance and criticism. Take, for instance, hand hygiene in health care. After Ignaz Semmelweis demonstrated that

---

1 J.A. Gette, A. K. Stevens, et al., 2021, "Individual and COVID-19-Specific Indicators of Compliance with Mask Use and Social Distancing: The Importance of Norms, Perceived Effectiveness, and State Response," *International Journal of Environmental Research and Public Health* 18 (16) doi: 10.3390/ijerph18168715

antiseptic agents could significantly reduce the transmission of germs in obstetric clinics, there was almost no adoption of his recommendation. This has been attributed to his failure to consult with colleagues and gave rise to the modern "recognize–explain–act" approach to successful infection control intervention.[2] But this kind of failure is certainly not the whole story for COVID-19.

In a recent review of the pandemic in connection to political polarization in the UK, Vlandas and Klymak conclude,

> *The first death had a negative effect on the perceptions of government handling of health among both Labour and Conservative voters, while Boris Johnson's hospitalisation improved perception among most voters for both dimensions. Lockdown improved perception of health handling but at the cost of lower perceptions of handling of economy among Conservative voters. It further led to a convergence in views about government handling of the economy towards a lower level among both Conservatives and Labour voters, whereas it led to partisan divergence in perceptions of its handling of health, despite leading to more positive views for all voter groups.*[3]

This complex causality makes a Rube Goldberg machine look as straightforward as a spoon. It would seem to be impossible to predict such oscillations and affinities of belief. Certainly, increased public morbidity might elicit distrust in government, but a rise in trust across the board based on the hospitalization of a prime minister who declared that he had "shaken hands with everybody" is less comprehensible.

In conclusion, I still believe in my original premise. I just no longer believe that society can act rationally under pressure. This places a great burden on science to be readily available and communicated well in times of stability. In other words, citizen-based medicine needs to become preventive medicine rather than treatment.

---

2 World Health Organization (WHO), 2009, *WHO Guidelines on Hand Hygiene in Health Care*, Genève, Switzerland: World Health Organization.

3 T. Vlandas and M. Klymak, 2021, "Pandemic and Partisan Polarisation: Voter Evaluation of UK Government Handling during Covid-19," *Swiss Political Science Review* 27 (2): 325–38, doi: 10.1111/spsr.12457

TRANSMISSION NO. 001

COMPLEX CRISES & THE INEVITABILITY OF ETHICALLY LOADED DECISIONS

DATE: *30 March 2020*

FROM: *David Kinney, Santa Fe Institute*

STRATEGIC INSIGHT: *In a complex crisis, scientists cannot avoid making value judgments.*

The rapidly unfolding COVID-19 pandemic has brought the interface between scientists and policymakers directly into the public eye. Examples include the role of Anthony Fauci in daily White House press briefings and the impact of a report by Neil Ferguson and his Imperial College London colleagues on decisions by several national governments. One might hope that this ongoing interaction between scientists and policymakers would respect a certain division of labor. Policy-facing scientists would provide politicians with decision-relevant facts, and in turn politicians would make decisions that require them to assess the value to society of different possible policy outcomes. As clean and compelling as this division of labor is, I don't believe that it is achievable, especially when dealing with a system as complex as a pandemic. In responding to this crisis, scientists must embrace the fact that they are being called upon to make ethically loaded decisions, including in cases where this may not be immediately obvious.

The idea that science ought to be free of value judgments has a rich history. As early as the nineteenth century, W.E.B. Du Bois (1898, cf. Bright 2018) argued that public trust in science could only be preserved if science was insulated from social and political concerns. In the twentieth century, the decision theorists Richard Jeffrey (1956) and Isaac Levi (1960) put forward mathematically precise frameworks for formalizing

---

OPPOSITE: *Pieter Brueghel the Younger, "The Village Lawyer's Office," 1626 (detail)*

the division of labor between scientists and policymakers. According to them, scientists should provide policymakers with an empirically supported assignment of probabilities to different relevant outcomes under a set of policy alternatives. It is then incumbent upon policymakers to do the value-laden work of evaluating the desirability and probability of each outcome under each policy, and formulating a decision rule that outputs an optimal policy. In the context of the current crisis, this division of labor would proceed as follows. Scientists would provide policymakers with an assignment of probabilities to the various possible public health and economic consequences of policies such as extreme social distancing, gradual de-quarantining, and the isolation of vulnerable populations. Elected policymakers would then use these probabilities, along with their own normative judgments, to arrive at a decision as to the optimal policy.

Scientists rarely possess an evidence base that allows them to be confident in a single assignment of probabilities to different possible outcomes.

However, as the Australian National University's Katie Steele argues in a 2012 paper, scientists rarely possess an evidence base that allows them to be confident in a single assignment of probabilities to different possible outcomes. Much more often, and especially in the face of significant uncertainty about the behavior of a system, the best that scientists can offer is a range of probabilities that a given outcome will occur. Here, scientists face a clear tradeoff. Wide ranges are much more likely to be correct, but can offer limited guidance to policy makers. Narrow ranges facilitate political decision-making, but are more likely to be wrong. Thus, when scientists decide how to report results to policymakers, they have to balance the need for action-guiding advice against the risk of their advice being wrong. These are value-laden decisions that cannot be outsourced to policymakers. Thus, as politicians continue to call on the expertise of scientists in order to respond to the current pandemic, scientists must embrace the fact that they are being asked to make ethical decisions. This may not be the ideal role for a scientist, but it is one that each epidemiologist, virologist, economist, and anyone else in a position to provide scientific advice to policymakers finds themselves in, like it or not. Likewise, the public must accept that even though scientific policy advisors have not been popularly elected, we have no choice but to grant them a certain level of value-laden decision-making power, or else abandon the idea of scientifically informed policymaking entirely.

REFERENCES

Bright, L.K. 2018. "Du Bois's Democratic Defence of the Value Free Ideal." *Synthese*, 195(5): 2227-2245, doi: 10.1007/s11229-017-1333-z

Du Bois, W.B. 1898. "The Study of the Negro Problems." *The Annals of the American Academy of Political and Social Science*, 1-23.

Jeffrey, R.C. 1956. "Valuation and Acceptance of Scientific Hypotheses." *Philosophy of Science*, 23(3): 237-246, doi: 10.1086/287489

Levi, I. 1960. "Must the Scientist Make Value Judgments?" *The Journal of Philosophy*, 57(11): 345-357, doi: 10.2307/2023504

Steele, K. 2012. "The Scientist qua Policy Advisor Makes Value Judgments." *Philosophy of Science*, 79(5): 893-904, doi: 10.1086/667842

REFLECTION

WHY WE CAN'T DEPOLITICIZE A PANDEMIC

DATE: *28 July 2021*

FROM: *David Kinney, Santa Fe Institute*

In my 2020 Transmission, entitled "In a Complex Crisis, Scientists Cannot Avoid Making Value Judgments," I stressed that scientists offering solutions aimed at mitigating the COVID-19 pandemic must recognize that their recommendations have an unavoidable ethical dimension. Since then, mitigation strategies for COVID-19 (e.g., masks, social distancing, or vaccines) have become deeply politicized. This is powerfully demonstrated by the fact that, as of July 2021, a map of US states indicating the percentage of adults in each state who have been vaccinated against COVID-19 very closely resembles the 2020 electoral map, with Biden-voting states tending to be more vaccinated and Trump-voting states having a lower percentage of vaccinated adults.[1]

In July 2021, Dr. Anthony Fauci called this political divide "unfortunate," and declared that US citizens needed to "put politics aside" to increase vaccination rates.[2] Indeed, we have good reason to believe that if Trump-voting adults were vaccinated at the rate of Biden-voting adults, then the current public health burden of COVID-19 would be less severe.

While politicization of the pandemic may be unfortunate, I will argue here that it is also inevitable. The pandemic asks each of us to arrive at a personal view of how COVID-19 works, and what we ought to do about it. Recent work in cognitive psychology suggests that we learn the causal structure of the world primarily through social interactions. Thus, each person's understanding of the world tends to be shaped by how those around them understand the world. Moreover, foundational

---

1 C. Cillizza, "2 Maps that Explain How Partisanship has Poisoned our Fight against COVID-19, *CNN*, July 10, 2021, https://www.cnn.com/2021/07/08/politics/electoral-map-vaccine-map-covid-19/index.html

2 S. Shahrigian, "Dr. Fauci Decries Nation's Political Divide over COVID Vaccinations," *New York Daily News*, July 11, 2021, https://www.nydailynews.com/news/politics/us-elections-government/ny-fauci-delta-variant-nasty-strain-20210711-ymir45yrpresrhfdtviqxvzmuu-story.html

work in philosophy argues forcefully that our understanding of any given phenomenon is shaped by how that phenomenon fits into our broader worldview. That is, we understand the world *holistically*. We can't just silo off pandemics and understand them using tools different from those that we use to understand other social phenomena, including politics. As a result, efforts to de-politicize the vaccination response to the COVID-19 pandemic amount to swimming upstream against the hard reality of human cognitive mechanisms.

In what follows, I will explain briefly how these two theses (i.e., the inherent sociality of causal learning, and the holistic nature of a person's worldview) jointly imply that it is unlikely that we will ever be able to truly de-politicize a pandemic at scale.

As very young children, we begin understanding the world in terms of cause and effect. We learn which interventions lead to what sorts of outcomes, and we start to strategize accordingly. This capacity for causal reasoning underwrites our ability to explain *why* things happen in a particular way in the world around us. Importantly, we do not build this causal apparatus on our own. As Legare, Sobel, and Callanan explained in a 2017 literature review, experimental and observational evidence from cognitive science strongly suggests that causal learning in early childhood is an inherently social affair.[3] Children use their parents and other adults as sounding boards for repeated "why questions" that inform their mental representation of the causal structure of the world, which in turn informs their understanding of why things happen in a particular way. As such, each person's understanding of the world is mediated through the contingent features of their unique social environment.

This inherently social nature of our understanding of the world continues throughout adulthood. As Sloman and Fernbach argued in their 2017 book, *The Knowledge Illusion*, all causal cognition involves outsourcing mechanistic knowledge to experts and black-boxing the fine-grained details of how many processes work.[4] This serves to reduce

3 C.H. Legare, D.M. Sobel, and M. Callanan, 2017, "Causal Learning is Collaborative: Examining Explanation and Exploration in Social Contexts," *Psychonomic Bulletin & Review* 24(5): 1548-1554, doi: 10.3758/s13423-017-1351-3

4 S. Sloman and P. Fernbach, 2017, *The Knowledge Illusion: Why We Never Think Alone*, New York, NY: Riverhead Books.

the cognitive load of understanding the world, but it is only possible because of our high degree of sociality; when we *do* need to understand something in granular detail, each of us has a cognitive social safety net that we can call on, by asking questions of those around us. Knowledge of the causal structure of the world doesn't really live inside each of our heads; it is distributed throughout our social network.

Almost no one is an epidemiologist or virologist, and so almost all of us must call upon our social network when we attempt to understand both how the COVID-19 pandemic works, and how to mitigate its negative impacts. Social networks are inherently biased along a number of axes, including political ones. Democrats are more likely to outsource their intellectual labor to fellow Democrats, and Republicans to fellow Republicans. Thus, once a certain amount of political bias is introduced into some individuals' understanding of how a pandemic works, these biases can quickly propagate and amplify, as more people turn to their social networks to help them understand the pandemic.

It is at this point that it becomes natural to insist on de-politicization. If each of us could see the pandemic through a strictly scientific lens, without political bias, then we would settle on a unified, and more accurate, understanding of how to mitigate its effects. However, the history of science teaches us that it is difficult to strictly separate the various aspects of our belief-forming apparatus. As Quine (1960) famously argued, it is always possible to hold on to some particular belief "come what may."[5] Consider the case of late medieval astronomers who, rather than accept a heliocentric model of the solar system, constructed increasingly complicated geocentric models. These astronomers were able to maintain one belief—that Earth was at the center of the solar system—by revising related beliefs about the laws of planetary motion. It is for this reason that philosophers of science like Carnap, Quine, and Kuhn have viewed each of us as having a fundamentally "holistic" worldview.[6] That is, each of us sees the world through a richly interconnected web

5 W.V.O. Quine, 1960, *Word and Object*, Cambridge, MA: MIT Press.

6 Although Quine set up Carnap as a foil for many of his arguments, and Kuhn explicitly viewed his project as an extension of Carnap's, Becker, 2002, "Kuhn's Vindication of Quine and Carnap," *History of Philosophy Quarterly* 19(2): 217-235, convincingly argued that all three are committed to a version of the epistemic holism described herein.

of particular beliefs about how the world works. In the face of data that contradict any particular part of that web, we can choose which parts to revise, in order to remain consistent with the data. Different agents may adopt different rules of revision, and it is not obvious that there is any rational basis for preferring one set of rules to another.

The history of science teaches us that it is difficult to strictly separate the various aspects of our belief-forming apparatus.

As argued above, cognitive science strongly suggests that our understanding of the world is inevitably informed by our social networks, which are themselves heavily politicized. Thus, our political and scientific views are deeply interconnected, and how we revise our understanding of the world in response to new data is invariably informed by our personal understanding of *both* science *and* politics. Short of a mass de-politicization of our social networks, it is unlikely that this close connection between science and politics will ever be eliminated.

This may seem pessimistic, but I do not believe that all hope is lost with respect to addressing wicked problems like COVID-19. Rather, I conclude only that efforts at de-politicization of the pandemic are likely to be in vain. Instead, and in keeping with the spirit of my first Transmission, I would urge scientists of all fields to embrace the inescapably political dimensions of their work. For instance, when scientists argue (with good reason) that vaccines are the most effective tool for curtailing the spread of COVID-19, they must recognize that, whether they like it or not, this message has a political valence. Thus, getting this message to the public requires scientists to embrace politics rather than shirk them. This is especially true when scientific recommendations bear so directly on the well-being of so many.

TRANSMISSION NO. 002

REDUCING CONFLICTING ADVICE ON ALLOWABLE GROUP SIZE

DATE: *30 March 2020*

FROM: *John Harte, University of California, Berkeley; Santa Fe Institute*

STRATEGIC INSIGHT: *Group size matters when it comes to how many people should gather in one place. Let's use mathematical models to pin down consistent guidelines for complicated situations.*

Beginning in early March, 2020, conflicting advice about COVID-19 emanated from local, state, and federal leaders, as well as from public health spokespeople. While there was unanimous agreement that some level of social distancing was critical to reducing the daily incidence of new COVID-19 cases, there was wide disagreement as to what group size should be allowed. Through the media, one learned that group size should be limited to 200, or maybe to fifty, or maybe to twenty, or maybe to ten. On the same day, you could learn that you could attend a large lecture but not a sports event or political rally, or you could go to a bar or restaurant but not a concert hall, or you attend a small dinner party with friends but not a restaurant. Eventually many regions of the US settled on home confinement, which implies group sizes of at most a handful.

So what is the effect of group size on the transmission rates of infectious disease? This question raises many secondary questions. How long does one stay within a group—perhaps two hours at a ball game, but all day in kids' classrooms—and how does that interact with group size? How thoroughly within a group does transmission occur? Surely somebody in the bleachers cannot directly infect someone in a box seat above home plate. And what about whether the group is indoors or outdoors; what about wind and humidity?

OPPOSITE: *Utagawa Hiroshige III (Ando Tokubei), "Progression during the Imperial Inspection at Ou, Matsushima," 1876 (detail)*

Clearly, it's complicated. But, to get at least a very simple insight, we can make some very simple assumptions and obtain a back-of-the-envelope result. Let us suppose that you are in a group of size n_0, and that there are N_0 such groups in a total population of size $n_0N_0 = P_0$. Moreover, we assume that if an infected individual happens to be in a particular group, then everyone in that group becomes infected. Finally, assume there is no mixing among groups. Both of those last two assumptions are readily altered, but let's look at this simplest case first.

We further assume that the group you are in, *A*, is comprised of your friends and/or family members. Hence, a simple but useful measure of your expected damage is the probability that an initially infected individual happens to be in your group multiplied by the number of people in the group.

Suppose that initially the population contains a single infected individual. That individual could be equally likely to be in each of the groups, and so the probability that it is in the group that you happen to be in will be proportional to 1/(number of groups). Multiplying by the number of individuals in the group, your expected damage is proportional to n_0/N_0.

It is imperative that, as we emerge from strict social distancing some months from now, we don't go straight from group sizes of two or three or four to unlimited group gatherings, lest we trigger a resurgence in infection.

How does n_0/N_0 depend on group size, n_0? Because $N_0 = P_0/n_0$, the expected damage to you varies as n_0^2. In other words, a doubling of allowable group size results in a four-fold increase in your expected damage. Group size matters a lot!

Suppose, instead of assuming that everybody in a group gets infected if one of the initially infected people is in the group, we assume that the number of infected in a group increases as the square root of the group size (the bleachers are far from home plate). Then it is easy to show that the damage to you varies as $n_0^{3/2}$.

Suppose we allow intergroup mixing. Then, depending on the rate of mixing, the infection rate, the duration of infectiousness in a person, and other factors having to do with the spatial pattern of mixing such as distance over which one mixes, the expected damage can become much larger, but the dependence on group size probably does not ever become steeper than the quadratic dependence derived above.

Clearly, more sophisticated modeling is needed here. It is imperative that, as we emerge from strict social distancing some months from now, we don't go straight from group sizes of two or three or four to unlimited group gatherings, lest we trigger a resurgence in infection.

So what's the magic number? There isn't a single answer. However, because group size matters a lot, the precautionary principle urges us to err on the side of small group-size restrictions. If mathematics informs our decisions, then as we eventually ramp up our sociality and return to some approximation of normality, we can do so with more clarity than was available when we went into quarantine.

REFLECTION

SOME COVID-19-TRIGGERED THOUGHTS ON COMPLEXITY

DATE: *22 May 2021*

FROM: *John Harte, University of California, Berkeley; Santa Fe Institute*

My March 2020 Transmission addressed a perplexing issue of scale, network structure, and risk. My essay was motivated by the bewildering variety of guidelines issued by public officials in the first months of the COVID-19 pandemic. The conflicting advice concerned the numbers of people that one could safely hang out with. There was well over an order of magnitude variation in safe group sizes advocated by experts, and often no distinction made between indoor and outdoor gatherings. I tried to shed a little light on the sources of ambiguity in epidemiological model predictions of the consequences of group size.

Similar ambiguities now arise in discussions of vaccination and herd immunity. A widely disseminated estimate is that the US could reach herd immunity when perhaps three-quarters of us are vaccinated. Using a standard epidemiological Susceptible-Infected-Recovered (SIR) model, such a value can be derived. But how the unvaccinated susceptibles are distributed over space and interact with the entire population matters. It is easy to imagine situations in which three-quarters of the population is immune and yet the network structure and spatial distribution of the susceptibles results in unacceptable death rates.

To complicate matters, human behavior, as reflected in patterns of social network structure, will both affect the outcome of a vaccination campaign and be altered by awareness of the degree of success of that campaign, just as it was altered by the pandemic. Given this complexity, the notion of herd immunity may be too simplistic a target of modeling; a broader analysis of the consequences of a massive vaccination campaign could allow us to better anticipate locations in which various interventions, including masking and social distancing, are still needed.

This is indeed a complex system, with a self-referential structure, in which outcomes both influence, and will be influenced by, policy interventions. Arthur[1] noted the importance of such structures in economics, particularly in non-steady-state economies. Is this fundamentally different from complex systems in physics, in which the human element is absent? In a provocative article, Goldenfeld and Woese[2] suggested that, while physics makes a clean separation between the state of a physical system and the equations that govern the time-evolution of the system, successful biological theory will inevitably be self-referential in the sense that the equations that govern dynamics will evolve with the predicted changing state of the system. In their words:

> *In condensed matter physics, there is a clear separation between the rules that govern the time evolution of the system and the state of the system itself. . . . (T)he governing equation does not depend on the solution of the equation. In biology, however, . . . we encounter a situation where the . . . update rules change during the time evolution of the system, and the way in which they change is a function of the state and thus the history of the system. To a physicist, this sounds strange and mysterious.*

This self-referential nature of theory described here is distinct from conventional feedback interactions that are found ubiquitously in both purely physical and also biological complex systems. Conventional feedback operates among the actual components of the system. In a sense Goldenfeld and Woese were also referring to feedback, but it is between the system operating rules (that is, the structure of the theory itself) and the state of the components of the system.

How does this play out in macroecology: the study of micro-level patterns in the distribution and abundance of species within ecosystems? Must ecological theory be self-referential or need it merely describe conventional feedback processes? To gain insight into this, it is useful to distinguish between relatively static, undisturbed ecosystems exhibiting patterns that at most fluctuate from year to year, versus systems in which

---

1 W.B. Arthur, 1999, "Complexity and the Economy," *Science* 284: 107–109, https://science.sciencemag.org/content/284/5411/107

2 N. Goldenfeld and C. Woese, 2011, "Life is Physics: Evolution as a Collective Phenomenon Far from Equilibrium," *Annual Review of Condensed Matter Physics* 2: 375–399, doi: 10.1146/annurev-conmatphys-062910-140509

natural or human-caused disturbances produce real trends, not just fluctuations, in the patterns. In analogy with pressure, volume, and temperature—the state variables of an ideal gas—we can consider the macro-level descriptors of the ecosystem, such as the number of species, the number of individuals, and the total metabolism, to be the state variables. We have shown[3] that in ecosystems in which these state variables are relatively constant in time, the patterns alluded to above can be quite accurately predicted using the maximum-information-entropy (MaxEnt) principle derived from information theory. This powerful inference procedure finds the least biased distributions over the micro-level variables that satisfy the constraints imposed by the macro-level state variables.

Society desperately needs to deal more intelligently with the next pandemic than it did with COVID-19, and, more generally, it needs to prevent catastrophic climate change, preserve biodiversity, ensure sustainable food and water supply, reverse the trend toward increasing inequity in the distribution of wealth and opportunity, and prevent war waged with weapons of mass destruction.

Viewed from the perspective of the rapidly developing new subfield of "disturbance ecology,"[4] however, the story above appears quite inadequate. In particular, ample data indicate that in ecosystems in which the state variables undergo secular change in response, for example, to

3 J. Harte, K. Umemura, and M. Brush, 2021, "DynaMETE: A Hybrid Max-Ent-plus-Mechanism Theory of Dynamic Macroecology," *Ecology Letters* 24(5): 935–949, doi:10.1111/ele.13714

4 E.A. Newman, 2019, "Disturbance Ecology in the Anthropocene," *Frontiers in Ecology and Evolution* 7: 147, https://www.frontiersin.org/articles/10.3389/fevo.2019.00147/full

anthropogenic disturbance, the patterns change and the purely information-theoretic predictions fail rather dramatically.

To remedy this, a theory of dynamic disturbed ecosystems can be constructed by hybridizing information-theory methods with explicit mechanisms causing disturbance. In this dynamic theory, as the state variables change over time, the form of the micro-level dynamics (e.g., birth, death, and ontogenic growth of individuals) is altered, and thus the probability distributions used to determine the consequences of the mechanisms that generate disturbance change over time. Thus, at each forward-in-time iteration of the theory, the distribution over which micro-level variables are averaged to update the macro-level constraints is itself updated in a way that depends on the constraints.[5] Or, as Goldenfeld and Woese described, the update rules change during the time evolution of the system, and the way in which they change is a function of the state of the system.

Society desperately needs to deal more intelligently with the next pandemic than it did with COVID-19, and, more generally, it needs to prevent catastrophic climate change, preserve biodiversity, ensure sustainable food and water supply, reverse the trend toward increasing inequity in the distribution of wealth and opportunity, and prevent war waged with weapons of mass destruction. To do so, we need to improve our capacity to understand complex, dynamic, disturbed coupled systems of all sorts, including ecologic, economic, and social, which not only are replete with feedback, but which, in addition, require theory that flexibly evolves with evolving systems' configurations. Arguably, such theory can help us avoid rigidly designed, inflexible interventions and point the way toward sustainable policies. Insights from ecological theory may help guide progress toward that goal.

5 Harte, Umemura, and Brush (2021).

TRANSMISSION NO. 003

MODELING THE EFFECT OF SOCIAL-DISTANCING MEASURES IN THE COVID-19 PANDEMIC[1]

DATE: *30 March 2020*

FROM: *Luu Hoang Duc, Max Planck Institute for Mathematics in the Sciences, Leipzig*
Jürgen Jost, Max Planck Institute for Mathematics in the Sciences, Leipzig; Santa Fe Institute

STRATEGIC INSIGHT: *To forecast the spread of the novel coronavirus, we must attend to the quality and consistency of the data*

The coronavirus disease 2019 (COVID-19), which apparently started in December 2019 in Wuhan in China under still-unclear conditions and spread worldwide in the first half of 2020, is caused by a small virus (SARS-CoV-2) which by now has diverged into several variants or mutants, and which affects the human population and human societies across the world. Thus, modeling it might naturally span a vast range of scales, from the molecular to the global. In fact, while the virus itself is only moderately complex, its dynamics depends on the complexity of human biology and society. Thus, analyzing the genome of the virus does not give a complete picture because its replication machinery depends, for its assembly and proliferation, on the properties of the host cells—human, in the case of this disease. (Although it apparently originated in other animals, we will not address that further here.) The virus is transmitted through the air, and we seem to have some rough understanding about its aerosol physics.

1 This Transmission has been updated from the original version published on March 30, 2020. To read the original Transmission, visit https://sfi-edu.s3.amazonaws.com/sfi-edu/production/uploads/ckeditor/2020/04/06/t-003-luu-jost.pdf

---

OPPOSITE: *Leonardo da Vinci, "Perspectival Study of the Adoration of Magi," c. 1481 (detail)*

Many issues remain unclear. For instance, how many virus particles are needed or can suffice for successful transmission between people? Transmission depends on the behavior of people; behavior can be modified by voluntary restraint or by various political measures, which people may or may not obey. The spreading of the pandemic, as well as the measures taken to constrain it, may have many psychological, social, political, and economic side effects. While the severity of the disease is strongly correlated with age, treatment in the medical system can mitigate the mortality risk, although the medical systems in many countries have had difficulties coping with this challenge. Also, the repercussions of interactions between scientific opinions and public controversies are not easy to model.

Should a good model incorporate as many of these scales and dynamics as possible, from the biochemistry to the global economy, from the scale of virus replication in a host to the long-term political instabilities? Perhaps not, as it may depend on too many parameters that cannot be reliably estimated, and it may become far too complicated, in any case. There may exist certain universal patterns underlying the

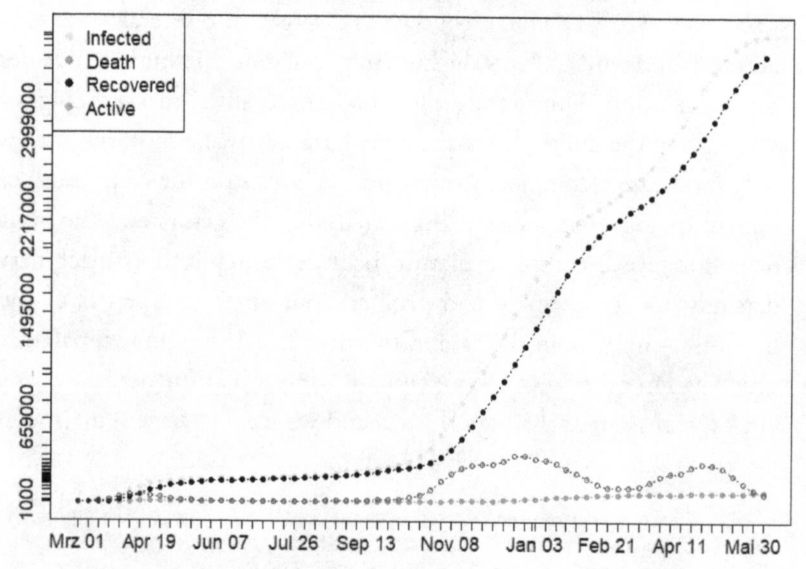

Figure 1: *COVID-19 case progression. Country: Germany.*

dynamics of this, and perhaps also other, pandemics. Such patterns can only be captured by simpler models that identify the essential aspects.

Now, there does exist a well-established model in mathematical epidemiology consisting of a few coupled, ordinary differential equations that depend on a small number of parameters. This is the SIR (Susceptible, Infected, Removed) model. Some variants or extensions of the basic model try to incorporate additional effects. As such, it assumes a homogeneous population, but one may also work with populations that are structured by age or other properties, or introduce an underlying network structure for the contact patterns between individuals. Also, the parameters need not necessarily be constant, but can fluctuate stochastically. This is also the theoretical setting that we adopt in this contribution. The key parameter is the contact rate because it depends on the behavior of people and therefore can be influenced. We build our model upon a careful analysis of an extensive body of data, although the quality and reliability of some of those data do have certain problems that we have to overcome.

EMPIRICAL EVIDENCE FROM THE COVID-19 PANDEMIC

COVID-19 came, and probably will continue to come, in waves. A wave triggers responses—of varying adequacy—in each country, and those, together with people's voluntary behavior, seem to have some effect. While not eradicating the disease, they dampen its spread and, importantly, ease the strain on the medical system. When the disease is believed to be at least under some control, these measures are typically relaxed. Then, a new, and typically more severe, wave arises (although experts dispute the extent to which such a relaxation contributes to a new wave, and even to what extent mitigation measures are effective in controlling the pandemic). The new wave triggers renewed countermeasures, which last until the wave is thought to be under control or until the population gets too impatient. Then a third wave may arise, and so on.

Different countries respond to the pandemic in their own ways and thus, their results are different. Some Asian countries preferred to maintain strict policies to eradicate the disease, while European countries continued a "flattening the curve" strategy to slow down the spread.

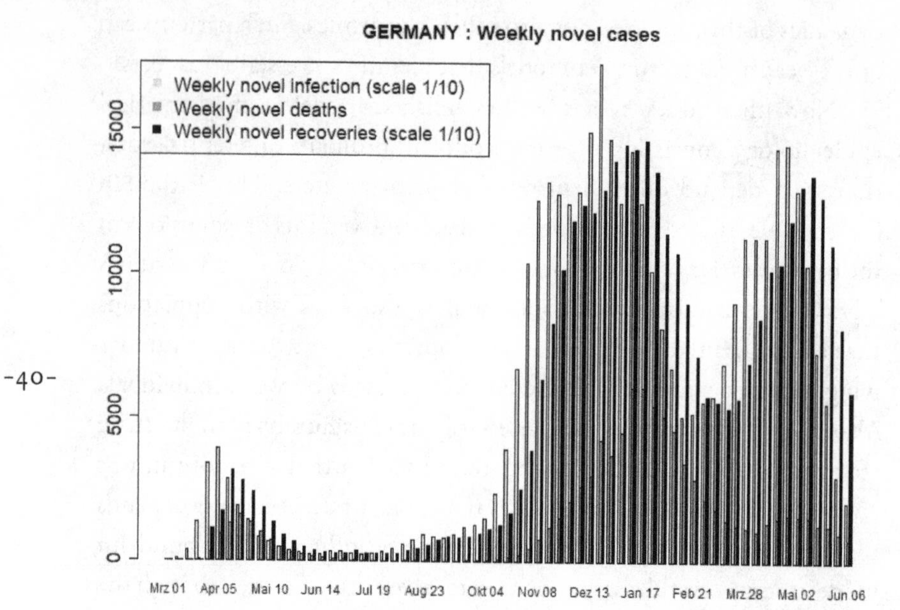

Figure 2: *Weekly novel cases. Country: Germany.*

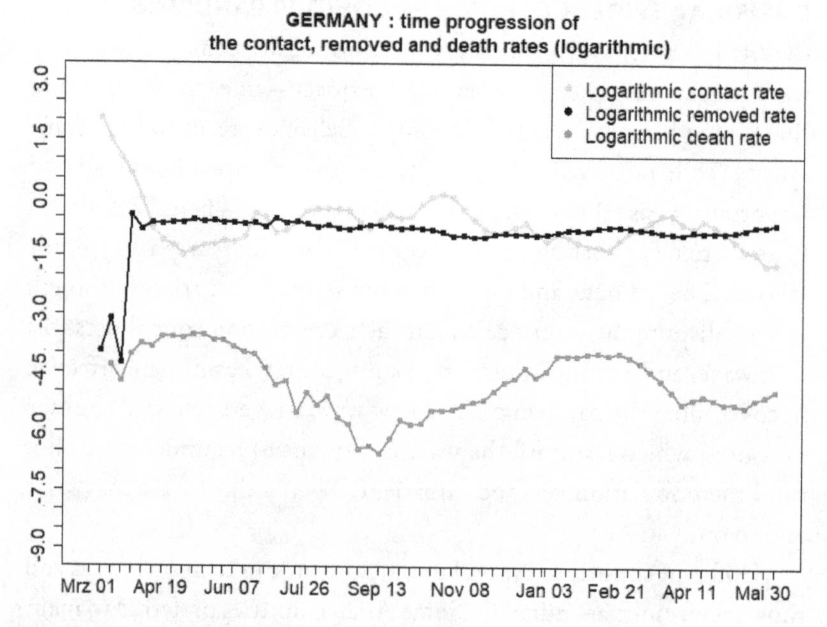

Figure 3: *The logarithmic contact, removed, and death rates. Country: Germany.*

Nevertheless, certain patterns seem to be rather universal, and we illustrate them here with the data from Germany.[2] (See figures 1 and 2 for the case progression and the weekly novel cases.) It is hoped that vaccination programs will soon get the pandemic under sufficient control so that life can return to normal.

A TIME-VARYING SIRD MODEL

Our main aim is to model the effect of social-distancing policies in reducing spread of the virus. The classical SIR model for epidemic spreading, which is our starting point, describes the dynamics of some global time-dependent quantities in a population. We consider the normalized fraction S_t of susceptibles, I_t of active infected cases, D_t of casualties, R_t of removed cases (i.e., the sum of the recovered and the deceased), $\gamma_t = I_t + R_t$ of total infections, as reported daily, in the total population. Thus, $S_t + I_t + R_t = S_t + \gamma_t = 1$. The model stipulates

$$\frac{dS_t}{dt} = -\beta_t S_t I_t, \quad \frac{dI_t}{dt} = \beta_t S_t I_t - \gamma_t I_t, \quad \frac{dR_t}{dt} = \gamma_t I_t, \quad \frac{dD_t}{dt} = \mu_t I_t, \tag{1}$$

where β_t, γ_t, μ_t are time-varying processes modeling the contact, removal, and death rates. The model assumes that recovered persons do not get infected again, and so, the pandemic may come to an end when it has swept through a sufficiently large fraction of the population, but because of the many casualties this may cause, this is generally not considered a viable option. Rather, one tries to control the virus by making the parameter β_t, the contact rate, smaller than the removal rate γ_t.

ESTIMATING THE CONTACT, REMOVED, AND DEATH RATES

The dynamical quantities in equation 1 can be measured, and many data are available, with the problem that those for R_t are not reported in many countries or are based on some rather rough and indirect estimates. But,

---

2 We evaluate the data provided by the Robert Koch Institute—Germany: https://www.rki.de/EN/Content/infections/epidemiology/outbreaks/COVID-19/COVID19.html, and the European CDC: https://www.ecdc.europa.eu/en/cases-2019-ncov-eueea. The figures are updated on a weekly basis in our blog on our institute's website www.mis.mpg.de/COVID19/COVID19-mpi-mis-leipzig-start.html. With regard to the quality and accuracy of the data: The number of cases reported may be significantly lower than the number of people actually infected, and these discrepancies may be different in different countries.

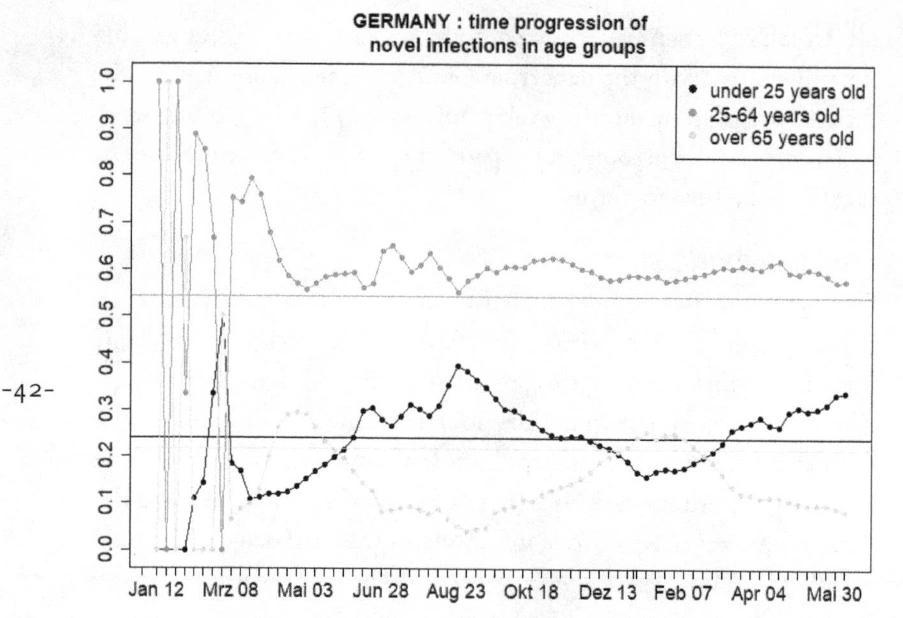

Figure 4: *Novel infections in age groups. Country: Germany.*

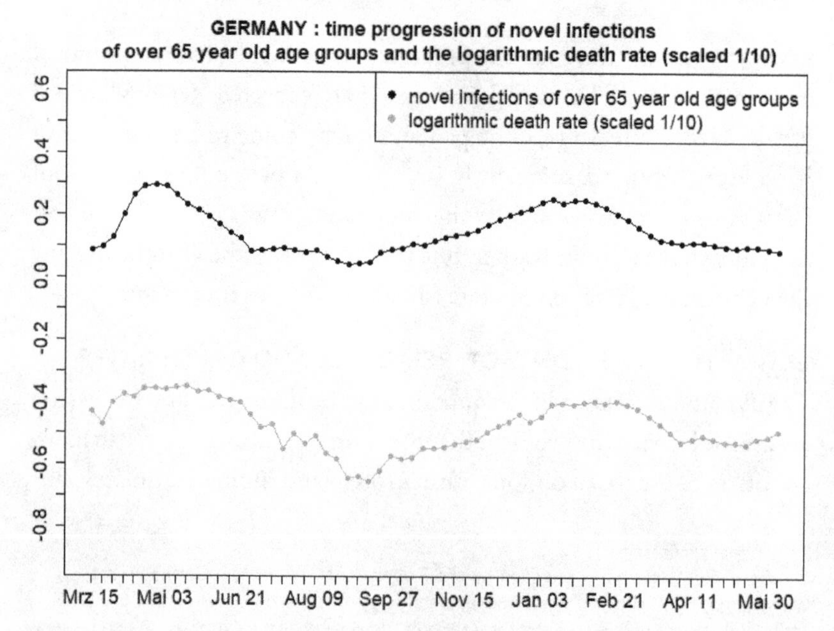

Figure 5: *Time progression of the novel infections in the over-65-year-old group and the logarithmic death rate (scaled 1/10). Country: Germany.*

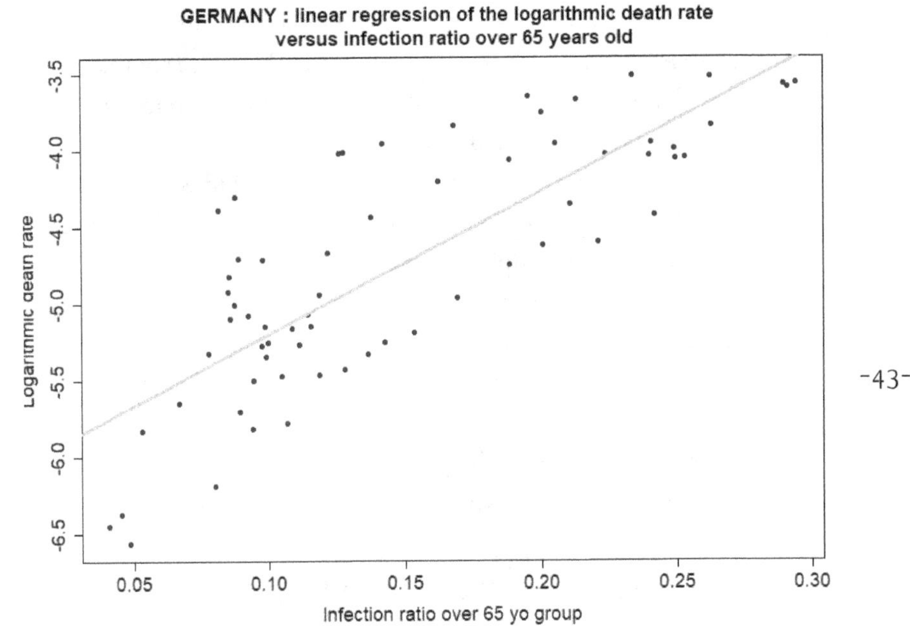

Figure 6: *Linear regression of the logarithmic death rate versus novel infections in the over-65-year-old group. Country: Germany.*

importantly, we can use them to estimate the crucial parameters. We use discrete equations for the original data to compute the rates:

$$\beta_t = \frac{\Gamma_{t+1} - \Gamma_t}{(1 - \Gamma_t)I_t}, \quad \gamma_t = \frac{R_{t+1} - R_t}{I_t}, \quad \mu_t = \frac{D_{t+1} - D_t}{I_t}. \tag{2}$$

With COVID-19, we observe that the logarithmic contact rate fluctuates in particular ways, creating trends corresponding to pandemic waves. As shown in figure 3, in the initial phase of a wave, the contact rate was high, but then started to decrease considerably and in a controllable way (in terms of low fluctuation of the residuals) to a low level in the second phase, possibly as a result of social-distancing policies. Figure 7 shows that during the time of control measures in the first wave of the pandemic in Germany, the logarithmic contact rate $\log \beta_t$ depends linearly and negatively on the total number of infections Γ_t. Towards the end of the wave, when the active cases had decreased to a level that was considered safe for the public health system, the social-distancing policies were partially lifted—for example, schools reopened and social gatherings were allowed with certain limitations. Thus, we see that the contact rate β_t bounced back gradually until crossing the removal rate, which may have caused the onset of a new wave.

In figure 4, we see the proportion of infected middle-aged people approaches their proportion in the population (straight horizontal line), while for the older people the relationship between these proportions eventually becomes lower. In particular, we see that, at least in Germany, the spread of the virus is not totally mixed, perhaps as a result of social-distancing policies that aim to protect the elderly who have a high mortality risk.

Figures 5 and 6 show that the logarithmic death rate and the novel infections among those over sixty-five are proportional. This suggests a relation,

$$\log \mu_t = \log\left(\frac{D_{t+1} - D_t}{I_t}\right) = a + b\frac{d\Gamma_t^o}{d\Gamma_t} \qquad (3)$$

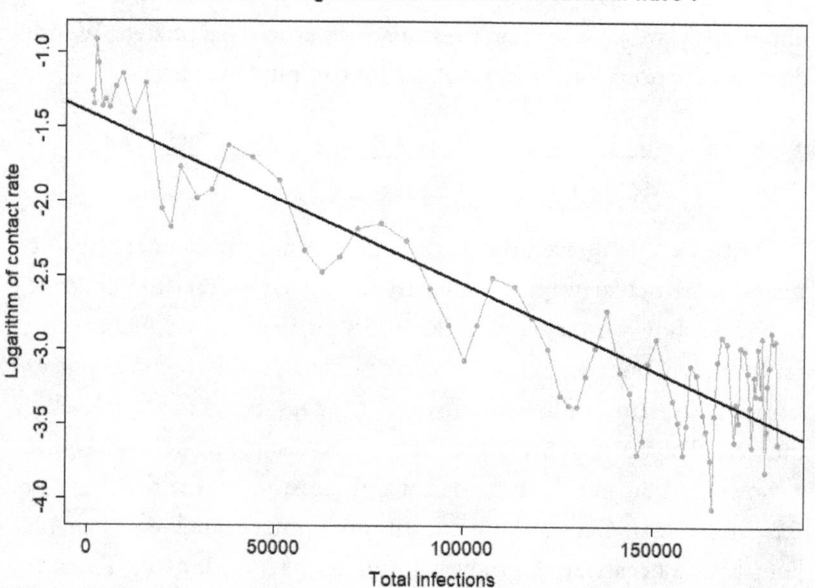

Figure 7: *Logarithmic contact versus total infections in the 1st wave. Country: Germany.*

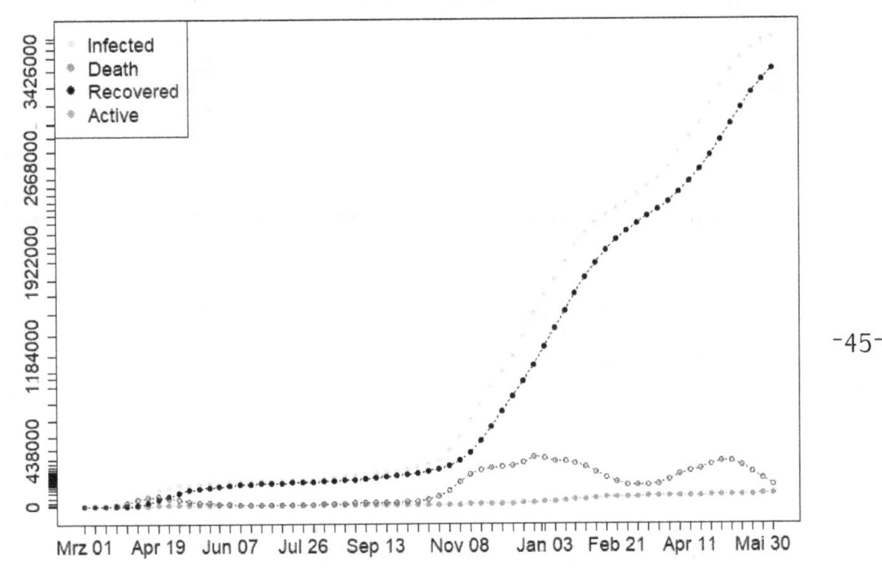

Figure 8: *Case progression using ECDC data. .*

with constants a and b, where $d\Gamma_t$ and $d\Gamma_t^o$ are the total novel infections and those among people over sixty-five. This may indicate that social-distance measures and/or behavior are implemented differently depending on age.

ESTIMATING THE REMOVED CASES USING TIME TO REMOVAL

We have already indicated the problem that one cannot get good data for the recovered cases, because they are either not systematically recorded, or because different countries have different criteria for recovery. We can, however, approximate the recovered cases, and thus the removed cases, based on an algorithm that assumes that the length of the period from the detection of a novel infected case to its removal (by recovery or death)—called the removal time τ_R—is a random variable with a lognormal distribution *Lognorm* (c, σ). The removed cases are approximated by

$$dR_t = Ed\Gamma_{t-\tau_R} = \sum_{i\geq 1} d\Gamma_{t-i}\,\mathbb{P}(\tau_R = i). \tag{4}$$

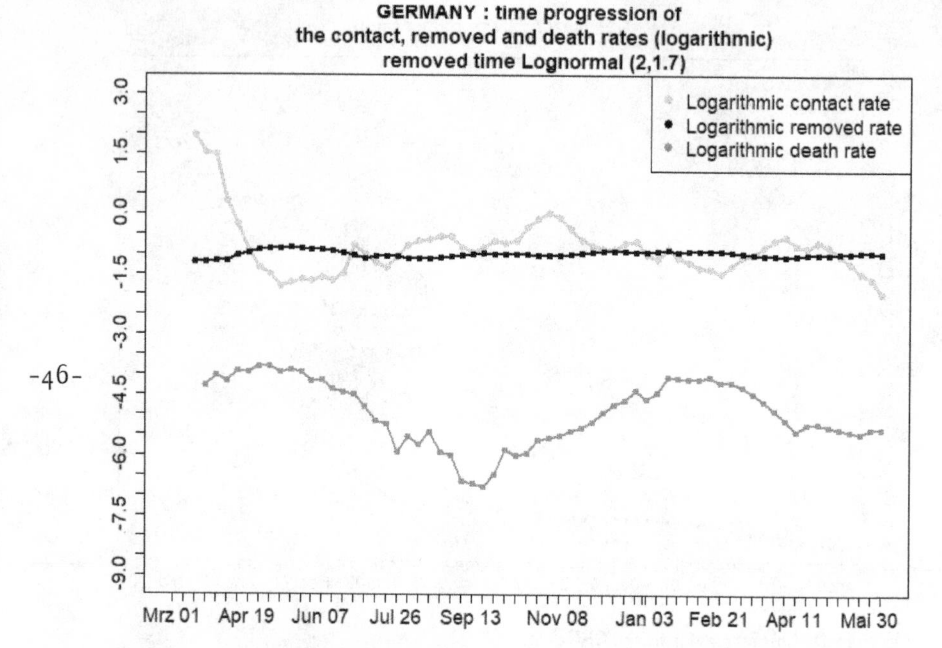

Figure 9: *The logarithmic contact, removed and death rates using ECDC data. Country: Germany.*

With COVID-19, we may model the distribution of the removal time based on the fact that a median time from onset of infection to recovery for patients with mild cases is approximately two weeks, and three to six weeks for patients with more severe symptoms—with 81% of cases estimated as mild to moderate, 14% as severe, and 5% as critical. Where cases of recovery are reported locally and are higher than the modeled value, the reported value is used. As such, one can choose, for example, $\tau_R \approx Lognorm\ (2, \sigma)$ for $\sigma \in (1.7, 2.3)$. Figures 8 and 9 plot the case progression and time progression of rates, but the computation uses only the data of novel infections and novel deaths provided by European Centres for Disease Control and Prevention (ECDC) and the assumption $\tau_R \approx Lognorm\ (2, 1.7)$. The precise parameters should be chosen cautiously depending on empirical clinical studies by curve fitting when full and complete data are provided, to avoid mis-estimation of the pandemic scales. As observed, there are only some minor differences between figures 1 and 8, and between figures 3 and 9.

Remarkably, the method using removal time detects the time of new pandemic waves as well as using the full-data approach.

CONCLUSIONS

The classical SIR model for the spreading of diseases in populations can be extended by making the parameters variable, structuring the population, and introducing a time delay. The most important parameter that needs to be controlled is the contact rate, and it will vary in the course of a pandemic because people will adapt their behavior, perhaps forced by political measures. Populations are not homogeneous. They are distributed across space and are socially structured, and the parameters, like the contact rate and the death rate, can vary within and between sub-populations. Finally, people do not become infectious or die immediately after having been infected, but there are time-delay factors that themselves fluctuate.

As there is a high correlation between the dynamics of novel infections in the elderly group (over 65 years) and the logarithmic death rate, the most important aim is to use social-distancing policies and control measures to avoid a full mixing in the spread of the novel coronavirus to that group.

While an encompassing model that includes all these effects simultaneously may be too complicated and have too many parameters to be estimated from the currently available data, we have analyzed several extensions that each included one of the effects just described. Still, not all parameters can currently be estimated with precision. In particular, for structured populations, we would need precise demographic data, and this would go beyond the scope of the present Transmission. Nevertheless, we have shown how contact and other rates can be estimated from the available data, and we have used that for predicting the trend of the pandemic.

As there is a high correlation between the dynamics of novel infections in the elderly group (over sixty-five years) and the logarithmic death rate, the most important aim is to use social-distancing policies and control measures to avoid a full mixing in the spread of the novel coronavirus to that group. The number (sixty-five in this example) might vary from country to country and might also be different with respect to various variants. In addition, those with underlying medical problems like cardiovascular disease, diabetes, chronic respiratory disease, and cancer are among the high-risk group, but we do not have such data for evaluation.

We also need to cope with missing or incomplete data. For instance, we do not have the precise number of recovered cases but only new infections and casualties. Hence, careful consideration of time-delay factors like incubation time, removal time, etc., is necessary to design algorithms to estimate recovered and removed cases.

In the long run, we expect the pandemic to develop in waves and become endemic before the vaccination campaigns show their effects.

In the long run, we expect the pandemic to develop in waves and become endemic before the vaccination campaigns show their effects. Hence, one needs to study the cost of control measures—social-distancing and mask-wearing policies; local and national lockdowns—and their economic impacts. A natural strategy seems to be to keep the contact rate as close to the removal rate as possible through gradual implementations and cautious lifts of control measures.

SUMMARY

It is obvious that the spread of an airborne virus like COVID-19 depends on the number and the structure of contacts that individuals have. Therefore, this contact rate is a key parameter in epidemiological models, and it is a main target of social distancing measures that have been implemented in various forms in different countries. We have studied these effects and incorporated them in models where the contact rate is varying in time, heterogeneous across populations structured by age or other parameters, or randomly fluctuating. During the time of social distancing measures the contact rate depends exponentially and disproportionally on the total number of infections. The removal rate fluctuates as a lognormally distributed random variable. In contrast, the logarithmic death rate depends linearly and proportionally on the percentage of the old age group in the novel infections.

REFLECTION

WHY DO WE NEED COMPLEX-SYSTEMS SCIENCE TO UNDERSTAND THE COVID-19 PANDEMIC?

DATE: *29 July 2021*

FROM: *Luu Hoang Duc, Max Planck Institute for Mathematics in the Sciences, Leipzig*
Jürgen Jost, Max Planck Institute for Mathematics in the Sciences, Leipzig; Santa Fe Institute

INTRODUCTION

The coronavirus disease 2019 (COVID-19) is caused by a small virus (SARS-CoV-2), which has diverged into several variants, and affects humans and societies across the world. Thus, modeling the disease might naturally span a vast range of scales, from the molecular to the global. This virus itself is only moderately complex, but its dynamics depend on the complexity of human biology and society. Analyzing the genome of the virus does not clarify these dynamics because its replication machinery depends, for its assembly and proliferation, on the properties of the host cells. The virus is transmitted through the air, and we know something about the aerosol physics behind transmission. The transmission depends on the behavior of people, and this may be modified by voluntary restraint or political measures, which are not always accepted and obeyed by people. Also, in many countries, vaccination campaigns encounter resistance from substantial parts of the population, while in other countries, there is a vaccine shortage. The spreading of the pandemic as well as the measures taken to constrain it may have many psychological, social, political, economic, and financial side effects and perhaps as-yet-unknown long-term consequences. Apparently, the severity of the disease is strongly correlated with age, but the availability and quality of treatment in the medical system can mitigate the mortality risk, although the medical systems in many countries are poorly equipped to cope with this challenge. Also, the interaction between scientific opinions and public controversies has repercussions, but is not easy to model.

Should a good model incorporate as many of these scales and dynamics as possible, from the biochemistry to the global economy, from the scale of virus replication in a host to the long-term political instabilities? Perhaps not, as such a model may depend on too many parameters that cannot be reliably estimated, and it may become far too complicated. However, there may exist certain universal patterns underlying the dynamics of this and, perhaps, other pandemics. Such patterns can only be captured by simpler models that identify essential aspects. But, while this valuable path may lead to important contributions, a profound scientific understanding needs to go beyond such models and incorporate complexity.

CONCEPTUAL ISSUES

As one of us (JJ) has advocated, the key feature of biological life is that a biological process can control and regulate other processes, and it improves that ability over time. This control can happen hierarchically and/or reciprocally. Thus, the information that a biological process needs to use concerns only the control, but not the content or the internal structure, of those processes. Those other processes can be—or rather, have to be—vastly more complex than the controlling process itself. Each biological process draws upon the complexity of its environment.

The novel coronavirus illustrates that thesis. From that perspective, we should conceptualize the virus not as a physical molecule, but as a dynamic process. Complex systems in general may consist of many interacting levels and scales. In fact, to understand the pandemic, it does not suffice to sequence the 26–32 kilobases of RNA of the virus. We rather need to understand the complexity of the human cells that the virus uses for its reproduction and the complexity of human societies that enable its transmission.

Complex systems are characteristically both vulnerable and resilient at the same time. In noncomplex systems, stochastic fluctuations usually average out, and small random events have only small consequences. In complex systems, by contrast, small and local random events may have large and global consequences. A single mutation of one copy of the virus may change the course of the global pandemic. The negligence of a single individual may have catastrophic consequences. The consequences may also be positive: the human immune system may adapt to the virus and learn to fight it off. Vaccinations may make people immune. A society may rearrange its interaction patterns to restrict physical contact between individuals. Through that, it may discover that certain rearrangements make it more efficient in other respects—for instance, in reducing unnecessary traffic and improving time management among home office workers.

-52- CONSEQUENCES

We need specialists in many disciplines. Knowing the genetic sequence of the virus and its mutated variants, we can apply our knowledge of human cell biology to understand how the virus uses the molecular machinery of human cells for its reproduction and how its copies can then invade other cells inside and across organisms. We can then try to interfere with various stages of the reproduction and transmission process by creating medicines or vaccines. Understanding the physics of transmission, we can propose contact restrictions that reduce the transmission chances. Analyzing the large-scale structure of social contacts, using concepts like network modularity or assortativity (i.e., to what extent the network consists of distinct modules that have only few connections between them, or whether highly connected individuals preferentially connect to other such individuals or rather stay away from them), can improve epidemiological models. In order to assess the effectiveness of countermeasures, we have to take into account psychological microfoundations of human behavior as well as cultural practices and differences. Contact restrictions emerge from interactions between political actors, scientific advice, public debates, and opinion dynamics in the population. Thus, insight into political processes and opinion dynamics can guide the scientific system in communicating its findings and formulating its advice efficiently, anticipating public reactions and possible countermeasures by individuals against political measures emerging from scientific advice. We also need to look at the scientific system from the outside and see how individual scientists or scientific institutions react to public pressure. We must anticipate how internal disagreement in the scientific system can be exploited by political agitators or interest groups to discredit science at large. We should also look at the long-term consequences, such as increasing social inequality, growing public debt, economic reorganizations, and education deficits for many schoolchildren and students.

The preceding seems rather obvious, but the point we want to make is that an understanding of the dynamics of the pandemic and of its potential long-term effects on our societies and economies requires that all of these factors and their interactions be considered. Pandemic models and advice based on just some of these factors—virology, epidemiology, aerosol physics, social contact structure, opinion dynamics, and collective behavior, or whatever—are almost surely inadequate. Of course, we need specialists in all those disciplines, but we also have to integrate their insights into a comprehensive understanding from a complex-systems perspective. This poses challenges, from analyzing the dynamics across vastly different scales and at many different levels, to coping with radical uncertainty.

We must anticipate how internal disagreement in the scientific system can be exploited by political agitators or interest groups to discredit science at large.

Building on fundamental research in various disciplines, complex-systems scientists need to integrate these findings and create appropriate models. The models themselves may be simple, but effective simplicity must be grounded in the complex details and qualitative insights those various disciplines have uncovered. That is, we need both specialized research and its integration to understand how the virus exploits the complexity of human biology and society and to devise and propose appropriate intervention strategies.

On one hand, we need to be humble. We cannot predict chance events. We do not know which mutations will occur and whether the virus will become more harmful or more benign in the future. We may, however, simulate potential mutations and their ability to control the molecular mechanisms in human cells. Which of those will actually be realized, we do not know, but we might be able to predict their consequences when they occur. We cannot predict when and where superspreading events will happen, but we may propose strategies to reduce their probability. We need to keep in mind the point expressed earlier about complex systems, that small random events need not average out, but may have large-scale consequences for the system. Such amplifications exploit particular vulnerabilities of a system, and we should try to identify those. Some of our predictions may be self-defeating, because

people get scared by that prediction and react appropriately.

On the other hand, we need to convince the general public and the politicians that scientific advice is the best advice we have. Science can provide insight into how the virus reproduces and spreads and point to short- and long-term consequences of the pandemic. Good complexity science can also integrate the individual findings and balance the effects at the various levels and scales.

We also need scientific insight to cope with radical uncertainty. In particular, when the pandemic first started to unfold in the spring of 2020, we did not know what consequences radical countermeasures or mere inactivity might have. In hindsight, of course, we know a lot more. But those decisions cannot and should not be held against the decision-makers. Rather, in order to make good decisions and act upon them, we must understand which psychological factors may help or hinder the exploitation of diffuse knowledge, the use of analogies with other situations of uncertainty, the assignment of appropriate weights to the many factors in play, and the adaptation of heuristics suitable for the challenges we face.

As complex-systems scientists, we try to build upon specific expertise and then integrate that expertise into the large-scale frame. Thus, we have analyzed and developed mathematical models for the spread of epidemics. The key parameter is the contact rate; because it depends on the behavior of people, it can therefore be influenced. We build our modeling upon a careful analysis of an extensive body of data, although the quality and reliability of some of those data do have certain problems that we have to overcome. The important point is that this parameter, the contact rate, can include stochastic fluctuations as extracted from the data, heterogeneity across populations, and systematic shifts resulting from the voluntary or enforced implementation of countermeasures. The model can thus readily include insight coming from social network science or social psychology.

> ... we need to convince the general public and the politicians that scientific advice is the best advice we have.

In turn, when the model is run with different parameter values, it can be used to assess the effect of political actions. Other parameters are less variable, but may be estimated on the basis of results from virology or from an understanding of the infection process. We also regularly update the data about the pandemic in most countries around the world to enable a comparison of the different dynamics and to correlate them with political measures and social, economic, and other factors. Please consult our website[1] for details.

1 https://www.mis.mpg.de/covid19/covid19-mpi-mis-leipzig-start.html

TRANSMISSION NO. 004

THINKING OUT OF EQUILIBRIUM

DATE: *30 March 2020*

FROM: *Simon DeDeo, Carnegie Mellon University; Santa Fe Institute*

STRATEGIC INSIGHT: *Getting the quarantine end game right means thinking about how to change thinking itself*

Our world mostly works. When you're leaving the airplane, don't think, follow: good design nudges you all the way to the taxi. Architect Christopher Alexander (Alexander et al. 1977) built a life's work on showing how something as simple as the design of a home's window seat has, over centuries, adjusted to a delicate balance of physical, psychological, and social needs. In equilibrium, good systems get you by on instinct.

Like the hiker who brought a can of espresso beans, however, many of us are now noticing how much of day-to-day mind-life has been cooked, not left raw. By choice, or by necessity, we're forced to think about things we've usually left to the environment. As I asked a friend who teaches philosophy: have you ever done this much thinking before?

There's a challenge down the road from today's mental jubilee. Call it the quarantine end game. We do, actually, want to get kids out of the house, throw a dinner party, or hear some jazz. When COVID is tamed, but not yet defeated, we'll need new norms for how we bring our worlds back together.

It's more than just hand-washing. It's whom, for example, we trust, and why. What reasons do we give when we turn someone down for a play date? What reasons can that person throw back? You can't hand them a printout of an article from the *Financial Times,* or a machine-learning

---

OPPOSITE: *Photograph of Arshile Gorky working on his "Modern Aviation: Evolution of Forms under Aerodynamic Limitations" mural at Newark Airport, 1937 (detail)*

GPS trace. It matters, because the explanations we give each other are buildings we come to live in, often for life.

Getting the quarantine end game right means thinking about how to change thinking itself.

We know quite a bit about how people go about those buildings, and it's more than just how it makes them feel, or whether "it makes sense." For an explanation to flourish, it often has to have a particular kind of feel, and make a particular kind of sense. We're predisposed, for example, to prefer "unifying" explanations about the world—something we call a consilience drive. The same preferences also go wrong, and those who tilt too far can end up on a conspiracy-theory forum, or in a mental ward. Parallel work shows that moral matters are just as complicated, and high-level concepts like justice, friendship, or dignity are more than window-dressing on a world of short-term contracts.

Getting the quarantine end game right means thinking about how to change thinking itself. You can't sell an idea the way you sell a can of soda. That's no silver lining, but if evolution did indeed make us thinkers, we might as well get back to getting good at it.

REFERENCES

Alexander C., S. Ishikawa, and M. Silverstein. 1977. *A Pattern Language: Towns, Buildings, Construction.* Oxford, UK: Oxford University Press.

REFLECTION

FROM VIRUS TO SYMPTOM

DATE: *10 August 2021*

FROM: *Simon DeDeo, Carnegie Mellon University; Santa Fe Institute*

After trauma, complex systems rarely return to where they began. It's a principle we can lament: consider, for example, the persistent disruption to a cell's gene regulatory network after brief exposure to a carcinogen. It's also a principle we've learned, in some cases, to make use of: in how, for example, the mental "noise" of a psychedelic experience can lead to long-lasting escape from life's mental ruts. You can't go home again, and hysteresis—the physicist's term for this effect—is a social law as much as a physical one.

One of the consequences of the COVID-19 trauma was to accelerate what my colleagues George Loewenstein and Nick Chater[1] have termed the drive for sense-making. Our tolerance for endless catalogs of disconnected facts is always low, and out of the avalanche of events in the last two years, we were mad to explain what was going on. What matters, in that sense-making process, is not accuracy, but more ambiguous values such as simplicity and unification.[2] We didn't just want to know how it all happened. Our minds wanted the explanation to feel right—and every day the "it" we wanted to explain grew larger.

In retrospect, many of the things we know about explanation-making held true. With so many out-of-scope events, people built explanations to account not just for the disease, but also for the accompanying—often incidental—political events, and even, at a meta level, for the distortions of our appointed explainers.

As time went on, these explanations expanded to encompass more than just the immediate progress of the virus. The virus was not simply something that caused the classic symptoms we were so vigilant for: it

---

1 N. Chater and G. Loewenstein. 2016. "The Under-Appreciated Drive for Sense-Making," *Journal of Economic Behavior & Organization* 126(B): 137–154. doi: 10.1016/j.jebo.2015.10.016

2 Z. Wojtowicz and S. DeDeo. 2020. "From Probability to Consilience: How Explanatory Values Implement Bayesian Reasoning," *Trends in Cognitive Sciences* 24(12): 981–993. doi: 10.1016/j.tics.2020.09.013

seemed to become itself a symptom of more fundamental grievances, fears, and hopes. The virus was used to diagnose everything from the wisdom of the managerial class to sustainability of globalism and the rising power—or hidden weakness—of China. Just as doctors do, we seemed to see beneath these symptoms—wailing sirens and quarantine breakers—to a world that made sense, where a biological virus could in turn be caused by a social one. But which one?

As the sirens fade, and vaccination rates slowly move toward herd immunity, we have not only traveled far from where we, personally, each began. We also, it appears, have traveled away from each other.

If our understanding of sense-making stumbled, it was a matter of underestimation. The semiotics of COVID-19 were far more complex than I could have imagined. It was less about keeping safe—a question we still have yet to answer well—and more about making sense of how safety could be so strangely taken away. The work in our laboratory at CMU sees this now in our synoptic, data-science studies of online conspiracy groups, and in the ways in which COVID-19 is quickly tangled into explanations of why, for example, liberal democracies are primed for implosion.

In the social domain, COVID-19's hysteretic release has revealed a far greater heterogeneity than we ever thought possible. As the sirens fade, and vaccination rates slowly move toward herd immunity, we have not only traveled far from where we, personally, each began. We also, it appears, have traveled away from each other. It's not just that our explanations for *the way things are* have been fundamentally altered by the nearly two chaotic years of COVID-19. It's that our explanations have been altered differently. In a newly maskless world, many of us—myself included—play our cards a little closer to the chest, lest we elicit a story even crazier than our own.

The transformation of a virus into a symptom in this fashion seems to be a part of a longer-term development in political life—one almost Hegelian in its scale. In retrospect, COVID-19 pushed our democracies further and further away from the coalitional politics of the twentieth century, and into a new *epistemic* realm. Our tribes may no longer be matters of common interest, but ones of common belief. Such a world is one where we recognize each other's dignity in our toleration—or not—of their explanations, rather than the validity of their needs. In such a world, politics is won and lost not in the material realm, but in that of ideas. And those ideas are only getting stranger.

Batch 1 Podcast

THE TRANSCRIPT OF *COMPLEXITY PODCAST* TRANSMISSION SERIES EP.1, DISCUSSING THE FOLLOWING TRANSMISSIONS

[Batch 1, released 30 March 2020]

T-000: Citizen-Based Medicine
David Krakauer

T-001: Complex Crises & the Inevitability of Ethically Loaded Decisions
David Kinney

T-002: Reducing Conflicting Advice on Allowable Group Size
John Harte

T-003: Modeling the Effect of Social-Distancing Measures in the COVID-19 Pandemic
Luu Hoang Duc & Jürgen Jost

T-004: Thinking out of Equilibrium
Simon DeDeo

This podcast transcript has been abridged for length and clarity. Find the full podcast at https://complexity.simplecast.com/episodes/26

COMPLEXITY PODCAST: TRANSMISSION SERIES NO. 001

RIGOROUS UNCERTAINTY: SCIENCE DURING COVID-19

DATE: *6 April 2020* **HOSTED BY:** *Michael Garfield, featuring David Krakauer*

MICHAEL GARFIELD: We're here to talk about this new essay series that the Santa Fe Institute is doing, the Transmission series, a set of complex-systems perspectives on the COVID-19 pandemic, and its sort of broader consequences and implications. Why don't we start by just explaining a little bit about the villain of this drama, the virus itself? Tell us a little bit about RNA viruses, and then what you had in mind for framing this essay series as an exploration of these broader issues.

DAVID KRAKAUER: These are trying times for all of us. We want to just try and get as many useful ideas on the table as possible. I thought I might just start by explaining what a virus is and what a coronavirus is, in particular, and what makes it so frightening. Viruses are parasites, and they're parasites that can only reproduce by gaining access to the cells of their hosts. Once in there, they hijack or hack the machinery of the cell, in order to copy themselves and continue their transmission.

Now, viruses come in a lot of different shapes and sizes. For some of them, DNA is their genome, like us, and some of them have RNA genomes, like the coronavirus. The coronavirus is a very big RNA virus by the standards of RNA viruses—about 30,000 bases long. The good news about that is that, unlike most RNA viruses, it's not very mutable. The reason why it's so big is to ensure that its replication is high fidelity. Most viruses that are small make tons of mistakes when they copy themselves, but the coronavirus is very large and has a machinery in place to ensure that's not true. It's good news for us because it means that once we have a vaccine, we won't have to constantly be reinventing it because the virus will be the same one, we hope, unlike flu, which is very mutable.

Coronaviruses have been around for a long time. They account for about 30% of common colds, along with other viruses like rhinoviruses, and usually we haven't been that concerned about them because they're not very virulent. They don't cause lots of pathology. This one does. This virus is particularly horrible because it hijacks a physiological mechanism that we are all dependent on—sometimes called the renin–angiotensin pathway—and this is the way that the body regulates various cardiovascular functions. The coronavirus—this particular one, COVID-19—binds to one of the crucial cell surface receptors that we need to regulate our cardiovascular system and gains entry into the cell.

A variety of different groups are more susceptible to the virus; diabetics, for example, because they have more of this receptor, so there's more for the virus to access and enter, and, people who are potentially having medication for cardiovascular disease might also suffer because that also interferes with this pathway. This is a particularly horrible virus because it's targeting something that we absolutely need, and exploiting that need in order to ensure that it replicates itself effectively.

The virus itself has lots of different proteins in it. Everyone's heard of the spike protein—the S protein—and this is that one that gives it its name—coronavirus—because they look like little sunspots, and it needs that to get into the cell. Current vaccine development is, in part, targeting that spike protein to prevent it getting in the cell. But the most frightening protein it has is the E protein. This sits in its membrane and elicits an immune response. It generates cytokines, which are small little peptides that lead to inflammation, little proteins that tell the immune system to trigger the inflammation response. It's the inflammation response that gives rise to the edema—the filling up the lungs with water—that give us our respiratory complications. So, there is this little protein that causes our immune system to go a bit haywire, leading to these horrible pathologies that we're all hearing about. It's not the spike protein that's necessarily causing the damage, but it's the spike protein that gives it access to the cell.

M. GARFIELD: So there we have the "Map of the Problematique" in miniature. Let's move on to this essay series and the broader motivation for doing this.

D. KRAKAUER: SFI has been working on virus evolution, immunology, and epidemiology for twenty-five years, and many of our researchers are on the front line of this disease. They're forecasting the progress of disease, providing that to the CDC[1] and the White House.[2] They're

1 https://www.cdc.gov

2 https://www.whitehouse.gov/priorities/covid-19/

working on tests, repurposing fundamental research labs in order to make them useful to society. That's actually something very interesting, that all of these labs that were doing what we would call basic science or fundamental science . . . it's taken almost no time in this period of great urgency for them to turn around and actually provide a service to society, which I think is a really interesting fact. It hasn't really been noted that these distinctions that we often make between basic and applied or fundamental aren't really real, because when it comes to crisis and when it comes to real need, scientists—fundamental scientists—are very prepared to turn that into an applied project very quickly. So that's been a big part of what we've been up to as a community.

But beyond that, there are all these questions that are the ramifying, cascading effects of this infection on behavior, on commerce, on transport, on ecology, human psychology, institutional credibility . . . these are the things that are making life so trying, above and beyond the primary infection; flu doesn't do this. And this series is an effort to explore why it is that a tiny bug that's thousands of times smaller than a grain of salt, that has none of the alarming features of an AI that we've been talking about kill switches for in the last several years, is actually bringing down the complex systems of the world. And we would like to understand that in order to forestall this happening ever again: what we need to put in place, what new ideas we need, what new models we need, what new social norms we should cultivate, in order to ensure this kind of tragedy doesn't repeat itself. That's what this series will be about. So, everything, all the layers of the onion above and beyond the primary biology of the virus and epidemiology.

M. GARFIELD: Before we started recording, you and I mapped out how the first five essays in this series are linked to one another conceptually. We're going to save Simon DeDeo's for the end, but I think it's worth noting, to connect to what you just said and to lead into your introductory piece, that this is one of those issues where we don't notice the systems upon which we depend until they break. So that very, very, very small virus, leading to a global systemic restructuring is really perfect in the way that it exploits our own physiological vulnerabilities to great effect. It's also an interesting invitation to a broader menu of really important, fundamental complex-systems concepts. The first concept that seems worth discussing is the name of the series itself: Transmission. Why don't you kick us off with a coarse outline of your article?[3]

D. KRAKAUER: What I tried to do is understand why it is that this particular outbreak has been so disruptive of so many complex systems. If you look at diseases like cancer or Alzheimer's or respiratory infection in general or cardiovascular disease, these things have been around for a long time, are terrifying, cause many, many deaths, and there's very little prospect of us curing them in the next decade or more. And yet, they haven't led to catastrophic failure of systems. I wanted to understand why.

One of the characteristics of these diseases is that they're truly complex in the following sense: If you look at cancer, there is no *one* gene that is responsible for the disease; there are many tens, if not hundreds, if not thousands. That's made treatment very hard, because you can't target one site in the genome. So, that's a negative from the point of view of intervention. On the other hand, that is a property of complex systems that's critical for robustness. Our body doesn't want to give one lever or one switch that controls the entire system to one function or one disease because it collapses the complexity into simplicity, and it allows for the possibility of complete disruption with a single point of change. Complexity is what we would call homeostatic. It's a good thing—it's hard to understand, but it means it is much harder to hijack.

Now, look at this particular outbreak. What was very clear to me and many others is that there was one basic dominant causal principle at all complex scales, and that's transmission. The virus transmits from cell to cell. It then transmits from body to body in social networks. It then transmits across cities, through commerce, and surfaces. It then transmits from city to city through transport networks, and then across the globe. It hijacks supply chains, in some sense, in the same way that it was hijacking the ACE[4] receptor. It's hijacking the transmission mechanisms of our world. And that does something strange: It leads to a paradoxical simplicity, which gives the virus huge power over a number of scales that wouldn't be true, for example, for Alzheimer's. That was the first

3 See T-000 on page 15

4 angiotensin-converting enzyme

point I wanted to make, which is: human culture has become vulnerable at the level of transmission because too many different mechanisms are aligned . Moving forward, we have to think about engineering misalignments, in some sense, such that this can't happen.

The second point I was trying to make is that quarantine and social isolation are actually collaboration in this world; by staying home, you're not just hiding from a menace; you're actively participating in misaligning one element of this complex system so as to interfere with the progress of the infection. This idea somehow that a citizen could actually play a part that was as important as a vaccine, but instead of preventing transmission of the virus into another cell at the ACE-receptor level, it's preventing transmission of the virus at the social-network level. So we're actually adopting a kind of behavioral vaccine policy, by voluntarily or otherwise self-isolating. I think that's a very important point for everyone to understand, and I actually argued that everyone should be awarded some fraction of the Nobel Prize in Medicine for the sacrifices they're making in order to minimize the transmission.

> Complexity is what we would call homeostatic. It's a good thing—it's hard to understand, but it means it is much harder to hijack.

M. GARFIELD: I think that the prize would be significantly less than proposed federal bailouts, but the fractional prestige would be a nice consolation. This seems linked to something that we've discussed a couple of times on the show in the last few weeks, talking to SFI-network epidemiologists about how there's another layer of transmission, which is the informational and behavioral contagions that you allude to here. How do we get people to adopt different behavioral regimes that are going to relate to the transmission of this virus in complex ways?

D. KRAKAUER: One of the things that we've been working on—our community, for the last several decades—is the mathematics of infectivity. We can write down equations or computational models to describe how things transmit. And one of the things that we discovered very early is that the same kinds of mathematics used to describe the transmission of disease could be used to describe the transmission of ideas. The same kinds of networks that promote or impede the transmission of disease promote or impede the transmission of ideas. So that's in part what I mean by the alignment.

One of the things that happened in this particular outbreak is the memetic transmission—the transmission of ideas—was much faster than the transmission of the virus. But the disappointment, I think, that many of us felt is we didn't act on that. We weren't leveraging the comparative advantage of idea infection over biological infection. But the common mathematical structure that applies to both cases and more, incidentally, is very valuable because it does give us insights into how we might be able to bring these systems out of alignment to potentially control them.

Human culture has become vulnerable at the level of transmission because too many different mechanisms are aligned.

M. GARFIELD: With that, we can frame the next four articles in this series as the beginning of an effort in a kind of informational contagion. Like a conceptual vaccine, maybe.

Let's start with David Kinney,[5] because I think his piece addresses something that's really critical here, something that we talked about when he was on the show a few weeks ago[6] about the difference between the way that science is actually practiced and the way that it's publicly perceived. Typically there's rhetoric around science that this is objective and purely quantitative: "Just give me the numbers." But, especially when you're dealing with models of complex scenarios that produce a

5 See T-001 on page 21

6 Complexity Podcast, episode 19, aired on February 19, 2020. https://complexity.simplecast.com/episodes/19

spectrum of possible futures, there is a hidden value judgment that scientists and scientific advisors have to make in communicating their findings, and that value judgment is driven by a kind of market dynamics about the estimation of risk in giving the broadest available spectrum of possibility, versus giving the clearest and most actionable advice.

D. KRAKAUER: Here we are in New Mexico, where Los Alamos National Laboratory is based, where the Manhattan Project was conducted. And, to my mind, that's one of the most salient examples of this fuzzy boundary line between basic science and policy. If you remember, General Leslie Groves appointed Robert Oppenheimer to basically oversee the production of the atom bomb. That first collaboration worked out, from their point of view, reasonably well. Oppenheimer was overseeing the science, and Groves was interfacing with government and society in a certain way.

But when it came to the hydrogen bomb, everything changed, and Oppenheimer was so demoralized by what had happened with the dropping of the two atom bombs, that he felt compelled to make policy remarks. And that led, as we all know, to various trials and conflicts between Edward Teller and Oppenheimer, where Oppenheimer was accused of being a communist spy. It's an extremely difficult problem, and scientists are, quite naturally, as a consequence, very cautious about making policy recommendations.

What David Kinney says in his Transmission article is that in the late '50s there were philosophers and ethicists who were advocating for a strict division of power. This is largely the work of Richard Jeffrey and Isaac Levi. They suggest that it was the role of the scientist to just give raw numbers to policymakers and politicians and allow them to make the decisions. What David points out is that what policymakers want is certainty. So, if I, as a scientist, provided you with probabilities for certain outcomes, if those probabilities are near 1 or 0, you will act on them. That's what you want—certainty. But they're much less likely to be correct.

On the other hand, if I give you nuances, subtlety, and say, "Well, maybe there's a 0.6 probability of this and 0.4 of that," you, as a policymaker, are very dissatisfied, but it's more likely to be closer to the true distribution of outcomes. There's a catch-22 here. Scientists want to present the true numbers, but the true numbers are unlikely to be popular with the policymakers. So we are forced to exaggerate the severity of outcomes in order to have them adopted. And I think this is actually a problem that can't be escaped.

M. GARFIELD: Yeah, and in a way, this is also a sort of outbreak of its own, at least conceptually, where we're no longer assigning the accountability or the authority of these values-laden decisions to . . . It's sort of the typical process to want to pass the buck: blame the politician, "Well, they were just acting on the advisor." And then, it's like, "Well, actually, we have bad data . . . "

D. KRAKAUER: Let me say one more point on this, actually, Michael, which I think is very important. That is: good scientists are comfortable with uncertainty; bad ones are not. This leads to a genuine conundrum, and it's why pseudoscience flourishes in periods of uncertainty in culture. Because if the real scientist is saying, "Look, we don't have the data to make the decision. To the best of my ability, these are the odds," and a bad scientist comes along and says, "No, I know exactly what's going on," that's why bad scientific ideas are adopted so readily and with such alacrity in periods of great stress. They appear to give certainty in periods of uncertainty.

It's something that the public should be very cautious of. When someone says, "I have a solution, I have a remedy," what they're really doing is exploiting this need for certainty, and I think it's a very desperately difficult situation to be in. But just like sport, where outcomes are not deterministic, the same is true of all complex systems, including disease. We have to cultivate our patience, I think, and our understanding that certainty is acquired very slowly, and be extremely suspicious of anyone who claims to have found a solution with inadequate data.

M. GARFIELD: When we had Rajiv Sethi on the show[7] and he was talking about stereotypes within the context of criminal justice, one of the things that really lingered with me after that discussion was how so much of this comes down to the ways that our brains are tuned to

---

7 Complexity Podcast, episode 7, aired on November 13, 2019. https://complexity.simplecast.com/episodes/7

reduce uncertainty by grouping what are determined to be equivalent experiences, equivalent phenomena.

Now, given enough time, the ominous stranger in a dark alley, you might be able to sit down and have dinner with them and get to know them and develop a model of that person as a unique individual. But one of the problems with stereotyping and criminal justice and also now, as you just mentioned, the desire to cleave to oversimplified explanations—explanations that in a longer timeframe are going to appear maladaptive—is that we don't have the time to collect the data that we need and form the truly rigorous and robust models because of various problems with disease surveillance or economic urgency. There's a balance that we have to strike. That leads into the next paper by Luu Hoang Duc and Jürgen Jost[8] on what do we do when we're left with only bad data and relatively little time to act on it.

D. KRAKAUER: Let me just clarify, in periods of uncertainty, we have to learn to live with it and try as best as we can to avoid those claiming to have certainty in the face of inadequate data. What Jürgen and Luu Hoang have done is make a very interesting point—and that is very paradoxical, I think—that when data is really bad, you should use the simplest model at hand. When data is very good, you can use complicated models. There's been a lot of conversation about this early prognostication that came out of the Imperial College model. It had, in retrospect, potentially wildly overestimated the number of fatalities, and now, of course, that model has been modified so as to reduce that number.

One of the problems with that model philosophically is it was vastly too complicated given the data that we had. There was a temptation, because of policy, in fact, to put everything in, so we're going to put in the number of schools, the age distribution, the household, the number of hospitals, their spatial locations, the position of airports. These are very complicated agent-based models, and these models are absolutely critical when you have really good data, like those of cities under normal conditions.

But what Jürgen and Luu Hoang are saying is, what if you don't? Well now, you should do the paradoxically opposite: use the simplest

8 See T-003 on page 37

model you possibly can because they're much less sensitive to fluctuations in the data; they don't "overfit"[9] the data. The last thing you want to do is overfit, parameterize a model based on sparse or bad data. And the particular idealization that they advocate is a linearization of the data. If one reads that paper, you'll see they do linear extrapolations based on a logarithmic transformation—they essentially linearize the data, and they try to say that this would be, given our current state of uncertainty, the best course of action.

Now, whether that's true or not remains to be seen, but from the point of view of us as a community and these listeners, the key philosophical point is that if someone comes to a complicated model with bad data, you should be very suspicious of it, and you should be more tolerant of simplicity in times of uncertainty.

You only want to put complexity in your model if it's justified by the empirical data. If it's not, leave it out, because it will underperform the simple model.

M. GARFIELD: Let me see if this analogy holds with you: we're approaching Earth and we don't have a map of the coastlines of the continents; we're operating with a very low-resolution camera deciding where to land. You don't try to come up with a map of the coastlines based on data that's too granular, that you don't actually possess. You basically just aim for the center of the continent and continue to update your maps as you get closer to it, as things come into focus.

D. KRAKAUER: It's reasonable. I think maybe a better example would be in an economic setting where we have basic laws of supply and demand. They're very simple models. If the price goes down, then you're more likely to purchase. If the interest rates go down, you're more

9 IBM says "Overfitting is a concept in data science, which occurs when a statistical model fits exactly against its training data. When this happens, the algorithm unfortunately cannot perform accurately against unseen data, defeating its purpose." https://www.ibm.com/cloud/learn/overfitting

likely to borrow. These are very simple models. And you can imagine a very complicated model that said, "I know exactly what Michael likes, I know what foodstuffs he purchases, I know what TV show he likes, and I'm going to put all of them into my model." But now imagine that data was all totally bogus. That model would do much less well than the simple macroeconomic model that just looks at supply and demand regularities. You only want to put complexity in your model if it's justified by the empirical data. If it's not, leave it out, because it will underperform the simple model.

M. GARFIELD: We have an example here. John Harte[10] gave a simple model to reduce conflicting advice on the allowable group size as we're going through school and business closures. This is something that I brought up with Laurent Hébert-Dufresne, when we had him on the show,[11] this question of where are we getting these numbers that say 200 people is too many, but twenty-five is acceptable? This sort of gets back to that David Kinney remark on having to sort of place a bet on where you draw the line. But John's got a model here that's simple enough to understand and doesn't rely on complicated extrapolations about the internal structure of these different kinds of groups, and so on.

D. KRAKAUER: We, like everyone else in the world who's being sensible now, are isolating and in quarantine. What we're doing essentially is reducing our exposure to others, reducing our group size. And what John asked was, "Look, what is a sensible, allowable number of individuals with whom you can interact?" And as we come out of quarantine, the last thing we want to do is resume normal behavior. We have to titrate it or iterate up slowly toward normalcy.

He gives a very interesting simple analysis. He has a model: you have a number of groups that could be schools, for example. Each school has a number of students in it. Typically, you have more students in each school than you have schools. What he shows is that if you double the size of a group, you have a four-fold increase in the transmission of the disease; this is called a superlinear scaling. In other

10 See T-002 on page 29

11 Complexity Podcast, episode 24, aired on March 25, 2020. https://complexity.simplecast.com/episodes/24

words, even small groups can have very, very high transmission rates. So, when we talk about these numbers, five, ten, one hundred, they're actually quite meaningful. I heard early quotes coming from, I won't name who, where they said, "Well, there's no real difference between ten and one hundred." There's a huge difference between ten and one hundred! The sensible thing for people to do would be to be in the single digits, that's what a sensible group size is. Because of the superlinearity of the scaling, a small increase in a group leads to a very large increase in transmission.

> We're moving into a world now where everyone should see themselves as a part of a solution. In a complex system, you are absolutely as important as, if not more important than, the ACE receptor.

I would hope that what comes out of those simple models, if people can internalize this message, is that, as we basically break free of the grip of this horrible virus, we actually don't rapidly or too precipitously return to normal, but we *gradually* return to normal. That would be the sensible thing to do. It would be the considerate thing to do for all the citizens of the world, and I think John's model makes that extremely clear. There is a difference between ten and one hundred. There's a difference between one and ten, there's a difference between ten and twenty. In some sense, add one or two at a time. That seems kind of preposterous at one level, but it would be the sensible thing to do.

M. GARFIELD: To link this back to your piece, this is the continued citizen-science experiment, that every time anybody leaves their house, they're increasing the size of that group, and then providing data to someone, hopefully not at great personal expense.

In the case of this novel coronavirus what we have is a situation where, as people like SFI External Professor Lauren Ancel Meyers have shown in their research, the interval between someone receiving it and actually developing symptoms is great—this is part of why our data is so patchy on this particular outbreak, because it's probably impossible to reconstruct the actual transmission network. There are so many sort of mysterious links too far separated in time and space.

And yet we've been relatively effective, at least by some measures, at containing it. Like you said earlier, this was a very rapid and emergent social mobilization. People were isolating themselves before they were ordered to do so by the governments. At least in that respect, it's like you said, "an antiviral flash anti-mob." Kudos to the internet and the citizens of the web for being able to coordinate our misalignments.

D. KRAKAUER: I think you made a very good point earlier, that this gets back to that notion of citizen-based medicine. We're very used to these interventions that are in some sense biochemical, and they give us a sense of security because they're basically reductionist. But we're moving into a world now where everyone should see themselves as a part of a solution. In a complex system, you are absolutely as important as, if not more important than, the ACE receptor.

This graduated return to normalcy is actually quite a sophisticated mathematical principle. If people could acquire these new kinds of habits, we'd make the world a much better place looking forward. What we would like to do, I think, is develop guidelines for sensible forms of behavior that don't feel oppressive, but feel empowering, that give people a sense of what they might be able to do under these circumstances. And they're not easy. They're actually, in some sense, more difficult than developing a vaccine because we're developing a social analog. So you're absolutely right. I think every citizen now is actually, in some sense, empowered to control this pandemic. And John's article is telling us, is giving us an insight into, how that should be engineered. What I would like everyone to do is develop a kind of sophisticated sense of how social networks and transmissions work, in order that they can make the best decisions for themselves and society in terms of how we roll back to normalcy.

M. GARFIELD: That brings us to Simon's piece, because whatever normalcy we roll back into is going to look very different from the normalcy that was disrupted at the beginning of this year. As Andy Dobson pointed out in our episode,[12] we've done such an excellent job at containing this, but very few of the models make it look like this is a one-wave infection. Very few of them make it look like we're going to have completely trounced this thing.

It seems very likely that there will be additional waves, that there will be additional injunctions to go back home after we have returned to work, to distance ourselves again, because we will not have infected enough people through our successful early containment efforts to have developed herd immunity for this. This leads us into Simon's piece: we're in a kind of interregnum here where we don't really know what the new normal is going to be.

D. KRAKAUER: Simon DeDeo's contribution[13] is really about the dangers of habitual thinking in relation to all that we've discussed before. Society has evolved over thousands of years so as to minimize the cognitive burden on individuals. And we call that minimization *habit formation*. We have rules of thumb—heuristics—that allow us to solve problems quickly because the environment in which we live is more or less constant. But when that environment changes, those habits become deleterious, they become maladaptive, they no longer fit.

And then we have to sit down and reason again. What Simon is arguing is now is the time for us to rethink *thought*, meaning rethink *habit*. There's a nice language for this. Daniel Kahneman wrote this book[14] on System 1 and System 2, where System 1 is the instinctual, the reflexive, and System 2 is the analytical, the more ponderous. And what's happened, in a sense, is the crisis has forced us to move behaviors that would normally sit in System 1 into System 2. It's almost as if all these years you've been playing chess, and someone came along and said, "Oh, by the way, the rook now can move on diagonals, and the king can move three squares on the horizontal and vertical, and the pawn actually now behaves like a bishop." Things that you would just have almost

12 Complexity Podcast, episode 23, aired on March 18, 2020. https://complexity.simplecast.com/episodes/23

13 See T-004 on page 57

14 D. Kahneman, 2011, *Thinking, Fast and Slow.* New York, NY: Farrar, Straus and Giroux.

done automatically now you have to rethink completely; it's a really sort of difficult thing to have to handle and confront.

I think what the pandemic has done is exactly that: things that we took for granted in society, things that are extraordinarily comforting for us, as human beings—human proximity and conversation and group living—have been challenged, and we have to rethink it. Hopefully, we will return to that again, but we might not, and I think that's Simon's deep point, which is that it's time to be analytical again and to reconstitute what becomes habit of the future. One of the things that we can all be doing, as we're all incarcerated in this horrible moment, is to think a little bit about that. I mean, what would I change given that the world might throw at me a catastrophe of this scale again? We don't want to become paranoid—this is not about stockpiling firearms, which seems to me absolutely ludicrous—but it is about being very thoughtful about a world where perturbations of this magnitude or slightly less might be more common than we had anticipated—the so-called long tail.

M. GARFIELD: As you've mentioned a few times in this, through your lens on the now suddenly obvious importance of individual citizens to a collective experimental process here . . . It's been interesting, I've been seeing a lot in the SFI-affiliated economics discussion about how this is changing the assumptions that we make about the balance between labor and capital, for instance. Recently, Suresh Naidu on *Quartz*[15] was talking about the gig economy and how these essential independent contractors are suddenly the stone upon which so much well-being is balanced right now. They have a negotiating power that they didn't have six months ago. American Public Media's Marketplace,[16] in their reading group that's going on right now for the CORE Econ textbook[17] that Suresh and Rajiv Sethi and Sam Bowles and Wendy Carlin worked on, was talking recently about how this is shifting the way that we understand healthcare from something that's conventionally understood as a private good to something that's understood as a public good. This crisis has made it much

15 M. Cheng, "Does Being Made 'Essential' Finally Give Gig Workers the Upper Hand?" *Quartz*, April 1, 2020, https://qz.com/1829003/do-essential-gig-workers-finally-have-the-upper-hand

16 https://www.marketplace.org

17 https://www.core-econ.org

more evident than it used to be that well-being is something that exists in a network—that if I get sick, it's going to affect your health, your economic well-being, and so on. What seems to me to be the trend here is a kind of democratization of the SFI way of thinking, really—an understanding of our individuality as something that is emergent and relational. How do you imagine that moving forward?

D. KRAKAUER: I'd like very much that society became more empirical, more analytical, more cooperative, more prosocial. These are things that we all hope for, but I'm not that optimistic. If you look what happened on 9/11, there was this extraordinary collegiality at the level of the globe, great sympathy, the prospect and possibility of global collaboration. But very quickly, it turned into rabid xenophobia and protectionism, and it all went horribly wrong. I don't think we can place the blame anywhere in particular, but human beings very quickly reverted to their usual selfish selves.

That's why I think this point about analysis versus habit is so crucial because what we have to do is make what is clear *when we reason* also clear *when we act* ; we have to turn thought into action by making analysis habit. Unless we do that, there'll just be a massive rebound to normalcy. I mean, we're all looking at oil prices, and it's extraordinary. We're also looking at the incredible rapid cleaning up of the environment, given the reduced traffic and human activity. But we all know what's going to happen: as soon as a green light goes on, humans will go out and say, "Look how cheap gas is, I'm going to take road trips," and it'll massively increase production. And all of those transient positives that only a little bit offset the terrible negatives of what's happening to people in their lives—in unemployment and health—will be eradicated.

So, I'm not the person who believes in this. I believe that, unless you move from something into a social norm and a habit, it's basically behaving like a Pollyanna to think that the decency that one observes under periods of crisis are maintained in periods of normalcy. We actually have to act in very thoughtful ways in order to ensure that that's true, instead of hoping that it will be true.

M. GARFIELD: Certainly, although were we to rush back into this, we would be running back into the same wall that nailed us the first time.

The normal that we would be rushing back into is a mirage, and the second lesson would be even more difficult than the first. I imagine, over time, running multiple iterations of our foolish and uninhibited return to the great American road trip and business as usual, and it seems like it would select out those behaviors over two or three runs.

Incremental change is real change; revolutionary change is illusion.

D. KRAKAUER: But you see, I'm much more sympathetic to the desire for normal. The example that I like to give is exercise. One of the things we've learned, I think, in the last several decades, is that everyone wants to be healthier, but it's extremely difficult to change your habits. And there was this perception, say, twenty years ago, that you'd have to sort of cut your caloric intake in half, and you'd have to start running twenty miles a day, and so forth. And no one could do it. Now we live in a much more humane world. You can say, "Look, if you take a ten-minute walk, that's fantastic. That's genuine progress. You're developing habits that you can then build upon." And I think the worst thing that could happen is an alternative ideology that doesn't like this one, proposes these draconian transformations of society that will be rejected wholesale.

The right thing to do is learn from fitness coaches and say, "Look, it's not a big change, maybe we should stop shaking hands, right? It's not a big change, we should be a little bit more considerate of individuals in the gig economy." Incremental change is real change; revolutionary change is illusion. One of the things we learn from complexity is that any attempt to treat the system as if it was simple and that there are one or two levers that will lead to transformation is a mistake, and we have to have a nuanced attitude toward tiny changes in many different places.

2

[BATCH 2, RELEASED 6 APRIL 2020]

Transmission need not **DISEASES**. *Culture and even franchised*— **INDUSTRY** *focused that convince people is* **ESSENTIAL** *to*

be confined to infectious

and ideas are transmitted

ADVERTISING *is an*

on transmitting ideas

that a **COMMODITY**

their quality of life.

—ANDREW DOBSON, Transmission 005

C.E.Orm.

TRANSMISSION NO. 005

THE NEED FOR DISEASE MODELS WHICH CAPTURE KEY COMPLEXITIES OF TRANSMISSION

DATE: *6 April 2020*

FROM: *Andrew Dobson, Princeton University; Santa Fe Institute*

STRATEGIC INSIGHT: *The disease models used to guide policy for the COVID-19 pandemic must capture key complexities of transmission.*

The study of infectious disease dynamics can be divided into two areas. The first is pathology, which focuses on the changes in the hosts due to the presence of the pathogen (e.g., dry cough, high fever, and acute respiratory syndrome). The second is transmission, how the pathogen moves from an infected individual to an uninfected one in order to initiate a new infection. In sharp contrast to pathology, we almost never see transmission occurring—we can only infer that it has occurred when a healthy individual develops signs of infection. We then need to trace back their activities in the hopes of identifying an individual with symptoms that led to the transmission event.

This post will supply an overview of the problems that beset epidemiologists when we try to measure transmission, while familiarizing the reader with some key models used to measure transmission, and prevent it.

Transmission need not be confined to infectious diseases. Culture and ideas are transmitted and even franchised—advertising is an industry focused on transmitting ideas that convince people a commodity is essential to their quality of life. Teaching may be the ultimate form of cultural transmission; from kindergarten right on through to graduate school and beyond, ideas are developed into presentations that transmit

---

OPPOSITE: *Gabriel Orm, "Two Bats Nailed to a Timber Wall, Knife and Quill Pen," 1738*
NATIONALMUSEUM, STOCKHOLM

knowledge from one individual to others. Similar mathematical frameworks are used to describe the ways in which ideas and pathogens are transmitted. Here I will focus on pathogens, and only occasionally indicate applications to cultural transmission.

Teaching may be the ultimate form of cultural transmission; from kindergarten right on through to graduate school and beyond, ideas are developed into presentations that transmit knowledge from one individual to others.

Models that are used to describe the dynamics of infectious diseases fall into four broad categories: a major split is into microparasites and macroparasites; a second split is into vector-borne diseases (VBDs), sexually transmitted diseases (STDs), and other infectious diseases (OIDs). The first division acknowledges that some pathogens are too small to be accurately counted and their dynamics are best described by a mathematical framework that divides the host population into Susceptible, Infected, and Recovered/Removed or Resistant hosts, depending on whether they died, recovered, and rejoined the susceptible hosts, or developed immunological resistance that protects them against future infection. The macroparasites are physically large enough to count: ticks, fleas, and worms. The pathology they create is a function of their abundance; having one or two ticks or fleas is an annoyance, but carrying around a large number of worms in your gut is a major constraint on your ability to grow physically and intellectually. Models for these pathogens need to consider the frequency distribution of parasites across the host population; these are inevitably aggregated, with a large parasite population relative to the host population, and most hosts have only minor infections. The macroparasites produce complex but short-lived immunological responses that only weakly protect them against future infection. However, they can modulate the efficiency of immune responses produced in hosts concomitantly infected with microparasites, and may also significantly reduce the efficacy of vaccines against these pathogens.

The second division, into VBDs, STDs, and OIDs, takes us back to transmission, with examples of each form known for both macro- and micro-parasites. Malaria is the poster child for vector-borne diseases. The pathogen sequentially uses mosquitoes and humans (or other vertebrates) as hosts, converting each of them from susceptible to infectious individuals. Technically it may be possible to identify which mosquito bite gave rise to an infection, but only a very small proportion of mosquitoes are infected. As a result, we often experience many mosquito bites before getting infected, and we are again stuck with the problem of who the mosquito bit to acquire its infection.

Sexually transmitted diseases have many similarities with vector-borne diseases, although transmission mainly occurs alternatively between each sex within the same species. The problem of when someone was infected and by whom again arises, and it is never apparent that one has been infected until after the incubation period, when symptoms appear. STDs can also be vectored. The smut fungi transmitted between the flowers of many plants are effectively STDs transmitted by pollinating insects. Heterogeneity in transmission is driven by the frequency distributions of numbers of new sexual partners; this variability may be fairly low in the case of insects transmitting smut fungi, but could be significant for human STDs in individuals with many partners. These heterogeneities hugely increase the rate of spread of the pathogen; a big early success in the AIDS epidemic was changing people's behavior by encouraging them to have few new partners. This reduced the variance and significantly reduced net transmission rates.

The OIDs are transmitted by free-living particles, expelled from the infectious hosts, that directly infect the susceptible hosts. The duration of time these infective stages last in the free-living stage is crucial—in the case of influenza and SARS-CoV-2, coughing and sneezing releases a cloud of infective particles into the air that may only be infectious for a couple of minutes. In contrast, bacteria such as anthrax produce spores that may survive in the soil for decades. The parasitic worms that live in the guts of most vertebrates produce eggs that can survive for weeks to months. This creates a bifurcation in transmission mode, when infectious stages are short lived, then transmission from one infected individual is a function of the density of susceptible hosts in their vicinity. You may infect many more people when you sneeze on a crowded subway car than when you sneeze in a nearly empty one. This form of transmission, usually called density-dependent transmission, is assumed to be linear, with each infected individual infecting a fixed proportion of the susceptible individuals in their vicinity. In contrast, pathogens with long lived free-living stages, as well as vector-borne and STDs, tend to have saturating transmission functions: mosquitoes have to digest between blood meals, and even Casanova had to rest or dine occasionally. This form of transmission is modeled by frequency-dependent functions with the product of susceptible and infected hosts appearing in the numerator, and the total population in the denominator.

How do pathogens get transmitted between different host species?

Most animal and plant species harbor a significant diversity of parasites and pathogens—it may be there are ten times as many parasitic and inquilinistic species as free-living species, possibly more. Although we focus significant energy on conserving charismatic free-living species, each time one of these goes extinct, it takes with it a significant number of species that depended upon it for their existence. Some of these pathogens will be specialists that are usually only found on that host species, while others are generalists that can also use closely related species, or ones with similar feeding habits. In the majority of cases the parasites only have a minor effect on their hosts and their pathology is relatively low unless the host becomes stressed due to other changes in its environment, such as a reduction in food supply or change in local climate.

Occasionally, the pathogen infects the wrong host. This can occur when a vector feeds on a novel host that is more abundant than its normal host (as occurs when domestic livestock or humans are more abundant than native species). Animals captured for food or the pet trade often become stressed, leading to them releasing significant numbers of infective stages that can contaminate humans involved in the trade. Usually, the pathogen is unable to survive in the new hosts, as the cells it needs to infect in order to replicate are absent. However, when a pathogen does manage to infect novel cells and start replicating, this can then lead to the emergence of a new disease. This seems to be what is happening with the COVID-19. Genetic evidence suggests its natural host is a bat species, or possibly a pangolin. These species have very different physiology from humans (and most of our domestic livestock species); we rarely see any overt pathology in bats infected with these pathogens. This can change dramatically when the pathogen finds itself in the wrong host.

... pathogens with long lived free-living stages, as well as vector-borne and STDs, tend to have saturating transmission functions: mosquitoes have to digest between blood meals, and even Casanova had to rest or dine occasionally.

Bats have very different immune systems from other mammals, likely as a consequence of their ability to fly. Humans and other non-volant mammals produce the B-cells of their immune system in their bone marrow. Because bats fly, they have hollow bones; the only place they have bone marrow is in their pelvises, so they produce B-cells at much lower rates than other mammals. Similarly, active flight raises their body temperature to levels akin to fever in non-volant mammals, possibly constraining viral growth. Bats also do not store fat as it compromises their aerodynamic ability. Instead, they can enter torpor to get through periods when food resources are low. These all act as constraints on viral pathogens that disappear when the pathogen finds itself in a novel host whose immune response may interact with that pathogen in ways that are detrimental to both the host and the pathogen.

The dynamics of generalist pathogens provide important insights into the transmission dynamics of pathogens in structured human populations. Consider the dynamic of a pathogen that can infect multiple host species, each of which has a different body size and thus different birth and death rates and population densities. The species with the smallest body size will have the highest birth and death rates and population density. The largest will have the opposite. If the pathogen follows simple dynamics, with within-species transmission far exceeding between-species transmission, then each host will interact independently with the pathogen and each will exhibit its own epidemic cycles: large and frequent outbreaks in hosts with low body mass, and slow, less dramatic cycles in larger hosts. As we increase the relative rates of between-species transmission, these cycles will die out. Additionally, any tendency for epidemic outbreaks to occur is buffered by the pathogen's constant jumps between host species, preventing any one species from becoming too abundant. If we increase between-species transmission to levels where it matches within-species transmission, then the small species can use the pathogen to drive the larger species extinct; small species are abundant and recover quickly from outbreaks, while rarer large species cannot recover from frequent epidemics. Ultimately, only the smallest species survive, and they revert to the epidemic behavior they exhibited when between-species transmission was rare.

This exercise suggests that understanding rates of between-species transmission is an additional, vital component of disease dynamics. To that end, Who Acquires Infection From Whom (WAIFW) matrices provide a framework to examine how the pathogen moves between different groups of hosts and allows us to identify which section of the population acts as a reservoir to maintain the infection and which is subject to spillover events.

The matrices were originally developed to study the transmission of pathogens such as measles and rubella between different age-classes in human populations. The population in any area can be divided into preschool children, kindergarteners, middle school, high school, college, etc. In some ways, it will resemble the matrix for our hypothetical multispecies example; most of the transmission is within the same age class, because most of our interactions are with people our own age. These interactions may become more diffuse as we get older, and off-diagonal elements will become important as children interact with their parents and grandparents. If we could quantify the structure of these matrices, we would know a lot about how pathogens spread in populations. However, a problem instantly appears. The data on disease exposure we have for each age class increase linearly with the number of age classes into which we divide the population. The number of transmission elements in the matrix increases as the square of this number, so we have created an unsolvable problem. Unsolvable, unless we can assume similar rates of transmission in different age classes, or if we can gather independent social behavior on rates of human interaction. Both approaches have been used. The latter was particularly instructive, as it gathered data across multiple European cultures with different approaches. There was a curious similarity in the structure of the interactions; it would be helpful to expand these surveys to nations where different sex- and age-dependent interactions might modify the structure of the matrices.

These matrices are central to the social distancing now being put in place for the COVID-19 epidemic. Essentially, we are trying to massively reduce the strength of interactions within each age class and completely remove the interactions between age classes in order to protect older people who seem to be more susceptible once infected. Quantifying the structure of these transmission matrices is crucial for understanding the size of the epidemic and how to control its spread.

REFLECTION

UNDERSTANDING THE COMPLEXITIES OF TRANSMISSION IS KEY TO CONTROLLING VIRAL PANDEMICS

DATE: *4 August 2021*

FROM: *Andrew Dobson, Princeton University; Santa Fe Institute*

COVID-19 has fulfilled the traditional Chinese definition of chaos: an event that creates both crisis and opportunity. For those interested in epidemiology and the dynamics of infectious agents, boundless opportunity has appeared: over fifty thousand COVID-19 papers have appeared since a December 30, 2019, report on ProMED described a cluster of patients around the Wuhan market in China with an unusual respiratory disease.[1] Since then, more than four million confirmed cases have been fatal, and more than two hundred million people have been infected. Perhaps ten times as many as this have actually been infected, and we will probably never know how many additional fatalities have occurred indirectly because of COVID-19. The spread of the virus has provided deep and vital insights into multiple complex systems, particularly the structure of all currently extant human societies and the transportation networks that facilitate connections between them. Papers referring to SEIR (Susceptible–Exposed–Infected–Removed, where Removed signifies recovered, resistant, or deceased) models have been published by epidemiologists, but also people with little background in public health, from physicists through mathematicians to almost everyone who can spell SEIR; this has created considerable insight and significant confusion. In my earlier Transmission, developed in the first months of COVID-19, I raised questions about what we already knew as epidemiologists that might be helpful. Here, I'd like to focus on other important considerations that are central to understanding more subtle complexities of the dynamics and evolution of COVID-19.

The study of infectious disease dynamics can be divided into two areas: (1) pathology, which focuses on the changes in the physiology and

1 ProMED-mail, "Undiagnosed Pneumonia - China (Hubei): Request for Information," December 30, 2019, https://promedmail.org/promed-post/?id=6864153

abundance of the pathogen's hosts due to its presence, and (2) transmission, which infers how the pathogen moves from an infected individual to an uninfected one. We almost never see transmission occurring; we can only infer retrospectively that it has occurred when a healthy individual develops signs of infection. We then need to trace back their activities in the hopes of identifying an individual, with or without symptoms, that led to the transmission of the virus to the known infected individual and potentially other individuals not yet showing symptoms. Understanding transmission matters.

The spread of the virus has provided deep and vital insights into multiple complex systems, particularly the structure of all currently extant human societies and the transportation networks that facilitate connections between them.

The traditional epidemiological framework is designed mainly to consider transmission rates between people of unspecified age. The emergence of HIV extended this framework to consider rates of interaction between people with different categories of sexual activity, as well as activities such as needle-sharing between intravenous drug users and potential transmission by blood transfusion in the early stage of the epidemic. Now, the models used to guide policy for the COVID-19 pandemic must capture key details of transmission; increasingly, we realize that these are driven by rates of interactions between people of different ages who operate in different sectors of the economy and interact at different social locations and at different intensities. Determining these rates of mixing and interaction will be a major challenge for the next generation of COVID-19 models. These epidemiological models will then need to be combined with economic models for a multisector economy; hybrid economic–SEIR models can then examine how to reduce activities in different sectors of the economy in order to reduce transmission at a minimum cost to economic productivity. Initial exploration of such models creates almost an entirely new set of interdisciplinary studies in econo-epidemiological complexity.[2]

WHO ACQUIRES INFECTION FROM WHOM?

WAIFW matrices (Who Acquires Infection From Whom) provide a framework to examine how a pathogen moves between different groups of hosts and allow us to identify which sections of the population act as a reservoir to maintain the infection and which are subject to spill-over events. The matrices were originally developed to study the transmission of pathogens such as measles between different age classes in human populations.[3,4] They were crucial in understanding how HIV moved between different sections of the population during the AIDS epidemic.[5] They will be central to understanding the efficacy of the social distancing introduced for the COVID-19 epidemic. Social distancing is, essentially, an effort to significantly reduce the strength of interactions across all age classes and completely remove the interactions between young and old age classes in order to protect people who seem to be more susceptible once infected. Quantifying the structure of these transmission matrices is crucial for understanding the size of the epidemic and how to control its spread.

Social mixing and disease transmission are strongly determined by contact rates between people working in different economic sectors. During the COVID-19 pandemic, a major challenge has been how to restructure WAIFW matrices to reflect those interactions. Age is less important than economic welfare in determining the structure of these matrices. For example, people working in service industries—places like

2 Boucekkine, Dobson, Loch-Temzenides, Ricci, Gozzi, Pascual; manuscript in preparation

3 B. Bolker and B.T. Grenfell, 1995. "Space, Persistence, and Dynamics of Measles Epidemics," P*hilosophical Transactions of the Royal Society B* 348: 309–320, doi: 10.1098/rstb.1995.0070

4 B.T. Grenfell and R.M. Anderson, 1985, "The Estimation of Age-Related Rates of Infection from Case Notifications and Serological Data," *Journal of Hygiene* 95: 419–436, doi: 10.1017/s0022172400062859

5 R.M. Anderson, T.W. Ng, et al., 1989, "The Influence of Different Sexual-Contact Patterns between Age Classes on the Predicted Demographic Impact of AIDS in Developing Countries," *Annals of the New York Academy of Sciences* 569: 240–274, doi: 10.1111/j.1749-6632.1989.tb27374.x

bars and restaurants—will be in contact with many more people than those who can readily work from home. Some agricultural workers will be relatively isolated, while others will be working intensively with others, often in crowded conditions. Children and students add additional levels of contact across these economic networks, both with each other and their teachers at school or college, and with their parents and grandparents at home. The dynamics of measles, mumps, and chicken pox are driven by the mixing activities of children until they are either vaccinated or infected and recover as immune hosts. COVID-19 (as well as other emergent pathogens) has very different dynamics, as everybody is initially susceptible and transmission can occur wherever people gather and interact. If we can quantify the background level of contact between people in different economic and educational sectors, we should be able to examine both the economic and epidemiological impact of partially closing down or constraining different sectors of the economy.

. . . if there is mixing between vaccinated and nonvaccinated hosts, it is likely that exposure of vaccinated hosts to infected, nonvaccinated hosts will boost the immunological protection of the vaccinated hosts, while concomitantly reducing the life expectancy of those not vaccinated.

In the simplest epidemiological models for a respiratory pathogen like COVID-19, we can divide the population into two parts: those who are willing to follow recommended public health guidelines and those who can not (or will not). We can also assume that rates of mixing within each of the two sections of the population are higher than between the two parts. Mask-wearing reduces transmission, so this will reduce infection numbers in the mask-wearing proportion of the population. Vaccination also reduces transmission and significantly reduces pathology and resultant mortality. In both cases, if there is any mortality associated with infection, the group that is mask-wearing and/or vaccinated will increase in relative abundance to those not undertaking either of these two actions. Over a four-year electoral cycle in a population that is initially evenly split between the two behaviors, this would eventually give mask-wearers and the vaccinated a significant numerical electoral advantage over those refusing to participate in these interventions. Why would their political leaders encourage them to avoid these activities? More subtly, if there is mixing between vaccinated and nonvaccinated hosts, it is likely that exposure of vaccinated hosts to infected, nonvaccinated hosts will boost the immunological protection of the vaccinated hosts, while concomitantly reducing the life expectancy of those not vaccinated. This is an unwitting and paradoxical form of altruism, with one group sacrificing their personal health to boost the immunity of others.

SUBPOPULATIONS OF HOSTS

COVID-19 dynamics have been subtly different in rural and urban areas, mainly because of differences in population density, but also because facilities to identify and treat people are much diminished in rural areas, particularly in sub-Saharan Africa, India, and South America. This generates weakly coupled sets of epidemics in areas and countries where subpopulations are coupled together at different levels of interaction. The dynamics of generalist pathogens provide important insights into the transmission dynamics of pathogens in structured human populations.[6] Consider a pathogen that can infect multiple host species, each of which has a different body size and thus different birth and death rates and population densities. The species with the smallest body size will have the highest birth and death rates and population density, while the one with the largest physical size will have low birth and death rates. If the pathogen follows simple SEIR dynamics and we initially assume that between-species transmission is much lower than within-species transmission ($<10^{-3}$), then each host will interact independently with the pathogen and each will exhibit its own epidemic cycles: large and frequent outbreaks in the hosts with low body mass, slow and less

6 A. Dobson, 2004, "Population Dynamics of Pathogens with Multiple Host Species," *American Naturalist* 164(5): S64–S78, doi: 10.1086/424681

dramatic cycles in the case of larger hosts.[7] As we increase the relative rates of between-species transmission (10^{-3}–10^{-1}), these cycles will die out and any tendency for epidemic outbreaks to occur will be buffered by the pathogen constantly jumping between host species and preventing any one species from becoming too abundant. If we increase between-species transmission to levels where it matches within-species transmission, then the small species can unwittingly use the pathogen to drive the larger species extinct; small species are abundant and quickly recover from outbreaks, while large species become rare and cannot recover from frequent epidemics. Ultimately only the smallest species persist, and they revert to the epidemic behavior they exhibited when between-species transmission was rare.

Similar dynamics are likely to be important in three areas of COVID-19 dynamics where heterogeneity creates weakly coupled structures within the host population.

1. Geoffrey West and Chris Kempes have suggested that city size is crucial in determining the dynamics of COVID-19 and other directly transmitted viral pathogens (this volume/these essays). I strongly suspect that the transmission dynamics of cities of different sizes are coupled together in similar ways as different species of hosts are coupled together. This would imply that small to intermediate levels of mixing by people commuting between different cities, towns, and villages would strongly buffer any innate tendency for cities to exhibit epidemic cycles at different frequencies. Instead, we would see a more homogeneous, low-level epidemic, with much longer persistence and no dominant epidemic frequency. Interestingly, this also suggests that when we observe differences in the epidemic dynamics of a pathogen in different-sized cities, this provides an alternative and indirect test of the efficacy of lockdown at each location. Similar logic applies to the dynamics of a pathogen in countries isolated from each other by transportation restrictions.

7 Ibid.

2. The rapid development of vaccines against COVID-19 has been a major scientific success. Logistical constraints, religious beliefs, and blatant misinformation have led to much lower rates of vaccination uptake than are needed to attain herd immunity. It now also seems that vaccinated hosts can acquire infection and transmit the virus, but for much shorter periods of time than those who have not been vaccinated. Their mortality rates are also significantly reduced. This will again generate heterogeneity in the epidemiological structure of the host population, buffering the tendency for epidemics in a fashion similar to that described for host species of different physical size, or cities of different size.

3. The physical size of the virus may also be important in determining key aspects of the dynamics of the epidemic. There is a curious and poorly explored relationship among the physical size of pathogens, the efficacy of vaccines, and the duration of immunity: measles seems to lie at a perfect "sweet spot" on a spectrum producing lifetime immunity from either a highly efficacious vaccine or natural infection. Larger viruses, bacteria, protozoa, and parasitic worms produce immunity of shorter durations, with vaccines of diminishing efficacy (Dobson et al., in prep). Ironically, viruses that are physically smaller than measles, such as influenza, produce prolonged immunity, but mutate so rapidly that new strains appear at rates that are increasingly less recognizable to the human immune system. Curiously, there is a 99% correlation between physical size and genome size for all parasitic organisms, suggesting that efficacy of immunity is mainly a function of the physical size of the pathogen and some scaling law underlies the relationship among virus size, vaccine efficacy, and duration of immunity. COVID-19 is a very large virus when compared to measles,[8] and the immunity of the commonly circulating coronaviruses is relatively transient (around a year). Although vaccinated and

---

8 E.C. Holmes, 2009, *The Evolution and Emergence of RNA Viruses*, Oxford, UK: Oxford University Press.

previously infected hosts do seem to recover more readily from challenge infections, it seems likely that vaccination against COVID-19 will need to be an annual event. This will require significant capacity building and infrastructure investment across the entire biomedical enterprise.

There is a curious and poorly explored relationship among the physical size of pathogens, the efficacy of vaccines, and the duration of immunity: measles seems to lie at a perfect "sweet spot" on a spectrum producing lifetime immunity from either a highly efficacious vaccine or natural infection.

EVOLUTION OF NEW STRAINS

The emergence of new strains of the virus was inevitable; viruses have a high mutation rate and constantly produce novel varieties, the vast majority of which are dysfunctional and fail to replicate. Occasionally, a functional novel variant will appear by mutation and it will quickly replace currently existing strains if it is either more transmissible or if the hosts it infects transmit the virus for longer periods of time. When levels of immunity in the host population are low, only novel strains with higher transmission success can persist and replace older strains.[9] As the appearance of these strains is essentially random, the rate at which novel strains appear will vary directly with the number of people infected. If you halve the number of people infected, then it will take twice as long for new strains to appear, while if you reduce infections by 90%, it will take ten times as long. Complications arise as levels of immunity to prior infection increase in the population; this creates selection

9 T. Day, S. Gandon, et al., 2020, "On the Evolutionary Epidemiology of SARS-CoV-2," *Current Biology* 30(15): R849–R857, doi: 10.1016/j.cub.2020.06.031

pressure on the virus to circumvent the host's immunity to prior infection. Potentially, this can occur at maximum rates when intermediate numbers of people are immune, and the virus is circulating in a large proportion of the remaining unvaccinated population.

Long-term research on an emerging bacterial pathogen of house finches in the US, which has very similar dynamics to COVID-19, suggests that selection for asymmetrical immunity (where one strain induces host immunity against itself and all other strains, while weaker strains only induce host immunity against themselves) is a powerful driver of virulence in emerging pathogens, with later strains generating a stronger immunological response that prevents reinfection of recovered hosts infected by earlier strains.[10] Mathematical models on the evolution of virulence in this system suggest that the ability of the pathogen to transmit before symptoms of virulence are apparent can lead to selection for significantly higher levels of virulence.[11] All these evolutionary considerations strongly underline the importance of getting as many people vaccinated as quickly as possible. Furthermore, the size considerations mentioned above suggest this will need to be done globally on an annual basis. This will require an increase of at least an order of magnitude in our global ability to produce and distribute vaccines.

THE BOTTOM LINE

Contact patterns between infected and susceptible people are what ultimately determine the dynamics of an epidemic. These contact patterns develop significant asymmetries within and across different sectors of the economy. They are further exasperated by people's resistance to wearing face masks, or politically based refusal to believe in either the pathogen or the benefits of vaccination. I remain shocked at how politically polarized discussion of these issues has become, particularly as it seems obvious that the last thing you want to do as a politician is reduce the health and life expectancy of your voters in ways that significantly reduce the probability that they will survive to reach the polling booth.

---

10 A.E. Fleming-Davies, P.D. Williams, et al., 2018, "Incomplete Host Immunity Favors the Evolution of Virulence in an Emergent Pathogen," *Science* 359(6379): 1030–1033, doi: 10.1126/science.aao2140

11 E.E. Osnas and A.P. Dobson, 2010, "Evolution of Virulence when Transmission Occurs before Disease," *Biology Letters* 6(4): 505–508, doi: 10.1098/rsbl.2009.1019

Welcome to the Covidocene! Annual vaccination against this virus will be the new normal. Unfortunately, COVID-19 will not be the last novel pathogen to invade humans—all the available evidence suggests we should see at least one or two more emergent viruses in the next decade.[12] This can be averted, or halted at an earlier stage of spread, but it will require considerable international cooperation.

12 A.P. Dobson, S.L. Pimm, et al., 2020, "Ecology and Economics for Pandemic Prevention," *Science* 369(6502): 379–381, doi: 10.1126/science.abc3189

TRANSMISSION NO. 006

USING SOCIAL MEDIA DATA TO DETECT SIGNATURES OF GLOBAL CRISES

DATE: *6 April 2020*

FROM: *Miguel Fuentes, Santa Fe Institute*

STRATEGIC INSIGHT: *We can use social media data to detect signatures of global crises, including early warning signs*

The oldest and strongest emotion of mankind is fear, and the oldest and strongest kind of fear is fear of the unknown. —H.P. LOVECRAFT

Our global crisis has produced a variety of personal and larger social responses, from social unrest to widespread fear. My colleague Juan Pablo Cardenas (Net-Works, Chile) and I have been studying the spread of fear and social unrest by analyzing social media in crisis contexts, particularly around the COVID-19 pandemic.

Our research examines the correlation of crisis-related behaviors with geographic isolation, group size, and other factors. This should give us an idea of the characteristics, perhaps universal, that occur in these types of tumultuous events—whether massive shifts in employment from one sector to another or political protests in response to a government's handling of a pandemic.

Few concepts have such diverse connotations as crisis. Current approaches that help to define the concept—from different fields such as epistemology and the social sciences—tend to deal with its negative connotations. Putting less emphasis on value and greater emphasis on quantification can contribute to a more general understanding of the phenomenon.

Our working hypothesis is that crises are manifestations of dynamics in all adaptive systems linked to the constant and spontaneous increase

---

OPPOSITE: *Blaine A. White, "The Argument," 2017*

of the system's complexity or internal information. We approach these concepts rigorously using large datasets and network science.

Current approaches that help to define the concept—from different fields such as epistemology and the social sciences—tend to deal with its negative connotations. Putting less emphasis on value and greater emphasis on quantification can contribute to a more general understanding of the phenomenon.

Based on the analysis of a large social dataset of communication events on Twitter, our analysis demonstrates that periods of greatest activity (a fully developed crisis) coincide with an increase in the number of nodes in a network of conversations as well as an increase in the number of components (maximal connected sub-graphs). Non-technically, this means that there are significantly more conversations and that these become increasingly structured around a common set of themes.

A crisis is therefore characterized on the one hand by a combination of more intense and more concentrated activity. On the other hand, tweets (in the case of social media data) during a fully developed crisis tend to be addressed to specific users—whereas in periods of low crisis, tweets tend to involve messages of lower specificity and greater reach (i.e., opinions are less binding).

These are just a couple of ways in which we can explore quantitative network properties that reveal features of crisis behavior, and that in some cases could provide early warning signs of an impending crisis. In a couple of our studies—"Social Crises: Signatures of Complexity in a Fast-Growing Economy"[1] and "Social Crises: A Network Model Approach"[2]—we found early warning signs more than a year in advance of the deep social crisis in Chile that began in October 2019.

With respect to early warning signs, we have been very interested in the larger question of how an ongoing social crisis like the one in Chile relates to the crisis of COVID-19. Another set of deep questions relates to how history influences outcomes. In a dynamic system, this is called *hysteresis*; we are only now exploring how history modifies future behavior.

By sharing research and information, we can begin to answer the many questions that remain in relation to how fear and social unrest interact across social media, and how to find the signal in all the noise.

ENDNOTES

1 J.P. Cárdenas, G. Vidal, et al., 2018, "Social Crises: Signatures of Complexity in a Fast-Growing Economy," *Complexity* 2018: Article ID 9343451, doi: 10.1155/2018/9343451

2 J.P. Cárdenas, G. Vidal, et al., 2018, "Social Crises: A Network Model Approach," *Physica A: Statistical Mechanics and its Applications* 505: 35-48, doi: 10.1016/j.physa.2018.03.031

REFLECTION

COVID-19, A PARTICLE ACCELERATOR

DATE: *25 August 2021*

FROM: *Miguel Fuentes, Santa Fe Institute*

Despite the fact that COVID-19 irrupted in every country and became an unquestioned and fundamental issue of global importance, the unknown aspects of this pandemic—its direct and indirect consequences—continue to emerge.

My lab's preliminary work on social crises presented in Transmission T-006 suggested possible slowed-down dynamics; government-imposed total or partial lockdowns (what is known as dynamic quarantines) would lead to strong blockages in social mobility. Despite the fact that the mobility restrictions hindered big demonstrations and street riots—events that put great pressure on the status quo—information continues flowing freely on different digital platforms. Thus, in Chile, where we have installed our virtual laboratory and case study, and in other places, COVID-19 has become a true particle accelerator; it has been one of the important actors accelerating the processes that lead to deep social changes.

In general, the intrinsic adaptive nature of a social system allows it to overcome not only the challenges imposed by the environment and external forces acting on it, but also its internal pressures. This adaptive process is arguably associated with an increase in complexity, which we can observe in a variety of innovative social artifacts, greater diversity of the system's components, and new forms of organization, among other transformations. However, the dynamical adaptations of social systems have a cost and must be managed; otherwise, they can trigger social unrest and deep crisis processes. Further, the emergence of the internet and other information technologies has accelerated and magnified the process of adaptation of social systems. These technologies have broken down geographical barriers, strengthened globalization, and significantly increased both social interaction and the flow of information. Although social crises have accompanied humanity since its origins, they have increased in magnitude and frequency in recent decades.

These contemporary social crises could have their origin in the emergence of the so-called "Networked Society." That is why online social networks have become a kind of projection of those traditional face-to-face human relationships and interactions, sharing signs and developing their own particularities, assuming an unusual role in the development, and even in the genesis, of some of these social-crisis events.[1]

> ...in Chile, where we have installed our virtual laboratory and case study, and in other places, COVID-19 has become a true particle accelerator; it has been one of the important actors accelerating the processes that lead to deep social changes.

We strongly believe that it is not enough to think about the virtues of "self-organization" in the social system at all scales, since in principle the speed of change and adaptation have very different timescales. In principle, this lack of governmental investment in a complexity and big-picture perspective has led to the emergence of an increasing number of social outbreaks around the globe in recent years.

In our recent work "The Structure of Online Information behind Social Crises," published in *Frontiers in Physics*, we explore the close relationship between adaptation, complexity, and crisis, showing the particular ways that complexity is expressed in a digital social environment.

As expected, we observe the polarization of the society and the negative sentiment of messages in times of crisis. However, our results also show that, during times of crisis, social organizations experience a loss of order. As a result, new complex and ephemeral information structures emerge, seemingly early warning signs of profound social transformations. In contrast to classical early warnings that reach a bifurcation or

1 These concepts were first laid out in J.P. Cárdenas, G. Olivares, et al., 2021, "The Structure of Online Information behind Social Crises," *Frontiers in Physics* 9: 650648, doi: 10.3389/fphy.2021.650648.

critical point, these early warnings are signals mounted on some underlying information with a lot of fluctuation and noise—they are signals within the crisis. COVID-19 has accelerated the appearance of these signals and the transformation processes of the system, a product of increased unemployment, health crisis, mobility, psychological impacts on society at all ages, etc.

In contrast to classical early warnings that reach a bifurcation or critical point, these early warnings are signals mounted on some underlying information with a lot of fluctuation and noise—they are signals within the crisis.

From the analytical point of view, it has been necessary to investigate new territories. Researchers have built new metrics to understand the particularities of digital systems in times of crisis and to learn from their subtle signals. In our case we have been able to create tools that clearly indicate possible threshold states in the social systems. We now have more information to propose plausible explanations, and thus increase the understanding of these complex systems. Even so, it would be an exaggeration to say that we fully or mostly understand the system. Returning to the beginning of this article, the consequences of COVID-19 have not fully unfolded. We have new data, intuitions, and predictive tools, but complete understanding, as always in science, is still open.

TRANSMISSION NO. 007

HOW TO REDUCE COVID-19 MORTALITY WHILE EASING ECONOMIC DECLINE

DATE: *6 April 2020*

FROM: *Danielle Allen, Harvard University*
E. Glen Weyl, Microsoft Office of the Chief Technology Officer; RadicalxChange Foundation
Rajiv Sethi, Barnard College; Columbia University; Santa Fe Institute

STRATEGIC INSIGHT: *A "mobilize and transition" strategy could reduce COVID-19 mortality while cushioning.*

While the human toll of the COVID-19 pandemic has been apparent for some time, the economic picture is now starting to come into greater focus. Initial unemployment claims in the United States jumped from 280,000 to almost 3.3 million for the week ending March 21,[1] then doubled to over 6.6 million for the following week.[2] By way of comparison, weekly unemployment claims have never previously exceeded 700,000 in the history of the recorded data. The S&P 500 Index reached a record high in mid-February, then lost a third of its value in a month. Congress passed a $2 trillion stimulus bill on March 27, one quarter of which allows for loans and grants to firms under the discretion of the secretary of the treasury. And the Federal Reserve invoked the "unusual and exigent circumstances" clause of Section 13(3) of the Federal Reserve Act to break out of its usual shackles and channel credit to (nonbank) firms, states, and municipalities.

The plan seems to be to drastically scale back economic and social activity and wait for the pandemic to pass, in the hope that it will do so in about two or three months, with a rapid return to normalcy thereafter. A statement[3] by Secretary of Labor Eugene Scalia on the

---

OPPOSITE: *Artist unknown, "The Ramu Setu Being Built by Monkeys and Bears," c. 1850 (detail)*

unemployment numbers exemplifies this thinking; he observes that the stimulus bill "provides incentives and funding for businesses to keep their workers on payroll, so that, as soon as possible, we can spring back to the strong economic conditions we enjoyed just weeks ago."

But what if three months is not enough,[4] and we see ebbs and flows in confirmed cases over one or two years, in concert with the relaxation and tightening of social distancing measures? The social, political, and economic implications of this would be dire. And what if changes in the composition of demand[5] for goods and services are enduring, with less expenditure on travel and lodging and more on public health and distance learning for years to come? Then a large-scale reallocation of workers and capital across sectors will be needed even after the threat has passed, and a return to the precrisis status quo will not be possible in any case.

An alternative paradigm,[6] which we call *mobilize and transition*, allows for a return to active participation in economic life for some portion of the population long before the pandemic has been fully contained. The goal is to use an initial period of aggressive social distancing of up to three months in order to build out the infrastructure of pandemic preparedness and management that countries like Taiwan and Singapore have used to maintain far greater control over COVID-19 than we have managed. We *mobilize* in order to *transition* to being a society with the kind of pandemic resilience that permits maximal mobility for as large a portion of the population as possible *even when the pandemic is ongoing*.

This strategy involves large-scale testing, on the order of several million individuals per day, in order to partition the population into those who are believed to be safe and those whose status remains undetermined. It involves two regimes, which we call *find the safe* and *find the virus*, with transitions between regimes being contingent on epidemic trajectories and economic conditions.

In the *find the safe* regime, those with a recent negative disease test or a serological test indicating immunity can return to the workforce, entering occupations for which there is intense demand. This will require credible verification of safe status. Testing would be focused on those in the health and care professions, first responders, sanitation workers, and those connected with production and delivery of food and other essential goods and services. And it would include those who are willing and able to enter or re-enter these occupations without the need for extensive training, and those who could provide such training as is deemed necessary.

The *find the virus* regime involves broad-based testing in order to find and isolate those who are infected, and to warn and recommend testing for their contacts. In order to make best use of scarce testing resources and personnel, those with greatest likelihood of carrying the virus should be prioritized for testing, and this determination could be based on location, occupation, demographic characteristics, and proximity to others who have recently tested positive.

There is a useful analogy here to the literature on police stops and searches, where differences across groups in contraband recovery rates (or *hit rates*) are viewed as a diagnostic test[7] for discrimination. A non-discriminatory police department seeking to maximize, say, weapon recovery should conduct searches in such a manner as to equalize marginal hit rates across identifiable subgroups of the population. That is, the likelihood of weapon recovery should be independent of group membership among those who are close to the threshold for a search. Translating this to the case of testing, the targeting of individuals should be such that the likelihood of testing positive is roughly equalized across locations, occupations, and demographic groups, at least among those who have been recently tested. If a location is turning up more positives than another at the margin, resources would be better used by shifting to the former at the expense of the latter. Such adjustments require extensive mobile testing capability.

Innovative use of mobile technologies can facilitate more finely targeted testing, while preserving privacy and civil liberties. For example, an app developed by MIT researchers[8] collects location data every five minutes and stores it locally without any identifying information. Anyone testing positive can transfer this data to a health professional, who can upload it to a central server, again with all identifying information redacted. This allows intersections to be traced, and people who have crossed paths with those who have tested positive to be warned, even though their personal data never leave their phone without their permission. Other related applications are under development.

One essential feature of the proposed strategy, and indeed any strategy under current conditions, is the universalization of mask use, subject to availability of supply. Widespread mask use has helped limit contagion in many Asian countries, but the practice remains far from universal in the United States. The meaning of mask use needs to be transformed through public messaging[9] by civic and political leaders, so that it is associated with altruism and civic responsibility instead of carrying the stigma of sickness or fearfulness.[10]

The meaning of mask use needs to be transformed through public messaging by civic and political leaders, so that it is associated with altruism and civic responsibility instead of carrying the stigma of sickness or fearfulness.

We believe that the mobilize and transition strategy will result in lower mortality from the disease while cushioning the decline in output and employment and leading to a more rapid recovery with a very different allocation of workers and capital across sectors relative to the prepandemic period. But the economic hardship will nevertheless be extraordinary, with double-digit unemployment rates for several months, if not years. Social support therefore has to be an essential component of the strategy. The stimulus bill allows for some cash payments to households, but this could be routinized in the form of a basic income and distributed through individual accounts at the Federal Reserve.[11] The latter would be a radical departure from current practice, but these are indeed "unusual and exigent circumstances" under which the central bank has the power—and indeed the responsibility—to take steps that may have been unimaginable in quieter times.

ENDNOTES

1 Department of Labor of the United States of America, "Unemployed Insurance Weekly Claims," news release, March 26, 2020, https://www.dol.gov/sites/dolgov/files/OPA/newsreleases/ui-claims/20200510.pdf

2 Department of Labor of the United States of America, "Unemployed Insurance Weekly Claims," news release, April 2, 2020, https://www.dol.gov/sites/dolgov/files/OPA/newsreleases/ui-claims/20200551.pdf

3 US Department of Labor, "Statement by Secretary of Labor Eugene Scalia on Unemployment Insurance Claims," news release, March 26, 2020, https://www.dol.gov/newsroom/releases/osec/osec20200326

4 D. Allen, L. Stanczyk, R. Sethi, and E.G. Weyl, "When Can We Go Out?: Evaluating Policy Paradigms for Responding to the COVID-19 Threat," Edmond J. Safra Center for Ethics, March 25, 2020, https://ethics.harvard.edu/when-can-we-go-out

5 L. Saad, "Americans Hesitant to Return to Normal in Short Term," *Gallup*, April 1, 2020, https://news.gallup.com/poll/306053/americans-hesitant-return-normal-short-term.aspx

6 E.G. Weyl and R. Sethi, "Mobilizing the Political Economy for COVID-19," Edmond J. Safra Center for Ethics, March 26, 2020, https://ethics.harvard.edu/mobilizing-political-economy

7 I. Ayres, "Outcome Tests of Racial Disparities in Police Practices," *Justice Research and Policy* 4(1–2): 131–142, doi: 10.3818/JRP.4.1.2002.131

8 http://safepaths.mit.edu/

9 A. Goodnough and K. Sheikh, "CDC Weighs Advising Everyone to Wear a Mask," *The New York Times*, March 31, 2020, https://www.nytimes.com/2020/03/31/health/cdc-masks-coronavirus.html

10 M. Zhou, Y. Yu, and A. Fang, "Asians in US Torn between Safety and Stigma over Face Masks," *Nikkei*, March 14, 2020, https://asia.nikkei.com/Spotlight/Coronavirus/Asians-in-US-torn-between-safety-and-stigma-over-face-masks

11 R. Sethi, "Social Support and Financial Architecture: A Proposal for Reform," *International Banker*, March 7, 2017, https://internationalbanker.com/finance/social-support-financial-architecture-proposal-reform.

REFLECTION

PREDICTION AND POLICY IN A COMPLEX SYSTEM

DATE: *30 July 2021*

FROM: *Danielle Allen, Harvard University*
E. Glen Weyl, Microsoft Office of the Chief Technology Officer; RadicalxChange Foundation
Rajiv Sethi, Barnard College; Columbia University; Santa Fe Institute

A number of events occurred in 2020 that were interconnected in complex and subtle ways that no single academic discipline is well-equipped to understand.

Two of these—initial claims for unemployment benefits (measured weekly) and new confirmed deaths from COVID-19 (measured daily and smoothed using a seven-day average)—are shown in figure 1, both for the United States. The two series track each other closely initially, with deaths lagging claims, and then start to diverge from July onward.

These two phenomena are clearly connected, and we could better predict and respond to each of them if their linkages were more deeply understood. However, it would have been virtually impossible to use historical data to build and calibrate models that provide an integrated analysis of this kind. For one thing, initial weekly unemployment claims reached a peak of almost seven million in April, after never having exceeded 700,000 in earlier years, even at the depths of the global financial crisis of 2008-2009. That is, the labor market data in figure 1 lie outside the range of all observed historical experience. Similarly, one needs to go back a century for a pandemic with comparable levels of mortality in the United States.

There were many other events in 2020 that were connected to these two, but in ways that are difficult to grasp based on traditional scientific methods and boundaries. The mass actions following the killing of George Floyd were quite possibly the largest in American history[1]

1 L. Buchanan, Q. Bui, and J.K. Patel, "Black Lives Matter May Be the Largest Movement in US History," *The New York Times*, July 3, 2020, https://www.nytimes.com/interactive/2020/07/03/us/george-floyd-protests-crowd-size.html

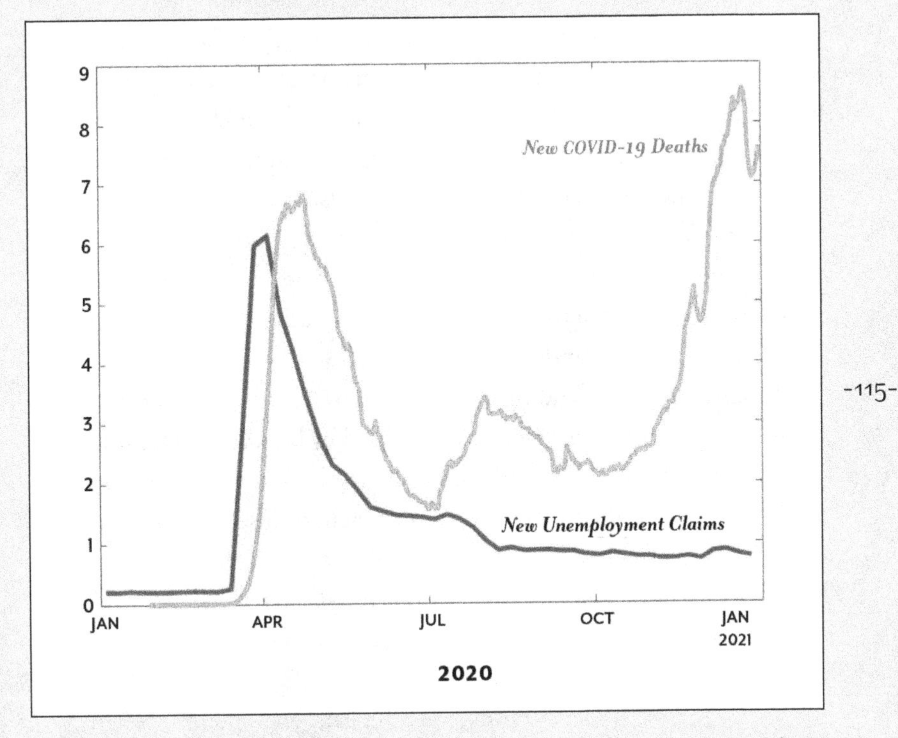

Figure 1. *Daily new confirmed COVID-19 deaths per million people (seven-day average) and weekly initial claims for unemployment insurance (millions of persons) in the United States throughout 2020. Sources: Our World in Data (https://ourworldindata.org/explorers/coronavirus-data-explorer) and Federal Reserve Economic Data (https://fred.stlouisfed.org/graph/), respectively.*

for a time; at peak in June there were a half million people protesting in more than five hundred cities. Mississippi removed the confederate battle emblem[2] from its state flag after a century and a quarter, which few would have imagined possible a year earlier. And Congress voted to override[3] a presidential veto of a defense bill that began the process of renaming military bases that honored Confederate leaders.

2 R. Rojas, "Mississippi Governor Signs Law to Remove Flag with Confederate Emblem," *The New York Times*, June 30, 2020, https://www.nytimes.com/2020/06/30/us/mississippi-flag.html

3 P. Ewing, "Congress Overturns Trump Veto on Defense Bill after Political Detour," *National Public Radio*, January 1, 2021, https://www.npr.org/2021/01/01/952450018/congress-overturns-trump-veto-on-defense-bill-after-political-detour

A veto-proof majority of the Minneapolis City Council pledged to dismantle[4] the city's police force, then began to reverse course[5] a few months later. New York City disbanded its plainclothes anticrime units,[6] which were responsible for a disproportionate share of abuse complaints and deadly force incidents, but also for significant numbers of gun confiscations. Many police departments responded to threats of defunding with slowdowns and pullbacks.[7] There was a rise in shootings and homicides[8] across towns and cities nationwide, the causes of which remain poorly understood.

Scientific understanding of such interconnected phenomena is hampered by the fact that they are addressed by largely separate research communities. Sociologists and criminologists look at crime and policing, epidemiologists at the spread and containment of disease, economists at financial and labor market movements, and so on. This specialization has both costs and benefits. It allows for deep focus on a key set of interactions and causal chains, at the cost of omitting factors that are (or can quickly become) highly relevant.

How might understanding, prediction, and policy design be improved by taking such interactive complexity into account?

One approach is to leverage big data, computational capacity, and recent advances in machine learning to build versatile and flexible models. The reliance on historical data is a limitation, however, because accurate prediction is most needed when one is outside the range of

---

4 D. Searcey and J. Eligon, "Minneapolis Will Dismantle Its Police Force, Council Members Pledge," *The New York Times*, June 7, 2020, https://www.nytimes.com/2020/06/07/us/minneapolis-police-abolish.html

5 A.W. Herndon, "How a Pledge to Dismantle the Minneapolis Police Collapsed," *The New York Times*, September 26, 2020, https://www.nytimes.com/2020/09/26/us/politics/minneapolis-defund-police.html

6 A. Watkins, "NYPD Disbands Plainclothes Units Involved in Many Shootings," *The New York Times*, June 15, 2020, https://www.nytimes.com/2020/06/15/nyregion/nypd-plainclothes-cops.html

7 A. MacGillis, "How to Stop a Police Pullback," *The Atlantic*, September 3, 2020, https://www.theatlantic.com/ideas/archive/2020/09/how-stop-police-pullback/615730/

8 D. Barrett, "2020 Saw an Unprecedented Spike in Homicides from Big Cities to Small Towns," *The Washington Post*, December 30, 2020, https://www.washingtonpost.com/national-security/reoord-spike-murders-2020/2020/12/30/1dcb057c-4ae5-11eb-839a-cf4ba7b7c48c_story.html

prior experience. Unique policy initiatives can make inference from past experience unreliable. The two series shown in figure 1 start to diverge in part because of the effects of large-scale policy interventions like the CARES Act,[9] which was signed into law in late March, and which had a much stronger impact on the economy over time than on the spread of the disease.

Purely data-driven technocratic approaches sometimes fail spectacularly because a lot depends on the data ontology, and relevant issues need to emerge through a bottom-up rather than top-down process. This suggests the need to explore participatory and democratic mechanisms that can be used to augment and improve upon top-down technocratic data analysis.

One approach to harnessing such decentralized knowledge, including information and insight held by nonexperts, is through market mechanisms. Prediction markets are exchanges on which state-contingent contracts can be traded; these contracts pay a specified amount if a referenced event occurs and nothing otherwise. Contracts that reference election outcomes have been traded on the Iowa Electronic Markets[10] for over thirty years, and other venues such as PredictIt[11] in the United States and Betfair[12] in the United Kingdom list heavily traded prediction market contracts. Kalshi,[13] a major new regulated exchange for the trading of such securities, launched in 2021.

With suitable caveats, the price of a prediction market contract (relative to the payment promised if the referenced event occurs) can be interpreted as the event's probability of occurrence, as judged by the market. Any feature that a trader considers relevant influences the contract price, so there is no *ex ante* restriction on the sources of information used for any particular predictive task.

But markets are just one mechanism for participatory information aggregation, and they have certain drawbacks. They tend to create a

---

9 Congress.gov, "Text - S.3548 - 116th Congress (2019-2020): CARES Act," June 3, 2020, https://www.congress.gov/bill/116th-congress/senate-bill/3548/text.

10 https://iem.uiowa.edu/iem/

11 https://www.predictit.org

12 https://www.betfair.com

13 https://kalshi.com

competitive—rather than participatory and communicative—dynamic and thus somewhat reduce the incentive to engage in nonlinear conversation. In addition, prices depend on the distribution of trader wealth, which need not be closely associated with good judgment. There can also be incentives for manipulation[14] when beliefs about an event can affect its objective probability of realization.

There exist alternative mechanisms for the aggregation of information and judgment that rely on persuasion and consensus building rather than strict competition. The pol.is[15] digital tool, for example, allows for the posting of policy statements by any user, which can be voted on by others, with the various positions being displayed on an interactive real-time map showing clusters of opinion. Users are incentivized to offer statements that can attract broad support, and the map can reveal the evolution of changing positions over time as the population of users moves toward agreement. This platform has been used in Taiwan to inform legislative action[16] on a number of issues.

During the early stages of the pandemic, the use of innovative digital tools[17] for information sharing in Taiwan helped to identify and reduce exposure without onerous restrictions, allowing the country to achieve lower case and death rates[18] compared to most democracies at comparable levels of prosperity and life expectancy.

---

14 D. Rothschild and R. Sethi, 2016, "Trading Strategies and Market Microstructure: Evidence from a Prediction Market," *The Journal of Prediction Markets* 10(11), doi: 10.5750/jpm.v10i1.1179

15 https://pol.is/home

16 A. Tang, "A Strong Democracy is a Digital Democracy," *The New York Times*, October 15, 2019, https://www.nytimes.com/2019/10/15/opinion/taiwan-digital-democracy.html; C. Horton, "The Simple but Ingenious System Taiwan Uses to Crowdsource its Laws," *MIT Technology Review*, August 21, 2018, https://www.technologyreview.com/2018/08/21/240284/the-simple-but-ingenious-system-taiwan-uses-to-crowdsource-its-laws/; C. Miller, "Taiwan is Making Democracy Work Again. It's Time We Paid Attention," *WIRED*, November 26, 2019, https://www.wired.co.uk/article/taiwan-democracy-social-media

17 J. Lanier and E.G. Weyl, "How Civic Technology Can Help Stop a Pandemic: Taiwan's Initial Success is a Model for the Rest of the World," Foreign Affairs, March 20, 2020, https://www.foreignaffairs.com/articles/asia/2020-03-20/how-civic-technology-can-help-stop-pandemic.

18 See, for example, Our World in Data (https://ourworldindata.org/explorers/coronavirus-data-explorer), comparing Taiwan's case counts and confirmed deaths to those of other nations.

The tools used to aggregate information for policy purposes could also, in principle, be used to improve scientific forecasting. Like prediction markets, such crowdsourced forecasts are not constrained by preselected variables or historical data in the way that most models are. Integrating conventional models, prediction markets, and nonmarket crowdsourced aggregation mechanisms remains an interesting challenge[19] for researchers.

But no matter how much better or well-integrated with other mechanisms the models get, there will always be spectacular failures of prediction. For this reason, it is important to design institutions and policies in such a manner as to be robust to limited understanding of the underlying science.

In the case of police use of deadly force, for instance, Lawrence Sherman has argued[20] that focusing on the individual culpability of officers is much less likely to lead to effective change than attention to the "complex organizational processes that recruited, hired, trained, supervised, disciplined, assigned, and dispatched the shooter before anyone faced a split-second decision to shoot." That is, most police homicides ought to be handled more like failures of air traffic control than crimes, resulting in an evaluation of organizational systems[21] alongside prosecution for unlawful conduct where appropriate.

And in the case of pandemics, a program of large-scale testing, tracing, and supported isolation[22] could allow for better suppression and economic resilience even if the connections between contagion, behavior, and economic conditions continue to remain poorly understood.

---

19 R. Sethi, J. Seager, et al., "Models, Markets, and the Forecasting of Elections," Social Science Research Network, January 16, 2021, doi: 10.2139/ssrn.3767544

20 L.W. Sherman, 2018, "Reducing Fatal Police Shootings as System Crashes: Research, Theory, and Practice," *Annual Review of Criminology* 1, doi: 10.1146/annurev-criminol-032317-092409

21 B. O'Flaherty and R. Sethi, "Policing as a Complex System," *The Bridge: Linking Engineering and Society* (National Academy of Engineering) Winter 2020.

22 "Roadmap to Pandemic Resilience," Edward J. Safra Center for Ethics at Harvard University, April 20, 2020, https://ethics.harvard.edu/Covid-Roadmap

TRANSMISSION NO. 008

THE IMPORTANCE OF TIMING IN RESTRICTIVE CONFINEMENT

DATE: *6 April 2020*

FROM: *Michael Hochberg, University of Montpellier, France; Santa Fe Institute*

STRATEGIC INSIGHT: *Physical distancing is necessary for reducing infections, but the timing of restrictive confinement makes all the difference.*

COVID-19 is revealing invisible ingredients in our contact networks—breath, touching, and physical surfaces. Understanding these networks is key to assessing how infectious diseases spread and how measures to extinguish epidemics are best deployed. The problem is that these same networks are a manifestation of our social freedoms and economy—and are both hard and dangerous to relinquish. This is basically why voluntary physical distancing is generally proving to be not enough to protect public health and health services.

Simple epidemiological "toy" models explain why, and to what degree, more restrictive confinement is necessary, and how communities and nations can be eased-out towards normalcy. These models can be revealing and can help with policy, but they come with operating instructions that even specialists can easily overlook.

The central concept in epidemiology found in toy models is the "basic reproductive number," usually referred to as R_0. R_0 is the expected number of new infections directly caused by a single infected person at the start of an epidemic (time $t = 0$), and as the epidemic proceeds, the analogous measure is the "effective reproduction number," or R(t) (hereafter R). R is an average. So, an R = 6 in a population of two infected and tens of susceptible individuals could mean that each infected six

---

OPPOSITE: *Jean-François Millet, "Shepherd Bringing Back his Flock at the Approach of the Storm," 1902 (detail)*

other people, or possibly that one infected zero and the other, twelve. Calculating R for real epidemics can be complex.[1] Notably, this key number may differ from one place to another, change during the course of an epidemic, and, fortunately, be reduced by disease-control measures.

Estimates of R_0 for COVID-19 are consistently about 2.5, and, given infectious periods and times from infection to symptoms, this results in the number of cases currently doubling about every three days.[2] The power of exponential growth is clear: after one month or ten doublings after the 100th case, about 100,000 confirmed cases would have occurred, had physical distancing and confinement measures not been put into place.

R is useful not only in understanding epidemics, but in developing control measures. Unsurprisingly, maintaining R below 1 for a sufficient period will progressively reduce case numbers to zero. The problem for COVID-19 is that not sufficiently lowering R below 1 will mean a more protracted epidemic, many people unnecessarily coming down with the disease, and a progressive strain on—if not the complete exhaustion of—health services. Such measures do indeed "flatten" the epidemic curve,[3] and, with ever-fewer susceptible individuals in the population, eventually reduce R to 0 associated with "herd immunity" (formally, this occurs when the susceptible fraction of the population is less than $1 - 1/R$).

Calculating R for real epidemics can be complex. Notably, this key number may differ from one place to another, change during the course of an epidemic, and, fortunately, be reduced by disease-control measures.

This simple insight is very powerful for yet another reason. Simply touting reductions in R as the ultimate objective of epidemic control overlooks the fact that when the epidemic measures begin will determine the effectiveness of a given reduction in R. As a general rule, R needs to be reduced early to prevent an epidemic, but should it take off (as seen for virtually every country where COVID-19 cases have occurred), strict measures are necessary to enact a reset to near zero cases, as has recently been reported from China.

For example, starting from 100 infected individuals in a population of 50 million, lowering R from 2.5 to 1 will result in about one hundred new infections about every two weeks, based on realistic epidemic parameters.[4,5] This would mean on the order of several thousand cases in a year. But the same measures starting with a population with 100,000 infected people will produce several million infections over the same period. This assumes that people mix freely—real contact networks will result in far fewer new cases.[6]

Nevertheless, the take-home message is that the size of the infectious population when measures are engaged is important in determining the degree of confinement, that is, the extent to which measures lower the baseline R_0. The near-complete lockdown in parts of China is case-in-point, where R approaching 0 was maintained for about two months, resulting in the virtual collapse of their epidemic. The question now for China—and soon for other nations easing out of strict confinement—is how physical distancing can be tuned to $R \approx 1$ or less, with the risk that failure to achieve this will produce a new epidemic.

REFERENCES

1 J.M. Heffernan, R.J. Smith, and L.M. Wahl, 2005, "Perspectives on the Basic Reproductive Ratio," *Journal of the Royal Society Interface* 2(4): 281–293, doi: 10.1098/rsif.2005.0042

2 https://www.ft.com/coronavirus-latest

3 R.M. Anderson, H. Heesterbeek, et al., 2020, "How Will Country-Based Mitigation Measures Influence the Course of the COVID-19 Epidemic?" *The Lancet* 395: 931–934, doi: 10.1016/S0140-6736(20)30567-5

4 O.N. Bjørnstad, 2018, *Epidemics: Models and Data Using R*, Cham, Switzerland: Springer.

5 http://gabgoh.github.io/COVID/index.html

6 M.J. Keeling and K.T.D. Eames, 2005, "Networks and Epidemic Models," *Journal of the Royal Society Interface* 2(4): 295-307, doi: 10.1098/rsif.2005.0051

REFLECTION

TWO EASY PIECES

DATE: *26 August 2021*

FROM: *Michael Hochberg, University of Montpellier, France; Santa Fe Institute*

When I wrote my Transmission essay in April 2020, scientists, particularly those with some background in virology and epidemiology, were just beginning to apply their knowledge to this never-before-seen situation. True, we were warned by Bill Gates. Also true, there had been pandemics before, but the two most notable in modern times—the 1918 Influenza and AIDS—*were* different than COVID-19. The Spanish Flu eventually killed tens of millions worldwide, in part because public education, communications, and epidemiological models were not what they are today. Major breakthroughs in epidemiological modeling[1] in the decades following the 1918 Influenza provided a basis for understanding subsequent influenza pandemics and the AIDS pandemic.

More than any infectious disease to date, COVID-19 stands alone in the massive mobilization of the mathematical modeling community. Models of all shapes and sizes were poised, and within weeks all baseline epidemiological parameters had been estimated for the disease and its causal virus, SARS-CoV-2. But perhaps most singularly, thanks to communication networks—social media, traditional media, and the lightning-fast dissemination of research preprints—information flowed freely, fostering progress. But information also flowed in *huge* quantities, which, along with inaccurate or poor communication between the scientific community and the public at large—generated confusion and mistrust.[2]

I've always championed first principles and parsimonious theory, and COVID-19 made clear that computational technology can go too

1 F. Brauer, C. Castillo-Chavez, and Z. Feng, 2019, "Introduction: A Prelude to Mathematical Epidemiology," in *Mathematical Models in Epidemiology. Texts in Applied Mathematics*, Vol 69, New York City, NY: Springer, doi: 10.1007/978-1-4939-9828-9_1

2 M. Hochberg, "COVID-19 in the Information Commons," *30000'* blog, March 1, 2021, https://mehochberg.wixsite.com/blog/post/covid-19-in-the-information-commons

far in the justification of data-driven, sometimes mind-bogglingly complicated models. John von Neumann said,[3] "With four parameters I can fit an elephant, and with five I can make him wiggle his trunk," which could be construed to imply that data justifies *truth*, which, in the end, is little more than a compilation of details.[4]

... COVID-19 made clear that computational technology can go too far in the justification of data-driven, sometimes mind-bogglingly complicated models.

Two central variables derived from the most basic epidemiological models go a long way toward understanding COVID-19. They are: the effective reproduction number R_{eff} and the fraction of infectiousness in the non-immune population I. True, R_{eff} can be decomposed into further parameters, foremost among them being the basic reproduction number R_0, the immune fraction of the population, and contact network heterogeneity. True too, the baseline epidemiological model has other important parameters (e.g., incubation and infectious periods, fatality ratio), but they wind up influencing one or both of R_{eff} and I.

This has both conceptual and practical implications, but above all, it is compellingly simple. In the early stages of an epidemic when the susceptible fraction is large, growth is R_{eff} and the population burden

3 Attributed to von Neumann by Enrico Fermi, as quoted by Freeman Dyson in "A meeting with Enrico Fermi" in *Nature* 427 (22 January 2004), p. 297.

4 "What a useful thing a pocket-map is!" I remarked.
"That's another thing we've learned from your Nation," said Mein Herr, "map-making. But we've carried it much further than you. What do *you* consider the *largest* map that would be really useful?"
"About six inches to the mile," I said.
"Only *six inches*!" exclaimed Mein Herr. "We very soon got to six yards to the mile. Then we tried a hundred yards to the mile. And then came the grandest idea of all! We actually made a map of the country, on the scale of a *mile to the mile*!"
"Have you used it much?" I enquired.
—Lewis Carroll, *Sylvie and Bruno Concluded*

is the product of I and R_{eff}. I cringe when I hear "this is *so important* because it's growing by X%." No! Or rather, yes, it can be: 100% interest on $1 is a pittance compared to the same on $1M, and 0.001% on the latter is still far more lucrative than 100% on the former. Two pieces are needed: growth rate and capital.[5] The same logic goes for COVID-19.

So, this is what you need to know:[6]

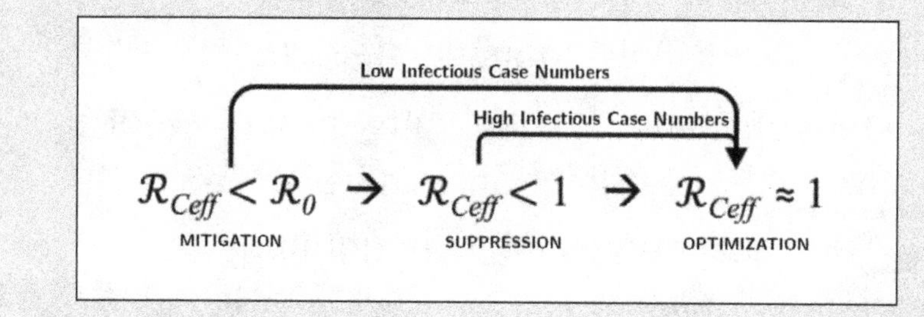

Data-driven statistical models now routinely estimate R_0 and R_{eff}. R_{Ceff} is the reduction in R_{eff} due to measures such as social distancing and lockdowns. Low numbers of infectious cases "buy time" in exponential growth. Higher numbers mean that health services are stressed and risk future collapse—lockdowns are necessary. Capping R_{Ceff} at 1.0 makes most sense if active case numbers are low (since otherwise although the curve is flattened, new case numbers remain high). But capping at low case numbers is problematic, because it stymies the growth of natural immunity. Capping is also a difficult sell for governments, whose constituents see that freedoms remain limited despite the virus apparently being under control.

---

5 Famously the basis of Thomas Piketty's 2014 book, *Capital in the Twenty-First Century*.

6 This schema shows the procession from initial attempts to mitigate outbreaks, either to optimization if successful, or suppression if unsuccessful. Once through a cycle, subsequent strategies will depend on active case numbers (and the correlated impact on health systems). Low case numbers are more likely to err toward baseline physical distancing and self-isolating, whereas high numbers meet with more restrictive curfews and lockdowns. Definitions of "low" and "high" and how "mitigation" is demarcated from "suppression" are somewhat arbitrary. "Low" and "high" depend on how decision-makers decide on thresholds of action, whereas mitigation and suppression are the packages of measures considered sufficient to either slow or reverse growth. See M.E. Hochberg, 2020, "Importance of Suppression and Mitigation Measures in Managing COVID-19 Outbreaks," *medRxiv*, https://doi.org/10.1101/2020.03.31.20048835; M.E. Hochberg, 2020, "Countries Should Aim to Lower the Reproduction to 1.0 for the Short-Term Mitigation of COVID-19 Outbreaks," *medRxiv*, doi: 10.1101/2020.04.14.20065268

Thus, importantly, there is a trade-off between minimizing morbidity and mortality and negative externalities to individuals and society.[7] A recent study[8] of optimal COVID-19 control is consistent with the above schema, and in particular the endpoint at $R_{Ceff} \approx 1$, which in turn is supported at least for some countries.[9] $R_{Ceff} \approx 1$ is, however, unsustainable due to the negative externalities associated with a long waiting time to achieve herd immunity. Until only very recently, vaccination in conjunction with physical distancing and natural immunity was knocking R_{Ceff} below 1 in many countries, promising local virus endemicity or extinction . . . but now as I write, the Delta variant is surging, suggesting that simple models are limited when it comes longer-term evolutionary dynamics.

Nevertheless, data-driven, computationally complex models are essential for determining what, when, and how much in decision-making. However, we shouldn't be blinkered into thinking that their "truth" makes them invariably superior to coarse-grained, toy models. Due to the importance of how science is communicated, and the centrality of collective behavior in both preventing the spread of SARS-CoV-2 and in vaccination campaigns, I believe that simple models will be pivotal in vanquishing the COVID-19 pandemic.

---

7 Prioritizing the former, as so many countries have, sacrifices the latter, and as it turns out, also generates negative externalities on social welfare, psychology, and the economy.

8 G. Li, S. Shivam, et al., 2020, "Disease-Dependent Interaction Policies to Support Health and Economic Outcomes during the COVID-19 Epidemic," available at SSRN, doi: 10.2139/ssrn.3709833

9 M.T. Sofonea, C. Boennec, et al., 2021, "Two Waves and a High Tide: The COVID-19 Epidemic in France," *Anaesthesia Critical Care & Pain Medicine* 40:100881, doi: 10.1016/j.accpm.2021.100881

TRANSMISSION NO. 009

HOW THE ANALOGIES WE LIVE BY SHAPE OUR THOUGHTS

DATE: *6 April 2020*

FROM: *Melanie Mitchell, Portland State University; Santa Fe Institute*

STRATEGIC INSIGHT: *The analogies we live by are shaping our thoughts about our current situation.*

I was struck by a recent headline on *BuzzFeed* about the current coronavirus pandemic: "The Disorienting Thing Is You Can't Compare this to Anything."[11] It's true that this pandemic is, literally speaking, like nothing we've ever experienced before. However, as humans, the only—and universal—way we make sense of novel situations is by making analogies to situations we have previously experienced. Contra *BuzzFeed*, coronavirus-related analogies are everywhere to be seen.

The way people conceptualize a situation drives their behavior in that situation. This means that the analogies we use to make sense of new situations can be powerful forces—for better or worse—in determining how we act. With this in mind, I've been collecting some of the analogies made by influential people about various aspects of the COVID-19 pandemic, and thinking about how these analogies are shaping our understanding and behavior in response to this "uncomparable" situation.

IS THIS JUST THE FLU?

On March 9, President Trump tweeted, "So last year 37,000 Americans died from the common Flu. It averages between 27,000 and 70,000 per year. Nothing is shut down, life & the economy go on. At this moment there are 546 confirmed cases of CoronaVirus, with 22 deaths. Think

---

OPPOSITE: *Paul Cézanne, "Nature Morte aux Pommes et aux Oranges," c.1895 (detail)*

about that!"[2] While a bad flu season might seem, on the surface, analogous to what we are experiencing, this analogy neglects some all-important facets of the novel coronavirus. Not only is it more infectious than typical flu viruses, it has at least ten times the fatality rate. We have a vaccine for flu, albeit an imperfect one. Most importantly, the illness caused by the novel coronavirus is often very different from flu. The scholar Zeynep Tufekci drove home this last fact using another vivid analogy: "By January 29, it was clear that COVID-19 caused severe primary pneumonia in its victims, unlike the flu, which tends to leave patients susceptible to opportunistic, secondary pneumonia. That's like the difference between a disease that drops you in the dangerous part of town late at night and one that does the mugging itself."[3]

The seasonal flu analogy has misled many people into minimizing our current situation, and has fostered serious misunderstandings of what the real perils are. As recently as March 21, the conservative pundit Bill Mitchell tweeted his corollary of Trump's analogy: "I do not understand how our hospitals can handle 670,000 flu cases every year, but a few thousand COVID-19 cases overwhelm the system. We have been dealing with massive flu outbreaks for decades. To me this is all part of the hype."[4] One reader replied with their own (sarcastic) analogy: "I do not understand how McDonald's can serve two billion hamburgers every year, but when I order five million at the drive-thru it overwhelms the system."[5]

IS THIS LIKE A NATURAL DISASTER?

One emergency room physician warned the *LA Times* that we need to make critical decisions "before we are in the throes of the tsunami."[6] Similarly, NPR reporter Mara Liasson warned, "This is a wave that hasn't crested yet."[7] The journalist Joy Reid noted, "the coronavirus hurricane is hitting all 50 states plus all the territories at once."[8] A newspaper in Tennessee told readers that "Coronavirus . . . will be like a flood. It will take time and action to subside."[9] *The New York Times* quoted one scientist's response as to when things will return to normal: "It's like asking a fireman when you can move back in, but your house is still on fire."[10] And even when the fire has subsided, we're told that "As long as the virus persists somewhere, there's a chance that one infected traveler will reignite fresh sparks in countries that have already extinguished their fires."[11]

The natural-disaster analogy is more legitimate than the flu analogy—it has the right urgency—but it can also mislead us. Natural disasters are times for panic buying, stockpiling supplies like pasta and toilet paper. This incorrect conception of our situation has led to unnecessary hoarding. Perhaps even worse, the thought of a natural disaster makes people want to flee to a safer location, to quite literally "head for the hills." One insightful tweeter noted the problem with this facet of the analogy: "I'm realizing that many people are responding to COVID-19 with the muscle memory of a natural disaster. After fires, we evacuate. Before hurricanes, we evacuate. It's time to realize: WE ARE THE HURRICANE. Running does nothing but bring the hurricane to others. Stay home."[12]

ARE WE AT WAR?

President Trump has called the pandemic "our big war," and referred to himself as "a wartime President."[13] Much has been written about the president's invoking "war powers" via the "Defense Production Act" to force American companies to manufacture medical equipment. Presidential candidate Joe Biden has also used a war analogy with respect to healthcare workers: "As president, I would never send an American soldier anywhere in the world without the equipment and protection they need. We should not do any less for the heroes on the front lines of the battle we're in now."[14]

Conceptualizing a situation as "war" can make people feel more nationalistic, xenophobic, willing to suspend civil rights, and susceptible to political exploitation . . .

The war analogy is appealing and in many ways apt, but again there are dangers lurking. Conceptualizing a situation as "war" can make people feel more nationalistic, xenophobic, willing to suspend civil rights, and susceptible to political exploitation, as was noted in a *New York Times* op-ed: "Amid all the signs of conflict—declarations of emergency, mobilizations of National Guard troops, the exercise of extraordinary powers—there is enduring constitutional danger in treating this crisis like a war."[15]

In making sense of our current difficult situation, we need to become more aware of how the analogies we live by are shaping our thoughts.

If you pay close attention, you will encounter myriad variations on these and other analogies, in television news programs, newspapers, social media, and even casual conversations. As I heard one radio commentator analogize about analogies: "We are in uncharted waters. We need some navigational buoys." Indeed, as humans we continually swim in a sea of analogies and metaphors that unconsciously but deeply shape our understanding of everything we encounter.[16,17] In making sense of our current difficult situation, we need to become more aware of how the analogies we live by are shaping our thoughts. Analogies are our buoys in troubled waters, our lighthouses in this dark storm, but we need to use them carefully to safely guide us to the other shore.

ENDNOTES

1 K. Miller, "The Disorienting Thing Is You Can't Compare This to Anything," *BuzzFeed News*, March 21, 2020, https://www.buzzfeednews.com/article/katherinemiller/corona-pandemic-economy-self-quarantine

2 D. Trump, Twitter post, March 9, 2020, https://twitter.com/realDonaldTrump/status/1237027356314869761

3 Z. Tufekci, "It Wasn't Just Trump Who Got It Wrong," *The Atlantic*, March 24, 2020, https://www.theatlantic.com/technology/archive/2020/03/what-really-doomed-americas-coronavirus-response/608596/

4 B. Mitchell, Twitter post, March 21, 2020, https://twitter.com/mitchellvii/status/1241340736504766465

5 "K. Trout," Twitter post, March 21, 2020, https://twitter.com/KT_So_It_Goes/status/1241465915486670848

6 S. Karlamangla and H. Ryan, "Coronavirus Hospitalizations Climbing Sharply in LA, Likely the Approaching Wave," *Los Angeles Times*, March 24, 2020, https://www.latimes.com/california/story/2020-03-24/coronavirus-hospitals-los-angeles-county

7 M. Liasson, "Update: Congress Negotiates a Stimulus Plan," *NPR*, March 22, 2020, https://www.npr.org/2020/03/22/819725424/update-congress-negotiates-a-stimulus-plan

8 J.-A. Reid, Twitter post, March 24, 2020, https://twitter.com/JoyAnnReid/status/1242576241997361152

9 B. Campbell, "Coronavirus in Nashville Will Be Like a Flood. It Will Take Time and Action to Subside," *Tennessean*, March 23, 2020, https://www.tennessean.com/story/opinion/2020/03/23/coronavirus-pandemic-nashville-flood-comparison/2902208001/

10 Bokat-Lindell, "How to Socially Distance and Stay Sane," *The New York Times*, March 17, 2020, https://www.nytimes.com/2020/03/17/opinion/coronavirus-social-distancing.html

11 E. Yong, "How the Pandemic Will End," *The Atlantic*, March 25, 2020, https://www.theatlantic.com/health/archive/2020/03/how-will-coronavirus-end/608719/

12 L. A. Smith, Twitter post, March 30, 2020, https://twitter.com/msLAS/status/1244847454035390464

13 B. Bennett and T. Berenson, "'Our Big War.' As Coronavirus Spreads, Trump Refashions Himself as a Wartime President," *Time*, March 19, 2020, https://time.com/5806657/donald-trump-coronavirus-war-china/

14 J. Biden, Twitter post, March 31, 2020, https://twitter.com/JoeBiden/status/1245066582792884224

15 G. Weiner, "Are We Sure We Want to Give Trump War Powers?" *The New York Times*, March 25, 2020, https://www.nytimes.com/2020/03/25/opinion/trump-executive-authority-covid.html

16 Hofstadter and E. Sander, 2013, *Surfaces and Essences: Analogy as the Fuel and Fire of Thinking*, New York, NY: Basic Books.

17 G. Lakoff and M. Johnson, 1980, *Metaphors We Live By*, Chicago, IL: University of Chicago Press.

REFLECTION

THE DOUBLE-EDGED SWORD OF IMPERFECT METAPHORS

DATE: *25 August 2021*

FROM: *Melanie Mitchell, Portland State University; Santa Fe Institute*

I wrote my Transmission in March 2020, just after most of the USA was put in lockdown due to the novel coronavirus pandemic. Our society was still trying to make sense of exactly what kind of situation we were in: was it more like a war, a hurricane, or just a bad flu season? Reflecting on this Transmission sixteen months later, I'm struck by how much we're still struggling to understand what is happening. But, as CDC scientists said in July 2021, "the war has changed," and the concepts we're grappling with have changed as well. The general public is sweating through an advanced course on virology and epidemiology that we're pitifully unprepared for, so we inevitably continue to rely on imperfect analogies and metaphors. Virus variants are like "a thief changing clothes"[1] or "sharks lurking beneath the surface."[2] mRNA vaccines "are like CD players that can play any kind of CD,"[3] and they won't cause permanent genetic changes because "the mRNA sends the instructions to your body's 'inbox' to interpret, after which it is erased."[4]

We still haven't found the right analogy to temper the expectations of the fully vaxxed. A vaccine is not a wall or a moat; vaccines "don't work like sunscreen, stopping the virus from entering your body. They

---

1 D. Stewart, "COVID Variants Are Like "a Thief Changing Clothes" – and Our Camera System Barely Exists," *Leaps.org*, April 2, 2021, https://leaps.org/covid-variants-are-like-a-thief-changing-clothes-and-our-camera-system-barely-exists.

2 K. Thomas, "Coronavirus updates, March 2: Montreal pharmacies to start administering vaccines," *Montreal Gazette*, March 2, 2021, https://montrealgazette.com/news/local-news/coronavirus-live-updates-quebec-reports-588-new-cases-sets-vaccination-record

3 Etobicoke Medical Centre Family Health Team, "COVID-19 Vaccination Questions and Answers," https://www.emcfht.ca/COVID-19-Vaccination-Questions-and-Answers.pdf

4 A. Pinder and K. V. Snyder, "Explaining mRNA Vaccines in Simple Terms," *Buffalo Healthy Living*, February 24, 2021, https://buffalohealthyliving.com/explaining-mrna-vaccines-in-simple-terms

fight for our cells."[5] Even harder is constructing that elusive analogy that will convince the half of the population that is still unvaccinated. Like refusing to wear a mask, eschewing vaccines is not like refusing to wear a seatbelt or a life jacket, since the decision affects not just you but your whole community.[6]

> The general public is sweating through an advanced course on virology and epidemiology that we're pitifully unprepared for, so we inevitably continue to rely on imperfect analogies and metaphors.

In fact, a lack of convincing analogies promoting collective action may be at the root of many of society's intractable problems. While our COVID-19 analogies are still a work in progress for understanding the complexities of this pandemic, it's worth contemplating the words of novelist Samuel Butler:[7] "Though analogy is often misleading, it is the least misleading thing we have."

5 A. Slavitt, Twitter post, August 14, 2021, 3:27 p.m., https://twitter.com/ASlavitt/status/1426656689432895489

6 https://www.theatlantic.com/science/archive/2021/08/covid-vaccine-analogies/619744/

7 Butler, S. 1912. *The Note-Books of Samuel Butler*, ed. Jones, H. F. London, UK: A. C. Fifield.

THE TRANSCRIPT OF *COMPLEXITY PODCAST* TRANSMISSION SERIES EP. 2, DISCUSSING THE FOLLOWING TRANSMISSIONS

[Batch 2, released 6 April 2020]

T-005: The Need for Disease Models which Capture Key Complexities of Transmission
Andrew Dobson

T-006: Using Social Media Data to Detect Signatures of Global Crises
Miguel Fuentes

T-007: How to Reduce COVID-19 Mortality while Easing Economic Decline
Danielle Allen, E. Glen Weyl & Rajiv Sethi

T-008: The Importance of Timing in Restrictive Confinement
Michael Hochberg

T-009: How the Analogies We Live By Shape Our Thoughts
Melanie Mitchell

This podcast transcript has been abridged for length and clarity. Find the full podcast at https://complexity.simplecast.com/episodes/27

COVID-19 & COMPLEX TIME IN BIOLOGY & ECONOMICS

DATE: *13 April 2020* **HOSTED BY:** *Michael Garfield, featuring David Krakauer*

MICHAEL GARFIELD: David, it's a pleasure to jump back into the complexity with you. Let's start on Andy Dobson's call[1] for disease models which "capture key complexities of transmission," because the origin story of this novel coronavirus in another animal is a really interesting point of the way that this is going to be remembered and a really key entry point for understanding some of the systems dynamics that have been revealed to us more popularly through this crisis.

D. KRAKAUER: Andy's was a very rich contribution, and he made three very deep points. The first point was, why is it that a disease like this one is so much worse in cities, whereas a disease like malaria is actually worse in rural communities? And it comes down to this technical insight of what the modelers call a "density-dependent transmission" versus a "frequency-dependent transmission." With density dependence, as the term would seem to imply, the higher the density, the more transmission there is. And that's why we're socially isolating. The more people there are, as there are in cities, the more the virus can spread. But if you look at a vector-borne disease like malaria, it's quite different. When the mosquito takes a blood meal that transmits this protozoan parasite, the malarial parasite, it has to stop and digest. If you present to the mosquito another potential susceptible host, there's nothing it can do because it's busy digesting. As you increase the number—the density, if you like, the number of people—that doesn't necessarily increase the transmissibility. In fact, as the population grows, the relative rate of transmission goes down. That was a very interesting point to make because people might be wondering why it is that not all transmissible diseases are worse in urban centers at high density.

So that was his first point. The second point I thought was really cool and had to do with the biomechanics of flying animals—in this case, bats. Bats have to have particularly light bones. A consequence of that

1 See T-005 on page 83

is that there's less of a volume of bone marrow. Bone marrow is where B cells of the immune system are synthesized, and the B cells create antibodies that create the inflammation that creates the pathology. And so bats, by virtue of being volant, flying mammals, are actually in some sense naturally more resistant to immune pathology.

He then went on to make another point about bats, which was every time a bat flies, in some sense it's generating a mini fever. And high body temperature, as everyone knows, is one way that physiology has invented to rid ourselves of infection. So, the bat is in some sense generating this fever every time it flies and killing off any viruses that it might be infected with.

... if human beings persist in the stupidity of engaging at a high rate, through whatever means of transmission, with nonhuman species such as the bat population, then there's a real prospect of a far worse entangling of the epidemic cycles, potentially leading to a very significant reduction in the human population.

The third thing about bats is that they enter into what's called "torpor." Torpor is a little mini hibernation where they have very reduced activity, and that naturally socially isolates the bat. In its natural repertoire of behaviors, it has built into it in some sense a natural means of combating viral infection that we do not. And that goes some way to explaining why bats don't suffer as much as humans.

Andy's final point is the most technical and most frightening. Every species that gets infected with a virus has a natural epidemic cycle. The factors that contribute to that cycle are things like how long they live, how quickly they reproduce, and so forth. Small mammals reproduce more quickly and have shorter lifespans, and large mammals reproduce slowly and have longer lifespans. That means our epidemic cycles are slow relative to a bat.

As long as we have separate, largely uncoupled life histories, that's fine. But if you increase the transmission between species to a level that you would observe within the species, something very alarming happens: the faster epidemic cycle dominates the slower one—which means that the bat species could drive the human species extinct, because we would be living in some sense on the timescale of a bat population. In this case, we kind of dodged a bullet because even though it's dreadful, there was in some sense one or a few points of contact, but if human beings persist in the stupidity of engaging at a high rate, through whatever means of transmission, with nonhuman species such as the bat population, then there's a real prospect of a far worse entangling of the epidemic cycles, potentially leading to a very significant reduction in the human population.

M. GARFIELD: This is an interesting place to enter the question you brought up last week, in terms of intentionally misaligning phenomena occurring at different scales, uncoupling systems that are operating at different scales, and this I think also touches on Danielle Allen, Glen Weyl, and Rajiv Sethi's Transmission, which we'll discuss later. As you know, we're looking at the epidemic timescale in bats, the epidemic timescale in humans, but then also the way that this epidemic is rippling through our economic systems and so on. There are a lot of areas where we either want to drive them apart or we are forced to align timescales to shift our economy in order to make it sort of work with the human epidemic timescale. But that's getting ahead of ourselves.

D. KRAKAUER: Well, no, I think it's a really important point you make. As I'm not sure everyone understands or knows, SFI has been for a long time fascinated with this topic of "complex time," which is how temporal phenomena play out across the scales of the complexity of Earth. And what evolution has done is, it's created this interesting coupling of all of these clocks. You can think of every species as being a little clock that ticks away at its own rate. We interact at some level, but we're not strongly coupled. If we strongly couple, then you get these strange synchronizations that appear. And in this case, the synchronization is dominated by the fastest clock. In the case of transmission, that could be a lethal synchronization.

M. GARFIELD: To get back to the core of this series and the dynamics of transmission, let's talk about Michael Hochberg's piece.[2]

D. KRAKAUER: Michael is in some sense is doing what John Harte did last week[3] in presenting us with the intuitions of a simple model. And the intuition that he's trying to instill, which we have a very hard time with, is exponential thinking. Human beings are basically by default linear thinkers. You double the size of that, you double the effect, you know. It takes twice as long, and it's twice the effort. But in the exponential regime, that's not true. These nonlinearities dominate. And the one that Michael focuses on is again this enigmatic quantity that everyone is talking about—that in England we call "R naught," but here we call "R sub zero"—which is the number of new infections caused by a primary infection. So, if I were infected, how many other people I would infect? We know that, for this particular virus, that value is about 2.5, so a primary infection would cause about 2.5 secondaries, which leads to a doubling of the virus about every three days.

What that means in large populations is really kind of staggering, and it bears on this whole question that we discussed on coming out of quarantine. So, if the initial size of the infected population is in the hundreds, in a population of millions as we live in, then thousands of people will be infected. But if the initial population of infected is in the hundreds of thousands, then the total number afflicted will be in the millions. This gets back to this crucial insight that because of the nonlinearity and this doubling effect, as we come out of isolation, it's crucial that we don't expose the world to a very large number of what might be asymptomatic infected people, because we're going to get these nonlinear effects, which lead to millions of people becoming infected.

M. GARFIELD: There's another number here, the effective reproduction number, R(t). People confuse these two numbers, and I think it's important to differentiate. This is what you're talking about—the R_0 gives you the transmissibility on day zero, but when we're looking at how

2 See T-008 on page 121

3 See T-002 on page 29

it lands in different populations, it's not just the size—as we talked about with Laurent Hébert-Dufresne[4] on an earlier episode—it's also the structure. Let's talk a little bit about how R(t) is different from R_0.

D. KRAKAUER: Essentially they're very similar concepts, but R_0, as you said, is in a way the concept that we worry about at $t = 0$ when the infection starts, but while the infection is moving along, there's also a reproductive rate of the virus, that's time-variant as the epidemic proceeds. It's not necessarily just what happens at the beginning that we should be concerned about, but what's happening as the infection unfolds. But they are, to all intents and purposes, very similar, because they all reflect this effect of the instantaneous current population size on the future epidemic size.

> If the initial size of the infected population is in the hundreds, in a population of millions as we live in, then thousands of people will be infected. But if the initial population of infected is in the hundreds of thousands, then the total number afflicted will be in the millions.

M. GARFIELD: It seems like one key point to take away from that, though, is that when people hear other people talking about lowering the R value for this virus, they're not talking about actually changing the virus itself to become less transmissible. They're talking about network dynamics.

D. KRAKAUER: Exactly. That's absolutely right. They all play into that calculation of R_0 or R, and as we said last week, there's been so much emphasis on the molecular biology that we've forgotten that the social-distancing measures are an essential parameter in the epidemic modeling.

4 Complexity Podcast episode 24 aired on March 25, 2020. https://complexity.simplecast.com/episodes/24

M. GARFIELD: We have a conversation with Caroline Buckee[5] coming out, and she's published quite a bit recently on disease surveillance using modern techniques . You know, this is a modern pandemic and so it's a pandemic happening in a world of data surveillance, happening at multiple different scales, and there are a lot of really interesting both technical and ethical questions that come up from that. But one of the reasons that it's such a hot topic is because it would allow us to potentially detect these things a lot earlier than we did this time, or to be able to respond to them in a much more precise and granular way. So that brings us to Miguel Fuentes's piece[6] on the use of social media data to detect signatures of global crises. What'd you think about this one?

D. KRAKAUER: This was very interesting, and it had that spooky epigraph from H.P. Lovecraft, which is, "The oldest and strongest emotion of mankind is fear, and the oldest and strongest kind of fear is fear of the unknown."

M. GARFIELD: Which speaks to last week's conversation on rigorous uncertainty.

"The oldest and strongest emotion of mankind is fear, and the oldest and strongest kind of fear is fear of the unknown."

—H.P. LOVECRAFT

D. KRAKAUER: It absolutely does—and our desire to do anything in our power to pretend that the unknown is not unknown. What Miguel and his colleague Juan Pablo Cardenas's contribution is, what they've been doing in fact for the last several years, is studying social unrest in their own home nation of Chile. And the question that they ask themselves is, what is the signature of social crisis? And they're interested in that largely because they want to detect it before it unfolds. And they've

---

5 Complexity Podcast episode 28 aired April 16, 2020. https://complexity.simplecast.com/episodes/28

6 See T-006 on page 101

been studying social media, in particular tweets and Twitter feeds, and they made an interesting discovery. In retrospect I think it's pretty obvious, but I think it might not have been before the analysis was performed. And that is, during periods of high crisis—during marches, during demonstration, during periods of significant civil unrest—if you look at the social graph or the social network that you can reconstruct from tweets—who sends messages to whom and so forth—you observe that, during crisis, that graph has a very particular topological structure. And that structure is that messages are very cliquish and they center around a very small number of dominant themes, dominant terms, of course, or keywords or tags. And so the graph kind of fragments and the particular reach of any particular person in that social media graph is limited. As you move into periods of low crisis, the number of terms or tags becomes more diverse and its reach increases. It extends over longer distances in the graph. And what was intriguing, I think, about their analysis of this particular dataset in Chile during the civil unrest is that those signatures were present before people were out in the streets.

It gets to our point last week that technology allows for mimetic transmission to be in advance of behavioral and biological transmission, and in just the way that we fail to act with alacrity in the presence of the data which we discovered today, for example—much earlier, perhaps even in January, we knew this ... that there are similar questions about social unrest and it's a double-edged sword. On the one hand, yes, you might be able to see this. On the other hand, in whose hands is that appropriate information and how would we use it responsibly?

M. GARFIELD: This sounds a lot like the way that Francisco Varela[7] described the 1973 Chilean revolution, when he spoke of turning on the radio and one radio station would say that it's sunny outside and another radio station would say it was raining. It was like a fragmentation of reality. And there's been a lot of research at SFI, including work by Joshua Garland, on polarization of networks on social media.[8] Reading Fuentes's

---

7 F. Varela, "The Cultural Contradictions of Power," Summer 1978 Fellows' Meeting, New York City, audio, 22:31:44, Internet Archive courtesy Schumacher Center for a New Economics, https://archive.org/details/FranciscoVarela

8 https://santafe.edu/applied-complexity/fellows#cyberhate

piece almost reminds me of the meiotic spindle and the polarization of chromosomal pairs at the moment that we're seeing this transition—a cell's reproductive cycle. Without suggesting that there's any kind of developmental telos here, I'm curious, in relation to your earlier remarks on multiple different timescales and the way that an epidemic in a smaller organism can drive disease dynamics in a larger organism, whether you think that what Fuentes and his colleagues are observing is the network coupling of two phenomena[9] that are going on at different scales, at different paces, and attempting to adapt to accommodate that.

D. KRAKAUER: I think you're right. Of course, we're all now obsessed over this virus, and it's rather dangerous actually. We're becoming a memetic monoculture, the news that should be reaching us is not, and our cliquishness, with respect to ideas, is shrinking, and I think it's a good example of that phenomenon. What's surprising, I think, when one performs these analyses of the kind that Miguel has, and Josh and others, is that what is clear at the level of our social lives also manifests at the population level, at the national level, and that's surprising. I think it's very difficult for us to project out our own psychological insights into the collective—and what the social media data is allowing us to see is these echoes, if you like, of the micro in the macro. And it needed to be that way. But in this case it does seem that as we become, in some sense, more narrow in our obsessions, so does culture at large.

M. GARFIELD: To explore this a little bit further, I just read a really fascinating essay[10] about the Thirty Years' War: the Anabaptists holing up in Münster fighting the Catholic Church, and the way that the fragmentation of Christianity across Europe was driven by the circulation of pamphlets due to the recently invented printing press. This was an early modern information explosion that led to a phenomenon that

---

9 See, for example, J. Garland, K. Ghazi-Zahedi, et al., 2020, "Countering hate on Social Media: Large-Scale Classification of Hate and Counter Speech," arXiv:2006.01974 [cs.CY], https://arxiv.org/abs/2006.01974 and J. Garland, K. Ghazi-Zahedi, et al., 2020, "Impact and Dynamics of Hate and Counter Speech Online," arXiv:2009.08392 [cs.SI], https://arxiv.org/abs/2009.08392

10 J. Stantonian, 2020, "Apocalyptic Cults and the Early-Modern Information Explosion," *Medium*, https://medium.com/@jamiestantonian/apocalyptic-cults-and-the-early-modern-information-explosion-708ad3cfbb84

Fuentes would call a "crisis," in the way that it bears this kind of signature in its networks. We talked a lot in the first episode about the way that the social contagion dimensions lay over this and make this, again, like a distinctly information-age pandemic. That period in the sixteenth and seventeenth centuries . . . I'm curious to what degree you might find that particular period of polarization and loss of collective narrative coherence as a useful analogy or guide for what's going on now—and then moving into Danielle, Glen, and Rajiv's piece, what does that suggest in terms of our options for recovery?

D. KRAKAUER: If we look at the article by Danielle, Glen, and Rajiv, what this is saying at the broadest level is that we have to be equally as sophisticated in the treatment of the socioeconomic networks as we are coming to be in the treatment of the social epidemiological networks. Again, it gets to last week—we're so accustomed to these reductionist forms of thought, where we look for these magic bullets at the genetic level, that we've become blinded to the fact that they're by and large useless at the level of the collective. We need new insights, which is in part what complexity tries to do. What they are proposing in their article is something in some sense, in retrospect, very obvious—which is that we need a multiple-timescale approach to this crisis that makes most effective use of confinement testing and mobilization. And they call their strategy "mobilize and transition."

The basic concept of mobilize and transition is while we are in confinement is the time when we should be maximizing virological and serological testing in order to move either the resistant or immune subset of the population back into the economy. So you don't wait for the wave of the pandemic to subside; you actually are actively mobilizing now while in confinement. And the point that they make, which I think is the most important, has to do with this question of technological certification—because everyone's asking this question: "If I tested negative and I returned into work, how do you know that I haven't been infected in the meantime? Because I've been exposed to any number of people by coming back to work and we can't be absolutely sure everyone is still negative." Now with serological testing, with antibodies, we're more secure because we hope you have persistent immunity and that's where technology comes in.

So, they are suggesting we need to develop new technologies, which certify negative-tested or serological positives that are reliable and can be accessed by other individuals. It's as if, Michael, we both came into a room and I looked at my phone and it said, "Yes, you can talk to Michael because he's a safe contact," but you can again note how frightening that might be with respect to security. I think the bigger question that they're raising is, yes, we need the sophisticated schedule of social change. We need a technological means of certification and we need to be sophisticated about the robustness of that certification.

M. GARFIELD: There's another piece here which ties into—if people heard Rajiv's episode of this podcast[11] when he was here talking about stereotyping and criminal justice, and he was talking about the imbalanced distribution of stop and search that certain groups are targeted due to ethnicity for discriminatory police action. And he uses that as an interesting metaphor here, to explain what we want to tune this process of testing for, in terms of optimizing for a best distribution of testing. Why don't you talk about that a little bit?

D. KRAKAUER: Well, now, this just gets to this point—again, it's very general. I'm not sure we have a good answer to this, which is what you don't want, right? Are all of these biases built into your process because you think you know who the highest-risk individuals are? In some sense you have to spread your survey much wider than your prior beliefs might lead you to. And my takeaway from Rajiv's point on that is that we have to be extremely aware of how we use historical bias to inform future survey, and that the best thing we could possibly do is know it in order to adopt a more uniform sampling strategy.

M. GARFIELD: He says, "If a location is turning up more positives than another at the margin, resources would be better used by shifting to the former at the expense of the latter." So, this is about reallocation of testing through a dynamic process that's provisional and under constant revision. You know, something else that came up around this

11 Complexity Podcast episode 7 aired November 13, 2019. https://complexity.simplecast.com/episodes/7.

paper on Twitter is people's discomfort around the idea. Nowhere in this paper does it say that people will be forcibly reallocated into different labor sectors. But, of course, when you're operating on a fraction of the workforce and there's been a massive shift in the composition of goods and services in economic demand, then it's clear that reallocation is going to happen. But it's going to happen, as they make clear in this, among those who are willing and able. And so this is another piece that I'd like to hear you speak to, which is how this kind of process is going to emerge naturally, rather than being forced in a kind of top-down way . . . you can't do that with a system like this.

D. KRAKAUER: But this gets to David Kinney's contribution[12] of last week having to do with the limitations of scientists' insight into policy advice, which is that what Rajiv is saying—very correctly—is there are statistical anomalies that come from these biased samples, and that there are techniques to actively avoid them. In his case, do the inverse, which is one way to do it. But then it leads to your question, which is a policy question, which is how do we then use that kind of insight?

It's certainly above my pay grade. I would say it would be my job and the job of our community to be as rigorous as we can with the limitations of the survey and suggesting improvements to harvesting the true information. At that point, we really have to turn to democratically elected officials to make difficult decisions, and to the citizens of the country who elect them. I'm always very careful about stepping into that domain because I wasn't elected for my sociopolitical wisdom. I was elected for my scientific capability.

M. GARFIELD: That is a fair caution. It seems, though, that there is a scientific opportunity here to investigate how the timescale at which people can be retrained suits the timescale at which people are going to be re-entering the workforce, and gives us some insight . . . . You know, like when, Maria del Rio-Chanona, Doyne Farmer's PhD student, came out for the symposium last year and was discussing looking at networks of skills rather than at job descriptions,[13] and looking at how we might

12 See T-001 on page 21

13 W. B. Arthur, E. Beinhocker, and A. Stanger, eds., 2020, *Complexity Economics: Dialogues of the Applied Complexity Network I*, Santa Fe, NM: SFI Press, chapter 9.

be able to help people see this in a fresh way by unbundling all of the different capacities that people have. Now that kind of work seems like it would be really useful in advising policymakers on recommendations for retraining and that kind of thing.

D. KRAKAUER: I do think that one thing we've learned in pedagogy is how vastly more flexible human minds are than we thought. All of us, when motivated, have an extraordinary latent capability to learn. And so, on that side of the equation, I think yes. But, as you are, I'm extraordinarily concerned about despotic moves during periods of crisis. They're very hard to undo. And so I would be extremely cautious in giving individuals in political power that instrument to dictate what people should be doing with their lives.

M. GARFIELD: To the point on plasticity and people being able to change the way that we think about something, that leads us very gracefully into Melanie Mitchell's piece on analogy.[14]

I'm extraordinarily concerned about despotic moves during periods of crisis. They're very hard to undo. And so I would be extremely cautious in giving individuals in political power that instrument to dictate what people should be doing with their lives.

D. KRAKAUER: This was really a wonderful, thoughtful, and very witty contribution having to do with how we struggle to conceptualize a moment like this, and how we're forced to reach for analogies to make sense of it because it seems so unprecedented. And of course, Melanie had worked on artificial intelligence systems with Doug Hofstadter[15] to try and draw analogies, so she has a really interesting background in this

14 See T-009 on page 129

15 D.R. Hofstadter and M. Mitchell, 1994, "The Copycat Project: A Model of Mental Fluidity and Analogy-Making," *Analogical Connections* 501: 31–112.

topic. The first analogy that Melanie mentioned is to flu. And many of us actually were quite confused by the situation because flu, as most people know, infects hundreds of thousands of us citizens a year. That leads to tens of thousands of deaths. Why is it that the coronavirus had a much stronger systemic effect than the flu virus? And one obvious answer to that is, because we knew the flu virus and society doesn't like uncertainty, as we have established. Just the novelty alone was enough to lead to a strong reaction.

But more importantly, other details of its biology—that the coronavirus is more virulent, it's more transmissible—mean that the rate at which society is being perturbed is higher. And this led to one interesting and amusing quote from someone with a comedic capability, in her article, who said facetiously, "I do not understand how McDonald's can serve two billion hamburgers every year, but when I order five million at the drive thru, it overwhelms the system." Right? It's a very important point being made there, and it's the difference between number, and number over time, and rate. And the problem—the difference between flu and coronavirus—is not the number. The problem is the rate.

And she then goes on to talk about these other analogies that we draw, you know, that we're in the throes of a tsunami. This is a wave that hasn't crested yet. And the question is, what's behind that analogy? I think what's behind that one is our awareness of something with immense power that we're not aware of until it's too late. If you unpack what is meant by a tsunami, that's what's meant, in contrast to a quote that she has from *The New York Times* where someone says it's like asking a fireman when you can move back into your house, but your house is still on fire. And that's completely different, because that's an inadequate grasp of the risks of a common event. And what that analogy is saying is, "Get real."

She goes on to develop all of these analogies. And the one that's, in some sense, most alarming in reference to our earlier conversation, is the analogy to war—that the pandemic is our war, the war of our time, perhaps World War Three, and with leaders, including presidents and prime ministers, referring to themselves as wartime leaders, as wartime presidents. What is behind that analogy? Well, on the one hand there's a positive: that we all have to cooperate. We all have to work together to solve this. We all have to make sacrifices. But there's also a very significant negative. And that is that during a war there is a reduction in civil liberties, and the question we're asking ourselves is, during this so-called war, are we cooperating or are we being controlled? And that's the question we should ask. The analogy is not benign. It carries with it both positives and negatives. Melanie's deep point is, we should all be very thoughtful about both of those, because they're going to change our behavior in the long run.

M. GARFIELD: There's an entry through this article to a deeper and broader topic in complex-systems thinking generally, which is that these analogies are informal verbal models—you know, SFI External Professor Scott E. Page and his book *The Model Thinker*.[16] It seems like a kind of adaptation to the information explosion that we were talking about earlier in this call. To try and fortify that analogy with the printing press, this was something that gave people the capacity for peer review. It was a technological innovation that enabled modern science, by a lot of historical accounts, and in some way the press was the cause of and solution to the information crisis.

D. KRAKAUER: You make a really important point, which is that in some sense, what we do when we build mathematical models, or models of any kind, is make analogies. If you look at the early theories of the solar system and the cosmos, they were based on clocks, right? In other words, mechanical devices we used to model the motions of the planets, and so hence we have armillary spheres and astrolabes and all of these extraordinary devices that are in some sense embodied physical analogies. You're right to say that we should understand that, because what your brain is doing when someone says "This is like a tsunami," is in some folkloric sense invoking fluid dynamics. You're saying, "I know what a wave is. I know how they're propagated. I have an understanding of the force behind them. I understand what happens when they reach shallow ground and a wave breaks." So, you have in your mind a mechanical picture of a dynamical system. It's just that you are not writing it down rigorously in mathematical terms. But your brain

---

16 S.E. Page, 2018, *The Model Thinker: What You Need to Know to Make Data Work for You*, New York, NY: Basic Books.

is using a similar understanding of mechanics, and it's a very natural extension in fact from the verbal analogy to the mathematical model. It's actually one of the very powerful things that the brain does, and it allows us to arrive at insights without doing a considerable amount of formal work.

M. GARFIELD: The question that Melanie's calling us to hear, I think, is really important, which is that not all models are created equal. There may be moments where the complexity of the situation calls for a kind of holding together of multiple models to get a stereoscopic view, to see things knowing that each model is hiding crucial aspects of the situation, and suggesting things that we don't want, like you mentioned, that wartime footing tends to lead to xenophobia and so on. So, these are different stories in a sense that are inhabiting a kind of evolutionary landscape. Some of these models are really robust and ancient, and some aren't. In communicating them, some of them are simple enough to be effectively communicated, but, like we were talking about in the last episode, the most encompassing, complete model may be dependent on data that we don't have or may overfit a rapidly changing situation. I think this really speaks to the issue of careful science communication here.

D. KRAKAUER: Well, this analogy concept and the relationship to models is very deep. I mean we could all sit down—it might be an interesting exercise, in fact—and say, okay, let's take the three analogies that are in play here, "war," "tsunami," and "fire." And if you were to sit down and just write down a list of what those things implied and which one we thought was the best fit to our circumstance, it's quite interesting, right? War typically has relatively equally matched adversaries who have strategic interests that are being in some sense reached for through conflict, typically territorial expansion. Does that fit this circumstance or not? I think it's a very interesting exercise, because language is so seductive and so infectious, and sentences are so easy to repeat, we often forget that they carry specific meanings. I think that Melanie's contribution is asking us to consider what those meanings really are.

3

[BATCH 3, RELEASED 13 APRIL 2020]

Beyond our response itself lie the longer-term **OPPORTUNITIES**—*economic. The deeper the subsequent opportunities,* **POSSIBILITY** *of shaping*

to the **PANDEMIC**

effects, including new

social, political,

CRISIS, *the greater the*

and the greater the

these opportunities.

—DOUG ERWIN, Transmission 014

TRANSMISSION NO. 010

INVESTMENT STRATEGIES IN TIMES OF CRISIS

DATE: *13 April 2020*

FROM: *William H. (Bill) Miller III, Miller Value Partners*

STRATEGIC INSIGHT: *Typical recession and recovery economic behavior offers great stock market buying opportunities.*

There have been four great stock market buying opportunities in my lifetime. This is the fifth.

The first was in 1973–74. There was war in the Middle East in October of 1973; there was war in Vietnam; we had soaring oil prices, double-digit inflation, double-digit interest rates, recession, and a constitutional crisis in Watergate culminating in President Nixon's resignation. I was a young first lieutenant in the Army making around $400 per month and putting $25 per month into the Templeton Growth Fund. In the fall of 1974, I went into the Merrill Lynch office in Munich, Germany, and bought shares near the market bottom.

The second was in the summer of 1982. We had started our mutual fund in the spring of '82 and kept it mostly in cash as the market continued to decline. Mexico defaulted on its debt, the market fell sharply, and we got fully invested in July. In August, Paul Volcker cut rates and the great bull market was underway.

The third was in October of 1987—the market crashed 22.6% on October 22. We had raised cash in the summer as the market got very expensive relative to bonds. Stocks peaked in August and began to decline. By the end of September, thirty-year bonds yielded over 9%, a cash return roughly equal to the average annual return of stocks since 1926. The market collapsed and we put all of our 25% cash position into

---

OPPOSITE: *Jean-Léon Gérôme, "The Tulip Folly," 1882 (detail)*

stocks. Our fund was the single best-performing fund of 1988 as most others maintained high-cash positions due to fears about the health of the economy.

The fourth was 2008–09. We did not navigate that period well, with the Opportunity Trust Fund being down over 60% in 2008. However, we remained fully invested and stayed that way throughout the 10-year bull market. The fund was in the top 1% since the market low in March 2009 through the end of 2019.

We have been hit about as hard as anyone in this decline because we have been overweight in higher-beta names, believing (correctly) that, since the financial crisis, people have been risk- and volatility-phobic, and that perceived risk has consistently been greater than real risk. Going into 2020, I thought that economic risk was low and that, if the market was going to decline, it would be because of either geopolitical events or some exogenous shock to global aggregate demand or aggregate supply. We got that exogenous event in the form of a global pandemic that took stocks from all-time highs to a bear market decline of almost 30% in the shortest time in history.

As is typical of these sorts of egregious declines, we are down a lot more than the market in all of our products. In times such as these, I am reminded of what John Maynard Keynes wrote to his board when he was managing money during the 1937 market collapse. They had been urging him to sell as the market went down, and he refused. He told them, "It is the duty of every serious investor to suffer grievous losses with great equanimity." He went on to note that their advice to sell more as stocks went down, if practiced by everyone, would be economically disastrous for the country, and that he wanted to be fully invested at the bottom and not out of the market when it recovered, which he would be if he followed their advice.

The market's behavior in the last week of March 2020 gives a strong confirmation that this recovery from a near-certain recession will follow the pattern of every other one since at least 1973–74. The leaders will be low-P/E, cyclical names with operating or financial leverage, viz., the exact names that were clobbered the most in the decline. Those names were also the worst performers when recession fears were high—fall of 2018, first six weeks of 2016—yet where no recession materialized. The reason why is fairly straightforward: prices and valuations are highly sensitive to the marginal return on invested capital and to business risk. When the economy is declining, or there are fears that it will, valuations on those companies whose return on invested capital (ROIC) is most sensitive to economic change, mostly traditional cyclicals, will decline more than those that are more resistant, such as consumer staples, utilities, bond proxies, and many recurring revenue businesses. Companies with high debt leverage and economic sensitivity fare the worst as the market discounts the possibility they will experience financial distress. When the market sees a recovery, the exact reverse occurs, which is what we saw March 24–25.

Going into 2020, I thought that economic risk was low and that, if the market was going to decline, it would be because of either geopolitical events or some exogenous shock to global aggregate demand or aggregate supply. We got that exogenous event in the form of a global pandemic that took stocks from all-time highs to a bear market decline of almost 30% in the shortest time in history.

If, as most strategists think, we will have a short, sharp snap-back rally that takes us up 20% or more, which will be followed by a retest of the lows, then that retest will see the reverse, and again the cyclicals will lag, but not as much as they did on the initial collapse. I am agnostic on that, as my ability to forecast the market's short-term path rounds to zero. So does theirs, by the way.

Last week gave a good indication of where the leaders and laggards are likely to be: names that have held up relatively well, such as AMZN, GOOG, FB, NFLX, were all down while most other stocks were up, led by cyclicals and high beta. The commonly offered advice in a steep market decline, such as we are experiencing, to "upgrade your portfolio" and to "buy quality names on sale" is a great prescription for underperformance in a recovery.

Sir John Templeton advocated buying at the point of maximum pessimism. The problem is that point is only known in retrospect. There is much pessimism and little optimism evident now, and it is impossible to tell whether stocks have declined enough to discount what the future holds with regard to the economic damage that the pandemic will inflict. In October 2008, Warren Buffett wrote an op-ed saying he was buying US stocks and urging others to do so as well. A few years later, he was asked how he knew that was the time to buy. He said he did not know the time, but he did know the price at which stocks were a bargain. They were a bargain then and, in my opinion, they are a bargain now. The market may have bottomed at an interim low on March 23, 2020—or it may not have. I do believe that shares bought at these prices will prove to be quite rewarding over the next year, and perhaps a lot sooner. If you missed the other four great buying opportunities, the fifth one is now front and center.

ADDITIONAL COMMENTARY

Miller Value Partners (3/16/20): How Markets as Complex Adaptive Systems Process COVID-19: https://millervalue.com/markets-complex-adaptive-systems-process-covid-19/

CNBC (3/18/20):

Clip 1: https://www.cnbc.com/video/2020/03/18/buy-amazon-in-response-to-coronavirus-sell-off-bill-miller.html

Clip 2: https://www.cnbc.com/2020/03/18/investor-bill-miller-one-of-the-best-buying-opportunities-of-his-life.html

Full interview (paywall) https://www.cnbc.com/video/2020/03/18/full-interview-with-bill-miller.html

Ameritrade Network (3/26/20): https://tdameritradenetwork.com/video/rB4AoXD6FmuBcRhYlikNzg

REFLECTION

BETTING ON THE FUTURE

DATE: *25 August 2021*

FROM: *William H. (Bill) Miller III, Miller Value Partners*

"The real trouble with this world of ours is not that it is an unreasonable world, nor even that it is a reasonable one. The commonest kind of trouble is that it is nearly reasonable, but not quite. Life is not an illogicality; yet it is a trap for logicians. It looks just a little more mathematical and regular than it is; its exactitude is obvious, but its inexactitude is hidden; its wildness lies in wait."[1] So wrote G. K. Chesterton in 1908, ten years before last great pandemic to strike the world prior to the one in which we are still engulfed.

Capital markets are quintessential complex adaptive systems: they contain millions of agents constantly acting under their own local rules, with differing time horizons, adjusting their behavior as circumstances or expectations change, all in an effort to earn returns for themselves or for their clients that are sufficient to ensure their survival. The first rule of the money game is to be able to stay in the game; if you don't have any money to invest, you can't play the game. Warren Buffett has said the first rule of investing is "don't lose money," and the second rule is to remember the first one. That isn't quite correct, though. Everyone in the game loses money from time to time; everyone makes mistakes. The legendary speculator Jesse Livermore once noted that if you were never wrong in the markets, eventually you would have all the money and if you were never right, you would go broke. George Soros was closer to the mark when he observed that it is not how often you are right or wrong that's important; it is how much money you make when you are right, less how much money you lose when you are wrong that matters.

Capital markets, rephrasing Chesterton a bit, exhibit both structure and wildness. There is enough structure to enable one to think useful predictions can be made about future prices and enough wildness to vitiate those beliefs. Because the payoff for getting the future right, or mostly so,

1 G.K.Chesterton, 1908, *Orthodoxy*, London, UK: Bodley Head.

is high, making predictions about prices is endemic. One popular strategy is to look at history, at events and circumstances that are analogous to the present and see how markets reacted during similar times.

> The first rule of the money game is to be able to stay in the game; if you don't have any money to invest, you can't play the game.

When this pandemic was compared to its closest analogue, the disastrous 1918 flu, the results of that exercise were not helpful. During the 1918–20 pandemic, the stock market was never more than 5% lower than when the pandemic began. Nine months after that pandemic got started, and during its second and deadliest wave, the market was actually up 11%. This time, after reaching an all-time high in late February 2020, the market had its biggest collapse ever in a four-week period, falling 37% to its low on March 23. Looking at similarities may be helpful, as long as one remembers Wittgenstein's quip: "I will teach you differences."[2] The difference between this pandemic and the one in 1918 was that state and local governments shut down vast parts of the US economy in 2020, whereas in 1918 businesses continued to be open. The result of that difference was a decline of US GDP in the second quarter of last year of 33%, the steepest single-quarter drop in history.

When I wrote in April of 2020 that the collapse in the market put investors "front and center" in the fifth great opportunity to buy stocks in my adult lifetime, I was making a judgment based on a few observations that had been helpful in forty-plus years of surviving in the game (encompassing a couple of close calls) and being involved at SFI for over twenty-five years. First, since no one has privileged access to the future, trying to predict the path of a complex adaptive system with any degree of granularity, particularly over short timescales, is a

2 Wittgenstein once told his friend Maurice O'Connor Drury, "I was thinking of using as a motto for my book a quotation from King Lear: 'I'll teach you differences.'" See M. O'C. Drury, 1984, "Conversations with Wittgenstein," in R. Rhees, *Recollections of Wittgenstein*, Oxford, UK: Oxford University Press.

very low-probability endeavor. Second, losing money is twice as painful as making money is pleasurable, as the psychologists have demonstrated, which makes fear dominant over greed, particularly in the short run. Third, the stock market goes up most of the time, about 70% of the years in the postwar period, so believing the market will be higher in the next year or so is a good bet. Finally, as Jesse Livermore also noted, the big money is made (or lost) in the big moves, and the big move down in 2020 from late February to the March 23 low was one for the record books. The odds seemed tilted in favor of the next big move being up, as it was in the previous four great buying opportunities that occurred after steep market declines, and so it has turned out in the fifth.

TRANSMISSION NO. 011

A COMPLEX-SYSTEMS PERSPECTIVE OF VIRUSES

DATE: *13 April 2020*

FROM: *Santiago F. Elena, Spanish National Research Council; Santa Fe Institute*

STRATEGIC INSIGHT: *A complex-systems perspective of viruses offers insight for controlling SARS-CoV-2 and future emerging viruses.*

Numerous questions emerge when considering the nature and relevance of viruses: What exactly are they? Are they even alive? How and when did viruses first evolve, and are viruses unavoidable consequences of biological replicators? Why are they so diverse in genomic architecture, yet so limited in lifestyles? Why are there so many emerging viruses, and what are the ecological and genetic drivers of that emergence? Why do they become pathogenic, and what is the nature of the complex molecular interaction they establish with their hosts?

Viruses are complex systems spanning orders of magnitude in size. The study of their behavior and structure, particularly using multidisciplinary frameworks, has revealed a number of universal patterns of organization. RNA viruses display high mutation rates that push them to the edge of disorder, where high instability, but also adaptability, occurs. As with many phenomena in virology, this edge is well described by phase transitions—where small changes in mutation rates can lead to large changes in structure—analogous to a liquid transitioning to a gas with a small shift in temperature.

Viruses have influenced evolution at all levels of biological organization, from cells and organisms to populations, ecosystems, and even the entire

---

OPPOSITE: *Wellcome Historical Medical Museum, Wigmore Street, London, "Reconstruction of a Sixteenth-Century Alchemist's Laboratory" (detail)*

COURTESY WELLCOME TRUST

biosphere. Their dynamics involve nonlinear phenomena, tipping points, and self-organization. Viruses offer unique experimental and theoretical windows into the origin and evolution of complex systems.

Viruses are complex systems spanning orders of magnitude in size. The study of their behavior and structure, particularly using multidisciplinary frameworks, has revealed a number of universal patterns of organization.

For thirty years, I have combined experimental evolution, molecular genetics, systems biology, molecular epidemiology, mathematical modeling, and computer simulations to pursue the enigma of the virus, using different animal and plant systems. In this search for basic knowledge, we have acquired many small pieces of information that, together, delineate a picture of how viruses evolve. This picture offers insight for our current pandemic.

We now know that the mutations required for an emerging virus to adapt to a novel host come with a cost, and we know that this cost is relaxed if the virus circulates among different host types. We have also learned that high mutation rates allow RNA viruses to achieve the highest possible fitness and replication rates. And, we have identified the molecular mechanisms by which some viruses, such as influenza A, infect a wide range of hosts, whereas others, like mumps, are highly specialized to particular hosts. To generalist viruses, hosts are more or less the same on the cellular level. The range of hosts for these viruses all have common, highly conserved, elements in their genetic regulatory networks. Specialist viruses, on the other hand, interact with unique, or non-conserved, elements within their host's cells.

The emergence and pandemic spread of SARS-CoV-2 has raised many questions, hitherto of interest to fundamental, theoretical science, to the level of immediate practice. The pressure to give rapid support to authorities in our public health systems and the urgent need to find new specific antiviral treatments has moved us to rapidly repurpose the lab and to focus our research toward more urgent needs.

For example, our know-how in virus detection and purification can now be applied to diagnostics. By moving our quantitative polymerase chain reaction machines from the lab bench into biohazard secure rooms and adopting protective measures, our research protocols transform into medical diagnostic procedures.

Our past experimental and mathematical analyses of the interactions between viruses and so-called defective interfering particles (DIPs)—basically, the parts of a virus that are incapable of self-replication—are inspiring a research project aimed at finding new SARS-CoV-2 antivirals. Because DIPs interfere with the replication, accumulation, and transmission of the wild-type virus, we can think of them as antivirals. DIPs offer several benefits over conventional antiviral approaches:

- They are transmissible in the presence of the full virus and hence can spread with it to target infected cells;
- DIP-mediated protection is effective immediately, unlike classic vaccines that work via priming the adaptive innate immune pathways; and
- DIPs cannot replicate and are transcriptionally defective in the absence of the full virus, thus limiting possible side effects.

DIPs are just one example of a potential new antiviral inspired by the analysis of viruses as complex replicative systems. My colleagues and I are sure that many other biomedical applications will emerge from this multidisciplinary perspective. Basic science is not just future applied science; it is our applied reserve, available to be repurposed when society calls.

REFLECTION

A YEAR OF PROGRESS AND OF NEW COMPLEX CHALLENGES

DATE: *3 June 2021*

FROM: *Santiago F. Elena, Spanish National Research Council; Santa Fe Institute*

In my Transmission, "A complex systems perspective of viruses offers insight for controlling SARS-CoV-2 and future emerging viruses," I described how many years of experience in basic and theoretical virology research placed us in the situation to quickly move into applied SARS-CoV-2 research. The diverse range of problems included the biological effect of mutations circulating among infected people and the identification of defective interfering particles (DIPs) that can be selected and further engineered as potential antiviral agents.

From a purely scientific perspective, this last year has been an exciting time of discoveries for virologists and evolutionary biologists. The amount of data accumulated about SARS-CoV-2 origins, biology, and epidemiology has been growing daily with no precedent in the history of biomedical sciences. In fact, the process of developing several working vaccines in a record time, moving them from lab assays to clinical trials and, finally, beginning to vaccinate the population (of rich countries) on a very wide scale must be qualified as a herculean task. Nonetheless, many challenges still remain. Some are undoubtedly associated with the virus biology, but others are popping up as a consequence of the existing economic, sociological, and educational inequalities across countries. Both are tightly linked.

As an RNA virus, SARS-CoV-2 has an error-prone replication machinery that results in the generation of a large number of genetic variants within infected individuals. Not surprisingly, we have been witnessing the sweeping of variants of concern (VOC in the World Health Organization jargon) into the virus population worldwide. These VOCs so far all affect the spike (S) protein. S interacts with the ACE-2 cellular receptor protein and facilitates entrance into the target cells. By now, many readers will have heard of the UK (lineage B.1.1.7 or Alpha), South

African (lineage B.1.351 or Beta), Brazil (lineage P.1 or Gamma), and even the most recently described Indian (lineage B.1.617.2 or Delta) and Californian (lineage B.1.427 or Epsilon) variants. All these variants show increased transmissibility, induce stronger symptoms, and may even jeopardize the effectiveness of vaccines. Luckily, most of these variants share a limited repertoire of mutations (e.g., L452R, E484K and E484Q, N501Y, D614G, P681R), suggesting that there are only a few evolutionary paths that the virus can take to adapt from its likely bat reservoir into the new human host. And here is where a first challenge appears. Once the virus is well adapted to human hosts, and as the frequency of immunized hosts increases until we reach herd immunity levels, a new selective force will enter into play: immune pressure. Host immunity will favor the virus escape variants that would probably fix a different panoply of mutations, not necessarily in the S protein but in other proteins that form the capsid.

The amount of data accumulated about SARS-CoV-2 origins, biology, and epidemiology has been growing daily with no precedent in the history of biomedical sciences.

Unfortunately, vaccination campaigns are not progressing at a similar pace in all countries, with 75% of all vaccines dispensed in only ten countries. This heterogeneity creates new challenges. First, and most important, it creates large reservoirs of susceptible hosts in which the virus continues evolving and adapting and, eventually, generating standing genetic variation that might contain future VOCs. Second, the effectiveness of current vaccines would diminish as new VOCs escaping from vaccine-elicited antibodies sweep into the viral population across the world. Third, obviously vaccine failures would impose the necessity of periodically reformulating the vaccines and re-vaccinating people. If we, as a society, really want to get back into our "normal" luxury lives, we must push our governments to enter into vaccination programs for the less wealthy countries and set up effective surveillance systems of reservoirs and intermediate animal hosts that may allow for predicting future waves of the virus. We will likely not eradicate SARS-CoV-2, and this is far from being our goal, but we must learn to live with it as we do now (more or less) with influenza A virus. We should learn a number of lessons from this pandemic and prepare ourselves for the next one, which will for sure occur.

If we, as a society, really want to get back into our "normal" luxury lives, we must push our governments to enter into vaccination programs for the less wealthy countries and set up effective surveillance systems of reservoirs and intermediate animal hosts that may allow for predicting future waves of the virus.

Another challenge is to develop new and more efficient antiviral treatments. In my Transmission, I mentioned the possible usage of DIPs as an anti-SARS-CoV-2 evolving and adaptable therapy. In 2021 we have actively explored this possibility. We now have compelling evidence that large amounts of defective viral genomes (DVG) are found in multiple patients, with some forms being preferentially accumulated during the infection process, and pervasively found in different individuals. This suggests that DVGs may play some yet unknown role in virus replication and accumulation, as well as in its interaction with the host's immune response. We are now synthesizing in vitro some of the most commonly found DVG variants and testing their interfering role (as DIPs). In addition, we are also developing machine learning algorithms to associate the presence and abundance of these DVGs with clinical symptoms of

COVID-19, generation of neutralizing antibodies, and the progression and duration of infection.

Finally, in my previous Transmission I also advocated for a multi-disciplinary approach to tackling the COVID-19 pandemic. This year has shown that this is, indeed, an absolute necessity. As mentioned above, astonishing amounts of different types of data are being produced that are not well integrated into a common predictive framework. Multilayer network approaches would provide the theoretical framework and mathematical tools to integrate information from the most elemental level of virus biology (i.e., the frequency and nature of genetic variants within infected individuals), the distributions of variants among individuals from the same population, the movements of individuals among populations and even between countries (despite quarantine measures), up to transport networks at continental levels. Objects in each layer should incorporate information relevant for this particular layer (e.g., clinical symptoms or even social networking (mis)information in the layer connecting individuals). Together, this multilayer integrative approach would provide a fresh look at all of the accumulated data and help us to better understand the ongoing processes and perhaps to predict future behaviors.

TRANSMISSION NO. 012

HOW EVERY CRISIS IS AN OPPORTUNITY

DATE: *6 April 2020*

FROM: *Manfred D. Laubichler, Arizona State University; Santa Fe Institute*

STRATEGIC INSIGHT: *Every crisis is also an opportunity.*

Can the COVID-19 pandemic trigger pathways toward sustainable global futures? It seems like a long time ago, but at the beginning of 2020 our biggest threat was the failure of world leaders to agree on a common framework to combat climate change. The collapse of the Madrid summit, which was supposed to deliver tangible plans to stay within the goals of the Paris accord, again demonstrated how difficult it is to change the trajectories of highly interconnected complex systems with multiple actors.

Complex systems, such as organisms, social institutions, or the economy, all have their own histories. They develop over time and accumulate certain features that become increasingly difficult to change. We refer to those features as path-dependencies and constraints. For example, in the early evolution of multicellular organisms, certain genetic circuits performing specific functions emerged. Subsequent innovations then depended on these older parts of the genome, which, in turn, made it increasingly difficult to change them. They became, in a sense, locked in. Such constraints then limit the future possibilities of evolving complex systems.

As a consequence, these genes are highly conserved, as most mutations would have large ripple effects throughout the whole organismal system. We can actually use the conservation of certain gene sequences to reconstruct the history of life and establish the genealogical

---

OPPOSITE: *Alphons Mucha, "The Hussite King Jiří of Podebrady and Kunstat," 1923 (detail)*

relationships among organisms. But it is not just gene sequences that are conserved. Structural features of the anatomy of organisms, which are the product of highly complex developmental processes that unfold as these organisms emerge from a single cell—the fertilized egg—are also extremely stable. This conservation of structures, or complex relations among parts, to use the language of complex systems science, explains one of the most intriguing puzzles of comparative biology: why, despite the hundreds of millions of species, living and extinct, are there fewer than fifty basic body plans, or recognizable designs, most of which are hundreds of millions of years old?

Major long-term historical developments, including the scientific and, later, industrial revolutions, had their origins in major upheavals, such as the plague, that radically changed entrenched structures and constraints (both internal and external) . . .

Actually, there are two types of constraints: internal ones resulting from the architecture of complex systems and their networks of interdependent parts, and external ones that are a consequence of rules governing the dynamics of complex systems, such as natural selection and optimization. The latter is captured by the idea of a fitness landscape with multiple peaks, or local optima, separated by valleys. Natural selection implies that better variants will increase. Under normal circumstances, every population will eventually approach the closest local optimum, but any other, even better, optimum is out of reach as the valleys of lower fitness cannot be crossed. Similarly, there are also countless examples in the history of technology where better solutions did not win in the marketplace, as there was no viable path for them to overcome initial periods of lower fitness.

Figure 1 shows a simplified fitness landscape with three optima. Depending on where a population starts, it can only reach the closest peak, as natural selection will always increase the mean fitness of a population.

Radically new technologies as well as large innovations within organisms or in societies have always required a corresponding transformation of the underlying fit-

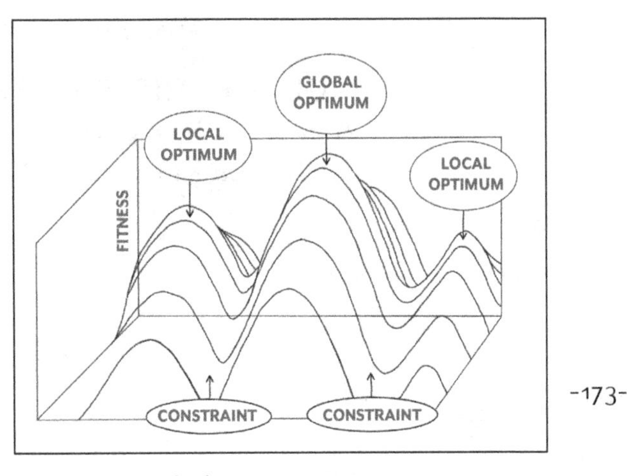

Figure 1: *Fitness landscape*

ness landscape. The iPhone was successful because it did not compete as a phone, where it would have lost, but as a new platform for a whole ecosystem of apps. Similarly, major long-term historical developments, including the scientific and, later, industrial revolutions, had their origins in major upheavals, such as the plague, that radically changed entrenched structures and constraints (both internal and external).

Early indications signal that COVID-19 may be such a disruptive event. At a time when it seemed almost impossible to accomplish major systems transformations needed to combat climate change and related challenges—not because of a lack of knowledge, but because of multiple constraints connected with globally interconnected systems—the current shock might open up a window of opportunity. The actions needed to combat the crisis are changing existing external constraints and also rewiring long-entrenched internal relations. This time of disruption is also one of opportunity.

REFLECTION

SCIENCE AS CRISIS RESPONSE: A NEW FRONTIER FOR COMPLEXITY SCIENCE

DATE: *26 August 2021*

FROM: *Manfred D. Laubichler, Arizona State University; Santa Fe Institute*

Major crises including pandemics, natural catastrophes, civil and international conflicts, and technological disasters are inevitable, as are their cascading consequences across social, economic, and environmental systems. Due to the increased interconnectedness of global systems, we now confront a whole new category of risk, often identified as global systemic risks, for which well-established practices of risk management and scientific analysis no longer suffice.

Managing global systemic risks requires scientific insights and guidance, but as the COVID-19 pandemic is revealing, traditional scientific practices are ill-equipped to provide those in a timely manner. A July 24, 2020 *Science* news feature by Paul Voosen concerning more accurate climate models perfectly summarizes this predicament.[1] A scientist involved in the study to narrow the bounds of climate sensitivity stated that we are today "light-years ahead of where we were in 1979," but Voosen concludes that "unfortunately, the years of work needed to attain that certainty came with a cost: four decades of additional emissions and global warming, unabated." In the context of COVID-19, we have observed efforts to accelerate the scientific response, mindful of the human cost in waiting to achieve conventional certainty. But what does this mean when science is uncertain?

The scientific community has been pushed beyond the traditional role of science and into a mode of science as crisis response, which includes shaping rapid policy decisions as well as accelerating scientific activities within traditional research tracks like vaccine and drug development. Traditionally, scientific research is iterative and methodical,

1 P. Voosen, 2020, "Earth's Climate Destiny Finally Seen More Clearly," *Science*, 369(6502): 354–355, doi: 10.1126/science.369.6502.354

using peer review to vet contributions. In contrast, science during a crisis is accelerated, based on limited information and in many instances published ahead of the rigor and scrutiny of the peer-review process. In addition, the nature of the problems underlying these crises does not map onto well-established scientific disciplines and their narrowly defined conceptual and methodological structures. Rather, all these problems represent challenges of increasing complexity and require a scientific framework capable of addressing these issues.

Complexity science is uniquely equipped to become the foundation of science as crisis response. But in order to do so, we need to confront a number of epistemological and organizational challenges.

COVID-19 serves as a preview of what is sure to become a more prominent mode of science as natural and social crises amplify in our complex, interdependent world and the time windows for necessary science-based responses shrink. The global interconnectedness of the world makes it more urgent to respond in real time and the global interconnectedness of science makes that possible. Modern communication tools facilitate rapid dissemination of ideas and results, and misinformation, and social media add a new dimension, including challenges to scientific authority.

Complexity science is uniquely equipped to become the foundation of science as crisis response. But in order to do so, we need to confront a number of epistemological and organizational challenges: How does acceleration lead to increasing levels of uncertainty? What criteria are we using to assess results? How are dealing with the fact that any intervention based on incomplete knowledge will change the intrinsic dynamic of the system? What are the patterns and consequences of this type of science? How has the communication and dissemination of knowledge changed? Has the rate of scientific convergence or divergence changed due to the nature of the crisis response? Who are the scientists that are most listened to and who have the biggest influence in shaping responses to the crisis? Are these the same who advance knowledge? How is scientific disagreement handled during a crisis?

The inevitable shift towards science as crisis response is a call to arms for complexity science.

Answering these questions will help us gain a better understanding of the structure and dynamics of science as crisis response. What we are confronting here is actually a threefold (minimally) problem of complexity:

1. **The complexity of the underlying problems.** As we have seen, COVID-19 is not just a pandemic, it has revealed complex interdependencies between molecular, immunological, ecological, evolutionary, economic, social, behavioral, cultural, and political processes, which we are only beginning to comprehend;

2. **The complexity of the crisis response.** The sheer magnitude and urgency of the COVID-19 pandemic has revealed serious flaws and challenges in our systems of crisis response as well as how unprepared all relevant institutions (including science) were, despite frequent warning signs that something like COVID-19 was likely to happen; and

3. **The complexities of the new science as crisis response.** Knowledge as well as the scientific structures and practices that generate it are themselves complex systems. In order to optimize these processes in light of ongoing and future challenges, we need to better understand the various elements of science as crisis response.

To conclude, the inevitable shift towards science as crisis response is a call to arms for complexity science. How well we will be able to meet these challenges will determine the future path of humanity. It is already obvious that small incremental changes will not be sufficient and more radical transformations of knowledge, action, and the structures that support both are needed. And for all of those SFI is uniquely positioned.

Robert Hooke dining with friends at Gresham College

TRANSMISSION NO. 013

OPPORTUNITIES FOR SCIENCE COMMUNICATORS

DATE: *13 April 2020*

FROM: *Mirta Galesic, Santa Fe Institute*
Henrik Olsson, University of Warwick; Santa Fe Institute

STRATEGIC INSIGHT: *The current spike in public trust in science gives science communicators an opportunity to reach new audiences.*

Our social circles are a crucial source of information about many things. Whom to trust? What behaviors are useful in the current environment? What is expected and what is possible?

Relying on our family, friends, and colleagues to answer such questions is usually a good idea. Using the collective wisdom of people who are similar to us and who live in similar circumstances can be a quick shortcut to solving a variety of problems in our own daily lives.[1] Moreover, we tend to seek people who are similar to us, as this typically enhances coordination and cooperation and helps to avoid costly conflict.[2,3,4] We also influence each other—ever more strongly and widely due to social media—further contributing to the homogenization of our own societal pockets.[5,6]

It is therefore not surprising that different sections of our society have very different beliefs about how dangerous COVID-19 is, what the appropriate individual and societal actions are, and what figures of authority should be trusted.[7,8] Our reliance on social circles for our judgments and decisions[9] has an unfortunate corollary: it is difficult for people to change their minds and also keep their social networks intact. Science educators know this all too well.[10] The immediate social costs of not being aligned with one's social circle might appear much larger

---

OPPOSITE: *Rita Greer, "Robert Hooke Dining with Friends at Gresham College," 2012 (detail)*

than the costs of not being correct, especially when a risk is perceived to be distant and personally irrelevant.[11,12]

Scientists are not a part of most peoples' social circles. Scientific judgment is not readily trusted as unbiased by significant parts of society,[13] and scientists can be perceived as a part of an elite that is not aligned with "our" own best interests. Significant sections of the US population have beliefs that do not align with the current scientific evidence.[14] Lack of basic scientific knowledge is likely both a consequence and a cause of further distrust that prevents acceptance of science facts—a cycle that is becoming hard to break[15] as it impacts people's decisions about which policies and politicians are worthy of their support.

But, in all its epidemic darkness, the current moment provides scientists with a unique opening: confidence in medical professionals is very high even though trust in other authorities can falter.[16] Scientists routinely appear on national televisions and in a variety of online settings to brief the public about the current progress of the disease. Furthermore, the general public has both more time (involuntarily) and more interest in hearing what scientists have to say.

Because of the current spike in trust and interest in science, this is *the* moment for science communicators to make a difference. Sending another tweet to our usual followers will likely not persuade anyone who is not already in our own social circle. But reaching out to audiences who otherwise have little exposure to scientific ideas and reasoning—and who may now be more receptive—can be a real game-changer.

Scientists can use these times to start a dialogue with the general public not only about the epidemics, but also about the many related broader issues ranging from how the scientific process works, what constitutes scientific evidence, and how scientists check their findings, to the societal value of many scientific discoveries. Science is not the answer to everything, but it now at least has a chance to be heard.

ENDNOTES

1 R. Boyd, P.J. Richerson, and J. Henrich, 2011, "The Cultural Niche: Why Social Learning is Essential for Human Adaptation," *Proceedings of the National Academy of Sciences* 108 (Supplement 2): 10918-10925, https://www.pnas.org/content/pnas/108/Supplement_2/10918.full.pdf

2 P.E. Smaldino, T.J., Flamson, and R. McElreath, 2018, "The Evolution of Covert Signaling," *Scientific Reports* 8(1): 1-10, https://www.nature.com/articles/s41598-018-22926-1

3 M. McPherson, L. Smith-Lovin, and J.M. Cook, 2001, "Birds of a Feather: Homophily in Social Networks." *Annual Review of Sociology* 27: 415-444, doi: 10.1146/annurev.soc.27.1.415

4 P. Richerson and J. Henrich, 2009, "Tribal Social Instincts and the Cultural Evolution of Institutions to Solve Collective Action Problems," *Cliodynamics* 3: 38–80, https://escholarship.org/content/qt981121t8/qt981121t8.pdf

5 W. Quattrociocchi, A. Scala, and C.R. Sunstein, 2016, "Echo Chambers on Facebook," SSRN, https://papers.ssrn.com/sol3/papers.cfm?abstract_id=2795110

6 I. Von Behr, A. Reding, et al., 2013, "Radicalisation in the Digital Era: The Use of the Internet in 15 Cases of Terrorism and Extremism," *RAND*, https://www.rand.org/content/dam/rand/pubs/research_reports/RR400/RR453/RAND_RR453.pdf

7 G.A. Smith, "Most White Evangelicals Satisfied with Trump's Initial Response to the COVID-19 Outbreak," *Pew Research*, March 19, 2020, https://www.pewresearch.org/fact-tank/2020/03/19/most-white-evangelicals-satisfied-with-trumps-initial-response-to-the-covid-19-outbreak/

8 J.M. Krogstad, A. Gonzalez-Barrera, and M.H. Lopez, "Hispanics More Likely than Americans Overall to See Coronavirus as a Major Threat to Health and Finances," *Pew Research*, March 24, 2020, https://www.pewresearch.org/fact-tank/2020/03/24/hispanics-more-likely-than-americans-overall-to-see-coronavirus-as-a-major-threat-to-health-and-finances/

9 M. Galesic, H. Olsson, and J. Rieskamp, 2018, "A Sampling Model of Social Judgment," *Psychological Review* 125(3): 363, https://drive.google.com/file/d/1NmVlusD4M8slPPYjoixKNiU5Sr0wfxyh/view

10 B. Nyhan, J. Reifler, et al., 2014, "Effective Messages in Vaccine Promotion: A Randomized Trial," *Pediatrics* 133: e835-e842. https://www.researchgate.net/profile/Sean_Richey/publication/260485891_Effective_Messages_in_Vaccine_Promotion_A_Randomized_Trial/links/5720aa4308aefa64889ec347.pdf

11 P.S. Hart and E.C. Nisbet, 2012, "Boomerang Effects in Science Communication: How Motivated Reasoning and Identity Cues Amplify Opinion Polarization about Climate Mitigation Policies," *Communication Research* 39: 701-723, https://research.fit.edu/media/site-specific/researchfitedu/coast-climate-adaptation-library/climate-communications/psychology-amp-behavior/Hart--Nisbet.-2012.-Reasoning--Identity-Cues-Amplify-Climate-Opinion-Polarization..pdf

12 R.I. McDonald, H.Y. Chai, and B.R. Newell, 2015, "Personal Experience and the 'Psychological Distance' of Climate Change: An Integrative Review," *Journal of Environmental Psychology* 44: 109-118. http://www.columbia.edu/~rim2114/publications/McDonald-Chai-Newell-2015.pdf

13 C. Funk, M. Hefferon, et al., "Trust and Mistrust in Americans' Views of Scientific Experts," *Pew Research*, August 2, 2019, https://www.pewresearch.org/science/2019/08/02/trust-and-mistrust-in-americans-views-of-scientific-experts/

14 Pew Research, "Major Gaps between the Public, Scientists, on Key Issues," July 1, 2016, https://www.pewresearch.org/internet/interactives/public-scientists-opinion-gap/

15 National Science Board, 2018, *Science and Technology: Public Attitudes and Understanding*, https://www.nsf.gov/statistics/2018/nsb20181/report/sections/science-and-technology-public-attitudes-and-understanding/public-knowledge-about-s-t

16 K. Archer and I. R. Levey, "Trust in Government Lacking on COVID-19's Frontlines," *Gallup*, March 20, 2020, https://news.gallup.com/opinion/gallup/296594/trust-government-lacking-frontlines-covid.aspx

REFLECTION

SOCIAL COMPLEXITY RESEARCH AFTER COVID-19

DATE: *25 August 2021*

FROM: *Mirta Galesic, Santa Fe Institute*
Henrik Olsson, University of Warwick; Santa Fe Institute

The World Health Organization's May 2021 report from the Independent Panel for Pandemic Preparedness and Response stresses the need to invest in and coordinate risk-communication policies.[1] What is missing from the report, however, is a discussion of how these communication policies might interact with cognitive and social factors. For example, how is the effectiveness of science communication affected by mistrust in the government or pharmaceutical companies? How does numerical literacy in the general population and among doctors affect the understanding of scientific information? How does the acceptance of scientific facts interact with the structure of the social networks that people find themselves in? What is the impact of telling people to get vaccinated when all of their friends say that they should not?

A pandemic, like many other social phenomena, is part of a complex social system where cognitions interact in social networks to produce dynamically changing patterns on the individual and aggregate levels. This is recognized by many scientists and governmental officials, but in practice the complex-systems view has yet to become commonplace.

The complex-systems view of our social worlds is not only a mere abstract idea to guide scientific reasoning. It comes with a commitment to formal models. Without formal models, it is practically impossible to understand the behavior of a complex system and nudge it in a particular direction. Most models constructed to understand and predict the time-course of COVID-19 lack the cognitive and social factors that might impact the time-course of disease spread. In order to fully understand

---

1 The Independent Panel for Pandemic Preparedness & Response, 2021, *COVID-19: Make it the Last Pandemic*, Geneva, Switzerland: World Health Organization, https://theindependentpanel.org/wp-content/uploads/2021/05/COVID-19-Make-it-the-Last-Pandemic_final.pdf

what happened in 2020 and how to deal with it, we scientists need to incorporate these important factors into our models.[2]

In our Transmission from 2020, we pointed to the opportunity for science communicators to make a difference in times of the pandemic. While trust in scientists remains high,[3] our society is fragmenting regarding COVID-related beliefs, with some people wearing masks even when they don't have to and supporting obligatory vaccination and others completely rejecting both masks and vaccination.[4] No amount of communication disconnected from people's social context will help change minds. We have been studying this complex social system using models paired with real-world tests of these models.[5] We find that effective communication not only provides facts but also succeeds at minimizing the dissonance between these facts and people's individual beliefs (e.g., about who profits from vaccines) and the perceived beliefs of others.[6] For example, some people believe that pharmaceutical companies unduly profit from vaccines. When presented with information about safety of vaccines they might experience dissonance between this information and their prior beliefs. Rather than avoiding to talk about such dissonant beliefs, our research suggests that educators should highlight this dissonance and at the same time provide additional information (e.g., explaining that pharmaceutical companies would profit

2 J. Bedson, L.A. Skrip, et al., 2021, "A Review and Agenda for Integrated Disease Models including Social and Behavioural Factors," *Nature Human Behaviour* 5: 834–846.

3 3M, "No Matter the Challenge, People Believe Science Will Solve It, 3M Survey Reveals," May 18, 2021, https://news.3m.com/2021-05-18-No-matter-the-challenge,-people-believe-science-will-solve-it,-3M-survey-reveals

4 D. Ahrendt, M. Massimiliano, et al., "Living, Working and COVID-19 (Update April 2021): Mental Health and Trust Decline across EU as Pandemic Enters Another Year," *Eurofound*, May 10, 2021, https://www.eurofound.europa.eu/publications/report/2021/living-working-and-covid-19-update-april-2021-mental-health-and-trust-decline-across-eu-as-pandemic

5 M. Galesic, H. Olsson, et al., 2021, "Integrating Social and Cognitive Aspects of Belief Dynamics: Towards a Unifying Framework," *Journal of The Royal Society Interface* 18: rsif.2020.0857, doi: 10.1098/rsif.2020.0857

6 M. Galesic, W. Bruine de Bruin, et al., 2021, "Human Social Sensing is an Untapped Resource for Computational Social Science," *Nature* 595: 214–222, doi: 10.1038/s41586-021-03649-2; T. van der Does, D.L. Stein, et al., 2021, "Moral and Social Foundations of Beliefs about Scientific Issues: Predicting and Understanding Belief Change, OSF preprint, doi: 10.31219/osf.io/zs7dq

more by letting people get sick) to help reduce the dissonance and foster the acceptance of the information about vaccine safety.

> Rather than avoiding to talk about such dissonant beliefs, our research suggests that educators should highlight this dissonance and at the same time provide additional information . . .

With advancements in theoretical models and an unprecedented availability of data, social science can now step up to the challenge of helping our society remain sustainable while continuing to flourish. By integrating knowledge from many disciplines in complex-systems models, social scientists can make real progress in understanding, predicting, and preventing future pandemics and other disruptive social phenomena.

TRANSMISSION NO. 014

ON NOT LETTING A CRISIS GO TO WASTE

DATE: *13 April 2020*

FROM: *Doug Erwin, Smithsonian Institution; Santa Fe Institute*

STRATEGIC INSIGHT: *Beyond our response to the pandemic itself lie the longer-term effects, including new opportunities.*

When Rahm Emanuel, then President Obama's chief of staff, said, "You should never let a serious crisis go to waste," he was channeling Winston Churchill and the economist Joseph Schumpeter. The great mass extinctions and smaller biotic crises documented by the fossil record reveal that life long ago discovered the importance of not letting a crisis go to waste, if not in quite the same manner intended by Emanuel. My first paper in college was about the end-Permian mass extinction, the greatest of all mass extinctions, which wiped out perhaps 90% of all species in the oceans some 251 million years ago, and almost as many plants and animals on land. Yet the extinction itself really is not very interesting, at least not to me, because in the aftermath of that cataclysm came dinosaurs, mammals, modern insects, turtles, ichthyosaurs, and probably the first flowering plants, as well as most of the invertebrate clades populating modern oceans. The great question (and the one that first drew me to SFI) is, what sort of opportunities were generated by this crisis, and how did life respond?

I in no way want to diminish the very real fear that many feel now, nor the frightening days that lie ahead for many in this country and elsewhere around the world. Many will die, even more will suffer serious illnesses, and myriad businesses and institutions will disappear. But, as treatments and vaccines come online, we will get through this, and knowing that allows us to begin to consider what we can learn from this event. And I see questions similar to the ones that I have been asking

---

OPPOSITE: *Józef Chełmoński, "Partridges in the Snow," 1891 (detail)*

about the end-Permian episode: What factors favor survival, whether of people or of institutions? How do these factors relate to those favoring success prior to the pandemic?

The great mass extinctions and smaller biotic crises documented by the fossil record reveal that life long ago discovered the importance of not letting a crisis go to waste . . .

In many past crises, the factors favoring survival have had little to do with those that provide an advantage during "normal" times. Can we design models to evaluate alternative treatments that would modify these outcomes (aside from the obvious ones of more timely interventions at the outset)? Resilience has long been a central theme at SFI; like the 2008 financial crisis, this pandemic provides an opportunity to articulate the factors underlying robustness in social, economic, and other systems, and to distinguish resilience factors that may be unique, either to this event or to pandemics in general, from those factors that enhance resilience more generally. Hurricanes, earthquakes, megafires in the western US, and, of course, climate change are just a few of the events in which resilience plays a critical role, and which argue for strengthening our portfolio in this area.

Finally, the turtles—one of the oldest has been found near Ghost Ranch, just north of Santa Fe. There is no hint of an ichthyosaur or a turtle fossil predating the end-Permian mass extinction. But ichthyosaurs and turtles exploded in diversity in the aftermath of the extinction as they helped construct entirely different ecosystems from those before the extinction. Beyond our response to the pandemic itself lie the longer-term effects, including new opportunities—social, political, economic, and otherwise. The deeper the crisis, the greater the subsequent opportunities, and the greater the possibility of shaping these opportunities.

REFLECTION

PANDEMIC AND CLIMATE CHANGE: PULSE VERSUS PRESS

DATE: *25 August 2021*

FROM: *Doug Erwin, Smithsonian Institution; Santa Fe Institute*

As I write this, temperatures in Death Valley approached (and may have passed) 130°F (54°C), the eighty-three active wildfires across the country have burned some 1.3 million acres (still below the ten-year average of fires since the beginning of the season), and although it is only mid-July, many fires in the northwest are expected to burn until rain and snow come in the fall. Climate change is making the wildfire season in the US longer, and the fires larger and more severe, as in Russia, Australia, and many other parts of the world. As I looked back on the Transmission piece I wrote last year about resilience, I realized that many of the challenges we face with climate change are of a fundamentally different kind than those we have experienced through the COVID-19 pandemic.

Ecologists differentiate between pulse and press events, and many years ago I imported this distinction into the study of mass extinctions and similar biotic crises, where it has proven useful for thinking through the relationship between the dynamics of the extinction and the subsequent biotic recovery. For mass extinctions, pulse events are quick shocks that are so rapid that no adaptation is possible (at least for most species). Plants, animals, and other organisms live or die based on what they brought to the game. The asteroid impact that triggered the mass extinction at the end of the Cretaceous 66 million years ago, wiping out the (non-avian) dinosaurs, was a classic pulse extinction. Press events differ in being long-lasting, pervasive, and persistent, but the longer they last the more species have an opportunity to adjust and adapt to the changing circumstances. The difference between press and pulse events changes the dynamics of robustness and resilience, although in ways that I do not think have been fully explored.

Lately, I have begun to use this framework of pulse and press events to think about the challenges we face moving forward. This approach is similar to the focus of Jessica Flack and her colleagues on slow variables, with rapid events superimposed on much longer patterns of change, and to the discovery of slow earthquakes—seismic events that occur over hours or months.[1]

Building a society resilient to climate change will require far more fundamental societal shifts than have resulted from the pandemic.

Much of the developing world remains unvaccinated and the Delta variant of COVID-19 is ravaging unvaccinated populations in the United States and Europe, threatening what had been a remarkably resilient recovery from the early phases of the pandemic. The emergence of the Delta variant and newer variants has surprised no one with even a passing familiarity of how evolution works. And we face the prospect of newer variants emerging to which extant vaccines are less useful in reducing severe illness and death. But the construction, testing, and production of at least twenty different vaccines worldwide, and vaccination of hundreds of millions of people since the virus was isolated is a truly remarkable achievement. That achievement reflects many factors: scientific, technical, economic, political, and cultural. But that capacity is essentially what we brought to the table in early 2020. There have been many successes and failures over the past year and as I outlined in my initial Transmission piece, there are many questions we can now address about the response to the COVID-19 crisis. Assuming researchers can safely put aside partisan wrangling, we should be able to develop a better sense of what worked and what did not work. Testing alternative scenarios will help us learn what steps could have been taken to reduce the human suffering and economic cost. Studies from social scientists may help identify scenarios that would increase acceptance of mitigation efforts. Such efforts represent contributions to generating more resilient social and political systems.

1 J. Flack, D.H. Erwin, et al., 2013. "Evolution and Construction of Slow Variables and the Emergence of Aggregates," in *Cooperation and its Evolution*, ed. K. Sterelny, R. Joyce, et al., Cambridge, MA: MIT Press, 45–74.

Let us assume that we mount a collective effort to identify the best lessons from the pandemic crisis, the events of 2008–2009, the aftermath of Hurricane Katrina, and other similar crises. Let's even assume that we implement at least some of them and manage the political will to keep such efforts in place as memories fade (living outside of Washington, DC, I do not really believe this is possible, but we can always hope). But in each of these cases the events are relatively short-term, pulse events. Would such lessons be relevant for the challenges facing us with climate change? Climate change is not going to go away. In fact, it will only get worse over the lifespan of everyone now alive, and for generations into the future. Building a society resilient to climate change will require far more fundamental societal shifts than have resulted from the pandemic; these shifts will not be just a function of what we have at hand today, but rather the adaptations and responses that we will develop in the future.

There are many at the Santa Fe Institute (and elsewhere) who have been deeply involved in addressing these problems, and as I expressed in the Transmissions piece, I hope that in coming years a greater understanding of resilience on both short and longer time spans will emerge. Work on power grids, studies by ecologists of resilience in natural ecosystems, and the resilience of native communities are all areas where SFI researchers have already made important contributions. While directly influencing climate change legislation may not be our first priority as a research institute, we must continue to generate knowledge of the features of resilient systems over a range of timescales, paving the way for a greater understanding of resilience.

THE TRANSCRIPT OF *COMPLEXITY PODCAST* TRANSMISSION SERIES EP. 3, DISCUSSING THE FOLLOWING TRANSMISSIONS

[Batch 3, released 13 April 2020]

T-010: Investment Strategies in Times of Crisis
Bill Miller

T-011: A Complex-Systems Perspective of Viruses
Santiago F. Elena

T-012: How Every Crisis Is an Opportunity
Manfred D. Laubichler

T-013: Opportunities for Science Communicators
Mirta Galesic & Henrik Olsson

T-014: On Not Letting a Crisis Go to Waste
Doug Erwin

This podcast transcript has been abridged for length and clarity. Find the full podcast at https://complexity.simplecast.com/episodes/29

COMPLEXITY PODCAST: TRANSMISSION SERIES NO. 003

CORONAVIRUS, CRISIS & CREATIVE OPPORTUNITY

DATE: *20 April 2020* **HOSTED BY:** *Michael Garfield, featuring David Krakauer*

MICHAEL GARFIELD: So we're back for week three. The first two episodes in this miniseries have really set us up to dig into this latest set of articles and find some jewels of insight in there.

DAVID KRAKAUER: This one is pretty much firmly in the complexity territories now.

M. GARFIELD: The first piece we have, which links back to your discussion of the viral action of the novel coronavirus in our first episode, is by Santiago Elena[1]. Why don't you open this one up for us?

D. KRAKAUER: This certainty touches all sorts of really intriguing ideas here. For those who don't know, Santiago basically does fundamental science using the virus as a model system. Santi is not primarily interested in the virus as an agent for disease, but he thinks about the virus the way, say, a physicist who is working in quantum mechanics would think about the hydrogen atom or a biologist would think about the worm *C. elegans*, in relation to development.

So, it's a model system, and in that respect it's rather like Darwin's finches, and I wanted to mention that. Folks may remember that, in 1835, Darwin gets to the Galapagos archipelago. He makes these observations on the finches, which, subsequently, once back in London, he realizes, working with a number of others, that the bill morphology of each of these finches varies according to the habitats in which they live, the islands on which they live, most notably the food that they eat—leaves, seeds, vampiric, insects, and so forth.

This became known as the Darwin finch story, and it's a story of adaptive radiation, and viruses are just like that. But instead of feeding on different plants and animals, they feed on different cells. So we talked about the coronavirus feeding on the cells in the pathway regulating cardiovascular function.

---

1 See T-011 on page 163

But you have viruses, like parvo viruses, that feed on heart cells, and herpes viruses that replicate in epithelial cells or hepatitis viruses in liver cells. In that respect, they look a lot like large, charismatic birds and mammals. But then there's one very crucial difference and that is that unlike large, multicellular creatures where we describe them as being members of a species, a virus is really a member of what Eigen and Schuster called a "quasi-species."[2] So, instead of thinking of a point in space which would be a species, a type, they occupy a cloud, because they're so mutable that what we think of as the lineage is a much more amorphous population of variants rather than a single category of biological variety.

M. GARFIELD: He makes a distinction with RNA viruses and the mutation rates.

D. KRAKAUER: Yes. So now we have this rather interesting difference between a regular species and a viral quasi-species. And the great thinker in this field was one of the early frequent visitors to the Santa Fe Institute, the Nobel Prize winner, Manfred Eigen. And he had been Werner Heisenberg's student, as in the Heisenberg uncertainty principle, which states that if you know more about position, then you know much less about its conjugate variable, momentum. This is very famous, and Eigen was very inspired by that way of thinking and looked for something like it in living systems. And what he discovered was the error threshold—that if a living lineage has a very high mutation rate, there is a possibility that it can cross a threshold and vaporize itself. So, think about water being heated up and becoming water vapor. If you take a virus and you start increasing its mutation rate, that cloud I described grows, and at a critical point called the error threshold, that cloud diffuses into free space and the virus disappears.

And in a paper in 2002, Manfred Eigen suggested[3] that one way that we could eliminate a virus infection is by doing something completely counterintuitive. Instead of creating a vaccine, say, that prevented its infection, we're actually going to increase its mutation rate. And you'd

2 M. Eigen and P. Schuster, 1988, "Molecular Quasi-Species," *The Journal of Physical Chemistry A* 92(24): 6881–6891, doi: 10.1021/j100335a010

3 M. Eigen, 2002, "Error Catastrophe and Antiviral Strategy," *Proceedings of the National Academy of Sciences of the United States of America* 99: 13374-13376, doi: 10.1073/pnas.212514799

think that would make it more adaptable. But at that critical point of the error threshold, the virus will just vaporize; it'll do what water does when you heat the temperature up above the boiling point. And this is something that Santiago has been very interested in, and in this article, he suggests a form of viral treatment based on what are called defective interfering particles or DIPs.

When every virus infects a cell or when every quasi-species infects a cell, it generates a ton of defective viruses because even though a coronavirus, as we said, has a relatively low mutation rate, it still generates a lot of error. Any given cell produces viruses that are viable—that could infect future cells—and viruses that are nonviable. He wants to sort of use those nonviable viruses as decoys and stick them in cells in order that most of the viruses appropriate defective parts. This insight of Manfred Eigen's, that you could push a virus over the threshold, can form the basis of a treatment in terms of these defective interfering particles.

M. GARFIELD: This issue of the error threshold ties in a lot from last week's conversation to pretty much everything that we're discussing today. I'll be calling back to Miguel Fuentes's piece[4] on the way that they were able to detect changes in the structure of social networks through Twitter data on an impending crisis before the crisis even happens. What I see there is something like this, like as society itself approaches this error threshold in communication. There seems like in a discussion of the opportunity of crisis, a deep relationship between the opportunity and the crisis, because there is something about this escalation of novelty, whether it's genetic or mimetic, is involved in these catastrophes. It's key to understanding why they even occur.

D. KRAKAUER: I think there is definitely running through today's conversation this idea of error and opportunity, collapse and rebirth, and whether it's a virus or the fossil record or the behavior of financial markets, and it is extremely interesting to try and understand the nature of that nexus. Santiago's Transmission gives one beautiful example. Viruses need to mutate to evolve in order to infect new cells and to evade the host immune system. But there is a critical value of mutation where they obliterate themselves, and it has exactly that character you're

4 See T-006 on page 101

describing. The question is, "Can we come to understand that and intervene?" The last thing we'd want to do—and it's one of the reasons, by the way, that people never used Eigen's original proposal for viral therapy, because it looked as if we didn't know where that threshold was and we would be inadvertently making it more evolvable as opposed to pushing it, if you like, to collapse.

The other one I do want to mention, though, since we are on the Santiago contribution is, we talked about this also before, which is fundamental science versus applied science. These fundamental ideas, which seem really quite esoteric—you know, go back to Heisenberg and uncertainty and mutability and selection—that actually become the basis of therapy. But it's also in practice because Santiago's lab very quickly moved all of its polymerase chain reaction machines from their lab into biohazard-secure rooms. And so the machinery of basic science could be mutated into application also. And I think that's really intriguing that these events are not only opportunities to rethink basic science, but to rethink the way that science is conducted.

M. GARFIELD: Which brings us to Doug Erwin's piece[5] about not letting a crisis go to waste. He says that, in past mass extinctions, "the factors favoring survival have had little to do with those that provide an advantage during 'normal' times." When you think about evolution making use of what is lying around, it seems like the opportunities here are, in this lovely Stephen Jay Gould word, the *exaptation*[6] —the appropriation of existing parts into new uses. So, let's talk about Doug's paleontological, big-picture view on this.

D. KRAKAUER: So paleontologists, as you well know, Michael, are a sort of a different species. They think about things on much longer timescales. While we're worrying about hundreds, thousands of years, they're worrying about millions, if not billions, of years. It's a very different mindset. Doug, interestingly, in a foreshadowing of Bill Miller's Transmission,[7] is talking about mass extinctions. Just as paleontologists

---

5 See T-014 on page 187

6 S.J. Gould and E.S. Vrba, 1982, "Exaptation—A Missing Term in the Science of Form," *Paleobiology* 8(1): 4–15, doi: 10.1017/S0094837300004310

7 See T-010 on page 155

recognize, very coarsely, five major mass extinctions, Bill mentions in his own lifetime five major stock market crashes. That was rather interesting. Most of us are familiar with the extinction of the dinosaurs, but Doug focuses on the End-Permian extinction event, the so-called Permian-Triassic extinction event about 250 million years ago, which was much, much larger with about 90% of marine species and about 70% of land species going extinct. It was a truly catastrophic event in Earth history.

> . . . these events are not only opportunities to rethink basic science, but to rethink the way that science is conducted.

But Doug makes this point that it's not the extinction itself that interests him, but what happened afterwards. As he points out, what happened afterwards were dinosaurs and insects and mammals and turtles and perhaps even the first flowering plants. There was something about that cataclysmic event that sort of cleared the board and allowed for the possibility of entirely new forms of life to emerge. And the question is, what are those characteristics that lead a clade or group to go extinct? What are those characteristics that allow it to survive? And then what are the characteristics which allow new species in groups to come into existence after the extinction event? Most people are familiar with trilobites, those little arthropods that you find in the fossil record that survive an awfully long time, from over 500 million years ago up until this event and then they were disappeared completely—as opposed to things like bryozoans, which are little invertebrates that look like corals, that survived and prospered, and all those new forms that we mentioned before that came into existence. And I think what Doug is suggesting is borrowing an insight from the Austrian economist Joseph Schumpeter, that mass extinction events are periods of creative destruction, and the aftermath of creative destruction is typically innovation. And the reason for that is, by and large, because all of the competitive trophic networks that had constrained the growth of lineages are now lifted, allowing for the possibility of entirely new forms to emerge and dominate.

M. GARFIELD: I'm especially fond, in a much smaller framing, of thinking about this in an ecological sense about a mature forest canopy and a tree falling and then suddenly new light reaching the forest floor. There's one way of looking at this, that it's about available resources—there's less competition and in some sense more to go around. Now, obviously this is kind of a morbid paleontological angle on what does "more to go around" mean? It means everything that has just died.

D. KRAKAUER: You're absolutely right. It's very strange that way because most people will be familiar with this from companies that in some sense go extinct, you know, like Research In Motion with their BlackBerry phone[8] that in some sense allowed or established the foundations for the selective context that became the adoption of the iPhone. It's very much a part of economic life and it's certainly a part of the life of ideas, but it's difficult to reconcile with the timescales that we live in. And that's what makes the paleontological view particularly interesting, because it points out that if you integrate over long enough periods of time, you realize that these events are crucial to the kind of Earth we've become familiar with.

M. GARFIELD: Doug leaves us with this question at the end of his piece, which is, why did these particular organisms, the ichthyosaurs, the turtles, the dinosaurs, why did they succeed in the wake of the Permian extinction? I think starting to answer that question carries us into Manfred Laubichler's piece[9] on the evolutionary fitness landscape, which is in a sense a sort of abstraction of the channels through which evolution can flow and reach opportunity. Although I guess that's sort of an inversion of the rugged landscape as an image.

D. KRAKAUER: Manfred is interested in stasis and change. Evolutionary biologists following Gould and Eldredge[10] and others slowly came to realize that non-change required as much explanation as change. There

8 J. McNish and S. Silcoff, 2015, *Losing the Signal: The Untold Story Behind the Extraordinary Rise and Spectacular Fall of BlackBerry*, New York, NY: Macmillan.

9 See T-012 on page 171

10 S.J. Gould and N. Eldredge, 1997, "Punctuated Equilibria: The Tempo and Mode of Evolution Reconsidered," *Paleobiology* 3: 115–151, doi: 10.1017/S0094837300005224

are essentially two sources of constraint, if you like, that Manfred introduces us to. One is external and that's what we think of as fitness or natural selection. In the 1930s, Sewell Wright introduced the idea of the fitness landscape.[11] Think of a landscape with topographical features like peaks and valleys, and think of a well-adapted species as living at the top of a peak and the valley as being maladaptive intermediates. What evolution does is over the course of time shift that landscape such that different species come to prosper at different points in time. A mass extinction event is precisely that event that completely changes the topography of Wright's fitness landscape.

So that would be an external constraint. But in the 1940s, about a decade after Sewell Wright, Conrad Waddington introduced the idea of the epigenetic landscape,[12] and that's not an external constraint. It's an internal one. It says that development from egg to embryo to adult morphology follows a path which is very canalized or very restricted. You can't all of a sudden, for example, develop three extra limbs that grow out of your back. It's very, very improbable. And the reason for that is that the trajectory of the phenotype from egg to embryo to adult is very constrained—what he called canalized. This has nothing to do with fitness per se because it operates on a much shorter timescale, but to do with the constraints, not of trophic networks but of regulatory networks, gene regulatory networks, and they constrain the number of possible forms that can be realized. And Manfred's question I think in this article is, how do we consider the relationship between the internal epigenetic and external fitness landscapes at times of crisis? How does one influence the other? And I think it's more a question than an answer because we don't know why a change in the topography of the fitness landscape liberates regulatory networks to explore new body types.

M. GARFIELD: I was just reading last night about research on the origins of multicellularity and some experimental work done with yeasts where they were able, over the course of a year—this was published in

---

11 S. Wright, 1932, "The Roles of Mutation, Inbreeding, Crossbreeding, and Selection in Evolution," in *Proceedings of the Sixth International Congress on Genetics*. 1(8): 355–366, http://www.esp.org/books/6th-congress/facsimile/contents/6th-cong-p356-wright.pdf

12 C.H. Waddington, 1957, *The Strategy of the Genes*, London, UK: Allen & Unwin.

Science[13]—to compel yeasts, through predation, to a form multicellular "snowflake yeast." And one of the things that snowflake, that very prototypical multicellular organism would do.... the oldest cells at the edges of the organism, as it grows from the center out, had the highest rate of mutation. And so they started to see early specialization in cell types because the cells at the base of those stalks of the snowflake would commit programmed cell death and release that mutated stalk out to grow its own separate colony. I wonder whether at the intersection between those two different ideas about the fitness landscape, more or less internal developmental and external evolutionary, are basically the same thing but operating at different scales. That brings us back to the way that slow and fast variables are related in the last conversation.

D. KRAKAUER: It's not known. I mean there are some simple explanations. For example, if the environment is fixed that's called normalizing selection, so rare mutations that don't align with the selection pressures tend to disappear quickly. They don't get a foothold. It's not that they're not there; it's just that they're largely invisible. When the landscape changes, all of this so-called cryptic variation becomes visible. But there are more interesting mechanisms, along the lines of perhaps the yeast study you mentioned, associated with error-repair proteins like chaperone proteins. These are proteins that help other proteins fold because not all proteins fold into their right tertiary configurations that give them their function. If you remove those chaperone proteins or down-regulate them, all sorts of protein folding patterns appear that weren't there before, because they're not all being folded into the same shape. This allows for the possibility of new function. There is some data to suggest that when the environment changes, these chaperone proteins are down-regulated in order to allow more protein variation to exist, which might turn out to be beneficial. These are very new results that suggest that there are pathways connecting the external to the internal, but we don't fully understand how they work, presumably through some means of epigenetics—that is, changes to DNA or protein that's non-heritable.

---

13 E. Pennisi, 2018, "The Momentous Transition to Multicellular Life May Not Have Been So Hard after All," *Science*, doi: 10.1126/science.aau5806

M. GARFIELD: So this is linked to Andreas Wagner 's work,[14] right? On applying selection pressures in the laboratory, and watching the bacterial cultures flail about for a creative solution?

D. KRAKAUER: Yeah. But with bacteria, because they evolve so quickly, you can never be sure that the variation wasn't there to begin with, but just at very low frequencies, whereas these epigenetic mechanisms are a direct causal line from some form of sensor in the world into the cell. And I should correct myself, because of course with epigenetic modifications of the genome, they are heritable. Epigenetic modification of chromatin is not. But that's in some sense the experimental challenge. Are we seeing regular evolution where we don't have to invoke a new connection between the fitness and the epigenetic landscapes, or is it something different? That's been very difficult to tease apart.

M. GARFIELD: So there's a kind of a Hail Mary pass I want to throw to you to get us to Mirta Galesic and Henrik Olsson's piece,[15] which is about affordances. What is possible in that epigenetic interrelating between the organism and its environment? It seems like it starts to possibly answer the question that Doug leaves us with about who survives the mass extinction to flourish after. Which is something in Santiago Elena's piece about the distinction between generalist and specialist viruses. Specialist viruses are sort of, as you put it, more canalized. They are narrower in their specificity and therefore more subject to disruption and that's what you see in the record of extinctions. When I was at the University of Kansas, our head of paleontology said, if you want to survive a mass extinction, live on insects, be as much of a generalist as you can, soak up what resources are available rather than subsisting on a very specific resource. With Mirta and Henrik's piece, this kind of thinking, you look at communication modalities and cross-cultural communication and this issue of scientists communicating with new audiences. If we only speak business to business or scientist to scientist, then we are less resilient in the face of disruption to our social networks.

---

14 J. Zhing, N. Guo, and A. Wagner, 2020, "Selection Enhances Protein Evolvability by Increasing Mutational Robustness and Foldability," *Nature* 370(6521): eabb5962, doi: 10.1126/science.abb5962

15 See T-013 on page 179

D. KRAKAUER: I think there are definitely connections, one of which is, what does a crisis like the one that we're facing do to science communication—what role does it play? How willing are people to accept facts as opposed to resist them? And the other one has to do with the internal and external constraints on credibility. Let me just mention them both. The first point that I think connects back is this observation that Mirta and her colleagues have been making, which is that people are not only influenced by reputable, centralized sources of discovery, but by their neighbors, by their friends, by their family and community. In fact, in many cases, when forced to make a decision between a fact as presented to you by a reputable newspaper, for example, versus your neighbors, you'll choose the opinion of your neighbor.

The reasons are obvious: the consequences of disagreeing with the newspaper are nearly zero, whereas the consequences of disagreeing with your friends and family can be rather catastrophic. That is very close to the external constraints of the fitness landscape versus the internal constraints of the epigenetic landscape, and scientists would be well advised to understand that. I think we've all been seeing this in other catastrophic areas like climate change. Just presenting people with the facts and hoping that they will be in some sense absorbed and acted on is never enough, because there are other forces at play that constrain people's abilities to accept them. And the big connection to this particular episode is what happens in crisis, because in crisis those internal epigenetic constraints are relaxed and all of a sudden more people are trusting the authority of scientists, as they point out in their article.

Medical practitioners now are considered reputable sources of information for very good reasons. The bigger push is, "Look, if this is true, now is the time not to become despondent about our inability to deal with the challenge, but actually to present the facts as best as we know them and ask everyone to collaborate in a rational empirical debate about how to move forward, and to extend it beyond the coronavirus to other issues like climate, which for related reasons people have been resistant to accept." Again, this is that "crisis in opportunity." What happens post-cataclysm? And I think they're being rather optimistic, which is nice, and considering the possibility that we'll have a different attitude toward reason.

M. GARFIELD: There was something in Mirta and Henrik's piece about abstraction and how the shift in the costs of disagreeing with your community versus disagreeing with scientific authority are about how the phenomenon becomes evident to people. It's a shifting in the external fitness landscape. But you had another point—you said there were two pieces.

In fact, in many cases, when forced to make a decision between a fact as presented to you by a reputable newspaper, for example, versus your neighbors, you'll choose the opinion of your neighbor.

D. KRAKAUER: One piece was this question of who is a reliable source of information, and why do local forces dominate global forces? In other words, the beliefs of your neighbor, family, and friends over the beliefs of a community of putative experts. The second point was why, during a crisis, the balance of forces are shifted such that the external source of information could come to dominate. I think that's what we've seen now. You make the point, Michael, that the reason that could happen is because what felt like a very remote consideration was domesticated, because our friends and family are suffering from this disease. We can see it at the local level. It's become an "epigenetic" phenomenon in that metaphorical sense, and now that opens up the opportunity for us to think in a much more rational, empirical way about how best to proceed.

M. GARFIELD: I'm curious how you see this linking to Bill Miller's piece because Mirta Galesic and Henrik Olsson are talking about a conversation that we're observing fold over itself. In an important way it's distinct—again, the Miguel Fuentes piece,[16] part of what we're seeing is the information explosion. The trust in scientific authority seems to be a kind of innovation at the edge of chaos like we were talking about with Santiago Elena and viral evolution. There's this threshold of social coherence. How do you see that linked to Bill's?

16 See T-006 on page 101

D. KRAKAUER: If you look at this catastrophe, it has many different elements. For most of us it's dominated by two factors: concerns about our health and concerns about our economic prosperity. Somehow the world of a virus, the microscopic world, has become aligned with the macroscopic market, and very few people have any understanding of how that connection works. Of course, the history of our scholarship is a history of silos and departments and disciplines. It's so rare to see an economist talking with an epidemiologist, and I think it's again one of those creative destructions that we've seen with the coronavirus, that that is now happening. It's something of course that we've been doing at SFI for a long time and I'm extremely gratified to see it. Now this contribution from Bill is about that, the market and trying to understand what's going on now.

Bill was the former chairman of our board, so someone we know very well. I've often thought of Bill as the Alex Honnold, the free solo climber, of investment, in the sense that he can rise faster and further than anyone else, but he can also take very dramatic plunges. He has a unique attitude toward risk, and he's pointed out many times that the market is a machine. It's a machine for pricing goods and services, and in order for the machine to work, it has to have the right inputs. Those are the costs of manufacture and the values that we assign to these goods and services. But, when faced with uncertainty, the market does something very strange. It always assumes the worst and that produces very rapid downturns. But it also means that as soon as that input, which is that information it wants, becomes available, you can get very rapid upturns. And that in part explains why the market is so volatile.

Bill makes the point that in the course of his own career, he's seen five, including this one, major stock market crashes, which he in the Schumpeterian paleontological mode refers to as opportunities. The first in '73–'74 during war in the Middle East, during the Vietnam War, another in 1982, another in 1987, a fourth in 2008–2009, which for many of us experienced, and now of course the most recent. The first is an insight into Bill as a very special kind of species. He was at the time a lieutenant in the army in Munich, in Germany. When the market was at its worst, he went into a local Merrill Lynch office—this was long before he was in any sense involved with investment—and bought shares near the market bottom. He was doing exactly what his hero, John Maynard

Keynes, suggested, which is that every serious investor has to learn how "to suffer grievous losses with great equanimity."

His view was when the market is at its worst is when the investor should get up and become involved. He makes this point that, just as in the fossil record there are new species that emerge after a mass extinction event, it is also true that, in markets, subsequent to a stock market crash, you see extraordinary innovation in the production of goods and services. So, he introduces us to this notion of "high beta." I have to ask everyone to sort of visualize the following graph. If you were to plot, over some appropriate unit of time, on the *y*-axis the returns in some stock, and on the *x*-axis the returns to the market in general—if they were perfectly correlated and you fitted a regression line through that data—the slope of that line would be 1 and your stock would behave, just like the market.

. . . just as in the fossil record there are new species that emerge after a mass extinction event, it is also true that, in markets, subsequent to a stock market crash, you see extraordinary innovation in the production of goods and services.

But if your stock is more volatile than the market, then that beta is greater than 1. A small change in the market corresponds to a very big change in your particular stocks. And Bill describes himself as a high-beta investor, which means that he will pick typically cyclical stock, which has certain characteristics that make them rather volatile in times of uncertainty, but with the possibility that they'll do extraordinarily well. He's one of these people that has the ability to consider the long timescale. Not all of us can do that—I'm not in that position. But if you do and you can see through to the end of the cataclysm, typically you're positioned to do extraordinarily well.

This is an insight that John Templeton made when he said you always buy at the point of maximum pessimism, which feels counter to our instinct, but is the adaptive thing in the cultural setting to do. I think Bill is presenting us with a kind of an insight into the mind of a free-solo investor who has done extraordinarily well by understanding fast and slow timescales and the behavior of markets and how things that might do badly during a crisis might do extraordinarily well after it. In some sense, we want to develop this long temporal vision to sustain these periods of uncertainty, and consider what we imagine might do well in an entirely new context.

M. GARFIELD: Bill says, "Going into 2020, I thought economic risk was low and that, if the market was going to decline, it would be because of either geopolitical events or some exogenous shock," and that "we got the exogenous event in the form of a global pandemic." This gets us back to this question of the internal and external fitness landscapes. It is an attack from within the body, but it's an attack from beyond the models that a lot of people had.

D. KRAKAUER: That's right, Michael. I think people have pointed out that there are significant differences between perturbations that reveal true endogenous structural deficiencies in markets—like, for example, subprime mortgages—versus exogenous perturbations. And there is this rather interesting empirical dataset that suggests that perturbations that are endogenous tend to persist longer than perturbations that are exogenous. That would be analogous to a mass-extinction event that was primarily driven by some change in the environment versus some deficiency in the genetic architecture of life itself that meant that it was extremely difficult to now adapt back to the new conditions of existence.

M. GARFIELD: When I hear Bill say, for example, "The commonly offered advice in a steep market decline, such as we are experiencing, . . . to 'buy quality names on sale' is a great prescription for underperformance in a recovery." What I hear there is, don't overfit your model. This is, again, possibly a prescription for generalist strategies evolutionarily. You don't want to overspecialize in one sector. You want to hedge the bets and that's a bad strategy when things are normal, like Doug was saying: if you try to fill every niche in a relatively mature, robust ecosystem, you're going to be out-competed by specialists who are able to do it more efficiently. But if you're a raccoon or a cockroach, if you're willing to sort of throw darts all over the board, then that strategy which kept the mammals under the feet of dinosaurs for 150 million years ended up being what brought them to prominence after the dinosaurs went out.

D. KRAKAUER: What you've described is the microbial mode of living, the quasi-species that we started with, which is the better description of microbial species than the concept of a species—that is the cloud versus the point or the particle. It's exactly this high-mutation, high-risk, high-return strategy. Viruses do very well in a crisis. I think what we're hearing from everyone in this particular episode—whether it's science communication, the way that we mobilize laboratories, what we learned from the history of life—is that you sort of have to adapt in that way and explore the possibility of variant ideas at a much higher temperature than you would during periods of stability. And I think that's an important message. It's extremely difficult to do, but it is an opportunity for a society to be more experimental than it otherwise would be.

M. GARFIELD: To double down on Mirta and Henrik's optimism, I'm pleased to see that this seems to be the way that society is responding to this. Everyone I see is talking about the creative opportunity that this provides. And I know that you and I have stressed, it's important not to be Pollyanna-ish about this. It's important to understand long-term optimism is not in competition with short-term pessimism, but at least people are generally displaying a sort of native, intuitive understanding that now is the time to try something fresh.

D. KRAKAUER: And I think that where we started the whole series—with citizen-based medicine and citizen-based science. . . . If science is anything, it's the rigorous pursuit of new ideas. The idea that the world now is entertaining a variety of new ideas is extraordinarily refreshing and necessary in order to be, as you say, a long-term optimist.

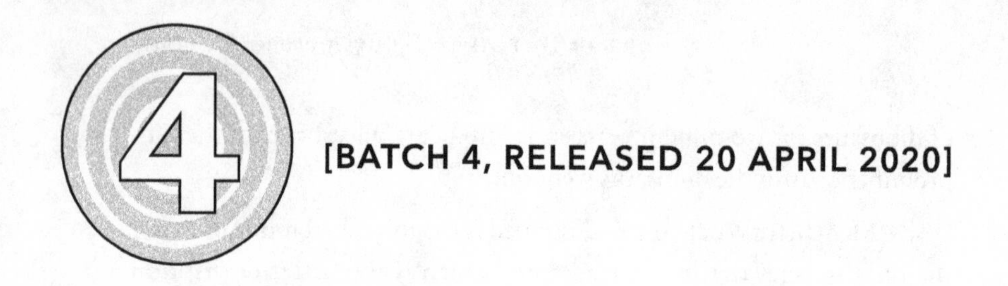

Higher education holds

propagate **EXISTING**

can thus be a powerful

SUSTAIN *society's*

PROPAGATE

the **POWER** *to*

STRUCTURES, *and*

TOOL *to promote and*

values and ideals—or to

OPPRESSION.

—CARRIE COWAN, **Reflection 016**

VIVE LA
FANFARES

TRANSMISSION NO. 015

FEDERALISM IN THE TIME OF PANDEMIC

DATE: *20 April 2020*

FROM: *Anthony Eagan, Santa Fe Institute*

STRATEGIC INSIGHT: *It is important to keep in mind that as agents we maintain bottom-up control, even if we lack decisive power.*

One of the central points of contention during the framing and ratification of the US Constitution involved the question of federalism, or the division of power between the state and central governments. The question was novel in the history of nation-building, in part because of the size and diversity of the American geographical and social landscape. In many ways, the richness of the debate demonstrated a prescient grasp of our germinal nation as a complex system, with questions of agency, robustness, decentralization versus centralization, economy and trade, public safety, and national versus subnational networks at its heart. This conflict can be understood as one between the urge for national power and the counter-urge for local control.

The Federalists advocated a strong national government, the primary role of which would entail overseeing the state and local governments across our large domain. This can be likened to a top-down approach to robustness. The Anti-Federalists advocated against consolidated central power and for greater local and state sovereignty: the bottom-up approach. The debate took for granted that any central authority was potentially both beneficial and detrimental. The primary benefit is the ability to safeguard liberty and protect the populace, whereas the primary detriment is the ability to threaten liberty in times when it would otherwise be unthreatened by external or internal dangers.

OPPOSITE: *James Ensor, "Christ's Entry into Brussels in 1889," 1888 (detail)*

Fearing that ratification was in jeopardy, Alexander Hamilton, James Madison, and John Jay wrote a series of eighty-five letters under the title of *The Federalist*, in which they attempted to explain how the Constitution amounted to, among other things, a compromise between local and centralized needs. In the thirty-ninth letter, Madison wrote the following:

> *But if the government be national with regard to the operation of its powers, it changes its aspect again in relation to the extent of its powers. The idea of a national government involves in it not only an authority over the individual citizens, but an indefinite supremacy over all persons and things, so far as they are the objects of a lawful government.... In this relation, then, the proposed government cannot be deemed a national one; since its jurisdiction extends to certain enumerated objects only, and leaves to the several states a residuary and inviolable sovereignty over all other subjects.*

National action has the obvious advantage of a quicker and more unified and consistent response; local action has the advantage of specialized adaptation to the needs of the community as well as the flexibility of innovative thinking appropriate for the milieu in question. We are seeing today how important these questions and tensions remain, as the American people have sought more rapid, more consolidated, and more direct responses from the federal government, particularly the executive branch. Yet simultaneously the same population has shown fear that unprecedented measures enacted in catastrophic times will inevitably amount to greater authoritarian power in Washington, not only in the near future, but also down the road.

We are currently trying to juggle different considerations: the health of our populace, the needs of our economy, and our political and social future. The obvious fear is that prioritization of one may cause incalculable and unforeseeable damage to the others. We worry that, for instance, a centralized and authoritative response may sacrifice the future for the present, leading to an economically desperate population which is hungry, out of work, resentful, radicalized, and ready to cede power to increasingly authoritarian political figures. Simultaneously, we know we cannot sacrifice the present for the future, since the death toll is rising, hospitals are overwhelmed, the virus is spreading, and our medical-care professionals and other essential workers are risking their lives on a daily basis.

As we wait and observe the ongoing governmental responses at the municipal, state, and national levels, it is important to keep in mind that as agents we maintain bottom-up control, even if we lack decisive power.

As we wait and observe the ongoing governmental responses at the municipal, state, and national levels, it is important to keep in mind that as agents we maintain bottom-up control, even if we lack decisive power. Whether one is a Federalist or an Anti-Federalist, as the COVID-19 pandemic continues, it remains the case that the most local response possible falls upon the individual agents constituting the population. The more uniformly and selflessly we act, the more control we have. In this way, the robustness of our nation may prove to be bottom-up. Health experts tell us to wash our hands and stay at home. We should do these things literally, but not figuratively. It is the best way to maintain control.

REFLECTION

THE COVID-19-INDUCED EXPLOSION OF BOUTIQUE NARRATIVES

DATE: *20 August 2021*

FROM: *Anthony Eagan, Santa Fe Institute*

Writing to his fellow novelist Paul Auster in 2011, J.M. Coetzee bemoaned the threat posed to narrative fiction by the proliferation of new information-exchange technologies. "So much of the mechanics of novel writing, past and present, is taken up with making information available to characters or keeping it from them, with getting people together in the same room or holding them apart. If, all of a sudden, everyone has access to more or less everyone else—electronic access, that is—what becomes of all that plotting?" A moment later, he adds: "The default situation has become that, save in extraordinary circumstances, [character] B is always contactable by [character] A."[1] Coetzee feared that the once-dependable conflict of information differential was rapidly disappearing.

In many respects, Coetzee was correct. Yet it is worth imagining what his letter to Auster would look like had he written it in 2021, when communication technologies had grown so extreme and our dependence on them so high that, though isolated from one another by quarantines, lockdowns, and travel restrictions, B and A were holding audiovisual meetings together with D and C, all the while separated by arbitrary physical distances. On the one hand, the basis of Coetzee's concerns has intensified. On the other hand, what Coetzee may not have foreseen was a different sort of proliferation, namely that of what I will call boutique narratives. In this scenario, B and A (along with D and C) hold divergent accounts of the state of the world despite having access to the same wealth of information.

Whereas an author, Coetzee continued, "used to get pages and pages out of the nonexistence of the telegraph and telephone (yet to be invented) and the consequent need for messages to be borne by hand or even memorized at one end and recited at another (example: the man

1 P. Auster and J.M. Coetzee, 2013, *Here and Now: Letters 2008-2011*, Letter dated August 29, 2011, 224–225.

who had to race from Marathon to Athens)," now an author might get pages and pages out of the conflict that arises in the narrative differential between B and A, who receive their information from different media platforms and publishers and from differently biased interlocutors. What in former times would have amounted to a shared narrative has become a conflict of narratives. In other words, literature might nowadays replace the conflict of information differential with the conflict of narrative differential (arising from divergent boutique narratives).

If, for example, Romeo and Juliet had had access to present-day technologies, their ironic double suicide would have been avoided. Instead, they might have discovered, not long after their successful elopement and erotic consummation, but just after the arrival of a pandemic, that, given their opposing boutique narratives, one believed the lab-leak hypothesis while the other maintained that a pangolin was the source of the outbreak. While one politicized mask-wearing, the other blamed every acronymic institution for prolonging the pandemic through the systematic suppression of ivermectin. Perhaps, in a sort of historical vertigo, they even disagreed on when and by whom their home country was founded. One saw the killing of a black individual by a white policeman as an isolated incident (or at least a statistically rare phenomenon) while the other saw it as indicative of the complete corruption of all police officers.

At a deeper level, the two failed to recognize the disparity in their grasp of the concepts under discussion, given their differing perspectives on the historical trajectory. Each had a different sense of concepts such as identity, discrimination, gender, offense, diversity, justice, protest, and so on. Given their lack of categorial accuracy, they failed to understand one another despite the illusion that they communicated in a common language. In this case, Romeo and Juliet's peril would escalate not because they could not see eye-to-eye on certain well-grounded and debatable issues, but because they could not even agree on the facts of the issue under discussion, lacking as they did any shared narrative context.

In the developing cultural anxiety arising from mortal fear, economic alarm, transparently spectacle-laden and even childish governance from both political parties, and the dissipation of world comprehension, the need for a salient account of things only intensifies, and thus one more urgently clings to the convictions arising from his or her extant boutique narrative. Instead of an unintended double suicide, the young lovers might have committed an intentional mutual homicide, with Juliet poisoning her newly antagonistic Romeo, and with Romeo stabbing the loathsome Juliet before he fully asphyxiated. We can classify this entire process as a metaphorical autoimmunity, where agents within an ostensibly cooperative system misfire because of insufficient or incorrect information and begin attacking one another rather than addressing the actual pathogen.

> Clearly, boutique narratives can be generated by the selective bias of an individual . . .

Clearly, boutique narratives can be generated by the selective bias of an individual, such as when a person deliberately reads the opinions of only one or two columnists, distrusting all the others because of minor or major differences of opinion, or when a person watches only CNN or only Fox News, failing to triangulate between multiple journalistic outlets to find a more objective thread running through the tendentious lines most sources implicitly or explicitly follow. But equally likely is that an algorithm, designed to generate clicks and to feed on our pre-existing tendency for selective bias and our attraction to narrative simplicity, has grown ever more personalized to the individual user's browsing data, so that in addition to bespoke advertisements, the user receives bespoke headlines and bespoke accounts about all local, national, and global circumstances beyond his or her empirical purview. The foreseeable result is a feedback loop of confirmation bias and informational feeds—an echo chamber between a self-conscious human and unselfconscious encroachment of code.

Why mention this in reference to my Transmission on Federalism in the time of pandemic? In that piece, while highlighting the difference between bottom-up control and top-down power, I emphasized that the more we assumed personal accountability in the attempt to slow transmission of the disease, the less concerned we would need to be with respect to the anticipated problem of centralized power grabs in our time of crisis. Like Coetzee, however, but in a different manner, I failed to appreciate the ways in which a global crisis and the consequent panic would coincide with recent advances in communication technologies to make us more dependent on applications that appear to bring us together but actually fragment us further.

Proper accountability requires proper understanding as grounds for action. The implied point of my Transmission was that quarantines and lockdowns only work if a certain high percentage of individuals follow the guidelines. Failing such voluntary cooperation, the quarantine is prolonged, hospitalizations and deaths continue unabated, and all the attendant economic, social, physical, and mental consequences are extended and amplified.

But if one group promotes herd immunity while another believes that claims about the dangers of COVID-19 are exaggerated, while still another politicizes the crisis by enumerating the errors the authorities have made and suggesting they and their constituents can never be trusted, while still another remains ignorant either willfully or through unfortunate circumstances, while still another is told that infection occurs through fomite transmission when in fact it is spread by respiratory droplets and aerosols, while still another is convinced it is all a hoax, etc., then the possibility of well-meaning accountability is severely diminished. As agents, in other words, we can only maintain local control with a view to the prevention of both transmission and tyrannical rule if we have a shared understanding of what is going on and how we might act in a selfless and yet responsible manner. We cannot align our behavior without aligning our sense-making apparatuses.

Among the innumerable branching effects of the COVID-19 pandemic has been its radical exposure of the boutique narrative complex. We have learned with greater clarity that, contrary to Coetzee's suggestion, global, remote "contactability" does not resolve the conflicts arising from information exchange.

If the pandemic revealed the dangers of global physical contact, so too has it more slowly revealed the dangers of technologies that appear to reduce the need for such contact.

TRANSMISSION NO. 016

THE FUTURE OF EDUCATION

DATE: *20 April 2020*

FROM: *Carrie Cowan, Santa Fe Institute*

STRATEGIC INSIGHT: *American higher education must think outside the academy in a postpandemic world.*

Can we talk about the kids? Well, college students, really. The COVID-19 pandemic is inadvertently prompting an experiment in higher education. More precisely, it's challenging the value of traditional educational models. The unwitting participants in this experiment are, first and foremost, the kids.

COVID-19 will leave a mark on higher education. At a minimum, students can expect more online content and the use of class time for deeper discussion and collaborative work, a shift that has already been occurring over the past decades. But why not more? The pandemic highlights a globally interconnected world and a changing economy. It's time to prepare students.

Coincidentally, the last transformation of education in the United States occurred around the time of the 1918 influenza pandemic, the most deadly pandemic the country has experienced. The pandemic did not cause educational reform; the economy did. Yet, in the decades before the pandemic, the majority of Americans completed no more than eight grades; only about 10% went on to graduate from high school. By the mid-1920s, the primary school to high school to college path had become the goal for many.

The supply and demand for higher education was driven by America's westward expansion. Taxation revenue generated by prosperous agriculture provided the funds. As mechanization expanded

---

OPPOSITE: *Albert Bierstadt, "The Oregon Trail," 1869 (detail)*

and immigrant laborers increasingly performed the heavy, dirty work of agriculture and manufacturing, more of the population was eager to use high school and college to prepare for the demands of a newly possible professional life. Thus, the current version of American education was born. Shortly thereafter, it began to stagnate.

It's hard to imagine that many students (or faculty) today are satisfied with the constraints of the American higher education system, with university bureaucracies, prohibitive tuitions, siloed departments, and historic methods of learning and evaluation imposed upon them. The challenges of the world require input from diverse participants, unconstrained by how much money or privilege they might have. Real-world problems—and the jobs that exist to solve them—demand more than a knowledge of political science or physics or art history. They require a new style of thought that emphasizes connection and commonalities—mutualism across disciplines. How otherwise to reconcile the enormity of social, behavioral, biological, and physical factors that contribute to multiscale recovery after a pandemic? Or a recession? Or ecological collapse? The world is not getting simpler.

American high school education mandates a Common Core curriculum and standardized testing, preparing students to be quite skilled at passing tests but not necessarily adept at dealing with the ambiguities and intersectionality of real-world problems. Then we hope they will figure it out in college. Sure, this is a generalization, but couldn't education better prepare students for the complexity of the world? Why shouldn't students define what their education looks like, what they want to learn, and how they will know when they've gotten there? The availability of no-cost online learning resources now makes individualized, self-directed curricula accessible to everyone. The two largest platforms, Coursera and edX, offer more than 2,750 classes. On Friday, March 13, 2020, as COVID-19's grip on the US became clear, both platforms saw a surge in Google searches. However, that search volume paled in comparison to another trend on the same day: Minecraft.

Minecraft is monumental. It's the highest-selling video game of all time, played by more than 100 million people each month. Players build sewers and cities, explore worlds, engage in an economy, embark on strategic partnerships, resolve disagreements, and, yes, prevent pandemics.

The game is hosted in a distributed manner, meaning that there are hundreds of parallel worlds evolving simultaneously, which provides an opportunity for us to see how small changes in social structure—for instance, more trading than fighting—impact health and prosperity. Minecraft is experiential and versatile learning. Why isn't it valued as such?

If the online experience prompted by the pandemic shows that we can educate students in a fully virtual setting, how does that change the role of costly and exclusive academic institutions? Beyond facilitating science labs, arts, and sports, what exactly is the unique value of in-person education? There has been a tendency to equate in-person education with higher merit. Online higher education has struggled with image because of the dominance of for-profit colleges in that space, but that's about to change as traditional non-profit institutions adapt postpandemic.

Today, as in 1918, pandemics don't change education. But such a crisis does inspire reflection about what the future will look like, what problems society will have to solve, and whether our education system is fostering the curiosity, ambition, and intellectual adventurousness that will be required. If anything, we know that our postpandemic world is going to be different. For the kids' sakes, education should be different, too.

REFERENCES

T.D. Snyder, 1993, "120 Years of American Education: A Statistical Portrait," US Department of Education, Office of Educational Research and Improvement, National Center for Education Statistics, https://nces.ed.gov/pubs93/93442.pdf

C. Goldin, 1994, "How America Graduated from High School: 1910 to 1960," *National Bureau of Economic Research*, no. w4762, https://www.nber.org/papers/w4762

Google Trends: Coursera, edX, Minecraft, https://trends.google.com/trends/explore?date=2020-01-01%202020-03-31&q=Coursera,edX,Minecraft

S. Protopsaltis and S. Baum, 2019, "Does Online Education Live up to Its Promise? A Look at the Evidence and Implications for Federal Policy," Center for Educational Policy Evaluation, https://mason.gmu.edu/~sprotops/OnlineEd.pdf

REFLECTION

HAS HIGHER ED BECOME SELF-SERVING?

DATE: *16 August 2021*

FROM: *Carrie Cowan, Santa Fe Institute*

The COVID-19 pandemic's effects on higher education appeared immediately. Colleges and universities rushed to adopt online teaching and learning while simultaneously emphasizing the necessity of in-person, on-campus experiences. Conversely, the effects of higher education on the pandemic emerged more subtly throughout the year.

The US population split on key issues stemming from the pandemic. That division fell along lines of educational attainment. In summer 2020, US adults without a college degree were four times more likely to report never wearing a mask compared to college graduates.[1] As of February 2021, US adults with college degrees were almost a third more likely to have been, or intend to be, vaccinated against COVID-19 than adults with no college education.[2] US adults who graduated from college were nearly a third more likely to report "a great deal of confidence" in medical scientists compared to adults without college education.[3] These responses aren't likely to reflect the college curriculum per se; the details of N95 technology and aerosol physics are not emphasized in a typical university course load, nor are diagnostic qPCR, mRNA-based vaccines, or the role of helper T cells. Only a specialized few will learn these things as part of their undergraduate education or career preparation.

---

1 M. Brenan, "Americans' Face Mask Usage Varies Greatly by Demographics," *Gallup*, July 13, 2020, https://news.gallup.com/poll/315590/americans-face-mask-usage-varies-greatly-demographics.aspx

2 Pew Research Center, "Majority of Black Adults Now Say They Plan to Get—or Have Already Received—a COVID-19 Vaccine," March 3, 2021, https://www.pewresearch.org/science/2021/03/05/growing-share-of-americans-say-they-plan-to-get-a-covid-19-vaccine-or-already-have/ps_2021-03-05_covid-19-vaccines_00-019/

3 Pew Research Center, "Americans with More Education Have Greater Confidence in Medical Scientists to Act in the Public Interest, as Do Democrats," May 19, 2020, https://www.pewresearch.org/science/2020/05/21/trust-in-medical-scientists-has-grown-in-u-s-but-mainly-among-democrats/ps_2020-05-21_trust-in-scientists_00-14/

As the pandemic progressed and new and varied crises confronted US society, division along the same axis persisted. Educational attainment was used to describe different views of racial inequity,[4] trust in police,[5] support for #BlackLivesMatter,[6] the legitimacy of the presidential election,[7] and views of the January 2021 Capitol riot.[8] Among Trump voters, 60% more college graduates viewed his conduct post-election unfavorably than those without college degrees.[9]

The higher education system apparently underlying these social and ideological divisions continued to face criticism from Republicans for its perceived liberal political tendencies, suppression of free speech, and cancel culture.[10] The US education system as a whole was the target of a Trump administration re-education proposal[11]—a revisionist history of the US and its role in the world to revive nationalist sentiment.

4 H. Lee, M. Esposito, et al., "The Demographics of Racial Inequality in the United States," *Brookings*, July 27, 2020. https://www.brookings.edu/blog/up-front/2020/07/27/the-demographics-of-racial-inequality-in-the-united-states/

5 J. Williams, "BLM and Policing Survey Identifies Broad Support for Common-Sense Reform," *Civis*, July 22, 2020. https://www.civisanalytics.com/blog/blm-policing-pulse-survey-analysis/

6 Trend poll, "Do You Support or Oppose the Black Lives Matter Movement?" *Civiqs*, https://civiqs.com/results/black_lives_matter

7 Muhlenberg College, "Pennsylvania 2020 Post-Election Poll: January 2021," https://www.muhlenberg.edu/aboutus/polling/surveys/election/pa2020post-election-poll/

8 P. Bump, "The Demographic Divergence that Helps Explain Perception of the Capitol Rioters," The Washington Post, January 14, 2021, https://www.washingtonpost.com/politics/2021/01/14/demographic-divergence-that-helps-explain-perceptions-capitol-rioters/

9 Pew Research Center, "Trump Voters are Divided in Views of His Conduct since the Presidential Election," January 14, 2021, https://www.pewresearch.org/politics/2021/01/15/views-of-biden-and-trump-during-the-transition/pp_2021-01-14_biden-trump-views_02-02/

10 K. Parker, "The Growing Partisan Divide in Views of Higher Education," *Pew Research Center*, January 30, 2019, https://www.pewresearch.org/social-trends/2019/08/19/the-growing-partisan-divide-in-views-of-higher-education-2/

11 M. Balingit and L. Meckler, "Trump Alleges 'Left-Wing Indoctrination' in Schools, Says He Will Create National Commission to Push More 'Pro-American' History," *The Washington Post*, September 17, 2020, https://www.washingtonpost.com/education/trump-history-education/2020/09/17/f40535ec-ee2c-11ea-ab4e-581edb849379_story.html

So what is the other learning that goes on in our higher education system that seems to shape how individuals respond to Dr. Fauci's recommendations or to political fanaticism? Over decades, through an emphasis on standardized testing and associated eschewal of creativity, the US higher education system evolved to reward compliance.[12] Higher education holds the power to propagate existing structures, and can thus be a powerful tool to promote and sustain society's values and ideals—or to propagate oppression.[13]

Higher education holds the power to propagate existing structures, and can thus be a powerful tool to promote and sustain society's values and ideals—or to propagate oppression.

Higher education is a system like any other, acting for its own perpetuation and growth. It perpetuates itself limitlessly, not because that was the initial goal of the education system, but because, like a successful virus, that has become the feature that matters.

According to the sociologist Randall Collins,[14] the higher education system now exists to produce credentials—degrees. Credentials, in turn, provide access to jobs. US higher education is subject to inflation: as more people acquire a college degree, the worth of a degree diminishes in the job market, demanding more, or more advanced, degrees to compete. Yet, according to Collins, the higher education system is founded on false pretenses: more education has not led to more equal opportunity, nor to better economic performance, nor to better jobs.

12 For example: S. Bowles and H. Gintis, 2011, *Schooling in Capitalist America: Educational Reform and the Contradictions of Economic Life*, Chicago, IL: Haymarket Books.

13 For example: P. Freire, 2013, *Pedagogy of the Oppressed*. New York, NY: Routledge.

14 R. Collins, 2019, *The Credential Society: An Historical Sociology of Education and Stratification*, New York, NY: Columbia University Press, http://cup.columbia.edu/book/the-credential-society/9780231192354

Has the higher education system ceased to serve its students and instead only serves itself?

As colleges and universities fought to justify onsite learning while temporarily moving to remote instruction early in the pandemic, things remained opaque as to precisely why in-person classes are better for achieving the oft-cited roles of post-secondary education: building knowledge and preparing for work. What about these activities requires a shared physical location? Enrollment declined at US colleges and universities in fall 2020,[15] reflecting many students' views that in-person classes and campus life are essential to their higher ed experience.

While online learning went better than expected,[16] the pandemic did not precipitate a massive collapse of our higher education system or a full transition to low-cost online learning for everyone, nor did it diminish the demand for on-campus life. Most colleges will survive;[17] fall 2021 enrollments are forecast to recover to prepandemic levels.[18]

The pandemic should, however, prompt a sustained discussion about the role of education in fostering a society that can robustly navigate a massive public health crisis—and racial injustice and insurrection—and that values facts, logic, and creativity to guide an equitable and fulfilling way forward.

15 National Student Clearinghouse Research Center, "Current Term Enrollment Estimates," https://nscresearchcenter.org/current-term-enrollment-estimates/

16 Tyton Partners in Partnership with Every Learner Everywhere, "Time for Class: COVID-19 Edition," January 2021. https://www.everylearnereverywhere.org/resources/time-for-class-covid-19-edition/#:~:text=This%20spring%2C%20higher%20education%20institutions,digital%20learning%20tools%20and%20techniques

17 M. Schifrin and H. Tucker, "College Financial Grades 2021: Will Your Alma Mater Survive COVID?" *Forbes*, February 22, 2021, https://www.forbes.com/sites/schifrin/2021/02/22/college-financial-grades-2021-will-your-alma-mater-survive-covid/?sh=467397454916

18 R. Garrett, "Higher Education Predictions for 2021, Part 1," *Encoura*, January 5, 2021, https://encoura.org/higher-education-predictions-for-2021-part-1/

TRANSMISSION NO. 017

PRIVACY CONCERNS THAT ARISE WITH THE PANDEMIC

DATE: *20 April 2020*

FROM: *Stephanie Forrest, Arizona State University Biodesign Institute; Santa Fe Institute*

STRATEGIC INSIGHT: *Complexity science and computer algorithms can help us address privacy concerns that arise with the pandemic.*

Pervasive surveillance is quickly becoming the new normal, whether the surveillance infrastructure is developed by governments for control (e.g., in China and many other authoritarian countries) or by large tech companies for profit, as we see in the US and other western countries. In the age of coronavirus, these are tempting tools to turn to, either to stamp out misinformation (censorship) or for efficient contact tracing of infected individuals or detection of social distance cheaters (surveillance). One recent example is the company Kinsa, which markets an Internet-enabled thermometer and made headlines in March when it announced that it had used data from 1 million thermometers to create a national map of fever levels and had spotted a downward trend in fevers, ahead of data reported by public health agencies.

Many of us would like to contribute to databases such as these, which can play an important role in the current epidemic; however, many are wary of Internet-enabled surveillance and wish there were a way to participate without sacrificing anonymity and privacy.

Complex-systems thinking can provide new ways of thinking about these problems. Several years ago, we studied how the immune system learns to distinguish self (its own naturally expressed proteins) from other (cells and molecules associated with invading pathogens). Put

---

OPPOSITE: *João Carrilho da Graça (architect), "Rubin's Optical Illusion," Pavilhão do Conhecimento, Lisbon, Portugal, 1998 (detail)*

very simply, the immune system uses a trick that is reminiscent of classic figure versus ground examples, such as the famous picture that is both an urn and two faces. In effect, the immune system builds a map of self (the urn shape) by constructing many small detectors that represent nonself (the two faces). In immunology, this is known as "negative selection" or "clonal deletion." This simple idea can be mathematized and coded as a computer algorithm in an approach that its inventor, Fernando Esponda, calls "negative surveys."

In terms of the thermometer example, suppose each thermometer recorded the correct temperature but reported a value to the database that was different from the actual value. With sufficient data, it is a relatively straightforward calculation to recreate a histogram of the original temperature frequencies. A similar trick can be used to disguise the location from which the temperature was recorded.[1,2]

Cranium Head Optical Illusion

Such an approach would work well for problems like crowd-sourcing fever maps, but contact tracing poses a greater threat. Epidemiologists argue that the most effective way to control epidemic spread through a population is with rigorous tracing of contacts. In today's world, the most efficient way to accomplish that is by inspecting cell phone data, since almost everyone carries a phone almost everywhere they go. And some countries are indeed taking this route—for example, China, South Korea, and Singapore. In the US there are already calls for us to adopt this approach, and we are seeing an explosion of "privacy preserving" apps for contact tracing. Because cell phone data reveal much more information than that which is required to alert and test all at-risk contacts, the potential for abuse is high.[3]

Suppose that I am secretly meeting with a potential new employer, keeping an appointment with my mental health provider, or perhaps having an affair that I don't want my spouse to know about. In these circumstances, if one of my contacts becomes infected, we would like a computation that can identify every person whose location data intersects with that of the infected contact, and we don't need to know who the infected person is or what locations exposed me or my other contacts. This problem is known as set intersection, and Ni Trieu[4] and many others have developed private set intersection algorithms that use cryptographically secure methods to compute set intersections without revealing members of different sets to one another.[5] Despite the urgency of the current situation, this is the time to insist on strong guarantees (on both data collection and use), and on secure methods for computing and alerting contacts.

These are just two examples of how complexity science and computer algorithms can help us address the many privacy concerns that arise with the pandemic. In the past, we have gone to war to defend the principles of a free and open democratic society. Putting in place poorly thought-out massive surveillance schemes puts these principles at risk. Instead, we should develop and deploy methods, many of which already exist, that allow us to recover the information we need in order to preserve public health without jeopardizing our right to be free of unreasonable search and seizure.

REFERENCES

1 M.M. Groat, B. Edwards, et al., 2012, "Enhancing Privacy in Participatory Sensing Applications with Multidimensional Data," *2012 IEEE International Conference on Pervasive Computing and Communications, Lugano (19–23 March 2012)*, https://forrest.biodesign.asu.edu/data/publications/2012-percomm-neg-survey.pdf

2 M.M. Groat, W. He, and S. Forrest, 2011, "KIPDA: *k*-Indistinguishable Privacy-Preserving Data Aggregation in Wireless Sensor Networks," *2011 Proceedings IEEE INFOCOM*: 2024–2032, https://forrest.biodesign.asu.edu/data/publications/2011-kipda-infocom.pdf

3 R. Anderson, "Contact Tracing in the Real World," *LightBlue Touchpaper* blog, April 12, 2020, https://www.lightbluetouchpaper.org/2020/04/12/contact-tracing-in-the-real-world

4 https://nitrieu.github.io/

5 B. Pinkas, M. Rosulek, et al., 2019, "SpOT-Light: Lightweight Private Set Intersection from Sparse OT Extension," in *Advances in Cryptology — CRYPTO 2019 — 39th Annual International Cryptology Conference Proceedings*, edited by D. Micciancio and A. Boldyreva, 401–431, https://eprint.iacr.org/2019/634.pdf

REFLECTION

COMPLEXITY THINKING FOR COMPUTATIONAL SYSTEMS

DATE: *August 20, 2021*

FROM: *Stephanie Forrest, Arizona State University Biodesign Institute; Santa Fe Institute*

As a computer scientist I was proud of how the technology developed over the span of my career enabled society to continue functioning in the face of a global pandemic: online education, remote social interactions, video meetings, online shopping, and telemedicine, just to name a few. At the same time, the pandemic threw into sharp relief many issues that my field has not addressed and probably cannot address on our own.

Many of these are complex-systems problems that involve interactions among sociopolitical systems, biology, and computational systems. For example, remote learning is much more effective when students have a stable high-speed internet connection, a quiet place to work, and a functioning computer; for most students online learning is still a pale facsimile of traditional in-person education. Similarly, the shortcomings of virtual work became painfully apparent as we staggered out of our home offices each evening suffering various forms of Zoom fatigue, mental fog, and other maladies, which highlighted the many subtle and mysterious ways that digital interfaces interact with our minds and bodies. Moving so much of routine life online so quickly also accelerated a trend that had been underway for some time, in which vast amounts of personal data are accumulating in the hands of for-profit companies and government agencies.

It will be many years before we fully understand the consequences of this trend, but if "data is power" as many suggest, then we might expect even more extreme concentrations of economic and political power in the hands of fewer and fewer actors, and we might ask what kinds of technical, institutional, and social innovations are required to mitigate these effects. The field of complex systems does not yet have answers for questions about social inequity, the differences between an in-person interaction and a high-quality video chat, or what role information and data will play in future economic and political shifts, but the pandemic brought them to the fore.

My Transmission suggested ways that complex-systems thinking might lead to new methods of privacy-preserving information capture, and it discussed phone-based contact tracing, which is an essential tool for controlling the spread of disease through human populations. Contact tracing apps record when two or more individuals are physically near each other, identify individuals who have been in close proximity to an infected individual, and notify them of exposures. In today's world, smartphone technology is quickly replacing earlier methods as the method of choice for recovering an infected individual's history of recent physical contacts and automatically sending out notifications. Yet existing apps do not provide sufficient coverage to be broadly useful in an epidemiological setting, and in the wrong hands they could provide new surveillance tools.

. . . the shortcomings of virtual work became painfully apparent as we staggered out of our home offices each evening suffering various forms of Zoom fatigue, mental fog, and other maladies . . .

Over the past year, my ASU colleague Ni Trieu and I have investigated existing contact tracing apps on smart devices and identified several shortcomings, including: lack of integration across different systems, heavy bandwidth requirements for connectivity and high computational requirements on client devices, and vulnerability to privacy and security attacks.[1] Integration and compatibility issues

1 See, for instance: S. Habib, N. Trieu, et al., 2021, "SoK: Eunomia: A Structured and Comprehensive Scoring System for the Assessment of Privacy Policies in Global COVID-19 Contact Tracing Applications," Submitted to 43rd IEEE Symposium on Security and Privacy; T. Duong, D. H. Phan, and N. Trieu, 2020, "Catalic: Delegated PSI Cardinality with Applications to Contact Tracing," in *26th Annual International Conference on the Theory and Application of Cryptology and Information Security*, Asiacrypt; O. Nevo, N. Trieu, and A. Yanai, 2021, "Simple, Fast Malicious Multiparty

arise when, for example, the user of one system—say, COVID Alert NY—travels to another state or country where a different app—say, CovidWatch—is favored. If the traveler happens to become infected and contagious before returning home, the inconsistency among these apps would likely prevent prompt and accurate notification of exposed individuals. Contact-tracing apps today place most of the computational load on the user's device, which in turn requires large data downloads. In many places, including tribal communities in the US, users may not own a state-of-the-art smartphone or have access to high-speed internet. We recently submitted an NSF proposal to develop an end-to-end unified contact tracing application with strong privacy and lightweight cost.[2]

> Meanwhile, the pandemic has only reinforced my view that complex-systems thinking and methods are sorely needed to understand the computational systems we have already devised and to shape the design of future systems so they enhance humanity rather than degrade it.

Even if we should succeed in our quest for lightweight, secure, ubiquitous contact tracing, other mechanisms will be required to address the larger questions posed by the wholesale integration of computing

Private Set Intersection," in *28th ACM Conference on Computer and Communications Security*, CCS; M. Rosulek and N. Trieu, 2021, "Compact and Malicious Private Set Intersection for Small Sets," in *28th ACM Conference on Computer and Communications Security*, CCS; and G. Garimella, B. Pinkas, et al., 2021, "Oblivious Key-Value Stores and Amplification for Private Set Intersection," the 41st International Cryptology Conference.

2 Our Unicon app would be deployed as a plug-in for existing apps, supporting integration across different systems, and allowing data to be shared secretly and securely via untrusted Unicon servers, similar to the distributed voluntary system of Domain Name Service (DNS) servers that supports internet communications. It would also allow users to opt in to the Unicon system even if their app provider does not.

and information with society. Others at SFI are concerned about similar issues, and during my annual summer visit we began batting around ideas for how to tackle them. Meanwhile, the pandemic has only reinforced my view that complex-systems thinking and methods are sorely needed to understand the computational systems we have already devised and to shape the design of future systems so they enhance humanity rather than degrade it.

TRANSMISSION NO. 018

QUANTITATIVE WAYS TO CONSIDER THE ECONOMIC IMPACT OF COVID-19

DATE: *20 April 2020*

FROM: *Sidney Redner, Santa Fe Institute*

STRATEGIC INSIGHT: *Common-sense estimates provide quantitative ways to think about the economic impact of COVID-19 in Italy.*

Is there a principled way of calculating the burden of disease on the economics of a nation? And doing so when the quality of information is heterogeneous and our knowledge of the disease improves from one moment to the next? The great economist John Maynard Keynes in a 1938 letter wrote, "The object of a model is to segregate the semi-permanent or relatively constant factors from those which are transitory or fluctuating so as to develop a logical way of thinking about the latter, and of understanding the time sequences to which they give rise in particular cases."

In the spirit of Keynes, it's an instructive exercise to quantitatively estimate the impact of COVID-19 hospitalizations and deaths by using simple common-sense estimates. These estimates make use of more or less constant factors (in this case, national death rates and the costs of medical care) in order to best estimate factors about which we know far less, or, as Keynes put it, transitory factors (in this case, COVID-19 hospitalizations and costs).

Let's consider the case of Italy. For this country, the current life expectancy is 82.8 years.[1] This number translates to one person dying every 30,222 days, or a death rate of $1/27{,}375 \approx 0.0000331$ people per day. Equivalently, the death rate is 0.0121 per year, or 12.1 deaths per year per 1,000 people. This naive estimate is reasonably close to the currently

---

OPPOSITE: *Jan Provoost, "Death and the Miser," c. 1515–1521*

documented value of 10.6 annual deaths per 1,000 people in Italy.[2] Using this documented death rate and ≈ 60,000,000 for the population of Italy,[3] roughly 2,200 people die in Italy each day from all causes. In recent years, the leading causes of death were ischemic heart diseases, cerebrovascular diseases, and other heart diseases; these accounted for nearly 30% of all deaths.

The number of COVID-19 deaths in the past few weeks in Italy has been:

DAILY AND TOTAL COVID-19 DEATHS IN ITALY[4]

DATE	DAILY DEATHS	TOTAL DEATHS
3/16/20	349	2,158
3/17/20	344	2,503
3/18/20	475	2,978
3/19/20	427	3,404
3/20/20	627	4,023
3/21/20	793	4,825
3/22/20	651	5,476
3/23/20	601	6,077
3/24/20	743	6,820
3/25/20	683	7,503
3/26/20	712	8,125
3/27/20	919	9,134
3/28/20	889	10,023
3/29/20	756	10,779
3/30/20	812	11,591
3/31/20	837	12,428
4/01/20	727	13,155
4/02/20	760	13,915
4/03/20	766	14,681

Table continues on next page

DATE	DAILY DEATHS	TOTAL DEATHS
4/04/20	681	15,362
4/05/20	525	15,887
4/06/20	636	16,523
4/07/20	604	17,127
4/08/20	542	17,669
4/09/20	610	18,279
4/10/20	570	18,849
4/11/20	619	19,469
4/12/20	431	19,899
4/13/20	566	20,465
4/14/20	602	21,067
4/15/20	578	21,645
4/16/20	525	22,170

The peak value of the daily number of deaths in Italy due to COVID-19 (at the end of March) represents an additional ≈ 35% on top of the total number of 2,200 daily deaths.

Let's use the above numbers to get a sense of the impact of the epidemic on hospitals and on the economy. For COVID-19, presumably a substantial fraction of patient deaths occur in an acute hospital setting, such as an intensive care unit (ICU). It also seems reasonable to assume that almost all patients who die from COVID-19 spend time in the ICU before expiring. In the US, roughly 3,000,000 people die each year,[5] with roughly 700,000 in-hospital deaths[6] and approximately 500,000 of these deaths (≈ 17%) in an ICU.[7] Corresponding data for Italy does not appear to be available, but it is a reasonable assumption to use the same fractions for Italy; roughly one-fourth of all deaths occur in a hospital and roughly one-sixth of all deaths occur in the ICU. If we accept these fractions, then roughly 350 Italians die in ICUs daily. Over the nearly one-month period when the epidemic has been at its peak, the number of daily deaths in ICUs due to COVID-19 is at least 2.5 times larger than the steady-state number of 350. The stress on hospital staff who are working in an ICU setting during this peak epidemic time is hard to imagine.

The monetary costs of COVID-19 medical care are significant. Using the current number of deaths in Italy (≈ 22,000) and assuming another factor of two for patients who were gravely ill but survived,[8] there are roughly 44,000 patients who have needed or are currently needing ICU care. Multiple sources indicate that the typical length of stay of a COVID-19 patient in an ICU is around ten days.[9] Multiplying by the daily Italian ICU cost of roughly $1,600 gives an estimate of $700 million over the past month to tend to critically ill patients. It is difficult to obtain reliable numbers for the total cost of COVID-19-related medical care during the peak of the epidemic, but it is not unreasonable to make the guesstimate that the total cost is three times larger than the critical care cost. This gives an estimated direct medical cost of ≈ $2 billion that has been incurred over a period of approximately one month.

The gross domestic product (GDP) of Italy in 2019 was roughly $2 trillion,[11] which converts to roughly $166 billion per month. Thus, these direct medical costs represent a bit more than 1% of Italy's GDP for the past month. To give some perspective, let's translate these numbers to the scale of the US. In 2019, the GDP of the US was roughly $21 trillion.[12] An expenditure of roughly 1% of the US. GDP per month corresponds to approximately $17 billion dollars per month. This is a substantial spending rate; for comparison, the annual budget of the National Science Foundation (NSF) is $7.8 billion.[13] That is, Italian COVID-19 medical costs, when scaled to the size of the US, correspond to two annual NSF budgets being spent in a single month.

Another perspective comes from comparing COVID-19 medical costs with total medical costs. For Italy, health-care spending is roughly $3,000 per person annually or $250 per person monthly.[14] The $2 billion estimate for acute-care COVID-19 spending corresponds to an expenditure of roughly $33 per person over the past month. Thus, the direct costs of acute COVID-19 care represent an additional 13% charge to the medical costs for every person in Italy; the final cost will clearly be much higher.

By using estimates about which we can be fairly certain to make projections into domains of high uncertainty, this Transmission provides a quantitative way to think about the impact of the COVID-19 epidemic on Italy.

REFERENCES

1 World Health Rankings, 2018, "Italy: Life Expectancy," https://www.worldlifeexpectancy.com/italy-life-expectancy

2 Macrotrends, 2019, "Italy Death Rate 1950–2021," https://www.macrotrends.net/countries/ITA/italy/death-rate

3 Worldometer, 2021, "Italy Population," https://www.worldometers.info/world-population/italy-population

4 Worldometer, 2021, "Coronavirus: Italy," https://www.worldometers.info/coronavirus/country/italy

5 Centers for Disease Control and Prevention, 2019, "Deaths and Mortality," https://www.cdc.gov/nchs/fastats/deaths.htm

6 M.J. Hall, S. Levant, and C.J. DeFrances, 2013, "Trends in Inpatient Hospital Deaths: National Hospital Discharge Survey, 2000–2010," *Centers for Disease Control and Prevention*, https://www.cdc.gov/nchs/products/databriefs/db118.htm

7 UCSF Philip R. Lee Institute for Health Policy Studies, "ICU Outcomes," https://healthpolicy.ucsf.edu/icu-outcomes

8 S. Begley, "Coronavirus Model Shows Individual Hospitals What to Expect in the Coming Weeks," *Stat*, March 16, 2020, https://www.statnews.com/2020/03/16/coronavirus-model-shows-hospitals-what-to-expect

9 Centers for Disease Control and Prevention, "Interim Clinical Guidance for Management of Patients with Confirmed Coronavirus Disease (COVID-19)," updated February 16, 2021, https://www.cdc.gov/coronavirus/2019-ncov/hcp/clinical-guidance-management-patients.html

10 S.S. Tan, J. Martin, et al., 2008, "Microcosting Study of ICU Costs in Three European Countries," *Critical Care* 12: P526, https://ccforum.biomedcentral.comm/articles/10.1186/cc6747

11 Trading Economics, "Italy GDP," https://tradingeconomics.com/italy/gdp

12 Trading Economics, "United States GDP," https://tradingeconomics.com/united-states/gdp

13 National Science Foundation, "NSF Budget Requests to Congress and Annual Appropriations," https://www.nsf.gov/about/budget

14 Macrotrends, 2021, "Italy Healthcare Spending 2000–2021, https://www.macrotrends.net/countries/ITA/italy/healthcare-spending

REFLECTION

MODELING THE PANDEMIC: HINDSIGHT IS 2020.

DATE: *17 August 2021*

FROM: *Sidney Redner, Santa Fe Institute*

I based my Transmission essay about the folly of forecasting the course of the COVID-19 pandemic on an idealized but necessarily unrealistic model for the epidemic dynamics: random multiplication. Briefly stated, the model had two elements. In the early stage, there is no mitigation and the number of new infections grows exponentially with time. When the number of infections reaches a threshold where a societal alarm occurs, mitigation is imposed that gradually, but with some fits and starts, reduces the reproduction number R_0 to below 1. Reaching this point defines the end of the pandemic. The outcome of this toy model is that, although the time to extinguish the epidemic does not show much variance for different realizations of the pandemic, the ultimate size of the outbreak could vary by many orders of magnitude! This striking result suggests that forecasting the number of casualties over the course of an epidemic could be next to useless because this number is exquisitely sensitive to the details of the mitigation. The model also shows that early mitigation is crucial to reducing the outbreak size.

My essay was both prescient and quite misguided.

It was prescient because we are now witnessing the impossibility of making any quantitative predictions about the course of the epidemic. While sophisticated epidemiological models for the dynamics of the pandemic have been developed over the past year, essentially all of them have failed to account for the "unknown unknowns." Although we have long been aware that viruses mutate, we don't know how to incorporate their role in the COVID-19 pandemic. Will new mutations transmit more effectively or less effectively? Are they more virulent or less virulent? Which age groups are more affected? How effective will the antibodies from current vaccinations be against new mutations? How similar or different are the symptoms of the new variants? What other unknown unknowns might be lurking in the future?

My essay was also misguided because it did not account for many of the social behaviors that I couldn't have envisioned at the start of the pandemic. One lesson that I've come to learn, based both on this essay and on my experience over the past eighteen months, is to simply not trust any mathematical models that are based on mechanistic differential movement of fractions of people between various health categories (Susceptible, Exposed, Infected, Recovered, Dead, etc.) according to models of the SIR type. Much more careful modeling is needed that also accounts for behavioral features more faithfully. For example, models would need to account for anti-social behaviors, such as deliberately breaking quarantine and worse.

As someone who believes that science can solve many problems, it is disheartening to see the birth of a subculture in which ignorance is celebrated and alleged individual freedom infringes on the rights of the population as a whole.

Another aspect that I never could have imagined is the virulence (pun intended) of the anti-vax movement. While anti-vaccine sentiment has been around ever since the time that Edward Jenner invented a vaccine against smallpox, it's hard for me as a scientist to observe how it mutated into a toxic movement that has attracted such a large following. As someone who believes that science can solve many problems, it is disheartening to see the birth of a subculture in which ignorance is celebrated and alleged individual freedom infringes on the rights of the population as a whole.

TRANSMISSION NO. 019

STATISTICAL TOOLS FOR MAKING PANDEMIC PREDICTIONS

DATE: *20 April 2020*

FROM: *David Wolpert, Santa Fe Institute*

STRATEGIC INSIGHT: *There's no free lunch when it comes to making predictions about the COVID-19 pandemic.*

Human society has a lot of very, very hard decisions to make in the days ahead. These will require us to make a host of predictions: How will the epidemic spread if we do *this* versus *that*? How will the economy be affected if we follow *that* course of action rather than *this* one?

One of the major challenges in making these predictions is that they require us to specify the *dynamic processes* involved. Some of the models one can use to do this are based on equations, which are (typically) then approximated on a computer. Some models are instead based on massive simulations called "agent-based models," which were pioneered, in large part, at the Santa Fe Institute.

Whatever model we use to predict the future, we have to specify the initial condition of the variables in those models. We need to specify the current state of affairs, quantified with numbers ranging from the value of R_0 for the SARS-CoV-2 virus, to how many people have been furloughed rather than fired. In turn, to get those initial condition numbers, we need to convert some "noisy" data that we have gathered into a probability distribution over the initial condition numbers.

To illustrate the great challenge that we face, I'm going to describe why even just finding the distribution over the initial condition numbers for our models—never mind using those models to make the excruciating choices that await us—is fraught, with no right or wrong answer.

OPPOSITE: *Michelangelo Buonarroti, "Night" from the tomb of Giuliano de' Medici, 1523 (detail)*

Converting noisy data into a probability distribution is the subject of the field of statistics. Broadly speaking, there are two branches of statistics, and they provide different guidance for how to form such a probability distribution. To understand the older (and recently resurgent) of the two, a little algebra helps.

Suppose we have two random variables, A and B. The probability that those variables take the values a and b simultaneously is $P(A = a, B = b)$. What is the probability that $A = a$, no matter what value B has? This is called the "marginal distribution" of A, and if you think about it, it is just the sum of $P(A = a, B = b)$ over all possible values b:

$$P(A = a) = \sum_b P(A = a, B = b).$$

Similarly, the marginal distribution for values of B is

$$P(B = b) = \sum_a P(A = a, B = b).$$

What is the probability that B will have the value b, given that A has the value a? If you (again) think about it a bit, this "conditional distribution" is just

$$P(B = b \mid A = a) = \frac{P(A = a, B = b)}{P(A = a)}$$

Just like the quadratic equation holds, just like the sum of any two odd numbers is an even number, just like the product of two odd numbers is an odd number, the equations above mean that

$$P(B = b \mid A = a) = \frac{P(A = a \mid B = b)\, P(B = b)}{P(A = a)}$$

The left-hand side of the equation is the same. And the denominator of the right-hand side is the same. All I have done is substituted the joint distribution formula in the numerator of the right-hand side with an equivalent distribution—the probability of A given B for all values of B, which is just another way of writing the joint distribution.

This simple formula for converting a conditional distribution (A given B) into its "opposite" (B given A) is known as Bayes's theorem. (Berger 2013). To illustrate it, suppose that there is a blood test for

COVID-19 that ideally would say "+" if and only if one has the virus. Suppose it is 90% accurate, in the sense that the table for the conditional distribution P(test result | health status) is:

+	0.9	0.1
-	0.1	0.9
	Sick	Well

This table can be summarized by saying that the false-positive rate is 0.1 and the false-negative rate is 0.1.

Suppose you get tested—and are positive. How scared should you be? According to P (test | health), you might think that there's a 90% chance that you have the virus. But the truth is otherwise, and this is where the Bayes equation comes in.

Suppose that only 1% of the population is infected, so that—everything else being equal—the "prior" probability that you are sick, P (health = sick), is 1%. So, according to Bayes, the associated table for *what you're interested in*, P (health status | test result) is (approximately)

Sick	0.001	0.1
Well	0.999	0.9
	-	+

For example, P (well | +) / P (sick | +) = P (+ | well) x P (well) / P (+ | sick) x P(sick) = 11, so P (well | +) .9. So there's actually only a 10% chance that you're sick—still not good, but certainly less frightening.

Bayes's theorem has been elevated to the status of the scientific deities, as either the source of all truth and light, or of unending evil. Why?

Note that to apply Bayes's theorem we needed to know the prior. And in the case of the COVID-19 pandemic this is one of those estimates that we do not have; we do not know how many people in the population are infected. That is not just true in the example of blood tests; it is also true when (for example) using current data to set the initial condition numbers for our models for predicting the future course of the pandemic. Where do we get *that* prior from? In the case of blood tests it was relatively simple. But in more complicated scenarios—like formulating the probability distribution of the initial condition numbers for our models of how the pandemic will unfold—it can be a very difficult question. Answering this question, and using our answers to calculate what we want to know, is called "Bayesian statistics."

End of story? Not quite. Bayes's theorem embodies one of the deepest truths of life: *garbage in, garbage out.* Adopt a stupid prior, and you get a stupid answer. Not surprisingly then, Bayesian statistics was badly misused in the past, and produced many horrible results. Frustration with these results led people to create the main competitor to Bayesian statistics, called "frequentist statistics."

Bayes's theorem embodies one of the deepest truths of life: *garbage in, garbage out.* Adopt a stupid prior, and you get a stupid answer.

Can we justify frequentist techniques as actually being Bayesian, just for some implicit prior? If so, might frequentist techniques actually be a way to generate implicit priors, without violating the laws of math? Well, no. Even one of the most reliable, most widely used of frequentist statistics tools—the "bootstrap"—can be proven not to agree with a Bayesian analysis for any prior (Wolpert 1996).

This does not mean that we "should" use Bayesian statistics, in any normative sense, when we come up with the numbers to put into our models of the next year. (I myself am a great fan of the frequentist technique of bootstrap.) Even if rather than feeding garbage into Bayes's theorem we feed it ambrosia, we will still be making an assumption. If the virus—if our global economy—doesn't happen to agree with our Bayesian assumption, it does not matter whether our mathematics is correct. There is no free lunch.

REFERENCES

J. O. Berger, 2013, *Statistical Decision Theory and Bayesian Analysis*, New York, NY: Springer Science+Business Media.

D.H. Wolpert, 1996, "The Bootstrap is Inconsistent with Probability Theory," in *Maximum Entropy and Bayesian Methods*, ed. K. Hanson and R. Silver, Dordrecht, Netherlands: Kluwer Academic Press.

REFLECTION

LOOKING THROUGH SCIENCE-TINTED GLASSES

DATE: *8 August 2021*

FROM: *David Wolpert, Santa Fe Institute*

At the current stage of development of the scientific enterprise, it is divided into a set of many different scientific fields. Each scientific field comes equipped with several different pairs of "glasses" with which to examine the natural world. If you wear such a pair of glasses and look around you, you will see a limited set of features in the landscape highlighted. Each scientific pair of glasses highlights a different set of features in the natural world.

As an example, if a scientist is looking at the world while wearing "Newton's Laws" glasses, then every system around them is distorted to highlight the forces on that system, how fast that system is moving, and its rate of acceleration. In particular, if they look at a running horse, they see the forces the horse exerts on the ground (equal and opposite to the ones the ground exerts on the horse), the varying position, momentum, and acceleration of the horse, etc. As another example, if a scientist is looking at the world while wearing "Darwinian Selection" glasses, then every system around them is distorted to highlight the ancestry of that system, that is, previous systems that can be seen as the progenitors of the system. These glasses will also highlight how those previous instances of the system they're looking at may have differed from the current instance, and what may have caused them to evolve into the current system. In particular, if someone wearing such glasses looks at a running horse, they see the phylogenetic tree behind the horse, the way that horses have evolved over tens of millions of years, etc.

Finally, it's important to note that there are "coarse-grained" pairs of glasses that can be viewed as enveloping many of the other glasses. These coarse-grained glasses embody an entire scientific field. For example, there is a pair of glasses called "High-Energy Physics Theory" that embodies the scientific field of the same name. If a scientist wears those glasses, then they see all the features highlighted by the Newton's

Laws glasses—along with features highlighted by the "Relativity" glasses, features highlighted by the "Quantum Mechanics" glasses, and so on.

At the current stage of development of the scientific enterprise, there's a lot to be gained by having individuals who walk, breathe, eat, and sleep all while wearing one specific field's pair of glasses, and never wearing any other field's pair of glasses. These people eventually become what are sometimes called "domain experts" in that scientific field. This means they are completely comfortable navigating the world while wearing that field's pair of glasses. Such experts are able to see very subtle features in some of the objects in the natural world, and report on those subtle features back to the rest of us. In this way, we all gain from the dedication these people make of their entire intellectual lives to one particular, restricted pair of glasses.

Each scientific field comes equipped with several different pairs of "glasses" with which to examine the natural world. If you wear such a pair of glasses and look around you, you will see a limited set of features in the landscape highlighted. Each scientific pair of glasses highlights a different set of features in the natural world.

However, there is also much to be gained by flipping back and forth among many pairs of glasses, viewing the world in many ways. Such flipping among pairs of scientific glasses is what "multidisciplinary science" is supposed to be all about. Indeed, in what is essentially an evolutionary process, new "offspring" pairs of scientific glasses are often produced by flipping back and forth between some "parent" pairs of glasses very quickly. (As a completely gratuitous aside, incessant flipping back and forth among many different scientific glasses is the central feature of the Santa Fe Institute.)

Everything above was written while wearing the glasses called "Sociology/History of Science." Now, let me take those glasses off, and start flipping among many different ones.

In my Transmission, I first considered the COVID-19 virus while wearing a recently invented pair of scientific glasses called "Thermodynamics of Computation." If you wear these glasses, all dynamic processes you see around you have two central features. The first feature is the process' thermodynamic behavior, that is, how quickly the process uses energy as it implements its dynamics. The second feature is what precise computation is implemented by that dynamics. Wearing Thermodynamics of Computation glasses, all the dynamic processes you see around you involve an interplay between their rate of energy usage and the computation they are implementing. Indeed, the thermodynamics of computation teaches us that those two quantities are intimately related—in order for a dynamical system to implement more computation, it must pay for it, with a greater rate of energy usage. As an example, looking at a running horse while wearing these glasses, one notices the computation the horse is continually doing in both its brain and its limbs, of how to run, along with the energetic cost of that running.

After this preamble involving the Thermodynamics of Computation glasses, I replaced them with a different pair of scientific glasses, called the "Extended Phenotype" glasses. When you wear the Extended Phenotype glasses, the line between living organisms and the world outside of them gets blurry. Any organism will affect the dynamics of the part of the universe that contains its cells, that is, that contains small bags of protoplasm enveloping copies of its DNA. But any organism will also affect the part of the universe that does not contain its cells, the part of the universe typically called the "environment" of the organism. The "Conventional Phenotype" glasses are concerned with how the organism affects the more restricted part of the universe, containing its cells. The Extended Phenotype glasses enlarge that to consider how the organism affects the entire universe, including the environment of the organism.

One of the most prominent features one notices when wearing the Extended Phenotype glasses is how an organism affects its environment in ways that ultimately benefit that organism's genome, either directly, in the present, or indirectly, in the future. A standard example is how a beaver (organism) building a beaver dam (environment affected by the organism) affects the beaver's genome (helping the beaver survive). Another example is how the leader of a social group (organism) accumulates a large harem (environment affected by the organism) and so produces many progeny (the future genome of the organism).

The Extended Phenotype glasses can also highlight how an entire population, or species, affects its environment. (In this case, one considers the "aggregate" genome defining the population as a whole rather than the precise genome defining a specific organism in that population.)

When one sees this feature, one is often led to put the Darwinian Selection glasses on over the Extended Phenotype glasses, to consider the possible adaptive fitness value of the extended phenotype. This is particularly compelling if one is seeing this feature in an entire population rather than just a single organism in that population. (And yes, wearing one pair of glasses over another pair of glasses can be awkward, sometimes resulting in both pairs of glasses falling off—nobody said multidisciplinarity would be easy!)

We humans are part of the environment of the virus, and the virus affects us. In fact, it induces our immune systems to perform the computation (and bear the associated energetic cost) of replicating the virus. It then gets human society as a whole to perform the subsequent computation . . . of figuring how to spread the virus to new hosts.

So, what do you see when you wear the Extended Phenotype glasses on top of the Thermodynamics of Computation glasses, rather than flipping between them? You see populations that are constrained in how much computation they can do by themselves, due to the associated energetic costs, and that affect the environment by changing the computation that the environment does, with the associated energetic costs borne by the environment. One of the most striking examples of this kind of computational extended phenotype is exhibited by the COVID-19 virus (with the epigenome of the virus playing the role of an aggregate genome). We humans are part of the environment of the virus, and the virus affects us. In fact, it induces our immune systems to perform the computation (and bear the associated energetic cost) of replicating the virus. It then gets human society as a whole to perform the subsequent computation (and again bear the associated energetic cost) of figuring how to spread the virus to new hosts. This is obviously beneficial to the (epi)genome of the virus. So, the whole story still holds together when we put the Darwinian Selection glasses on as well, on top of the Extended Phenotype glasses and the Thermodynamics of Computation glasses. (The practitioners of multidisciplinarity look very odd to other people, wearing so many glasses at once, but we practitioners don't mind.)

This was the theme of my Transmission. However, I realized after writing that essay that there was a bit of a mystery concerning this example of COVID-19 offloading computations and associated energetic costs on its environment. Almost all real-world computational systems are modular and hierarchical, whether designed by humans (e.g., digital circuits) or constructed via natural selection (e.g., brains, genetic networks). There are many reasons for this. Both computers designed by humans and those constructed via natural selection benefit from the fact that modularity helps minimize the costs (both energetic costs and material costs) of communication among the subsystems of the computer. In addition, both types of computers benefit from the "evolvability" of hierarchical, modular design, as elaborated in arguments stretching back to Herb Simon.[1] Other benefits include robustness against noisy components/component failure (particularly important for human-designed computers). In addition, one can argue that hierarchy is actually almost inevitable in computers, since it often arises by

---

1 H. Simon, 1962, "The Architecture of Complexity," *Proceedings of the American Philosophical Society* 106(6): 467-482.

frozen accidents, especially in natural selection (so-called "accretional software construction").

At first I thought that the COVID-19 computation, offloaded onto us convenient humans, is an exception to this rule, that there is no sense in which the computation of propagating the virus' epigenome is being implemented in a hierarchical, modular system. Once one thinks about it though, by offloading the computation of propagating itself onto individual humans, the virus exploits the finer-grained modularity and hierarchy of the components of that human's immune system. But the lowest level of the hierarchy that forms an individual human's immune system is the individual cells in that immune system. Each of those cells comprises a set of interacting organelles, performing a joint computation. So, at a yet finer-grained level, the computation of a human's immune system that is doing the bidding of COVID-19 is constructed on top of the hierarchical, modular computation among the organelles within each cell in the human immune system. We can also go in the other direction, up to more coarse-grained levels than individual humans and their immune systems. After all, the virus also exploits the modularity and hierarchy of human social systems—the hierarchies built on top of the individual humans—to help it spread even further.

. . . by offloading the computation of propagating itself onto individual humans, the virus exploits the finer-grained modularity and hierarchy of the components of that human's immune system.

That's the COVID-19 extended phenotype computer: an aggregate hierarchical modular computer. That computer has organelles inside of individual cells at its lowest, most fine-grained level, at its smallest physical scale (the same physical scale as an individual COVID-19 virus). Those cell-computers are then aggregated into the components of the immune systems of individual humans at a higher level of hierarchy, to form a higher-level computational system. And those individual human computers are in turn aggregated into the components of the human socio-political hierarchy at even higher, more coarse-grained levels. Astonishing!

Part of my current research concerns precisely the thermodynamics of computation in hierarchical, modular systems. By contemplating the amazing extended phenotype of COVID-19, I have gained a completely new perspective of this issue. This is one of the ways that the pandemic has affected my research.

THE TRANSCRIPT OF *COMPLEXITY PODCAST* TRANSMISSION SERIES EP. 4, DISCUSSING THE FOLLOWING TRANSMISSIONS

[Batch 4, released 20 April 2020]

T-015: Federalism in the Time of Pandemic
Anthony Eagan

T-016: The Future of Education
Carrie Cowan

T-017: Privacy Concerns that Arise with the Pandemic
Stephanie Forrest

T-018: Quantitative Ways to Consider the Economic Impact of COVID-19
Sidney Redner

T-019: Statistical Tools for Making Pandemic Predictions
David Wolpert

This podcast transcript has been abridged for length and clarity. Find the full podcast at https://complexity.simplecast.com/episodes/30

RETHINKING OUR ASSUMPTIONS DURING THE COVID-19 CRISIS

DATE: *27 April 2020* **HOSTED BY:** *Michael Garfield, featuring David Krakauer*

MICHAEL GARFIELD: Well, David, we are in the mix for week four of the Transmission series and this week is kind of a grab bag. But with the diversity of entries for this one, I think we're starting to see something that taps into themes from all three of the last weeks' episodes.

DAVID KRAKAUER: I think there is a thread in this labyrinth, having to do with authority and power and its best use and control. The logical place to begin, as we have in the past, is at the micro and work to the macro, and I think David Wolpert's article[1] on Bayesian statistics is where we should start.

M. GARFIELD: David does a really good job with this piece in introducing the outline of Bayes's theorem. Not assuming our audience knows what this is, why don't you lay this out for us?

D. KRAKAUER: David starts by making the point that our best decisions are based on appropriate data and typically estimates of datasets, which are incomplete. We've talked a lot in this series about R_0—the reproductive rate during an epidemic—and these are things that we have to estimate from very noisy data. The point that David makes, which we all know, of course, is that our prognostications and projections and predictions are only as good as those estimates. Those are changing dynamically, and so we're constantly having to go back to the data to make better estimates. He introduces us to what is really a very foundational idea—it's about 200 years old, and it was derived by an English minister, Thomas Bayes—and it's the appropriate way to think about *conditional* probabilities.

Most of us growing up learned about probabilities in terms of fractions. If 50 out of 100 people are female, then the probability of being female in a population is a half. And that's how most of us think about probabilities or odds. But there is a more complicated idea, which is not

---

1 See T-019 on page 243

a straight probability, but what's called a conditional probability. For example, what was the probability of being British if your favorite food is a scotch egg? Well, I think it would probably be quite high, right? Because no one in their right mind who wasn't would ever go near the thing. And this is not a simple fraction. To calculate a conditional probability—what is the probability of *A* given *B*—you need to use an equation and that equation is known as Bayes's rule. Why does that matter? David gives a very nice example: because if you use fractional intuition for probabilities, you make terrible mistakes.

He gives the example of a test. Let's say it's a test to determine whether or not you are positive you possess the COVID-19 virus. Let's say that it's been established that the test is 90% accurate in its control group. So, we have a conditional probability here. What is the probability that you have the disease if you test positive? Now, most people, thinking "fractionally" if you like, would say, "Well look, you've told me that the test is 90% accurate. If I'm positive, there's a probability of 0.9 or 90% chance that I have the disease." And that would be absolutely wrong because Bayes's rule—the equation—says it's not enough to know the success of the test. You have to know the incidence of the disease in the population, what is sometimes called the prior. David says, let's imagine that we know that only about 1% of the population is infected.

So, now we have this test which we've established in our control group is 90% accurate. You plug it into the equation and it tells you what the odds are of being sick if you test positive. And it's about 0.1, or 10% chance. So not 90%, but 10%. That is because the equation says that the incidence in the population gives you a sense of the true population-wide false-positive rate. It's really important, because people are going to be taking tests. Those tests will be accurate. But if that prior, if that background statistic, is inaccurate—in other words, how many people are truly infected in the population?—then that reliability of the test could really titrate between almost 0 and 100%.

The second point—if you like, conundrum—is, how do you arrive at an accurate measure of what that prior is? I mean, what is the rate of infection in the population at large? Well, to know that you have to use Bayes's theorem, again. You have to test. But we've already established that to test you have to have a prior, and this gives rise to this whole field of what's called Bayesian updating or Bayesian particle filters, which is that you have to do it over and over again. So, you plug in a prior, which is approximate, you get what's called the posterior, and then you plug that back in as the prior, and you do this over and over again and you hope you converge on the accurate statistics. The big philosophical insight here is that your estimates of your chances of being infected are only as good as this data that you don't really possess. Even if the mathematics is absolutely watertight and correct, garbage in, garbage out, and it's extremely difficult to make those estimates.

M. GARFIELD: We're seeing this in practice now with the Stanford study[2] that dramatically recalculated the death rate that just came out this week. This links in two distinct ways with Mirta Galesic and Henrik Olsson's piece[3] and their work on social science and how all of us suffer local bias based on trying to infer global truths from the local sampling of our intimate social networks. Right? Because that's exactly what we're doing: we're assuming a sort of a marginal probability rather than a conditional probability. The condition being that we're only sampling, like, the five people who all want to see the same movie, or so on. That's a piece of it—having a little bit of forgiveness for the true and terrible complexity of making an accurate estimate on this issue.

D. KRAKAUER: Your mathematical formulae can be absolutely right—in other words, the logic can be correct—but if you feed that formula the wrong data, it will spit out the wrong answer. Of course, it's obvious, but it's not always obvious in debate because people will say, "your theory is wrong" or "the formula is wrong." No, it's not. What is wrong is that we don't have the empirical grounds with which to conclude, and I think that's something that's worth bearing in mind.

M. GARFIELD: It's like a point that Caroline Buckee brought up in a recent episode[4] when she was talking about the difference between weather forecasting and epidemiological modeling. She said, if you

---

2 E. Bendavid and J. Bhattacharya, "Is the Coronavirus as Deadly as They Say?", *The Wall Street Journal*, March 24, 2020, https://www.wsj.com/articles/is-the-coronavirus-as-deadly-as-they-say-11585088464

3 See T-013 on page 179

4 Complexity Podcast episode 28 aired on April 16, 2020. https://complexity.simplecast.com/episodes/28

predict that it's going to rain tomorrow and everyone goes out with an umbrella, it doesn't change the fact that it's going to rain. But if you say there's going to be a pandemic and everyone isolates and then there isn't a pandemic, people have a habit of blaming the model or blaming the scientist. I think that this is sort of a tricky condition. This is the other point with Mirta and Henrik's piece—that the opportunity to reestablish trust in science and scientific communication right now is contingent on people understanding this. I saw a kind of tragic cartoon that was talking about how, if we manage to do this well, if we manage to contain this and stave off the worst, if we have flattened the curve, then people are going to complain that the economic precautions that we have instituted weren't necessary. And so we face the possibility of a backlash against the people who were providing the very models that were used to protect us.

D. KRAKAUER: That's always been the problem with successful medicine—that it's a victim of its own success. If a vaccination policy is very successful, then you see very few infected and that leads some people to claim that there's no longer any risk. That's almost an insurmountable problem in the limits of human judgment. And the best you can do is show people, in some sense, in simulation, what the world would be like minus the vaccine. But that's a hard lesson to learn.

M. GARFIELD: There's another quantitative piece, Sid Redner's piece.[5] He's looking at this in a completely different way. I took this to be more back-of-the-envelope math. We were talking about, in the first episode, the importance of not using models that are more complex than the data that we have. This is a really good instance where you can ballpark an estimate of the economic impact of COVID-19. He looks at Italy specifically.

D. KRAKAUER: This is very nice. It's very simple arithmetic, starting with this quote from 1938 where John Maynard Keynes makes the point that what a model is trying to do is segregate what you know—what he calls *semipermanent*—from what you don't. It's a very important point about the philosophy of models. In the first episodes we talked about overfitting noise, as you just pointed out, and the virtues of simplicity in the face of uncertainty. This is another point, which is that a

5 See T-018 on page 235

good model minimizes novelty and restricts it to the hypothesis at hand. Meaning, if I write down a mathematical model for an epidemic, you don't want to check my math; you don't want to determine whether calculus is right or wrong. You want to determine whether my premise is right or wrong.

In any effort to theorize, you're always trying to take a framework where almost everything is known and can be verified in order to reduce the ambiguity around that one insight you're making, which is hard to verify. This is a very important philosophical point about model building and not everyone gets it, and so a lot of people try and do everything that's new. Of course, in the end, you're not sure what you're really evaluating: the mathematics, the logic, or the hypothesis. So, what Sid does is say, I'm going to take the simplest thing in the world—arithmetic, so there's no question yet—and I'm going to take data, which we sort of know to be true in the case of Italy. So here's the deal. We'll say, what do we know about Italy prior to COVID-19?

In any effort to theorize, you're always trying to take a framework where almost everything is known and can be verified in order to reduce the ambiguity around that one insight you're making, which is hard to verify.

Well, we know that life expectancy was on the order of 80 years, and we know that in a population of about 60 million people, just over 2,000 people die from a variety of causes. That's the Keynesian background, right? That's the thing we know. If you look at COVID-19 in late March, we can say right away that it's responsible for a 35% increase per day. This gets to that rate remark we were making in relation to Melanie Mitchell's analogies piece.[6] That's very significant. Now you can try and say what's the cost of that, and what do we know? Well, we know how much it costs to be in an ICU, but we don't know the Italian

---

6 See T-009 on page 129

ICU data. We know the American data. In the US, about 17% of daily fatalities end up in ICUs. If we go back to the Italian population, given the numbers we have in hand, we get about 350 expected ICU per day.

But what do we see under COVID? About 800 per day. That's 2.5 times baseline. So, now we can ask, okay, what's the cost of ICU? It's about $1.6 thousand per day. Under those assumptions, we get about 1% of Italy's GDP being accounted for by the medical costs associated with treating COVID-infected individuals. 1% of GDP. He makes the comparison to something like the National Science Foundation where he says, if you were to think about that in American terms, that 1% of GDP per month, by the way, would pay for the total annual budget of the NSF twice every month. These are very big numbers and it's simple linear arithmetic that he's using. And here's why I think it's important, and it's something he doesn't say explicitly, but I think it's a natural implication of his logic. If you look at the airline industry, the airline industry has lost on the order of tens of billions of dollars. The same goes for, say, the oil industry. The national economy is facing a loss of trillions of dollars. These are very large numbers. How much would a virus surveillance operation have cost preemptively? Probably millions. So, there's this interesting fact that industry—petrochemical, oil, airline, travel, service—if they had invested some sum of money in surveillance programs, they could have saved billions of dollars. I think that's the radical implication of these kinds of numbers: some preemptive expenditure with these kinds of risks in mind could have saved national economies trillions of dollars.

M. GARFIELD: Bending back to our last episode, this is exactly what we were talking about with bet hedging. That, in times of relative stability, this kind of thing should be obvious but is not because the economic competition, a blind evolutionary process, doesn't tend to regard this as a valuable expenditure because it's not thinking ahead on the kind of timescale that's going to integrate these costs.

D. KRAKAUER: Yes, that's absolutely true. But we've been hearing in the media very divergent opinions on this. Most of us are very aware that in the last decade, we've had comparable viral outbreaks, whether it's Ebola or SARS or MERS, and we were very fortunate that those viruses had lower transmissibility. If they had had higher transmissibility, it could have been worse than this situation. In just a decade we've seen multiple instances which should lead us to understand that this is not a black swan event. People have been citing Bill Gates, who pointed out, as many had, that something like this was inevitable. So, in the same way that actuaries who are consulting with an insurance company calculate plausible risks, this is a totally plausible risk. What was unanticipated was the shutting down of aggregate demand. But now, with this experience in hand, I think there's no excuse for industry and government not to take preventative measures and worry a little less about expenditures in the millions that might offset losses in the trillions.

The national economy is facing a loss of trillions of dollars. These are very large numbers. How much would a virus surveillance operation have cost preemptively? Probably millions.

M. GARFIELD: This is a possible link between—when we were talking about natural selection and epigenetics and the two sort of internal and external fitness landscapes—this is the kind of thing that we see in fine-graining when we look at the epigenetic impact of, like, a famine on up to something like fourteen generations of rat offspring. They find that environmental information is encoded so that they are remembering the condition of food scarcity. This is an instance where, luckily, our society will probably be prepared for another pandemic for as long as the various memory systems that the various ways that our civilization encodes the trauma of an event like this will actually last.

D. KRAKAUER: This is something that John Geanakoplos has been saying[7] in relation to the rapidity with which the government responded with stimulus to COVID-19 and a large part of that comes from the very recent memory of the 2008–2009 financial crisis. Most people in

7 https://www.santafe.edu/events/leverage-cycle-and-covid-economy

positions of authority experienced that event. That was only a decade ago. If that event had been several decades ago, it might be that there would have been enormous disagreement as to whether or not stimulus was a good idea, which would have been of course disastrous. So, I think you're absolutely right and we have to find mechanisms to allow for the persistence of memory. And that gets to exactly that point you were making earlier about vaccines, right? When a vaccine is too successful, people don't want to take them because they don't see the disease. We have to create in culture some means of maintaining an awareness of relatively infrequent events.

M. GARFIELD: This links us into Stephanie Forrest's piece,[8] because something that I've been seeing a lot of people talk about lately with respect to disease surveillance and the necessity of contact tracing and so on is the concern that's still very much alive in most of us about the way that privacy and surveillance were handled after 9/11. I think if 9/11 had not happened, then we would probably, as a whole in the United States, be much more willing to indulge in what we obviously now consider sort of draconian surveillance measures. But we're aware of the way that these opportunities are very difficult to roll back. And so, she's looking at how do we balance the need for surveillance with the need for privacy. And she's got some interesting insights from cybersecurity and immunology. You want to talk about this?

D. KRAKAUER: I think that's really interesting. But you know, that was novel—we didn't realize the extent to which that surveillance was possible. We now do and in consequence have been immunized, if you like, against that exploitation. What Stephanie is talking about is issues related to anonymity and privacy when we still can make use of these technologies to good effect. Stephanie has spent many years looking at parallels between biological evolution and immunology, and technological evolution and looking at computer viruses in particular. One of the things that was discovered several decades ago about the immune system, which might surprise everyone, is the following: how does the immune system mount a successful response to a pathogen it's never seen before? It does this by generating random variation in its

8 See T-017 on page 227

antibodies, in its immune response, in its T-cell receptors in particular, and this is enabled by a process called somatic hypermutation.

Now if it generates random responses, what stops it from attacking itself? That's called immunopathology and of course it does attack itself. It's developed a method of negative selection or clonal deletion whereby any element of the immune system that attacks self is tagged and deleted, and what you're left with is the negative complement of the self, which is everything that might infect you, right? In other words, you generate these random responses. You try and cover as much of the space as possible. You take out anything that attacks you, and in that space what is left is hopefully an adaptive response to a novel infection. Why is that important? Well, it's important partly because you're anonymizing yourself. You're not signaling what you are because the self is not present in that set, only the infection. This turns out to be really important for dealing with computer viruses, because you don't know in advance what bit of code the computer virus is going to use or which part of your operating system it's going to attack.

This is a very interesting idea that in some sense anonymizes the self, removes the self from the picture in order to remove non-self.

So, exactly the same principle can be applied. You generate a whole random set of possible interventions and you eliminate any that attack your own code, leaving only the response which attacks the foreign code. And so this is a very interesting idea that in some sense anonymizes the self, removes the self from the picture in order to remove non-self. That really is useful for dealing with things like malware or spam. But what about this issue of contact tracing? That requires a very different idea, and this is based on this notion of cryptographic set intersections, and it works like this. Let's imagine you and me, Michael, we get together in a brighter world where we can sit together at a bar or over a coffee, and we want to share our address books. We want to know which friends we have in common. We're looking for what's called the intersection of the set. What we want to do is come up with a method whereby we don't share everything with each other; we only share the friends we have in common. Everything else is encrypted. The only thing that's not encrypted is the intersection, what we share.

That is exactly the idea that Stephanie is calling for in relation to contact tracing. I want to know all of the people with whom an infected person has come into contact, and that means I'm going to have to take a huge survey of the entire population. But I'm only interested in this intersection, the intersection of that person with the population, not everybody. That's critical because, as Apple and Google and Facebook collect data, they're going to have to collect it all to calculate the intersection. But what Stephanie is saying is, everything that isn't the intersection gets encrypted, so that can't be used in the future for reasons that we never approved of.

The first experience most of us have in our lives of both authority and very strong anti-authoritarian feelings is education.

M. GARFIELD: I feel like we've moved into the domain where we're talking about the balance between a sort of central overview of the situation and a distributed overview. This is a dovetail into the last two pieces from Carrie Cowan and Anthony Eagan on the way that our system that we've inherited as we move into this crisis has structured governance, and has structured education, as a way of addressing the conditions of a world that, for all intents and purposes, in many ways, seems no longer to exist.

D. KRAKAUER: Right. So now we're moving back into this question of authority versus citizen. And the first experience most of us have in our lives of both authority and very strong anti-authoritarian feelings is education. Right? Speaking for myself. Carrie points out[9] that we're now experiencing one of the greatest large-scale experiments in education in

---

9 See T-016 on page 219

the history of education. That is, in particular, the experiment around digital distributed education and the use of digital classrooms and massively open online courses, or MOOCs, because we have to, because we're isolated. These are fairly new technologies at scale. The MOOC really started getting popular in the 2010s and it led to the whole notion of the flipped classroom—being this classroom where essentially you sort of do your prep in advance, you learn the facts and you use the social context to ask questions, as opposed to just listening to someone lecture, which you can do online anyway. I think the first point Carrie makes is this moment in our history is accelerating the adoption of these platforms that we had already grown weary of in the last ten years, but now are being forced to adopt because we have no choice. She makes this nice analogy to the flu of 1918 where that was a kind of pivotal moment, not only in society and demographics but in education. Prior to 1918 only about 10% of the population had graduated high school, but by the 1920s it was more than double that. And that education was then the aspiration that many people in society had.

She makes another point, which is that the economic burden of education is incredibly high. From 2008 to 2018, tuition increased by just under 40% and the net cost of engaging in education by over 20%. These are huge economic burdens and they're greater than the burden that many of us face from housing, for example. But there's also another burden, which is not economic but intellectual, which is that education has become so specialized and narrow. If the circumstance of this pandemic is telling us anything, it's that we now have to understand the entanglements in these complex systems and the educational system simply has to catch up with the reality of our circumstances. We need to emphasize connections, commonalities, what Carrie calls the mutualisms across the disciplines. Because if we don't, we're not going to really understand how to deal with the current circumstance.

Here's something really interesting that she says: when in mid-March we all realized that our lives had been upturned and that we were going to be living these more solitary existences, there was a huge increase in the demand for MOOC platforms like Coursera and edX—that is, online educational resources. But their increased popularity paled in comparison to another trend, and that was the popularity of

Minecraft. So, the game *Minecraft*, which is an online construction game, is extraordinarily popular. It's the highest-selling video game of all time. About 100 million people play it each month. And the point that she makes is that *Minecraft* is this deeply experiential, curiosity-driven, super-versatile environment to apply your mind to a whole multitude of projects that you and others have invented. It's in some sense the ideal classroom.

This is a world that I've been very involved in for quite a while. My colleagues in Madison when I was there, Kurt Squire and Constance Steinkuehler, investigated this world at length.[10] What do people learn when they play *World of Warcraft* that they don't learn in a classroom? How are they different? How can we transfer the insights from those games to the workplace? As you know, at the InterPlanetary Festival we've been talking to people like Tarn Adams who wrote *Dwarf Fortress*, and Jonathan Blow who wrote *The Witness*, about precisely those kinds of ideas.[11] I very much agree with Carrie's question, which is, if there are these very powerful, community-based, very free, playful environments in which people can apply their minds, how can they help us rethink education, in particular in this distributed sense? So, instead of recreating a classroom, which is what a MOOC basically does, what if you were to completely rethink it? I agree that there's a huge amount to be learned from the developers of those games.

M. GARFIELD: This is linked to the discussion we had about the way that a virus mutates so rapidly that it sort of occupies a cloud rather than a point. We're seeing society move into what you called last week "a microbial mode," where the educational process, the curricula, and the areas of specialization are much more bespoke and individually tailored to the local. Carrie mentions that the educational system we have now by and large emerged out of this early twentieth-century industry context, which was very standardized. It was in many ways like the adaptation of modern

---

10 See, for example, C. Martin, S. Chu, et al., 2011, "Ding! World of Warcraft Well Played, Well Researched," in *Well Played 3.0: Video Games, Value and Meaning*, ed. D. Davidson, Pennsylvania, US: Carnegie Mellon University ETC Press, 226-45.

11 For more on these conversations, see D.C. Krakauer and C.L. McShea's edited volume *InterPlanetary Transmissions: Stardust* (SFI Press, 2020) and video recordings at https://www.youtube.com/watch?v=_VeQy23qAgU.

time zones in order to coordinate train travel. We were looking at trying to make something consistent universally. In the last several decades, the surface area of society and of the information that's relevant has grown so much . . . You know, Adam Curtis's documentary, The *Century of the Self*, where he talks about lifestyle marketing and companies like Nike making it so you can design your own shoes on their website rather than just getting whatever they happen to sell you in the store.

This is also about an adaptation to the small and fast. To link this to Miguel Fuentes talking about where we see the signal of an impending social crisis in the breakdown of a consensus narrative and polarization.[12] But it's not just polarization—it seems like people are starting to split up and looking for local solutions to the problem as it emerges locally and regionally.

. . . *Minecraft* is this deeply experiential, curiosity-driven, super-versatile environment to apply your mind to a whole multitude of projects that you and others have invented. It's in some sense the ideal classroom.

D. KRAKAUER: I understand where you're going. I feel as if when I look at something like *Minecraft* and insights from gaming for education, I have a sort of list of what I think it's doing. One of them is freedom, which is moving away from the authoritarian top-down approach to education where there are received ideas that have to be imparted at very high fidelity and that's sort of the drill. That works for a very tiny fraction of people but not for the majority.

And that takes us to the second point: diversity. We have a far better understanding now of the diversity of peoples—in particular cognitive diversity—than we did have, let's say, twenty years ago, and we know that not everyone learns the same way, and we have to provide mechanisms

12 See T-006 on page 101

for, as you say, different timescales of learning and completely different ways of learning.

The third is collaboration. When I went to school, collaboration was not even a little bit on the table of the classroom. It was outside when we played, but not inside. It was you at your desk listening to someone bore you to death. Things have gotten a lot better, but the idea that the classroom itself could be deeply collaborative—which is what happens in primary school, much less in university, but it does happen—is something that we could massively scale up.

The final one is construction that you are being asked to contribute your world—your vision of the world—for others to scrutinize and inhabit. That's exactly what you do with a theory. If I told you my theory, Michael, of thermodynamic evolution, you'd say, "That's an interesting place to spend some time," or not, and you would critique it and give me feedback.

And *Minecraft* does all of this. It has freedom, it respects diversity, it enables collaboration, and it allows you to construct a world. And I think in a way that's what the best science is. We have been doing that historically, but I don't think the educational system, for various reasons, many of them I think understandable, has really had the tools to introduce those principles in a very effective way. The game environment is one way in which to think about that in a very principled way.

M. GARFIELD: I love that Carrie says, "Real-world problems—and the jobs that exist to solve them—demand more than a knowledge of political science or physics or art history. They require a new style of thought that emphasizes connection and commonalities—mutualism across disciplines.... The world is not getting simpler." I take this as a sort of a hopeful note, that what we're seeing here is that infrastructure such as *Minecraft* is going to enable a kind of cognitive complex organism, like a multicellular educational process.

D. KRAKAUER: I do feel very strongly that great classrooms, great research institutes and universities, have been doing this. In other words, these principles that we just discussed are not new. The problem has been that they've been restricted to a rather privileged few. They haven't been easy to scale up so that everyone can have access to that kind of free environment. And I think *Minecraft* in the domain of entertainment, very creative entertainment, has given us a keyhole to look through and see the possibility of a world where everyone, or many people, can have access to these principles. But there's a huge amount of work to be done. It's not as if we're just going to do science with *Minecraft*. Far from it. But we do need to learn from the joyousness and the freedom and the communitarian impulse that you see in those games to transform our educational system toward, as Carrie says, a more complex perspective.

M. GARFIELD: This takes us real neatly into Anthony Eagan's piece[13] because it seems that what we're talking about here—the modularity of the educational process through universal standards—brings us to the balance of power that he's talking about with federalism, and the way that we need to both lean on, but also question, the assumed priors of the United States Constitution and its articulation of this balance between bottom-up and top-down governance.

D. KRAKAUER: I think this piece touches on many of the previous pieces. The way I think about it, because I've worked on constitutions, is the rule systems that in some sense run societies—I sometimes call it law OS, the legal operating system. We have been watching the tensions arising recently between the president and state governors in relation to opening up states to allow for economic activities to take place, and how we trade off minimizing the risks of infection and maximizing productivity. And this tension was the tension at the very founding of this country, of course, in relation to national sovereignty and the monarchy, in Europe and Britain in particular.

Tony introduces this dichotomy that most people will be familiar with in this country between, on the one hand, the Federalists, who advocated for a rather top-down approach to robust governance, and the Anti-Federalists, who advocated for a more distributed local state-based sovereignty. Many of the principles that found their way into the Constitution in 1789 were articulated and explored in the Federalist Papers of 1788 that were written by Hamilton and Madison and Jay under their pseudonym Publius. They really tried to explain what is this thing, how do you compromise between localized versus highly

13 See T-015 on page 211

centralized needs? Tony cites the 39th Letter, where it's written that the proposed government cannot be deemed a national one since its jurisdiction extends to only certain enumerated objects. So, it has limited power and allows states to make certain residuary decisions of their own.

The deep point here is that that's all very well under normal circumstances. But what happens in crisis? There is a temporal dimension to the Constitution and the rules that it encodes in relation to individual liberties. Under crisis, people are willing to forgo freedoms—in order to benefit, for example, from centralized information—but in periods of peace and greater security, they want more autonomy. And there really isn't very good thinking about a dynamical constitution that would execute different rule systems according to the circumstances of the society.

. . . the shorter the constitution, in some sense, the freer you are in the interpretation of its principles, and the longer it is, the more it specifies how you should act in any given context.

I want to bring this to complexity science because there's a great deal that can be learned from the way that, for example, organisms are regulated. You could think of the genome as being a sort of constitution of a body, with the various cell types having greater or lesser freedoms in terms of pursuing their own metabolic or physiological functions. In 2010, actually, Mark Gerstein studied the structure of regulatory networks in five different species: bacteria, worms, flies, mammals, mice, and so on.[14] He found two very different modes of phenotypic governance, what he referred to as a more democratic structure, which is in some sense this sort of Anti-Federalist position where everything connected more or less to everything else very freely with very little evidence

14 N. Bhardwaj, K-K. Yan, and M.B Gerstein, 2010, "Analysis of Diverse Regulatory Networks in a Hierarchical Context Shows Consistent Tendencies for Collaboration in the Middle Levels," *Proceedings of the National Academy of Sciences of the United States of America* 107 (15): 6841–46, doi: 10.1073/pnas.0910867107

of hierarchy, and an autocratic structure where it was a much more sparse connection and much more top down.

The most important observation that the Gerstein lab made was that the more complex a phenotype—that is, the more parts it had, the more cell types, the larger it was—the more democratic it was in its regulatory governance.

This is really important, and it gets to another point, and I'll try and wrap them up together. In 2017 I wrote a paper with Dan Rockmore and Tom Ginsburg and Chen Fang and Nick Foti on mathematical analysis of the history of constitutions.[15] We looked at lots of different ideas, but we thought of the constitution as the rule system of a society, the playbook of a society, and how it changed through time. I want to make one point that relates to the Gerstein result and to Tony's insights, and that is that the US Constitution is quite remarkable. In addition to being the first written constitution, it's also one of the most parsimonious. It's the simplest model. It's the back-of-the-envelope physics model of how society should run. And that gives this country a huge amount of freedom in interpreting the Constitution. If you look at the Indian Constitution, it's huge. To put this in perspective, the American Constitution is about 8,000 words long. The Indian is about 150,000 words long. The next longest is Nigeria, about 60,000 words long. Monaco is the shortest. It's very intriguing that the shorter the constitution, in some sense, the freer you are in the interpretation of its principles, and the longer it is, the more it specifies how you should act in any given context. So, there's this nice complexity regulatory spin on this question of centralized and decentralized governance that's quite universal and it spans biological phenomena as well as cultural phenomena.

I think we can learn a great deal from that. I would actually argue that there might even be a case to be made for a dynamical constitution whose essential size varies according to the challenges that a society faces.

M. GARFIELD: It sounds like, at least in the case of the United States, we have a good deal of epigenetic flexibility with our constitutional code.

---

15 D.N. Rockmore, C. Fang, et al., 2018, "The Cultural Evolution of National Constitutions," *Journal of the Association for Information Science and Technology* 69 (3): 483–94, doi: 10.1002/asi.23971

D. KRAKAUER: Yes. It's sort of like saying it's the difference between, in another metaphor, an abstract painting—say, a Rothko—and a very realist—say, Hudson River School—painting. One of them doesn't give you much space to interpret. It's beautiful, but you know precisely what it's about. You know that's a tree and that's a human figure and that's a deer and so on. But when you go to an abstract painting, you play a bigger role in the interpretation. It's in some sense engaging with your brain in a more creative way. Not to say that realist art isn't creative.

There's something very deep there. I think these more parsimonious models of reality, whether they're mathematical models, whether they're abstract or realist paintings, or whether they're legal documents, titrate between specificity and generality, constraint and freedom. And I don't think society has learned enough from the way that regulation works in the biological world. It wasn't known, of course, until very recently to apply to the legal world and to the sociopolitical world.

There's no doubt in my mind that in the future, constitution-like objects will be algorithmic and dynamical

M. GARFIELD: To think of this in terms of the society as a type of organism, that reminds me of what we were discussing last week in terms of evolvability and how the more narrowly specified trophic niches an organism occupies, the more difficulty it's going to have navigating a period of extraordinary turbulence like this one. In some ways in the United States, our constitution affords us a sort of more of an insectivorous, or a raccoon approach.

D. KRAKAUER: I don't know. I think the right analogy was microbial. It's smaller and it's more of a generalist and that's why it's enduringly influential.

M. GARFIELD: So, to bring us back to the notion of authority and assumed priors, the question here really is, "At what timescale is the problem that we're facing operating?" In what ways is this a question for the aggregation of signals from throughout the social body and the intentional conscious decision to implement a coordinated central plan? In what ways is it a matter of reflex? You don't want to have to sit there and think about lifting your hand off a hot stove. Obviously, this is a problem that demands some balance of both. Where is that balance?

D. KRAKAUER: I think it gets to our research on the nature of complex time. The technologies that have allowed for knowledge to be transmitted have been by and large static objects: books, textual documents. And it's only recently that the possibility of dynamical objects has come into existence—like *Minecraft*, for example. There's no doubt in my mind that in the future, constitution-like objects will be algorithmic and dynamical, because we can now do that. We couldn't in the past. But we don't yet have the design principles, the engineering principles, to use to build such things, but we shouldn't assume that the constraints of history dictate the form of these objects into the future.

5

[BATCH 5, RELEASED 27 APRIL 2020]

We lionize Athens as **DEMOCRACY**, *but something as simple about the* **END** *of that*

the **CRADLE** *of*

are seldom taught that

as **GERMS** *brought*

democracy.

—STEFANI CRABTREE, Transmission 021

TRANSMISSION NO. 020

COVID SPIRALING FRAILTY SYNDROME

DATE: *27 April 2020*

FROM: *John W. Krakauer, The Johns Hopkins University School of Medicine; Santa Fe Institute*
Michelle C. Carlson, The Johns Hopkins University School of Medicine

STRATEGIC INSIGHT: *Exercise can make seniors less susceptible to frailty, and thus to COVID-19, but we need a systematic approach to making physical activity more enticing.*

COVID spiraling frailty syndrome. This was the term coined by the authors of a recent paper analyzing the unique vulnerability of the elderly to COVID-19 in Italy.[1] In the paper, they inform us that Italy is second only to Japan in terms of percentage of the population over sixty-five (22.4%). Along with the risk of increasing age, data from China and Italy implicate hypertension and type 2 diabetes as predictors of COVID-19 mortality, with these three factors defining the COVID spiraling frailty syndrome.

Frailty, outside the context of COVID-19, refers to a biological syndrome present in about 10% of Americans and indexed by three to five clinical components, including weakness from loss of muscle tissue (sarcopenia), slowed gait, low physical activity, sense of low energy or exhaustion, and unintentional weight loss.[2] The prevalence of frailty increases with age after sixty-five, and this increases the risk of infections, the need for hospitalization for a given medical illness, and the risk for falls and disabilities. The causal components of frailty—to the degree that they have been identified—include age-related decline in gonadal hormones, increases in inflammatory mediators, and impaired insulin sensitivity.[3,4] Evidence suggests that the non-additive interaction

---

OPPOSITE: *Alberto Giacometti, "Walking Quickly Under Rain," 1948*
COURTESY MOMA NEW YORK

of these failing systems leads to the emergence of the frailty phenotype, which has led it to be considered the consequence of disruption to a complex adaptive system. Critically, this dysregulated system lacks robustness, with the phenotype being unmasked and amplified by any external stressor or perturbation, e.g., infection by SARS-CoV-2. Notably, attempts to treat frailty with monotherapies—the one-system-at-a-time approach much beloved of mainstream medicine—have largely failed (for example, estrogen or testosterone replacement).[5,6]

In a recent article, Linda Fried, the investigator primarily responsible for first defining the frailty syndrome, makes a compelling case for physical activity as the kind of intervention needed to mitigate its effects. She states:

> *This is because physical activity simultaneously upregulates many systems that mutually regulate each other in combination. Thus, the whole organism could be re-tuned to a higher functional level. This offers a model intervention that matches well a complex system problem. If monotherapies are not sufficient, then finding an intervention that 'tunes' a critical mass of systems would be critical.*[7]

The benefits of physical activity extend well beyond the special case of frailty, with well-proven links between lack of physical activity and vulnerability to numerous chronic diseases. There is overwhelming evidence, for example, that the two chronic diseases mentioned at the beginning of this piece, hypertension and type 2 diabetes, are highly responsive to physical activity,[8] an intervention that can have effect sizes as large as those seen with medications. Neuroscientist Peter Sterling in his recent book, *What Is Health?*, has this to say about hypertension:

> *When the brain commands a rise in arterial pressure, blocking one peripheral mechanism leads the brain to drive the others harder, thereby requiring additional drugs. A therapeutic system based on blocking cerebral commands seems unlikely to succeed. A more promising strategy, we suggest, would concentrate on restoring social and psychological health, thereby reducing chronic conflicts between brain-centered and body-centered regulation. A therapeutic*

> *system based on brain-centered regulation would begin not with antihypertensive and antidiabetic polypharmacy (for example), but rather by enhancing sleep and exercise, both of which improve health.*[9]

Thus, a case can be made for reframing physical activity as either an emergent therapy or a complex medicine. If aging and chronic diseases are the consequences of a complex adaptive system gone awry, then perhaps a sustained complex perturbation is needed to fix it.

The high prevalence of frailty and chronic disease in the over-sixty-five population, which is growing faster worldwide than all other age groups, highlights the urgent need to address the global consequences of inadequate levels of physical activity in this population pre-COVID-19 and the exacerbating effect that can be expected by the imposition of further limitations on exercise and movement post-COVID-19. At the World Health Organization (WHO) website, physical activity is defined as any bodily movement produced by skeletal muscle,[10] and includes exercise as well as other activities which involve bodily movement and are done as part of playing, working, active transportation, household chores, and recreational activities. The benefits are not just physical but also extend to mental health, cognitive function, and social and emotional well-being.

If aging and chronic diseases are the consequences of a complex adaptive system gone awry, then perhaps a sustained complex perturbation is needed to fix it.

The WHO recommends that persons over sixty-five should do at least 150 minutes of moderate-intensity or seventy-five minutes of vigorous-intensity aerobic physical activity throughout the week (or an equivalent combination of the two).[11] However, for many adults, there are several functional and motivational barriers to engaging in regular physical activity, barriers that are exacerbated in the setting of aging and chronic disease. Barriers range from limited access to safe and fun environments that encourage physical activity (not everyone can afford a gym membership) to difficulty engaging in daily physical activity at all. For example, brisk walking is beneficial if you can balance safely and have somewhere to walk. Even if one does have access, the fact is that exercise is boring—making this kind of behavioral intervention hugely difficult to sustain. It is perhaps unsurprising then that, by age seventy-five, about one in three men and one in two women engage in no physical activity at all.[12] Efforts to promote modest increases in physical activity among at-risk sedentary adults have met with only limited success in delaying frailty risk.[13]

The one technology that we know keeps people on task is video-gaming.

The lockdown and self-quarantining recommendations being seen worldwide for COVID-19 and the lingering fear of going outside, even when these restrictions are lifted, will only exacerbate the myriad detrimental physical and mental consequences of inactivity on the aged. What is badly needed is a systematic approach that takes existing evidence and transforms it into an immersive, enjoyable, and gorgeous movement-based experience—human enrichment—for aging adults that is nevertheless realistic and scalable across locations and socioeconomic conditions. The one technology that we know keeps people on task is video-gaming. Indeed, about 50% of people over the age of fifty play video games.[14] We must take the impetus from the current crisis to accelerate innovation at the confluence of conventional video-gaming, virtual reality, real-time motion capture, and physical activity, with an emphasis on the over-sixty-five population. Given that COVID-19 is disproportionately targeting those at higher socioeconomic risk, the development of such interventions is particularly urgent for those with few opportunities for safe, regular, and accessible physical activity.

ENDNOTES

1 A.M. Abbatecola and R. Antonelli-Incalzi, 2020, "COVID-19 Spiraling of Frailty in Older Italian Patients," *The Journal of Nutrition, Health, and Aging* 24: 453–455.

2 L.P. Fried, C.M. Tangen, et al., 2001 "Frailty in Older Adults: Evidence for a Phenotype," *The Journals of Gerontology: Series A* 56(3): M146–M157.

3 Y. Chen, S. Liu, and S.X. Leng, 2019, "Chronic Low-Grade Inflammatory Phenotype (CLIP) and Senescent Immune Dysregulation," *Clinical Therapeutics.* 41(3): 400–409.

4 L. Ferrucci and E. Fabbri, 2018, "Inflammageing: Chronic Inflammation in Ageing, Cardiovascular Disease, and Frailty," *Nature Reviews Cardiol*ogy 15(9): 505-522.

5 A.M. Kenny, L. Dawson, et al., 2003, "Prevalence of Sarcopenia and Predictors of Skeletal Muscle Mass in Non-Obese Women Who are Long-Term Users of Estrogen-Replacement Therapy," *The Journals of Gerontology: Series A* 58(5): M436–M440.

6 P.J. Snyder, H. Peachey, et al., 1999, "Effect of Testosterone Treatment on Body Composition and Muscle Strength in Men over 65 Years of Age," *The Journal of Clinical Endocrinology & Metabolism* 84(8): 2647–2653.

7 L.P. Fried, 2016, "Interventions for Human Frailty: Physical Activity as a Model," *Cold Spring Harbor Perspectives in Medicine* 6(6):a025916.

8 S.R. Colberg SR, R.J. Sigal, et al., 2010, "Exercise and Type 2 Diabetes. The American College of Sports Medicine and the American Diabetes Association: Joint Position Statement," *Diabetes Care* 33(12): e147–e167.

9 P. Sterling, 2020, *What is Health?* Cambridge, MA: MIT Press.

10 World Health Organization, "Physical Activity," https://www.who.int/dietphysicalactivity/pa/en/

11 World Health Organization, "What is the Recommended Amount of Exercise?" http://www.emro.who.int/health-education/physical-activity/promoting-physical-activity/What-is-the-recommended-amount-of-exercise.html

12 The Center for Disease Control and Prevention, "Physical Activity and Health: A Report of the Surgeon General—Older Adults," https://www.cdc.gov/nccdphp/sgr/olderad.htm

13 R.M. Henderson, M.E. Miller, et al., 2018, "Maintenance of Physical Function 1 Year After Exercise Intervention in At-Risk Older Adults: Follow-up From the LIFE Study," *The Journals of Gerontology: Series A* 73(5): 688–694.

14 K. Terrell, "Video Games Score Big with Older Adults," *AARP*, December 16, 2019, https://www.aarp.org/home-family/personal-technology/info-2019/report-video-games.html

REFLECTION

MEDICAL GASLIGHTING & LONG COVID

DATE: *20 August 2021*

FROM: *John W. Krakauer, The Johns Hopkins University School of Medicine; Santa Fe Institute*
Michelle C. Carlson, The Johns Hopkins University School of Medicine

---

Our original Transmission was published in April 2020. Things have and have not changed since then. Despite the roll-out of several successful vaccines beginning in late 2020 and early 2021, the Delta variant has brought back the feeling of vulnerability that was prevalent early on, along with a renewed need for masks, distancing, and other restrictions. In our piece, we focused on the particular vulnerability to COVID-19 of the aged and those with chronic conditions such as hypertension, obesity, and diabetes. As one of us has recently written, this has been a "clash between an ancient family of viruses and the modern industrial world's diseases of lifestyle."[1] In this clash, we have fared far better against the virus than against chronic disease, and we suggested that new ways to promote physical activity are needed to make people more robust against morbidity and death from SARS-CoV-2 infection. What is really new and somewhat perverse is that the virus itself seems to have created a chronic disease all of its own, often affecting the young, called "long COVID," and this condition, along with its inherent medical complexity, has been accompanied by all the problems and prejudices the medical establishment has with chronic conditions, subjective complaints, and multi-organ effects, especially in women.

Long COVID is made up of a myriad of symptoms and signs that develop during or following a confirmed or suspected case of COVID-19, and which continue for more than twenty-eight days. In the largest online survey to date of long COVID (3762 patients), the patients were mainly women (approximately 80%) and the top three most debilitating

---

1 J.W. Krakauer, 2020, "A Theory for Curing the Diseases of Modernity," *Current Biology* 30: R1-R5.

symptoms listed were fatigue, breathing issues, and cognitive dysfunction. The probability of symptoms lasting beyond thirty-five weeks was greater than 90%, with no statistically significant difference between positively (diagnostic/antibody) tested patients and negatively tested groups.[2] This latter result is important because it has exacerbated what David Putrino, a neuroscientist and neurorehabilitation specialist at Mt. Sinai Hospital, calls "medical gaslighting." He elaborates this as meaning not accepting patients who are not "seropositive" for previous COVID-19 infection despite many problems with the sensitivities of the tests. This is accompanied by an unwillingness on the part of physicians to acknowledge physical symptoms in the absence of objective tests and therefore prematurely concluding that the problem is psychiatric.

. . . the virus itself seems to have created a chronic disease all of its own, often affecting the young, called "long COVID," and this condition, along with its inherent medical complexity, has been accompanied by all the problems and prejudices the medical establishment has with chronic conditions, subjective complaints, and multi-organ effects, especially in women.

In a fascinating piece in *The Atlantic* on patients with long COVID, known as "long-haulers," journalist Ed Yong comments, "The physical toll of long COVID almost always comes with an equally debilitating comorbidity of disbelief. Employers have told long-haulers that they couldn't possibly be sick for that long. Friends and family members

2 H.E. Davis, G.S. Assaf, et al., 2021 "Characterizing Long COVID in an International Cohort: 7 Months of Symptoms and Their Impact," *EClinicalMedicine*, doi: 10.1101/2020.12.24.20248802v3

accused them of being lazy. Doctors refused to believe that they had COVID-19."[3] Hence, long-haulers may represent a group with silent and persistent (potentially decades-long) neurocognitive and psychological symptoms, much like those with silent brain infarcts and TBI—but with one major difference: this pandemic-scale injury is affecting millions and associated healthcare systems simultaneously. Therefore, we need tools to anticipate, detect, and treat these cognitive and psychiatric symptoms in recovery rather than exacerbating them through lack of recognition.

The existence of long COVID,
like the chronic diseases highlighted
in the original Transmission,
has exposed again the inability of
the modern medical establishment
to cope with complex chronic conditions
that do not yield to simple lab tests
or mono therapies, but instead require
extended periods of multidisciplinary
care focused on behavioral interventions
for both body and mind.

So, we have come full circle. SARS-CoV-2 has created an "acute chronic" condition that hits the very age group that tends to avoid initial severe disease and hospitalization. They, however, develop a condition that may turn into prolonged disability.

The existence of long COVID, like the chronic diseases highlighted in the original Transmission, has exposed again the inability of the

3 E. Yong, "Long Haulers are Redefining COVID-19," *The Atlantic*, August 19, 2020, https://www.theatlantic.com/health/archive/2020/08/long-haulers-covid-19-recognition-support-groups-symptoms/615382/

modern medical establishment to cope with complex chronic conditions that do not yield to simple lab tests or monotherapies, but instead require extended periods of multidisciplinary care focused on behavioral interventions for both body and mind. One can also be certain, at least in the US, that it will be minorities who will suffer disproportionately from the inadequacies of healthcare delivery when it comes to long COVID, as they do for all chronic conditions.[4]

4 R. Marya and R. Patel, 2021, *Inflamed: Deep Medicine and the Anatomy of Injustice*, New York, NY: Farrar, Straus, and Giroux.

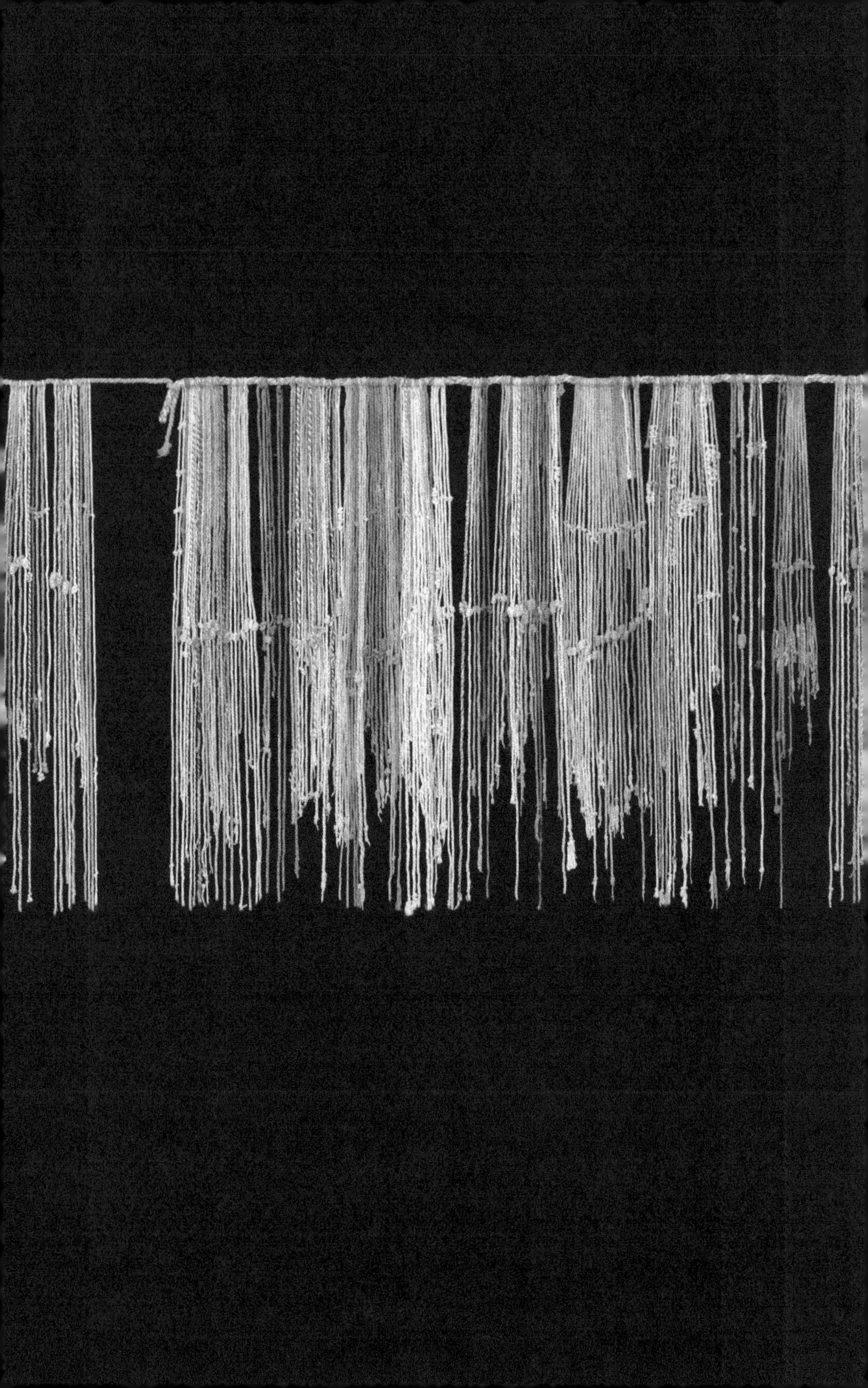

TRANSMISSION NO. 021

WHAT HISTORY CAN TEACH US ABOUT RESILIENCE

DATE: *27 April 2020*

FROM: *Stefani Crabtree, Utah State University; Santa Fe Institute*

STRATEGIC INSIGHT: *The archaeological record can teach us much about cultural resilience and how to adapt to exogenous threats.*

One thing humans are particularly good at is perceiving environmental cues, adapting our cultural practices around those cues, and encoding lessons for future generations on how to prevail. For example, the practice of quarantine comes from the fourteenth-century convention of keeping ships at anchor outside Venetian harbors to protect citizens from unknown pathogens; the etymology of the word comes from *quaranta giorni*, the forty-day period sailors had to wait before disembarking. Even Charles Darwin had to quarantine before exploring foreign shores as a naturalist. This behavior remains encoded in our own twenty-first-century cultural practice. Take comfort in knowing the boredom we face in our homes as we socially isolate was also felt by the sailors stuck at anchor outside the shining city of Venice in the 1300s.

Lessons can come from many societies worldwide, such as these premodern examples in the US Southwest. Among the Ancestral Pueblo people who lived in what is now Mesa Verde National Park and surrounding regions, challenges such as droughts and unrest led to innovations that helped the society thrive there for 700 years, though that society went through periods of prosperity and upheaval. Perhaps the most emblematic is the rise of Chaco Canyon, a large, highly centralized polity in northern New Mexico—and its rapid decline. As political hierarchies at Chaco grew, so too did increasingly large community

---

OPPOSITE: *Incan Quipu*
MUSEO DE LA NACIÓN (LIMA, PERU)

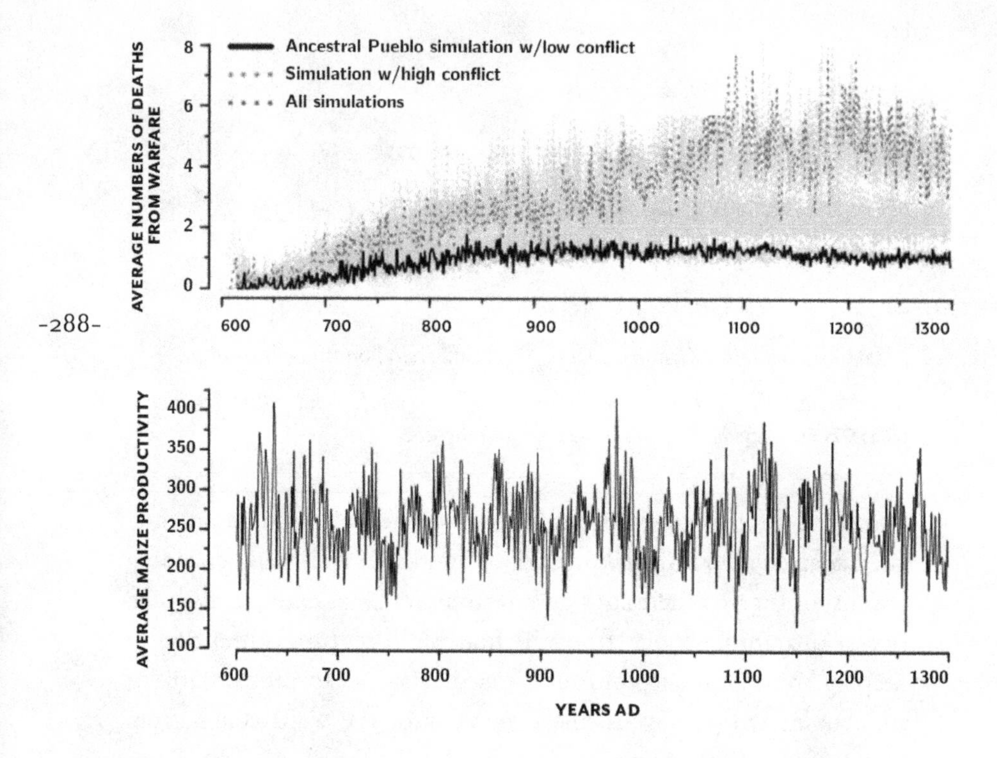

Figure 1. *Top: Simulations of Ancestral Pueblo society show that, as hierarchies grow and people fight over arable land to grow maize, warfare (and death from warfare) increases. This is juxtaposed against real average maize production by year for the whole region of Mesa Verde (bottom). The decline of the Chacoan hierarchy in the late 1100s corresponds to increases in violence and decreases in predictable food. Yet the Ancestral Pueblo society continued, reorganizing their society after the sharp exogenous shocks, and rebuilding away from Mesa Verde after 1300.*

structures built on the ability to control farming production. When productivity failed, the Chacoan hierarchical society fractured, violence increased, and society reorganized into regional communities. The archaeological record, bolstered by computer simulation (fig. 1), shows us how they reacted to societal upheaval; as maize farming became less predictable and violence escalated, hierarchy disintegrated.

When finally the communities left Mesa Verde, walking hundreds to thousands of miles to new regions, they brought with them the accumulated knowledge of successful society, demonstrating the resilience of culture in the face of exogenous stress. The descendent communities today demonstrate much about cultural resilience and how to adapt to exogenous threats.

This shared history and solidarity with the past suggests that we can learn from these difficult situations. The Plague of Athens (430–427 BC), which is estimated to have killed one-third of the city-state's population, would have been a traumatic experience to live through. Many have speculated that it caused the demise of Athenian democracy, but the writings of Thucydides offer lessons to help us understand how diseases can exploit weaknesses in society—and how we might overcome them. Public health issues like poor sanitation and overcrowding, he noted, led to increases in casualties from the disease. These could have been avoided had officials heeded concerns about crowding. Doctors and caregivers succumbed to the disease and died; little seemed to help those who contracted it to survive. A lack of understanding of the disease sparked fear that spread through the citizens as quickly as the Plague itself.

Societies that prevail encode these lessons through oral tradition, writing, songs, and artworks. We learn to adapt and innovate—a hallmark of our species.

Thankfully, these writings can help us identify commonalities between how citizens reacted in 430 BC and how they may react today. Bolstering citizen confidence via clarity of communication from officials (such as daily briefings from state governors like Washington's Jay Inslee), increasing our collective knowledge of the disease by applying open-science principles and sharing our models (as most scientists are doing), and thoughtful prosocial behavior by citizens (such as donating masks to healthcare workers) may help avoid the types of consequences Athenians experienced after the Plague ended.

Societies that prevail encode these lessons through oral tradition, writing, songs, and artworks. We learn to adapt and innovate—a hallmark of our species. From the first Oldowan stone tools to the development of complex communication platforms, we find ways to succeed, we persevere, we make do, as my great-great-great grandmother who lived through the Great Depression would say. Making do required focusing on what she had, not what she didn't have, and ensuring the health and survival of her family. Countless ancestors have had to make do through challenges in the past; it is, I think, one of the comforts of being an archaeologist: knowing that I am not alone in what I face.

We will innovate. Our descendants will learn from the lessons we are encoding now in our cultural memory.

And just like our forebears, so, too, will our society survive through this challenge. We will innovate. Our descendants will learn from the lessons we are encoding now in our cultural memory. Some 700 years ago Venetian sailors were stuck at harbor; 500 years later Charles Darwin also had to undergo quarantine before embarking on distant shores. In 700 years, our descendants will benefit from the innovations and lessons we are encoding today.

REFLECTION

REFLECTIONS ON ARCHAEOLOGY AND COVID-19, ONE YEAR LATER

DATE: *2 August 2021*

FROM: *Stefani Crabtree, Utah State University; Santa Fe Institute*

Nous entrons dans l'avenir à reculons. (We go into the future backwards.)

—PAUL VALÉRY

The decisions people made in the past have impacts on the present and the future. More than a year into a globe-spanning pandemic, we are witnessing the impacts of the past year's policy, public health, and messaging decisions. The result appears to be a bifurcation in the ways that citizens interpret the public health crisis. There are those who adhere to public health recommendations to be vaccinated or to wear a mask in public, and there are those who refuse the vaccine and refuse to wear masks properly or at all, thanks to rabid misinformation campaigns. Some officials are calling the current phase a "pandemic of the unvaccinated." The more the virus can spread among the unvaccinated, the longer the pandemic will go on.

. . . confronting a society-changing event poorly can lead to dramatic social unrest . . .

As I suggested in my Transmission, archaeology and history have lessons to offer from the past: effective leadership can instill confidence, thoughtful prosocial behavior can help quell the pandemic, and increasing collective knowledge of the disease can help with eradication efforts. On the other side, I would suggest that confronting a society-changing event poorly can lead to dramatic social unrest—something we have seen nationally and globally.

The Plague of Athens was not just a biological plague, but a social one as well. As the plague wore on, citizen confidence in the city-state's leadership plummeted as leaders failed to address the proximal cause of the spread of the pandemic. This loss of confidence in leaders led to overwhelmed doctors, an increase in fear, and panic. It seems that effective communication could perhaps have changed the tide for Athenians.

In the past year, places with clear communication from leaders have fared much better than places with poor communication or laissez-faire attitudes. Seattle was the first urban area to reach high vaccination, with strong leadership from the governor; New Mexico, likewise, had clear rules from the beginning that seemed to make the state come through the pandemic more rapidly than others. Clear and effective leadership that does not simply rely on a heterogenous population of citizens making the "right" decisions has allowed some places to fare well. Places with few rules have suffered. Unfortunately, we cannot relax our rules to benefit individuals at the expense of the group. In the fourteenth century, rules of quarantine were developed to prevent the spread of plagues, and breaking quarantine had dramatic repercussions. Minor personal inconveniences, like quarantine, can help the group.

We lionize Athens as the cradle of democracy, but are seldom taught that something as simple as germs brought about the end of that democracy. A society is made of individuals with varying opinions, beliefs, and practices. There will be heterogeneity in how people think leadership should work. There were people in Athens who believed there was no plague. There was disagreement on how to confront it. People became nihilistic, unafraid of the law, since they were certain of their untimely deaths. The plague resulted in stricter laws of citizenship and more draconian laws; eventually, due to the weakening of the democracy, Athens fell to Sparta.

Unfortunately, we cannot relax our rules to benefit individuals at the expense of the group.

How can we avoid the democracy-destroying effects of a plague such as those that Athenians experienced? There will be a long and drawn-out recovery from the COVID-19 pandemic, both nationally and globally. We must keep up communication about the efficacy of vaccines, we must help our global citizens with efforts to stop the spread, and we must continue to think of our neighbors as our neighbors, not our adversaries.

> How can we avoid the democracy-destroying effects of a plague such as those that Athenians experienced?

Yet even with the challenges come opportunities for positive societal change. The mRNA vaccines provide a paradigm shift in the way we can treat infectious disease, and habits of mask wearing or social distancing may, indeed, become cultural practices. These practices emerged with this pandemic, just as the practice of quarantine was new in Venice in the 1300s. And yet, they can have impacts reaching into our future. Our future is, after all, built upon the past. In this way we do go into the future backwards; it us up to us to ensure that we can look to the past for lessons that can help us build a better future.

TRANSMISSION NO. 022

THE INFORMATIONAL PITFALLS OF SELECTIVE TESTING

DATE: *27 April 2020*

FROM: *Van Savage, University of California, Los Angeles; Santa Fe Institute*

STRATEGIC INSIGHT: *Test kits cannot exponentiate at the same rate as the virus. Unless we ramp up to 500K, the curve will flatten due to artifact.*

Decisions about when and how to relax social distancing will ultimately come down to whether or not we think we're "flattening the curve"—slowing the growth rate for the spread of infection. But how do we know? The prevailing perception is that we can look at the curve's fit to reported new cases and deaths each day, but this might not be correct. If the true number of cases in the population is well beyond our maximum testing capacity (as it is in the US) and if we are primarily testing those with symptoms (as we are in the US), then in time the changes that we see might be dominated mostly by random noise. Because the true number of cases far exceeds the testing capacity, the signal is essentially saturated. Selectively testing the symptomatic cases is really testing what proportion of respiratory illnesses and fevers are due to COVID-19—not what percentage of the population has COVID-19—and that proportion could remain closer to constant as COVID-19 spreads. Currently, this proportion in the US varies by state, but is not changing much over time or as we increase testing capacity.

Consequently, the time series of new measured cases could simply reflect random fluctuations around an average that is given by the number of tests per day and the chances that someone with symptoms has COVID-19, as opposed to a different respiratory illness. Neither of

---

OPPOSITE: *H.C. Selous (illustrator) and Frederick Wentworth (engraver), "Edgar and Gloucester," c. 1865*

these is dependent on the current growth rate or trajectory of COVID-19 cases in the general population. For instance, estimating that the US is limited to conducting about 150,000 tests per day[1] and the positive rate for tests is about 20%, we expect 30,000 new cases being reported each day with random fluctuations based primarily on exactly how many tests are processed that particular day.

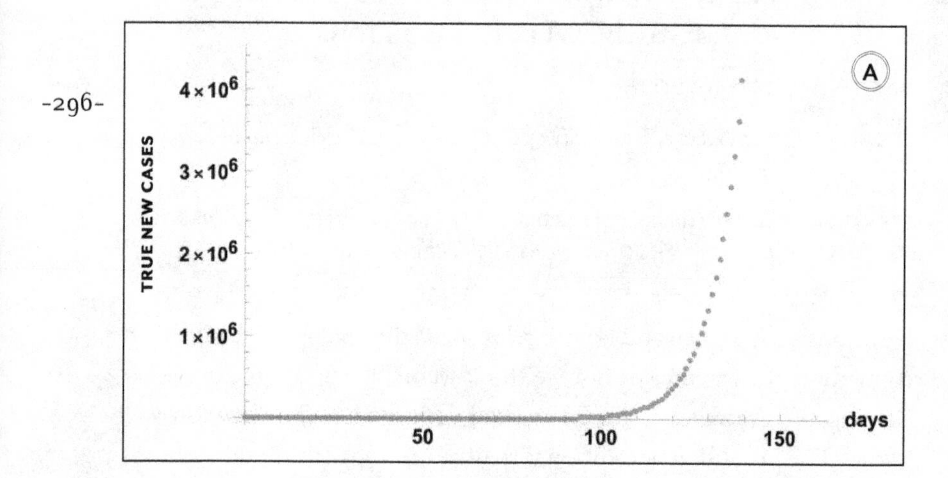

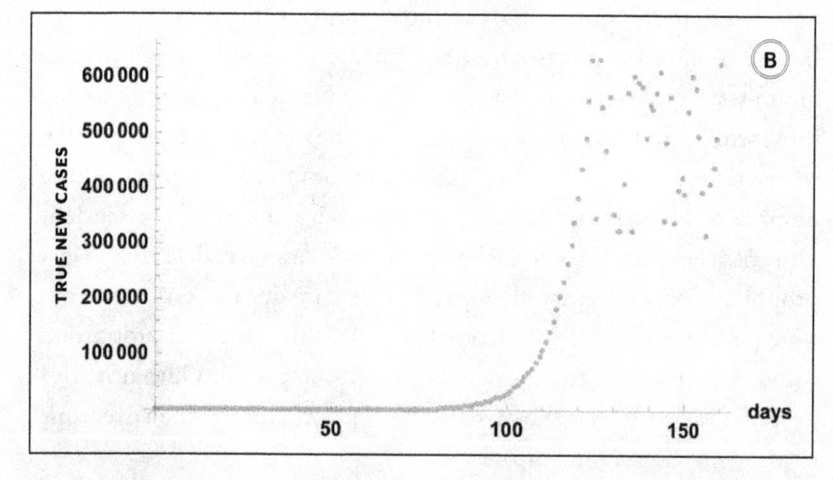

Figure 1. *Two scenarios for growth in true number of cases. A. Exponential growth for 160 days with an increase by a factor of e for every 8 days. B. Exponential growth for the first 125 days as in A. but a flattened curve for the last 35 days. For the flattened curve in the latter 35 days, the new cases per day randomly fluctuate around the number of new cases where exponential growth ceased. Parameter values are given in the Mathematica file linked in footnote 1. Changing parameter values has little effect on the results as long as the population and testing approach satisfy the two basic assumptions—true cases well beyond testing capacity and selective testing of population with respiratory distress, fever, etc. that has a roughly constant chance of having COVID-19.*

Early on when a disease is spreading, the number of cases will increase and look exponential either because the number of cases *is* increasing exponentially and can be adequately measured by tests, or the testing capacity is increasing exponentially and the positive test rate is roughly constant, or both. However, if the number of cases is growing exponentially, it will not take long for the true number of cases to reach millions or tens of millions. So, running approximately 100,000 tests per day can't possibly capture the true numbers. This still might be okay if the shape of our growth curve is the same—meaning the measured cases are a constant proportion of true cases. And if the population is being randomly sampled, this might actually work because the per-capita growth rate could still be captured, so that when the curve flattens, the percentage of tests that are positive will drop. However, this is not necessarily true for the percentage of tested symptomatic cases that are due to COVID-19, which will yield much higher rates of positive results than if tests were randomly sampled from the general population. This is because the percentage of positives from testing only symptomatic cases may remain closer to constant because those sick enough to come in have a reasonable chance of carrying COVID-19.

That is, we could be suffering from a double whammy in our approach. By having limited testing capacity, we can't track true numbers, and by testing only those with symptoms, we might not be tracking how the per-capita growth is actually changing. Because of this double whammy, we may be flying blind in terms of the growth rate of cases and not know whether we're flattening the curve.

To illustrate these points, Figures 1 and 2 present a very simple simulated example in which the true cases are growing exponentially for 160 days, versus a scenario where the true cases grow exponentially for 125 days but then flatten and new cases appear at a constant rate. For both scenarios, the total number of cases by day 125 is beyond the maximum testing capacity per day. (All parameter values are rough estimates. The percentage of asymptomatic cases has been varied from 40–80% and affects numerical values but not the overall conclusions.[1] I also assume

---

1 Shown in this Mathematica file: http://sfi-edu.s3.amazonaws.com/sfi-edu/production/uploads/ckeditor/2020/04/25/t-22-savage-mathematica.nb
PDF version here: https://sfi-edu.s3.amazonaws.com/sfi-edu/production/uploads/ckeditor/2020/04/29/savagetestmix3.pdf

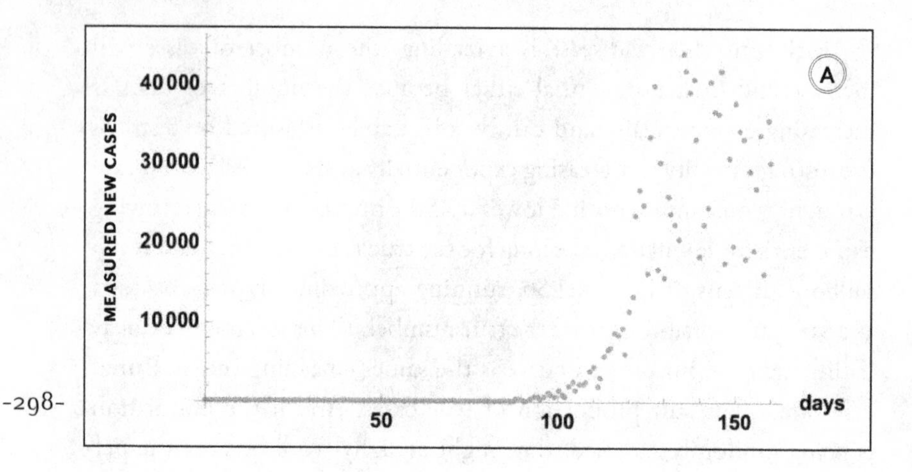

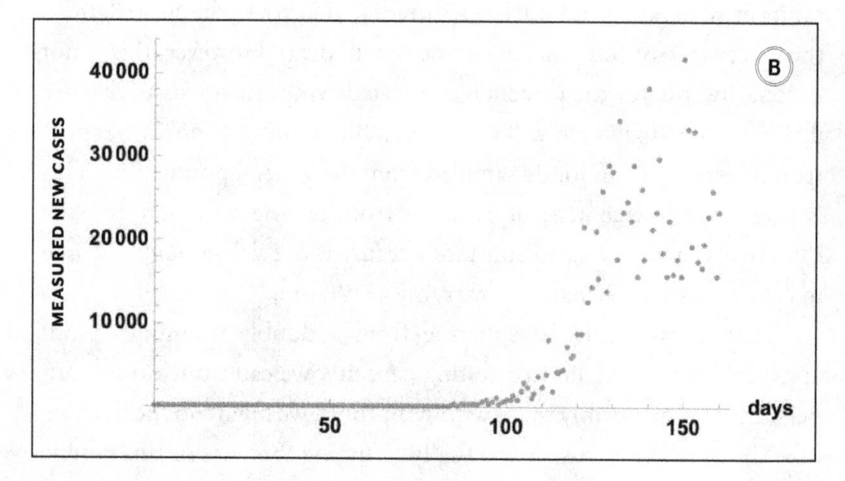

Figure 2. *Measured new cases for the two scenarios from Figure 1. These results are indistinguishable and both appear as if the curve has been flattened, even though one results from true cases that have pure and continued exponential growth (A), and the other (B) has true cases with a first period of exponential growth that is followed by a flattened curve with random fluctuations.*

testing is occurring only in some fraction of those with symptoms, and I estimate that people exhibiting generic symptoms of respiratory distress, fever, etc. have a 10% to 20% chance of testing positive for COVID-19. In terms of reported cases, both appear as if the curve is flattening, and it's not clear how to distinguish the dynamics of the true cases—exponential versus flat—using only the data for the measured number of new cases. Perhaps the range of values for new cases is slightly different, but the shape of the curve isn't.

Tracking the number of deaths will likely give similar conclusions unless presumed cases (not just those confirmed by testing) are included. This policy varies by state, but presumed cases are not included in official counts. Also, conclusions based on the number of deaths will be delayed by a few weeks compared with conclusions that could be based on actual case counts and won't help as much in anticipating or avoiding hospital overflow. Extreme backlogs and time delays in processing tests could also create short periods of a few days that look like the curve is flattened or spiked, even though new cases are still growing exponentially.

In summary, we must ramp up testing capacity and/or test more randomly. My rough calculations here also match prominent calls that 500,000 tests are needed to see whether we're flattening the curve.

ENDNOTES

1 R. Meyer and A. C. Madrigal, "A New Statistic Reveals Why America's COVID-19 Numbers Are Flat," *The Atlantic*, April 16, 2020, https://www.theatlantic.com/technology/archive/2020/04/us-coronavirus-outbreak-out-control-test-positivity-rate/610132/

2 J. Kashoek and M. Santillana, "COVID-19 Positive Cases, Evidence on the Time Evolution of the Epidemic or An Indicator of Local Testing Capabilities? A Case Study in the United States," *SSRN*, April 10, 2020, http://dx.doi.org/10.2139/ssrn.3574849

3 K. Collins, "Coronavirus Testing Needs to Triple Before the US Can Reopen, Experts Say," *The New York Times*, April 17, 2020, https://www.nytimes.com/interactive/2020/04/17/us/coronavirus-testing-states.html

REFLECTION

COMPLEXITY AFTER COVID-19

DATE: *18 August 2021*

FROM: *Van Savage, University of California, Los Angeles; Santa Fe Institute*

More than a year after writing about COVID-19, my views on the basic science of disease transmission—infectiousness, fatality rate, behavioral modifications to slow the rate of spread, the dangers of saturating and exceeding intensive care unit capacity—have changed very little. In direct contrast, however, my views have changed tremendously in terms of how to enact and encourage policies and behaviors that will help us—as a city, state, country, or planet—to slow the spread of COVID-19. This is necessary so that children can go back to school, people can socially and professionally engage in person, and the most vulnerable populations (people who are elderly, with complicating health conditions, with lack of access to healthcare, who can't stay home from work) can be protected.

On the first count of the basic science of disease spread, I was concerned over a year ago with how to reliably obtain and analyze data on new COVID-19 cases and new fatalities to infer when disease spread and risk are worsening versus improving. My Transmission was meant as a warning that we might have believed we were flattening the curve when in fact it was still growing exponentially, so that we were actually in worse shape than we recognized.

When I look back now, I think about the last day I was in my office and on campus before we went into full lockdown. One of my colleagues asked me if I thought COVID-19 was really that bad. I replied that the estimates of a fatality rate of 0.5–1.5% seemed believable to me based on a couple of pieces of independent data. Since there are about 320 million people in the United States, if you make a couple of assumptions—about half of people get the infection before disease transmission seriously slows, about half of people are asymptomatic or naturally immune—that reduces symptomatic cases to about 80 million people. So if 1% of those cases led to fatality, you'd guess there would be about 800,000 fatalities, and allowing for uncertainty in the range of estimates for fatality rate, you'd guess a rough range of about 500,000 to 1,000,000 deaths in the US. My colleague seemed genuinely surprised and replied that those were very large numbers, and that they didn't believe that would be allowed to happen. I responded that was the estimate if we did nothing, but that I hoped we—as a society—did something and could keep the numbers much lower than that.

As I'm writing this on August 18, 2021, the Delta variant is leading to new spikes in infections, and the current number of fatalities from COVID-19 in the United States is well over 600,000.[1] And this number ignores undercounting due to missed deaths—especially at the early stages of the pandemic—that could place the actual current number of COVID-19 deaths much closer to one million. That's now on par with the number of fatalities in the US Civil War—by far the bloodiest war in US history (but it occurred when overall population was one-tenth the size it is now).

. . . the current number of fatalities from COVID-19 in the United States is well over 600,000. . . . That's now on par with the number of fatalities in the US Civil War—by far the bloodiest war in US history . . .

Although none of my assumptions were completely correct, the order of magnitude and range of the final estimate were accurate. As a physicist, that's what we aim for in doing a back-of-the-envelope Fermi calculation.[2] A year later, I still believe that at the level of rough estimates and calculations—without needing either sophisticated models of dynamical systems or technical statistical analyses—much about this pandemic was relatively straightforward to predict.

1 U. Irfan, "How the World Missed More Than Half of All COVID-19 Deaths," *Vox*, May 7, 2021, https://www.vox.com/22422794/covid-19-death-numbers-total-us-vaccine-ihme

2 See V.M. Savage, J.F. Gillooly et al., 2004, "Effects of Body Size and Temperature on Population Growth," *The American Naturalist* 163(3), doi: 10.1086/381872

But even though I maintain the events that played out over the past year were quite predictable, a more compelling question is whether they were preventable. Indeed, when talking with my colleague, I said "if we did nothing." But it wouldn't be fair to say "we did nothing." At the personal level, my son didn't go to school, and my wife and I didn't go into work, for over a year. We wore masks when outside the home. We rarely saw friends and went a year and a half without seeing family. And in the grand scheme of things, I was extremely fortunate to have the flexibility and resources I had. At the level of the whole society,

we wore masks, shut down large public events and indoor gatherings, closed schools, imposed lockdowns, provided financial assistance, put a moratorium on evictions, developed multiple effective vaccines, and administered millions of vaccine doses.

In looking at our society as a complex system, it is now clearer to me than ever before that these social and behavioral issues are the bottleneck in our ability to help society and protect those we love.

So the question is: Why did these actions fail? Or to unpack that: Were these measures not followed closely enough by the public, not communicated well enough by leaders, not started quickly enough or over the right timescales, or not done in the right places? Was the failure because of political polarization that led people to take harmful actions and engage in dangerous behaviors, even when provided with reliable information and sound advice for modifying behavior? Was the failure an inability to enact these kinds of measures at the population level due to the demands of earning a living? Or is the number of actual deaths in the low range of my original estimate because our measures really did help avoid a few hundred thousand deaths? Is it some combination of these?

Having accurate information is necessary for making good decisions, but it is not sufficient. At the population level, we must also clearly communicate that information, work to provide people with the financial ability to follow recommendations, and implement best practices for ensuring that people are not making decisions in ways that are biased or stem from political polarization and misinformation. My Transmission was focused only on the accurate information. A year later what I think we need to focus on is all the rest—political polarization, communication, misinformation, and financial and health support. And I say this as someone who has lost four relatives to COVID-19 since I wrote the original piece, and as someone with relatives who have not yet chosen to get the vaccination (not counting my son, who's too young).

In looking at our society as a complex system, it is now clearer to me than ever before that these social and behavioral issues are the bottleneck in our ability to help society and protect those we love.

TRANSMISSION NO. 023

MAKING GOOD DECISIONS UNDER UNCERTAINTY

DATE: *27 April 2020*

FROM: *David Tuckett, University College London*
Lenny Smith, London School of Economics
Gerd Gigerenzer, Max Planck Institute for Human Development, Berlin
Jürgen Jost, Max Planck Institute Mathematics in the Sciences, Leipzig; Santa Fe Institute

STRATEGIC INSIGHT: *To make good decisions under uncertainty, decision-makers must act creatively to avoid paralysis, while recognizing the possibility of failure.*

The current COVID-19 pandemic presents decision-makers with situations where the range of possible actions and the probabilities of possible outcomes are not known or even imagined.

Because data are unavailable or contestable, using them to make sense of exactly what is going on or how the pandemic will play out is unreliable. Actions are untested, their acceptance by populations is not guaranteed, and long-term societal and economic impacts are unclear. Should we aim to eradicate, to slow, or accept deaths of the vulnerable and minimize the collateral economic damage?

Given our data, we may want to extrapolate a curve to make a prediction (fig. 1, below). It would be fantastic to have a perfect fit, but from limited statistical data, we cannot get that. Even worse, an optimal statistical fit (red line) is far inferior to a cone of possibilities conditioned on clearly stated assumptions. That is an important insight.

Radical Uncertainty (RU) is defined as a situation in which quantifying costs and consequences is contestable but we must choose.

OPPOSITE: *Matthias Stom, "The Judgment of Solomon," c. 1640 (detail)*

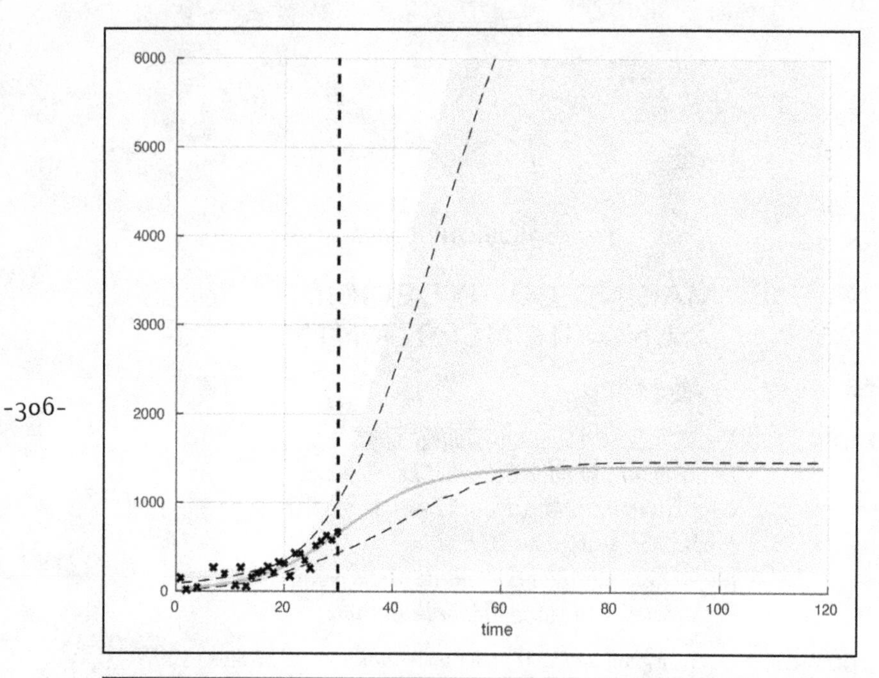

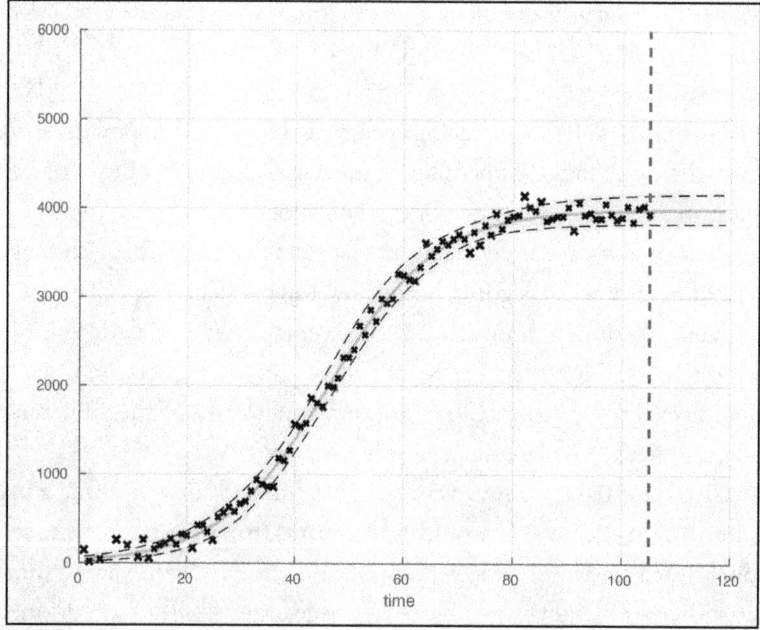

Figure 1. *Measuring (the small crosses) and modeling (the unbroken line) the progress of an epidemic through time. At the start, our uncertainty of the future remains unavoidably large (shaded area) but nevertheless an accurate measure of our true state of ignorance and more deeply informative than the confidence from the poor model (the unbroken line).*

As time passes, new data allow us to significantly reduce this cone of uncertainty. The dashed vertical line is the time at which measurements are made. Credit: Leonard Smith (@lynyrdsmyth).

Current scientific decision theory, based mostly on a rationality paradigm that aims at optimal decisions in contexts of known outcomes and their probabilities, is silent on what to do. The financial crisis, the climate challenge, and now COVID-19 emphasize the need to fill this gap.

Our collaboration focuses on how to aid decision-makers in selecting data enquiringly from diverse sources, to recognize structural instabilities, and to imagine possible Big Surprises. It emphasizes systematic ways to reduce complexity and to identify essential variables which have the largest effect and which we can control. Can we understand the relations between them, for instance, by checking the effects of small perturbations and their propagation? Can we reduce general vulnerability? Which heuristics are good enough to cut through details, and what kinds of narratives make sense of complicated developments?

Crucially, optimization in RU is dangerous. Its premises are not satisfied. It blinds us to the need to embrace diversity and experiment. It installs a blame rather than a creative culture and causes fragility rather than resilience—whether designing medical facilities or supply chains, for instance.

Decisively, RU necessarily evokes ambivalence—reasons for and against choices and the accompanying emotions of doubt or excitement. To make good decisions under uncertainty, decision-makers must act creatively to avoid paralysis, while recognizing the possibility of failure and accompanying anxiety. Decisions can work if supported by and communicated with a conviction narrative (CN). A CN generates confidence and cooperation in the public. Good CNs are the product of inquiring into and facing doubt transparently in conversation. Bad CNs are the outcome of dictatorial assertion, attraction to phantastic object solutions, and groupthink, creating "certainty."

Progress requires novel cross-disciplinary research combining mathematics, machine-learning, and physics with economics, engineering, social science, and psychology.

TRANSMISSION NO. 024

THE HEAVY TAIL OF OUTBREAKS

DATE: *27 April 2020*

FROM: *Cristopher Moore, Santa Fe Institute*

STRATEGIC INSIGHT: *R_0 is just an average: the transmission rate varies widely, and outbreaks can be surprisingly large even when the epidemic is subcritical.*

Much of the coverage of COVID-19 talks about R_0, the average number of people each sick person infects. If R_0 is bigger than 1, cases grow exponentially, and an epidemic spreads across the population. But if we can keep R_0 below 1, we can limit the disease to isolated outbreaks and keep it under control.

But R_0 is only an average. Your ability to practice social distancing depends on whether you are a first responder or healthcare worker, whether you have to work in close quarters, or whether you can work comfortably from home. (I'm one of the lucky few getting paid to work from my garden.) It depends on how seriously you take your government's warnings and how seriously your government takes the warnings of public health experts. And it depends on the structure of your family and your home.

As a result, R_0 varies wildly, not just from region to region, but across social space, as well. In New Mexico, Santa Fe has very few new cases, but there has been an explosion of cases in rural areas due to lack of running water, multigenerational homes, and other factors. As of April 26, 47% of our confirmed cases are in Native American communities, even though Native Americans make up only 11% of New Mexico's population. Clearly R_0 is larger in some parts of the state and of society than others.

OPPOSITE: *Jan Van Eyck, "The Ghent Altarpiece: Singing Angels," 1429 (detail)*

Even if $R_0 < 1$, outbreaks can be surprisingly large. Suppose you meet ten people while you are contagious, and you infect each one with a probability of 8%. The average number of people you infect is $10 \times 0.08 = 0.8$, less than 1. But those you infect may infect others in turn, and so on. If an outbreak starts with you, how many "descendants" will you have? A classic calculation shows that, if $R_0 = 0.8$, then the average number of people in this chain reaction is $1/(1 - 0.8) = 1/0.2 = 5$. But, like R_0 itself, this is only an average. Like earthquakes and forest fires, outbreaks have a "heavy tail" where large events are common.

Figure 1 shows a visualization of one hundred random outbreaks. The average size is indeed 5, and most outbreaks are small. But about 1% of those outbreaks have size 50 or more, ten times the average, and in this simulation the largest of these 100 outbreaks has size 82. This tail gets heavier if R_0 is just below the phase transition at $R_0 = 1$. If $R_0 = 0.9$, the average outbreak size is 10, but 1% have size 140 or more.

This tail has real effects. Imagine one hundred small towns, each with a hospital that can handle ten cases. If every town has the average number of cases, they can all ride out the storm. But there's a good

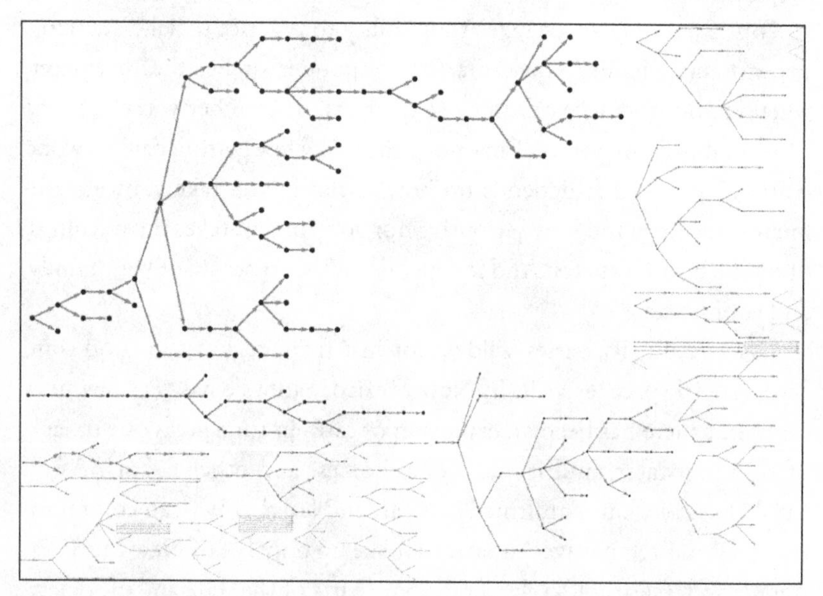

Figure 1. *A hundred random outbreaks in a scenario where each sick person interacts with ten others, and infects each one with probability 8%. Here $R_0 = 0.8$ and the average outbreak size is 5, but 1% of the outbreaks have size 50 or larger, and in this run the largest has size 82.*

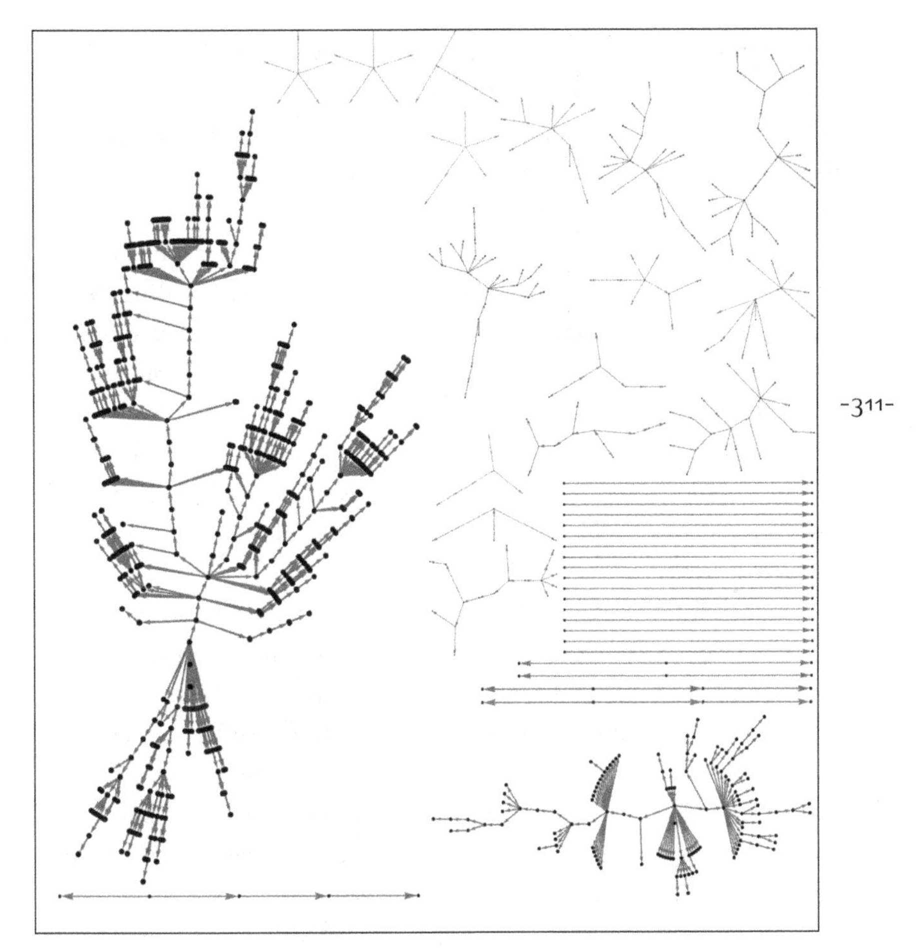

Figure 2. *A hundred random outbreaks in a scenario with superspreading, where 1% of the cases infect 20 others. As in Figure 1, we have $R_0 = 0.8$ and the average outbreak size is 5, but now the heavy tail of outbreaks is much heavier. In this run the largest outbreak has size 663.*

chance that one of them will have 50 or 100, creating a "hot spot" beyond their ability to respond.

The tail of large events gets even heavier if we add superspreading. We often talk of "superspreaders" as individuals with higher viral loads, or who by choice or necessity interact with many others. But it's more accurate to talk about superspreading events and situations—like the Biogen meeting, the chorus rehearsal, or the pork processing plant, as well as prisons and nursing homes—where the virus may have infected many of those present.

Suppose that 20% of cases generate 1 new case, 10% generate 2, 4% generate 5, and 1% "superspread" and generate 20 (and the remaining 65% infect no one). The average number of new cases is again $R_0 = 0.8$. Let's generate one hundred random outbreaks with this new scenario (fig. 2).

The average outbreak size is still 5, but now the tail is much heavier. If just one of the 100 original cases is involved in superspreading, we get a large outbreak. If there are several generations of superspreading, the size multiplies. As a result, large outbreaks are quite common, and the largest one in this simulation has 663 people in it.

What does all this mean? First, it can be misleading to look at statewide or national averages and celebrate if R_0 seems to be falling below 1. The epidemic could still be raging in particular places or among particular groups.

Second, even if R_0 is below 1, we need to prepare for hot spots. Even if the average outbreak is small, large outbreaks will occur due to superspreading or simply by chance. If we do a fantastic job at testing and contact tracing—using both technology and human effort—we will get this pandemic under control, but for the foreseeable future there will be times and places where it flares up and strains local resources. And through those flare-ups, we have to do our best to help each other, and hope that intelligent, generous voices prevail.

REFERENCES

L. Hébert-Dufresne, B.M. Althouse, et al., "Beyond R_0: Heterogeneity in Secondary Infections and Probabilistic Epidemic Forecasting," arXiv:2002.04004 [q-bio.PE], https://arxiv.org/abs/2002.04004

J. O. Lloyd-Smith, S. J. Schreiber, et al., 2005, "Superspreading and the Effect of Individual Variation on Disease Emergence," *Nature* 438: 355–359, doi: 10.1038/nature04153

REFLECTION

COMPLEXITY DOESN'T ALWAYS REQUIRE COMPLEX MODELS

DATE: *13 August 2021*

FROM: *Cristopher Moore, Santa Fe Institute*

A raging virus is devastating society, amplifying every form of inequality, and revealing the utter cravenness of our political leaders. What's a theorist to do?

I experienced most of the pandemic from the safety and peace of my garden. I myself have the freedom to radically reduce my human contact while continuing to feed my family. Transit workers, meatpackers, guards and inmates, single mothers, and people living in three-generation homes do not.[1] And, in addition to these heartbreaking inequities in employment and health, we have the infuriating fact that many others have the option to be safe but choose not to—and that this behavior is aided and abetted by some of the most powerful in our society.

Last year, despite being anguished over these inequities and idiocies, I was hesitant to say anything as a scientist about COVID-19. There was already so much noise, much of it coming from dilettantes who wanted to help but who didn't have the domain knowledge to really do so. We didn't need another theorist claiming to predict where and when the disease would peak. But then I was struck by a basic qualitative, pedagogical point that was worth making.

In the spring of 2020, the media seized on R_0—the average number of new cases generated by each case—as a measure of the spread of the virus. In the simplest models, if R_0 is bigger than 1, the disease grows exponentially; if R_0 is less than 1, it disappears. Websites popped up estimating R_0 in various states and countries. If we can push R_0 below 1, they said, everything will be okay!

But this is far from the truth. Like in so many complex systems, what matters are rare, large events, which were sadly not so rare:

---

1 See T-031 on page 401

superspreading in churches, workplaces, and jails. And there is no single value of R_0: it varies radically by location, employment, beliefs, race, and class.

Like in so many complex systems, what matters are rare, large events, which were sadly not so rare . . .

The Transmission series gave me a perfect opportunity to point this out. Using some basic math and simple simulations, I pointed out that large outbreaks can occur even if R_0 is less than 1. This happens by chance even without superspreading events, due to the multiplicative nature of epidemics, but it happens even more with them. It led to some media, peaking with a quote in *The Wall Street Journal* and a very brief appearance on Don Lemon's show on CNN. Although the same basic point appeared later in a number of venues, I felt I had done some good.

Looking back over the past year, it's clear that the right response to complexity is not always building a more complex model. Many attempts to build more realistic models for COVID-19 foundered on fundamental uncertainties about the disease and about society. The lack of effective testing and contact tracing meant that we were flying almost blind, making even simple models hard to validate. Friends of mine struggled to fit their model parameters to flawed and ever-changing data, and the effect of every intervention or re-opening was delayed by two to six weeks. It took months to understand that surfaces don't play much of a role in transmission, that aerosols do, and to finally learn that, yes, children can spread the disease.

While dire predictions can help spur action, demanding that predictions be precise and specific is a fool's errand. We don't, and can't, know precisely where a disease will spread, or how a hotter planet will change. But this is no excuse for inaction.

Unfortunately, intellectual humility and embracing uncertainty—the key values of science—do not play well with policy. Politicians complained that scientists' changing recommendations were contradictory.

They embraced models they liked, including amateurish data-fitting embarrassments like Kevin Hassett's "cubic model," and ignored the ones they didn't. Early entrants like the Institute for Health Metrics and Evaluations' (IHME) model were media successes but not scientific ones, and in reality provided little more than a cartoon. Someone should write a *post pestis* of the roles, both good and bad, that modeling played in the policy debate.

> Looking back over the past year, it's clear that the right response to complexity is not always building a more complex model.

In the absence of crystal balls, the basic message about COVID-19 remains the same: this thing is real, be kind to each other, be patient. Avoid leaders who whip up a mood of willful ignorance. Break the human habit of using every frustration and hardship as a chance to divide "us" and "them." Mobilize society's resources at every level to solve the problem, and support those for whom the intervening time is the most difficult.

Just like we need to do to address climate change.

THE TRANSCRIPT OF *COMPLEXITY PODCAST* TRANSMISSION SERIES EP. 5, DISCUSSING THE FOLLOWING TRANSMISSIONS

[Batch 5, released 27 April 2020]

T-020: COVID Spiraling Frailty Syndrome
John W. Krakauer & Michelle C. Carlson

T-021: What History Can Teach Us about Resilience
Stefani Crabtree

T-022: The Informational Pitfalls of Selective Testing
Van Savage

T-023: Making Good Decisions under Uncertainty
David Tuckett, Lenny Smith, Gerd Gigerenzer & Jürgen Jost

T-024: The Heavy Tail of Outbreaks
Cristopher Moore

This podcast transcript has been abridged for length and clarity. Find the full podcast at https://complexity.simplecast.com/episodes/31

COMPLEXITY PODCAST: TRANSMISSION SERIES NO. 005

EMBRACING COMPLEXITY FOR SYSTEMIC INTERVENTIONS

DATE: *4 May 2020* **HOSTED BY:** *Michael Garfield, featuring David Krakauer*

MICHAEL GARFIELD: All right, David. We're back for week five and, again, I think if we carry this from the micro scale to the macro scale, there's a lot in here that links to topics that we've discussed on previous podcasts, work that's been done by SFI faculty that has not shown up in the show yet. Let's start with Cris Moore's piece[1] because this one I think connects to a lot of work that's being done on network dynamics and social contagions and so on.

DAVID KRAKAUER: Yes. Cris's brings us right back to the start, to this whole question of R_0 again, which is this number that we don't seem to be able to escape from, which tells us how many secondary infections we expect a primary infection to induce. Cris makes this really important point. If you go online, for example, and look up R_0, you'll see an R_0 number associated with every virus, right? Different infections have different R_0 values. The first point to make, of course, is that R_0 is not just a property of a virus; it's a property of its host. So, it really doesn't make that much sense to assign an R_0 just to a pathogen, but you can because it's sort of averaged over the human population. And the whole point of Cris's contribution is to point out the dangers of averages.

Most of us know this already. I mean, you don't go into a clothes store and have them stock just the median size, right? The most common size. That would be useless, because they need sizes that are larger and smaller. You consider the variance as well as the mean. R_0 is a mean in that same sense, and Cris gives the following insight: Imagine that there are parts of the country where R_0 was greater than 1 and other parts of the country where R_0 was less than 1. Or imagine there are populations where it was greater than 1 and less than 1. You would still, when you present the average, conclude that the infection was subcritical. A supercritical infection refers to an infection that leads to exponential

1 See T-024 on page 309

growth, whereas subcritical is 1 that will peter out eventually. We've seen that ourselves in our own state here, Michael, because there are populations who might have preexisting conditions or there are all sorts of socioeconomic disparities that might increase the local R_0. McKinley County, where there are many Native Americans, has the highest rate in New Mexico, and so it's very dangerous to be reporting the average.

In his contribution Cris goes through a number of very simple models where he shows that, in the first instance, imagine you have an R_0 that's less than 1. I think in his particular case it's 0.8. Even with that, you expect to see a variance, a distribution of outbreak sizes. If you were to observe the outbreak independently and, say, in different localities, most of them would be small. About 1% of them would be about 50 and in one of those localities it may go up as high as 80 even though the expected numbers with an R_0 of 0.8 would be 5 on average. So, the average is dangerous because of regional variation. And then there's regional variation in individuality, the so-called superspreader, and the superspreader is an individual who infects many more people than the average. If you have a superspreader that can generate, say, 20 secondary infections, in those cases you can have, even though the average is still low, the average outbreak might still be, say, 5, you could have hundreds of individuals being infected. So, the story here for Cris is: beware of the average, consider the variance, which leads you to a contemplation of what's called the heavy tail in the probability distribution—outliers that are more common than you might have anticipated.

M. GARFIELD: This links to the conversation I had with David Kinney on bringing in economic or energetic considerations into the issue of granularity in scientific investigation.[2] At what level of resolution should we be asking the question? Or, to see it sort of inside out from that way, Mirta Galesic's work on the danger of drawing global conclusions from local samples and how, by only looking at our immediate networks, we generate extraordinary biases about the world at large. Something in Cris's piece that I really appreciate is how it draws our attention back to something I see discussed a lot at SFI about multiscale considerations.

2 Complexity Podcast episode 19 aired on February 19, 2020. https://complexity.simplecast.com/episodes/19

We've addressed this on the show a few times already. For example, the work that you and your co-authors just did on the information theory of individuality and the importance of being able to think of the individual as something that, you know, might be more conventionally understood, like a single person or perhaps an entire society. Like, how can we rigorously ask the question of whether this thing is bound as an individual in that way? When he talks about the difference between thinking of this superspreader individual versus superspreading events and situations, I think that starts to give a sense of the importance of the multiscale view in complex thinking as well as how that provides an entry point for asking, at what scale should we be considering this in order to make the most useful pronouncements?

D. KRAKAUER: It is very important to bear that in mind, because you'll often observe that there are discrepancies between what you observe in your own community versus what's being reported on the news, which tends to be an average. And I think we have to develop a sophistication that allows us to understand how those two things relate. The national average might not be what you're experiencing at all. You might have a much harder time. And Cris's contribution gives us a quantitative insight into why that is expected to be true.

M. GARFIELD: Let's move to the piece by Van Savage on testing capacity and the dangers of not being able to meet the actual spread of the disease with the number of tests we're administering.[3]

D. KRAKAUER: We discussed David Wolpert's contribution last week about the whole challenge of inferring what the true, false-positive–false-negative rates on tests are. But there's another issue in relation to testing that Van raises, relating to the other great phrase that we're all hearing regularly, which is "flattening the curve." Of course, "flattening the curve" refers to the time-dependent daily, weekly, monthly incidence or hospitalization rate and what contributes to that. We all understand that we're isolating so as to minimize the transmissibility of the pathogen and thereby its negative consequences, hence "flattening the curve." But how do we know? The way we know is by testing, and, Bayesian challenges aside, the point that Van makes is

3 See T-022 on page 295

that there's an even more profound rudimentary constraint, which is who you test and the number of tests that you possess.

He just makes this obvious point that if the true number of cases is beyond the maximum testing capacity—which it is—and you're testing people who present with symptoms—which we are—then what you're ultimately doing is reporting on noise, and the curve will appear to be flattening for pathological reasons. If the number of tests is much lower than the true number of infected, then what you're really reporting on is the number of tests, not the true underlying incidence. Take a sporting example—the Rose Bowl in Pasadena or Wembley Stadium in London. Both of those hold about 90,000 fans. Let's say that you filled the stadium. You wouldn't conclude that there were only 90,000 football fans in the US or UK—of course, they're different kinds of football fans—that you'd somehow reached the maximum popularity of their respective sports because they're saturating. And so the signal there is not truly informative of the true demand. It's limited supply. The only way you could truly estimate how popular those two forms of football were would be to take random samples across stadiums from all over the world. And that's precisely Van's point.

. . . if the true number of cases is beyond the maximum testing capacity—which it is—and you're testing people who present with symptoms—which we are—then what you're ultimately doing is reporting on noise, and the curve will appear to be flattening for pathological reasons.

M. GARFIELD: This is compounded by the issue that so many of the COVID-19 cases are asymptomatic, so it really does make filtering, trying to focus who and how we test, a much more complicated issue, and it really does bring us back around to the importance of random testing.

D. KRAKAUER: Exactly. The other point that Van makes is this issue of the exponentials. You know, a disease could be increasing exponentially and you have enough test kits to observe that. Or it could be that you have a huge number of infecteds that we hadn't realized—as you said, who were asymptomatic—but the testing capability is increasing exponentially, in which case it would look like exponential growth. Whereas, in fact, you'd already have saturated the numbers who are infected. There are these interesting perplexities of testing that you have to bear in mind. He goes on to present some simulations where he shows just how bad this problem is. You know, folks can read through those studies. But the key point here is there are only two options available to us: We either massively increase the testing capability or we test randomly.

This brings us to a general point I want to make again about the philosophy of the models, and theory. When Van presents his simple simulations to demonstrate the challenge that we face, what he's doing is generating what we would call a null model. That is, we don't really know how to interpret data unless we have an expectation, and the expectation is established by a rigorous simulation. And we've talked about, throughout this series, other kinds of considerations. We've talked about the prior criterion. We've talked about issues in Cris's contribution about what I call the structure criterion, that is, the assumption of the mean field or the average. And so we're getting a bit of an instruction, I think, over the course of this pandemic into how model building actually works and what we should consider reliable and what we should remain suspicious of.

M. GARFIELD: Definitely. You know, just to generalize this a bit, for people who are more familiar with similar problems and related domains, this kind of problem comes up a lot when we're asking about changes in the incidence of diseases like cancer over the last 150 years, or anything involving a spotty archaeological or paleontological, geological record—changes in species' distributions or volcanic eruptions over time. It brings us back to the same question that Van's asking about whether we actually are observing a difference in the frequency of these events or whether we're observing an improvement in our ability to actually measure them. As that becomes harder and harder to figure, it brings us into the next piece, on making good decisions under uncertainty.[4] This is a really cool piece because this is four authors, all of whom are focused on this particular issue of radical uncertainty.

D. KRAKAUER: This is a very active area of research, often associated with one of the authors here, Lenny Smith, on this question of what do you do when you really don't know. That's what radical uncertainty is meant to convey, where the quantification of a cost and a consequence is highly contestable. Nevertheless, you have to reach a conclusion and implement a policy. What they present in this contribution is the point that, very frequently, the best statistical model that you could generate in earnestness with the best data that you have to hand will be vastly less informative in a deep sense than the cone of all possibilities. In other words, it's a bit like Cris's point—the difference in the mean and the variance. You'd be better off knowing the variance than the mean. And we've seen this—yesterday, US corona-correlated deaths were about 60,000, and that was a number that surprised a lot of people because it wasn't expected to reach that level until about August.

And of course that's right, because we had radical uncertainty about the projected trajectory. The question then is, if you can't report on probabilities because the data just aren't good enough to support that, what can you do? This community has been arguing that, in addition to training people to get used to these presentations of the cone of uncertainty, the variances, and the higher moments of the distribution—but also to think about credible, what they call conviction narratives. Which is, you have to reach a conclusion, the data is not good enough, but a decision needs to be forthcoming: what are you going to do? Most of us in positions of leadership or having to make decisions do this all the time. We weigh to the best of our ability these factors, and then we contrive or invent a narrative that makes sense of them that doesn't appear irrational or totalitarian.

This style of reasoning has become very popular. Actually, this year a book was written by John Kay and Mervyn King—and John, of course, is a frequent visitor to the Santa Fe Institute, very prominent economist. They wrote a book called *Radical Uncertainty*. In that book they do a very good job in some sense of dethroning the reigning probabilistic,

4 See T-023 on page 305

quantitative school, when it comes to domains of uncertainty, but they don't do such a good job in replacing it. They also argue for conviction narratives. I'm very conscious of the fact that narratives can be very, very dangerous because they're very subject to cognitive biases. We've talked about these in the past—you know, attribution biases, confirmation biases, framing biases, and so forth. And so my question in reading this was, it's a very compelling piece; what else is there?

I think we can take some pages out of mathematics. There are ways of reaching certainty that have nothing to do with quantification, and people forget that. People often confound mathematical reasoning with quantitative reasoning. It's not. The Greek proofs in mathematics are very rarely quantitative; think about proofs in typology or geometry or logic. So, what this raises for me is really a conversation about how we reach reasonable conclusions where quantity doesn't dominate, where we recognize the limitations of narrative. Are there other formalisms we might explore which give us means of acting without assigning numbers? I think there are, actually, and this is, I guess, a conversation I would like to have with the authors.

People often confound mathematical reasoning with quantitative reasoning.

M. GARFIELD: In business culture, there's the famous OODA loop—"observe, orient, decide, act"—that was developed by a United States Air Force colonel, John Boyd. You know, this is how to make decisions rapidly under conditions of great uncertainty as a fighter pilot. And that's where he said, if you think about your narrative as the direction you're pointing the plane, right? Calling back to Manfred Laubichler's piece on fitness landscapes, the authors of this piece say, crucially, optimization in radical uncertainty is dangerous because its premises are not satisfied. In a period of cataclysmic change, the fitness landscape shifts, as we have discussed in previous episodes. There are valleys where mountains used to be, and if you continue to optimize for the mountain of your narrative, you're going to walk off a cliff.

D. KRAKAUER: It's very difficult, because we don't really have many alternatives. I was very convinced in about, I think it was 2018—I read a book called *The Tyranny of Metrics* by Jerry Muller. He's basically making this point that we've gone from measuring performance to fixating on measurement itself. Science is full of this. I mean, most of us are nauseated by this idea of the h-index, which is a means of measuring scientific popularity. And that's the key point: it measures scientific popularity, not scientific profundity, and these indices which have been proliferating across different sectors are essentially a substitute for judgment. I do think that this article raises a question that we have to address in all walks of life, where quantification is spurious rigor and I feel it's that important that we should, at the Institute, be thinking very carefully about it. I tend to like logical simulation environments generating the cone of uncertainty through a principled set of null models, which at least gives you a sense of the full space of counterfactuals which you can then use to navigate. But I think it's very open territory.

. . . we've gone from measuring performance to fixating on measurement itself. Science is full of this.

M. GARFIELD: As you did last week, calling to orthogonal discussions that we've held at InterPlanetary Festival—your piece on counterfactuals and null models is a callback to the world-building panel that we held last year with James S.A. Corey and Rebecca Roanhorse and Michael Drout on how the work of science-fiction authors helps us do this in a nonquantitative way. It allows us to speculate, and this is possibly why the work of science-fiction authors has become so useful in a very rapidly moving business environment in helping us map the full spectrum of possible scenarios. Right? That's why you've got Neal Stephenson on the list of former Miller Scholars.

D. KRAKAUER: That's exactly right. I've described narrative as ontology engines a number of times. I think that is actually what they mean, the authors, by narrative and conviction narratives. I think great authors do precisely that. And Neal is a great example of someone who has some of the most sophisticated reasoning about the near-term future of the economy, say, or technologies, because he's exploring that counterfactual space. But I think that there must be alternatives to that, that we could somehow conjoin the best of narrative with the best of modeling to create a real engine that allows us to generate alternative possibilities. And I don't think we have such a thing yet.

M. GARFIELD: Let's act as if we can. You know, moving from that, let's take a dog leg here. It's a little harder to find a clear through line into the work by your brother John Krakauer and Michelle Carlson. But it's a really interesting piece on spiraling frailty and the importance of physical exercise in countering some of the worst effects of this disease.[5]

D. KRAKAUER: Obviously, I'm somewhat biased. It's a very interesting article, partly because it was written by my brother and a colleague, but what they're raising is the question I think that many of us are asking, sitting at home: How do we remain active? How do we remain healthy? There are alarming observations that the COVID-19 virus, in addition to immobilizing very many of us, is actually attacking individuals who have conditions that are compounded through sedentary lifestyles. The Transmission begins with this discussion of frailty, which is a real biological syndrome that about 10% of Americans have, that's associated with various clinical components, including muscle tissue loss, a slowing down of gait, a significant reduction of physical activity, that feeling of always being low energy or being exhausted and so forth. This syndrome increases markedly over the age of sixty-five—and we can't but notice that the age of sixty-five is that kind of magic category where you become significantly more susceptible to the coronavirus.

It's the group of sixty-five up that are most vulnerable to this infection and are experiencing the highest hospitalization rates. So, it looks as if there is a synergy between these two things and the frailty is often unmasked or amplified by a perturbation such as this infection. In the medical community, John and Michelle make this point that there is a preference to treat conditions with monotherapies. We can attack that single system with a single drug, and the fact is that most of us know

5 See T-020 on page 277

it doesn't work. Right? In the case of energy, estrogen or testosterone replacement therapy would be a good example. They cite the work of Linda Fried, who defined frailty syndrome, and she makes this really interesting point that the extraordinary thing about physical activity is that it's not a monotherapy. It actually up-regulates many systems that are necessary to maintain health and they're particularly effective at ameliorating two chronic diseases, hypertension and type-2 diabetes, that are the primary comorbidities with coronavirus infection.

So, it's really quite deep. It's not just that you're better off exercising because it will reduce your susceptibility, but those groups that are most susceptible stand to benefit most. I think that point was really important. The other point that they make is how might we intervene on the system that is the entire-body system. And one point that they're making here is that there is a reluctance to engage in physical activity. It can be psychologically trying. You might feel as if you're not healthy enough even to begin. John has been a very strong advocate for game-like environments, which you can conduct without judgment, which are extraordinarily fun and can encourage relatively high levels of mobility inside the house. Think about things like Wii Sports or Beat Saber or Ring Fit Adventure. All of these are games which ask you to move. They're fun to engage in, the activation energy required to engage in them is relatively low, unlike going for a five- or six-mile run, and I think they're arguing that perhaps this kind of technology might be an important complement to regular forms of physical activity outside in times when you're being asked to socially isolate.

M. GARFIELD: This pandemic has driven so many social gatherings online kind of prematurely before we can really engage in a robust simulacrum of what it actually means to meet with one another in embodied physical spaces. It feels like we're right at the point here where this has happened to us early enough that we're still sort of climbing up from base camp in terms of the way that embodied physical computing—you know, the actual potential here. The plans that were popularized by films like *Minority Report* where you're not just sitting at a desk typing on a thing—as I and several of us at SFI are now trapped in a sedentary work environment and really have to force ourselves out of the seat—but, in addition to the games that you listed, I've been seeing ads for games like these balancing boogie board-type deals with video games built in, where the video game controller is your body. That's really exciting. But, to your first point, I feel like what they're getting at is something that I see brought up again and again in the underlying hidden patterns, the generalities that are opened up and revealed through complex-systems science, specifically in relationship to understanding things, again, as multiscale networks. And this piece in particular reminded me of the work of Jennifer Dunne on trophic networks and the conversations that we had about this way back in episodes five and six. If you think about the causal relationships in the body, like a food web, then physical exercise might be like the keystone species. If you remove it then the entire web collapses, and it's really important to identify it in, analogically, the conservation efforts that you apply to your body ecology.

It's not just that you're better off exercising because it will reduce your susceptibility, but those groups that are most susceptible stand to benefit most.

D. KRAKAUER: I think it's this very interesting shift of perspective that comes with familiarity with complexity—this notion of a systemic intervention. You know, in their article they cite this new book by Peter Sterling called *What is Health?* He makes this point that a complex therapeutic intervention—and he's interested in the case of hypertension or diabetes—is not that you begin with an antihypertensive or antidiabetic polypharmacological intervention, but you do exercise, right? You move. It seems to many people a little counterintuitive. And there is something that needs to be explained about our reluctance to engage in activities that are extraordinarily beneficial to health, which have very low costs. These are not pharmacological interventions. Why are we reluctant to take those, but we are so willing to spend huge sums of money on pills which do far, far less? I think one of the areas where this has been very obvious has been depression, right?

This is a real area of concern. We treat them with antidepressants, they have significant side effects, but one of the most effective means of treating depression is with exercise. I think that John and Michelle are getting at a really interesting point: are there technologies like gaming, for example, which would make exercise feel as legit as taking a drug? Somehow hanging them on an armature of engineering gives them a placebo-like salience that they might not have if we just step outside to go for a run. This is a psychological challenge for us that we all need to face.

M. GARFIELD: I wonder to what extent this is due to the fact that we have sort of eroded the topsoil, culturally, in terms of the ways that we're able to look to the lessons of human history for how people in other cultures have addressed comparable issues in their own times. We tend to think of our time as radically different than other moments in history, and so we leave questions about the innate benefits of the health and well-being to someone through gardening and the way that gardening helps cultivate a person's microbiome and exposes them to sunlight and keeps them physically active and so on. Because we're not economically motivated to draw those analogies or we're motivated to believe that our times are unique and call for modern, high-technological interventions.

Medicine will evolve by recognizing the complexity of the body and pursuing complex interventions. Physical exercise and sleep are two very good examples of that.

D. KRAKAUER: You know, it's interesting, Michael. I think it comes down to insufficient appreciation of complexity. I like to think about this in terms of causality and I think you raised it, which is, we're energy-minimizing cognitive systems. And if we can find dominant mono-causal explanations, we prefer them—and probably for good reasons. But that leads us astray and it does lend a cognitive bias to us which disposes us toward solutions that look as if they're single silver bullets. What we now know, I think, about health, about mental health, physical health, is that they are the convergence of many different systems and that they have to be treated with something as complex as they are. Right? It's a little bit like education that we discussed in relation to Carrie's Transmission.[6] You can pretend that learning is simple and have a correspondingly simple pedagogy, or you can recognize it's not and explore the full panoply of possibilities. And I think it's exactly the same here. Medicine will evolve by recognizing the complexity of the body and pursuing complex interventions. Physical exercise and sleep are two very good examples of that.

M. GARFIELD: There's a link here to the first piece we discussed in this call and Cris's caution around R_0, the average, is exactly what you're saying here. It appeals to our desire for a simple story and the link to David Kinney's work and his insistence on the inclusion of context in our philosophy of how we address specific scientific questions, that it's unpopular in philosophy to insist that there's a "yeah, but," that there's conditional considerations. People want to be able to make sweeping universal claims. And the reality is that that's insufficiently detailed, right? I mean, it's not going to get us there.

Let's move on to Stefani Crabtree's piece.[7] I think this is a really simple and inspiring statement that she is making about the importance of reflecting on historical precedent here.

D. KRAKAUER: We had already talked in relation to the frightening economic downturn that this is one depression among many—or one opportunity among many—and the same goes for the experience of plagues and epidemics and famines. Stefani opens her piece by making the observation that the practice of quarantine that we are now experiencing derives from a fourteenth-century Venetian convention of keeping ships at harbor to protect the citizens of the city-state from pathogens. She goes on to discuss the plague of Athens in the fifth century, which killed one-third of the city-state's population, and the value of narrative, of historical narrative, by Thucydides and others in helping us come to terms with that plague. How, in fact, he describes the spread of fear as moving more quickly than the spread of the plague.

6 See T-016 on page 219

7 See T-021 on page 287

It's something that we discussed in relation to the infection of memes as opposed to the infection of viral genes.

There is a historical context that all civilizations have experienced and learned to cope with. But I think where the article becomes more serious, which is something that we haven't yet discussed in as much detail as I'm sure we will, is in terms of the social upheavals and implications of these plagues. Stefani herself has worked for a long time on southwestern archaeology, in particular studying the Chaco culture. This is the culture right here, very close to us, which thrived between the tenth and the twelfth centuries, but experienced a very precipitous decline when the Ancestral Puebloans left. There are all sorts of hypotheses for why this is true, many of them having to do with drought and severe shortages in protein. But the point is, it led to significant escalations in warfare and violence and that connection, not just between disease and markets, but between disease and civil unrest, is something that I'm sure is at the back of many of our minds.

. . . if the climate catastrophe should begin to become more apparent on the heels of the coronavirus catastrophe, society is in for a real battle.

So that's a very important point. I was reminded, in fact, when reading through Stefani's Transmission, of a book that I read a couple of years ago by Kyle Harper called *The Fate of Rome*, and the great book on *The Decline and Fall of the Roman Empire* was written by Edward Gibbon. And in Gibbon's book, the empire declines for two reasons, really. One is self-destructive religious superstition, and the other is warfare at its boundaries. What Harper is arguing is that we've overstated those social factors and underestimated the importance of climate and epidemics. And of course he talks at length about the Antonine Plague, which was in the second century AD, where a thriving Roman Empire under Augustus was brought to its knees by smallpox. It's a little bit like our bull market that was brought to its knees by the coronavirus. So, that's another example where we tend to underestimate certain invisible factors, certain complex causes, in favor of the standard explanations. What Kyle Harper goes on to do, actually, in that book is show that, right after the Antonine plague, just as Roman society was recovering, Rome experienced what's called the Late Antique Little Ice Age, which was very, very unfavorable to agriculture. And so one catastrophe came on the heels of another and that compounding effect really brought the empire down, as opposed to the standard Gibbon explanation. Many of us are talking about climate now and if the climate catastrophe should begin to become more apparent on the heels of the coronavirus catastrophe, society is in for a real battle.

M. GARFIELD: It'll look somewhat similar to coronavirus infecting someone who's already suffering from spiraling frailty, right? For our listeners, Kyle Harper gave a presentation at our symposium last year on this book that was extremely fascinating and I think can be filed under a long list of rather prophetic presentations by people at SFI, including Lauren Ancel Meyers's work on preventing the next pandemic for last year's Ulam lectures. And you can find all of that stuff on our YouTube channel if you want to go deeper into that.[8]

D. KRAKAUER: Michael, I think that you make a really good point about nonobvious analogies between Kyle Harper's work and John and Michelle's points, which again relate to this complex-systems perspective—that it's very rarely the case that interesting or dramatic phenomena have simple causes. One of the dangers, I think, of the current crisis is this collapse of the dimension of causality to a single cause, that is, the virus. Forgetting that much of what we're observing is pre-existing inequalities of susceptibility or economic circumstance amplifying the effects of a virus. And that is important to bear in mind, and I'm hoping that this series is contributing to that kind of insight.

M. GARFIELD: You know, relatedly, to call back to Doug Erwin's commentary on mass extinctions,[9] I think most people similarly look for a silver bullet when we're trying to understand Earth history. And over the last fifteen years or so there's been a shift in the way that we

8 https://www.youtube.com/c/SFIScience/

9 See T-014 on page 187

understand mass extinctions as complex events that are caused by the compounding of factors like we're talking about here. Most people, thanks to Luis Walter Alvarez, have it in their minds that the end-Cretaceous mass extinction was triggered by an asteroid impact. But at the same time, this was concurrent with a two-million-year volcanic eruption in what is now India that caused climate change, that there were changes in atmospheric and ocean chemistry, there was the emergence of land bridges. You know, when I was studying this stuff, my mentor Robert Bakker talked a lot about the way that organisms were changing their migration patterns and he really insisted that pandemic disease transmission also probably played a very large role in the extinction of the dinosaurs. It's just yet another example where, like you said, we have to be careful about oversimplifying the story here.

D. KRAKAUER: Absolutely.

M. GARFIELD: There's one more link I'd like to make because you brought it up with Stefani's piece. When she's talking about the fall of the Chaco Canyon society, she mentions that, when productivity failed, the Chacoan hierarchical society fractured, violence increased, and society reorganized into regional communities that issued hierarchy. This calls back to Miguel Fuentes and his work on how social graphs disintegrate in crisis,[10] and Tony Eagan's piece on the balance between federal and state governance and disasters,[11] and how you suggested that maybe acclimating to the challenges that we're facing today might require dynamic constitutions that can become more or less strict in their regulation depending on the needs of the moment.

I feel like we're at a point in the series now where we can start to trace the outline of a more general systems way of thinking about these things. You know, that we can understand why it would be that the societies that came up after the fall of Rome or after the fall of Chaco or the sort of specter of Balkanization that's lingering over us now are actually not just a failure of a coherent system, but actually an adaptation to a crisis that's happening faster than the latencies of a massive and unified institutional response can address with respect to on-the-ground

10 See T-006 on page 101

11 See T-015 on page 211

conditions. It becomes very difficult for the general to conduct the war when something bizarre is happening rapidly on every front, more emergency local elections, more self-organizing duocracies that we see spring up in disaster relief. I'm curious how you understand this in terms of the insights that complexity science can offer us in adaptive strategies for social organization on the other side of this.

D. KRAKAUER: I think one useful mantra might be, complex causality requires complex control or complex intervention, and the history of thought is a history of low-hanging fruit. We start with the simplest things that we can intervene into in the simplest way. We see an apple fall from a tree and with remarkable ingenuity arrive at a theory of gravity. We develop engineering techniques of ballistics that are mirrors of that world. But when you get to the complex world, things don't work that way anymore. Things are connected in subtle ways. There are nonlinearities. There are different phases and critical transitions and we're just at the baby steps of understanding how to intervene into complex systems. The last thing we want to do is pretend that they're simple systems, and I think that's what we have been doing. A number of these contributions that we've been reading over the last few weeks are pointing out the challenges and dangers of doing that, and today we've arrived at this point where we're getting an insight into what systemic intervention and control might look like. The good news is it's available to us and available to us not at considerable cost, at least in relation to health.

6

[BATCH 6, RELEASED 4 MAY 2020]

To what extent can humans alone? Or is **PHYSICAL** *connectivity the* **VERY** *life,* **LIKE SLEEP: REGARDLESS** *of entertainment, or money*

live by the **INTERNET**

three-dimensional social

ESSENCE *of urban*

without it you die,

how much nutrition,

you have?

— GEOFFREY WEST & CHRISTOPHER KEMPES, **Transmission 025**

CLEAN

TRANSMISSION NO. 025

UNDERSTANDING CITIES TO RESPOND TO PANDEMICS

DATE: *4 May 2020*

FROM: *Geoffrey West, Santa Fe Institute*
Christopher P. Kempes, Santa Fe Institute

STRATEGIC INSIGHT: *Policies for responding to pandemics should be rooted in a scientific understanding of cities.*

It is hardly surprising that the vast majority of COVID-19 cases have occurred in cities; after all, more than 80% of the world's population lives in urban environments, and these are significantly denser than either suburban or rural communities. Consequently, understanding the detailed transmission, spread, and mitigation of the disease—and developing realistic pathways to long-term sustainable biological, social, and economic recovery—is intimately tied to developing a deeper understanding of the principles underlying the structure and dynamics of cities. This is critical at both a coarse-grained global scale as well as at a more fine-grained local level. Indeed, a much-needed conceptual perspective is one in which we view this as a quintessential complex adaptive system, where all components are inevitably interrelated, resulting in a plethora of "unintended consequences."

We should not be totally surprised that an accidental mutation of a virus in a city in China, when left unchecked, would potentially lead to huge unemployment in the US, the fall of global markets, no football games in Spain, less pollution in India, and a shortage of yeast in the UK: the ultimate "butterfly effect," where the subsequent time development is exponentially sensitive to the initial conditions, making detailed predictions extremely challenging. We are seeing that many dimensions of society are severely less robust than we might have

---

OPPOSITE: *George Bellows, "New York," 1911 (detail)*

thought. The spread of COVID-19 is a complex chaotic phenomenon in more ways than one!

The big question is, of course: to what extent, if any, can any of this be quantitatively predictable? It certainly involves considerably more than "just" traditional epidemiology, vaccinology, and health care. In the near term, these obviously play a dominant role, but as we move forward into the recovery phase, and more importantly, into developing a long-term sustainable phase, this needs to be holistically integrated with socioeconomic dynamics—such as finance, inequality, neighbor-

hood structure, and so on—coupled to the physical infrastructural organization of buildings, transport, etc. In a word, we need to develop and integrate our thinking about epidemics and similar "predictable" threats—such as earthquakes, tsunamis, and conflicts—across multiple time and spatial scales, with a quantitative "Science of Cities." It is a daunting challenge.

Cities are machines we evolved to facilitate, accelerate, amplify, and densify social interactions. The larger the city, the more the average individual interacts with other people in a multiplicative positive feedback process; by engaging in social discourse, exchanging ideas and information, making financial transactions, and, unfortunately, transmitting viruses!

Despite this, some baby steps have been made. Urban scaling, the theoretical framework quantifying how urban metrics change with size, reveals that cities are highly nonlinear yet share surprisingly "universal" commonalities.[1] For example, socioeconomic quantities (Y) like wages, patents, and wealth scale superlinearly with population size (N) following a classic simple power law whose exponent is ~ 1.15: i.e., as $N^{1.15}$.

In English, this means that their percentage increase is 15% larger than that of the population: i.e., $dY/Y = 1.15\ dN/N$. Consequently, the bigger the city, the more wealth, innovation, and social connectivity there is *per capita*—a major reason why cities are so attractive. This has its origins in the mathematics and dynamics of social networks. Cities are machines we evolved to facilitate, accelerate, amplify, and densify social interactions. The larger the city, the more the average individual interacts with other people in a multiplicative positive feedback process; by engaging in social discourse, exchanging ideas and information, making financial transactions, and, unfortunately, transmitting viruses![2]

Consequently, all of the benefits and attractions of larger cities that result from increased social connectivity have a dark side: more crime, greater inequality, more pollution, and more disease, all following superlinear scaling laws.[3] Not only are there systematically more cases but, equally importantly, their growth rate, like all socioeconomic urban phenomena, increases systematically faster. If the number of cases increases exponentially with time (t) as e^{rt}, then the rate parameter (r) is predicted to systematically increase with city size as $\sim N^{0.15}$. Consequently, a city of a million people will double the number of cases in approximately half the time a city of 10,000 (one hundredth of its size) will, since r changes by a factor $(100)^{0.15} \sim 1.99$.

Such superlinear scaling predictions have been confirmed with data from past AIDS epidemics,[1,4] so it is not surprising that the growth rate of COVID-19 has likewise followed similar laws, at least in its initial phase before intervention, as recently shown.[5] The underlying theory potentially provides a quantitative framework for mitigating the spread of the disease by appropriately pruning the social network (i.e., by decreasing connectivity, or "social distancing"). But, importantly, the city's social network is coupled with the physical constraints of infrastructural networks like transport and commerce. This necessary component is typically missing from epidemiological models, which are therefore blind to the strong dependence of social distancing on a city's size. Consequently, they are unable to incorporate or predict the inherently greater risk in larger cities, a fact which needs to be taken into account when formulating policy decisions.

There are a plethora of other problems and risks associated with the pandemic that are being hotly debated, many related to the economy, poverty, and quality of life, and therefore, by implication, to the role of cities.

How does an individual's risk of infection change based on the city, or the particular neighborhood, they live in? Can urban scaling theory help formulate optimal strategies for aiding cities of different sizes? What can we learn from urban scaling theory that might inform us about trajectories of recovery and what cities and urban life might look like post-pandemic? To what extent can humans live by the internet alone? Or is physical three-dimensional social connectivity, the very essence of urban life, like sleep: without it you die, regardless of how much nutrition, entertainment, or money you have? The short, but incomplete, answer to these questions is that cities are not monochromatic, and comprehensive policies for preparing responses to, and recovery from, a variety of crises should begin by at least taking into account the overall size and underlying network structures of cities.

ENDNOTES

1 L.M.A. Bettencourt, J. Lobo, et al., 2007, "Growth, Innovation, Scaling, and the Pace of Life in Cities," *Proceedings of the National Academy of Sciences* 104(17): 7301–7306, doi: 10.1073/pnas.0610172104

2 M. Schlapfer, L.M.A. Bettencourt, et al., 2014, "The Scaling of Human Interactions with City Size," *Journal of the Royal Society Interface* 11(98): 20130789, doi: 10.1098/rsif.2013.0789

3 L.M.A. Bettencourt and G.B. West, 2010, "A Unified Theory of Urban Living" *Nature* 467: 912–913, doi: 10.1038/467912a

4 F. Antonio, S. de Picoli, Jr., et al., 2017, "Spatial Patterns of Dengue Cases in Brazil," *PLoS One*. 2014; 9(10): e111015, 10.1371/journal.pone.0180715

5 A.J. Stier, M.G. Berman, and L.M.A.. Bettencourt, 2020, "COVID-19 Attack Rate Increases with City Size," arXiv preprint, https://arxiv.org/pdf/2003.10376.pdf

REFLECTION

SCALING AND COMPENSATING TIMESCALES OF RISK & RESPONSE

DATE: *2 September 2021*

FROM: *Geoffrey West, Santa Fe Institute*
Christopher P. Kempes, Santa Fe Institute

The 1918 flu pandemic took roughly two years to complete its three-wave progression of global infections. Since then, much has changed: we have come to understand the fundamentals of molecular biology, including how RNA viruses hijack cells for replication, developed a huge variety of sophisticated models of viral evolution and epidemiology, designed vaccines for a large number of diseases, and created the infrastructure for producing any mRNA or DNA sequence on demand and at low cost. There has been almost unimaginable progress in the biological and medical sciences in the intervening century. At the same time we have developed sophisticated methodologies for understanding social behavior, including the dynamics of social networks. And, with the advent of the IT revolution, we now have rapid communication across the globe and the ability to gather enormous amounts of data. None of these extraordinary advances was part of our toolkit a hundred years ago. Yet, despite all of this, and rather surprisingly, the first major pandemic to hit since then will take approximately the same length of time, namely about two years, to work through a similarly three-wave structure. Somehow all of the marvelous scientific advances of the last hundred years do not seem to have changed the timescale or temporal pattern of the pandemic.

How can this be? At first glance it is one of the more unexpected aspects of the present pandemic, but on closer inspection it is something that perhaps we might have anticipated.

In our original Transmission we described the need to understand how scale fundamentally affects the dynamical evolution of systems. For example, as cities grow and become more densely connected, with higher interaction rates between people, the transmission rate of any virus—or, for that matter, any idea—systematically increases.

Consequently, larger cities are generally more creative than smaller ones as manifested, for instance, in the systematic increase in the number of patents they produce per capita. However, the dark side of this increasing connectivity is that cities are fundamentally more susceptible to the rapid spread of disease between people.

As our advanced technological response and information-gathering and reporting abilities increased, they were effectively compensated for, or counteracted by, faster infection rates that reached all corners of the globe.

The answer to why the COVID-19 pandemic is taking roughly the same amount of time to play itself out as did the 1918 flu pandemic, despite all of the marvelous scientific advances, could be that the same societal infrastructure that has produced these innovations has also caused numerous commensurate shifts in human society. The same features and dynamics that increased the rate of innovation have also demanded a faster overall pace of life, resulting in increased global interactions through travel, upticks in human population and density, and radical surges in the rates at which information and misinformation spread. There is inherent feedback and interconnectivity between scientific innovations and social rates because, ultimately, they both originate in the dynamics of social networks, and so the two grow together. Thus, as our advanced technological response and information-gathering and reporting abilities increased, they were effectively compensated for, or counteracted by, faster infection rates that reached all corners of the globe.

Strikingly, when all of this is put together, an approximately invariant (that is, unchanging) timescale may have emerged. This mirrors many aspects of biological and social scaling theory, where certain invariants apply to diverse organisms of radically different size and evolutionary history. For example, the cost of repairing a unit of biomass is an invariant, as is the lifetime energy use per unit mass; this results in the number of heartbeats in a mammal's lifetime being approximately the same, whether a mouse, a giraffe, or an elephant. These are the results of severe evolutionary pressures that define a strict set of trade-offs that lead to invariants. The question for human society is whether we can escape such invariants. Are all future pandemics destined to take approximately two years, or are there fundamental shifts that can be made to allow innovations to outpace the processes that cause risk to propagate more quickly? We need more fundamental theories of trade-offs to answer these questions, but at present it would seem that the emergent dynamic compensates for the increasing rate of innovation with increased risks in certain dimensions.

However, it is important to note that other well-known characteristics of viruses affect the temporal dynamics of an epidemic, such as transmissibility, mortality rate, and percentage of asymptomatic and vaccinated individuals. After all, many smaller epidemics have ended faster than the two years that we have been discussing, such as the 2009 H1N1 flu or various Ebola outbreaks. Our suggestion here is that for viruses with similar properties, the timescales of risk and timescales of response may offset one another because of similar upstream generative mechanisms. Finally, it is worth emphasizing that though the length of the pandemic may be difficult to control, the huge biomedical scientific advances of the last hundred years have potentially mitigated many infections and deaths relative to an unchecked outbreak or longer timescale to produce a vaccine.

TRANSMISSION NO. 026

MECHANISM DESIGN FOR THE MARKET

DATE: *4 May 2020*

FROM: *Eric Maskin, Harvard University*

STRATEGIC INSIGHT: *Mechanism design can aid the market in meeting extraordinary needs under unusual circumstances.*[1]

Under normal circumstances, most goods and services are produced, bought, and sold through free markets. But in an emergency like a pandemic, markets may not suffice. Imagine, for example, that society suddenly needs to undertake tens or even hundreds of millions of virus tests a week (so that employers can put their employees back to work safely). To whom can we turn to produce the testing equipment? There may be many potential manufacturers, and how can we know who they all are? Even if we know their identities, how do we decide which ones should actually do the producing? How much should each produce? And what price should a producer receive to cover its costs?

If we had the luxury of time, the market might resolve all these questions: prices and quantities would adjust until supply and demand were brought into balance. But getting a new market of this size to equilibrate quickly is unrealistic. Furthermore, markets don't work well when there are concentrations of power on either the buying or selling side, as there might well be here. Fortunately, mechanism design can be enlisted to help.

---

1 This piece is based on a Santa Fe Institute webinar talk on the Complexity of COVID-19, April 14, 2020.

OPPOSITE: *Melton Prior, "Auctioneer Selling Fish from a Platform to an Excited Crowd," 1870 (detail)*

WELLCOME COLLECTION

MARKETS

Before getting to mechanism design, let's review why markets normally work so well. Suppose that there are many buyers and producers for some good. Suppose that buyer i enjoys (gross) benefit $b_i(x_i)$ from quantity x_i. Similarly, each producer j incurs cost $c_j(y_j)$ to produce y_j. Hence, society's net social benefit is:

$$\sum_i b_i(x_i) - \sum_j c_j(y_j). \tag{1}$$

At a social optimum, (1) is maximized subject to the constraint that supply equals demand:

$$\sum x_i = \sum y_j. \tag{2}$$

The solution to this constrained maximization is optimal in several senses:

(i) total production $\sum y_j$ and total consumption $\sum x_i$ are optimal;
(ii) y_j is optimal for each producer j; and
(iii) x_i is optimal for each buyer i.

Achieving all three optimalities may seem complicated, but the market provides a simple solution. If p is the price at which the good can be bought and sold, then each buyer i maximizes

$$b_i(x_i) - px_i \text{ (net benefit)} \tag{3}$$

and the first-order condition for this maximization is

$$b'_i(x_i) = p \text{ (} b' \text{ denotes the derivative of } b\text{).} \tag{4}$$

Similarly, each producer j maximizes

$$py_j - c_j(y_j) \text{ (profit)} \tag{5}$$

with first-order condition

$$p = c'_j(y_j). \tag{6}$$

But notice that (4) and (6) are also the first-order conditions for the problem of maximizing (1) subject to (2). And so the market outcome attains the social optimum as long as p is chosen so that (2) holds. (Mathematically, p is the Lagrange multiplier for (2).)

But how do we get the right choice of p? In a free market, p falls if supply exceeds demand and rises if demand exceeds supply. Eventually, the equilibrating price is found. But this process takes time. In the meantime, the price may be way too high, in which case, buyers who need tests are being "gouged," or too low, in which case there may be a serious shortage of tests.

There is an additional problem with the market solution: it relies on producers and buyers being "small" so that they can't individually affect the price. If some of these agents are big (e.g., if one of the equipment-producers supplies a significant fraction of demand), then the optimizations in (3) and (5) have to be modified and a social optimum no longer obtains. Moreover, by withholding supply, a big producer can distort the price-adjustment procedure and generate an outcome in which the price is too high and market supply is too low relative to the optimum. (A big buyer can do just the opposite.)

MECHANISM DESIGN TO THE RESCUE[2]

For both reasons, we now turn to mechanism design.[3] For now, let us assume that the government attaches (gross) benefit $b(\sum y_j)$ to total production $\sum y_j$. (In the next section we decompose $b(\sum y_j)$ into the underlying benefits $\{b_i(y_j)\}$of test-equipment users.)

The government is interested in maximizing the net social benefit

$$b(\sum_j y_j) - \sum_j c_j(y_j)$$

but it doesn't know the cost functions $\{c_j\}$ (and may not even know the full set of potential producers). We solve this difficulty using a variant of the Vickrey–Clarke–Groves mechanism.[4] Specifically, the government

2 This section and the next are a bit math-heavy. For a simple explanation, see page 349

3 An alternative to markets or mechanism design would be for government to simply order some company or companies to produce all the equipment. But this might be an extraordinarily inefficient outcome if these companies aren't up to the task or if there are other companies who could produce it much more cheaply (which the government is not likely to know in advance). Moreover, how does the government know equipment level is "right"?

4 See W. Vickrey, 1961, "Counterspeculation, Auctions, and Competitive Sealed Tenders," *Journal of Finance* 16(1): 8–37; E. Clarke, 1971, "Multipart Pricing of Public Goods," *Public Choice* 11: 17–33; and T. Groves, 1973, "Incentives in Teams," *Econometrica* 41(4): 617–631.

announces a call for test-equipment production and has each potential producer j submit a cost function $\hat{c}_j$. It then computes the production levels $\{\hat{y}_j\}$ that maximize the *apparent* net social benefit

$$b(\sum_j y_j) - \sum_j \hat{c}_j(y_j) \tag{7}$$

and has producer k produce $\hat{y}_k$ and gives producer k a payment:

$$\left[b(\sum_j \hat{y}_j) - \sum_{j \neq k} \hat{c}_j(\hat{y}_j)\right] - \left[b(\sum_{j \neq k} \hat{y}_j^*) - \sum_{j \neq k} \hat{c}_j(\hat{y}_j^*)\right], \tag{8}$$

where the levels $\{\hat{y}_j^*\}_{j \neq k}$ maximize $b(\sum_{j \neq k} y_j) - \sum_{j \neq k} \hat{c}_j(y_j)$.

Claim: Given that the government chooses $\{\hat{y}_j\}$ to maximize (7) and pays producer k the amount (8), it is optimal for producer k to report its costs *truthfully*, i.e., it will take $\hat{c}_k = c_k$.

Proof: The second expression in square brackets in (8) doesn't depend on $\hat{c}_k$ and so doesn't affect producer k's maximization. In effect, producer k maximizes

$$b(\sum_j \hat{y}_j) - \sum_{j \neq k} \hat{c}_j(\hat{y}_j) - c_k(\hat{y}_k). \tag{9}$$

But (9) is just net social benefit with cost functions c_k and $\{\hat{c}_j\}_{j \neq k}$, i.e., producer k's objective is the same as society's. Thus, the optimal choice of $\hat{c}_k$ is indeed c_k. Q.E.D.

BUYERS' BENEFITS

Let us now decompose $b(\cdot)$ into $\sum_i b_i(\cdot)$.

Because government doesn't know the benefit functions $\{b_i\}$, it will have buyers report $\{\hat{b}_i\}$ (as well as having producers report $\{\hat{c}_j\}$) and, instead of maximizing (7) it will choose $\{\hat{x}_i\}$ and $\{\hat{y}_j\}$ to maximize

$$\sum_i \hat{b}_i(\hat{x}_i) - \sum j\hat{c}_j(\hat{y}_j) \text{ subject to } \sum_i \hat{x}_i - \sum_j \hat{y}_j. \tag{10}$$

Buyer h then receives $\hat{x}_h$ and pays

$$\left[\sum_j \hat{c}_j(\hat{y}_j) - \sum_{i \neq h} \hat{b}_i(\hat{x}_i)\right] - \left[\sum_j \hat{c}_j(\hat{y}_j^*) - \sum_{i \neq h} \hat{b}_i(\hat{x}_i^*)\right],$$

where $\{\hat{x}_i^*\}$ and $\{\hat{y}_j^*\}$ maximize

$$\sum_{i \neq h} \hat{b}_i(x_i) - \sum_j \hat{c}_j(y_j). \tag{11}$$

By analogy with producer k's problem in section 2, it is optimal for buyer h in these circumstances to set $\hat{b}_h = b_h$.

SIMPLE EXAMPLE

Imagine that there is just a single buyer with benefit function $b(\cdot)$ and a single producer with cost function $c(\cdot)$. In that case, the government

(i) has the buyer report $\hat{b}(\cdot)$ and the producer report $\hat{c}(\cdot)$;
(ii) calculates z^* to maximize $\hat{b}(z) - \hat{c}(z)$;
(iii) has the producer produce z^* and deliver this to the buyer; and
(iv) pays the producer $\hat{b}(z^*)$ and taxes the buyer $\hat{c}(z^*)$.

Notice that the buyer's objective function is

$$b(z) - \hat{c}(z)$$

and the producer's is

$$\hat{b}(z) - c(z)$$

and so it is optimal for the buyer to report $\hat{b} = b$ and for the producer to report $\hat{c} = c$.

As usual in the mechanism design literature, the way to align social and individual goals is to give individual producers and buyers monetary transfers (either positive or negative) that transform their personal objective functions into the social objective function.

REFLECTION

FURTHER REFLECTIONS ON MECHANISM DESIGN AND THE PANDEMIC

DATE: *25 August 2021*

FROM: *Eric Maskin, Harvard University*

In May 2020, I wrote a Transmission based on my participation in an April SFI panel on complexity and the COVID-19 pandemic. My particular angle was to offer some suggestions about how mechanism design—the "reverse engineering" part of economics—might be useful in dealing with the pandemic.

I noted then that, in normal circumstances, ordinary markets do an excellent job of ensuring that the goods and services people want and need are produced and distributed. If, for example, there is currently greater demand for potatoes than available supply, we can expect the price of potatoes to rise. This will have two effects: (1) demand will be curtailed, but, more importantly, (2) potato growers will be induced to sell more potatoes, thus ameliorating the initial shortage.

However, as I discussed, there are at least two reasons why a laissez-faire approach is not likely to work very well for certain critical goods during a pandemic: (1) there may be no existing market for a good, yet it is needed right away; and (2) the good is, at least in part, a *public* good (its benefits go not just to the person using the good but to everyone else as well).

Suppose, for example, that a country needs to acquire millions of SARS-CoV-2 test kits quickly. This exact good hasn't been produced before, since the virus is new. Thus, there is no existing market (although there do exist companies producing similar products).

In principle, the country's government could leave matters to the market: suppliers that wish to produce kits would produce them and sell them to citizens (or hospitals or employers) who wish to buy.

But there are several problems with this solution. In particular, how is a supplier to know (at least at first) how many test kits to produce? After all, this is a new good and demand for it is uncertain. Furthermore, the supplier doesn't yet know who else will be producing test kits and how much they will produce. Under such circumstances, the supplier may be reluctant to incur the significant setup costs entailed in production until the uncertainties are resolved. Given time, the market would ultimately resolve them through the equilibration of supply and demand. But that process isn't instantaneous, and test kits are needed quickly.

Furthermore, given that supply can't be ramped up immediately, prices are likely to be high at first, which will disproportionately hurt poorer citizens and businesses (the very groups that are worst hit by the pandemic).

. . . in normal circumstances, ordinary markets do an excellent job of ensuring that the goods and services people want and need are produced and distributed.

And, finally, the market approach ignores the public good aspect of test kits. If I buy and use a test kit, I will get some benefit—I will know whether or not I have the virus and can take proper precautions and seek treatment if I do. But most of the benefit goes to *other* people, who will be protected from infection if I quarantine as a result of testing positive. Since I have little incentive to take into account those other benefits, I am likely to underpurchase test kits. And so the market system will result in too few kits being supplied and used.

At the opposite extreme, an alternative solution would be for the government to step in, pick some potential suppliers, and *order* them to produce test kits—that is, a command economy approach. Indeed, this approach was actually used to some extent in the United States for ventilators.

But it gives rise to some difficult questions. Which suppliers should the government choose? Clearly, it would like to choose the ones with the lowest production costs, but it doesn't know which ones those are. Indeed, the government might not even know the identities of all potential suppliers. Moreover, how many test kits should the government order? And how do the suppliers' costs get covered (if, in fact, they do)?

For all these reasons, I proposed a mechanism-design solution in my May 2020 Transmission. In this mechanism, the government first announces its intention to buy test kits and invites potential suppliers to furnish information about their costs. Then, on the basis of this information, the government gives each supplier a target output level (possibly zero if the supplier costs are too high) and a corresponding price that it is willing to pay for this output. After the kits are delivered, it then turns around and resells them to society for a very low or zero price. I showed that it is possible for the government to design the mechanism so that suppliers are induced to provide accurate cost information and the production targets maximize the net social benefit from test kits.

Our country has endured a staggering death toll—over 600,000 people lost already—due in large part to government mismanagement. And when the final reckoning is done, the absence of test kits and protective equipment is almost certain to be an important part of the story.

To what extent did the US government actually use a mechanism like this for critical pandemic goods? In the case of virus test kits and personal protective equipment, the answer appears to be: almost not at all. And, as a result, there were dangerous shortages of both, especially in the first year of the pandemic. Our country has endured a staggering death toll—over 600,000 people lost already—due in large part to government mismanagement. And when the final reckoning is done, the absence of test kits and protective equipment is almost certain to be an important part of the story.

The one bright spot for mechanism design was vaccine development. There, instead of leaving everything to the market, the Trump administration created Operation Warp Speed. In particular, the administration picked a number of pharma companies on the basis of their reputation or promise and covered a lot of their upfront development costs. It also offered them futures contracts: if they successfully developed a vaccine and got it approved (at least on an emergency basis), the government promised to buy a large number of doses at a specified price. And, indeed, we ended up with several vaccines in record time.

But Operation Warp Speed didn't go far enough, at least as far as the developing world is concerned. Although there have been enough vaccine doses for almost everyone who wants them in the US and Europe, only about 2% of Africans have been vaccinated so far. Furthermore, the fact that successful pharma companies retain patent rights over their vaccines is proving to be a major stumbling block to getting enough doses to the Third World.

A far preferable solution would have been for pharma companies to give up their intellectual property in exchange for a hefty buyout and for the vaccines to have been put in the public domain. That would have allowed doses to be manufactured on a far more massive scale.

Had it been used properly, mechanism design could have saved hundreds of thousands—perhaps millions—of lives. As it was, I would give the United States and the developed world a grade of D for their response to the pandemic. Not a complete failure, but not something to be proud of either.

TRANSMISSION NO. 027

STUDYING WILDLIFE IN EMPTY CITIES

DATE: *4 May 2020*

FROM: *Pamela Yeh, University of California, Los Angeles; Santa Fe Institute; Ian MacGregor-Fors, University of Helsinki*

STRATEGIC INSIGHT: *COVID-19 lockdowns provide a once-in-a-lifetime opportunity to study wildlife in empty cities.*

The shelter-in-place orders and the massive drop in human activity in our cities, designed to slow the spread of COVID-19, have given us surprising and unexpected sightings of wildlife species across cities around the world. But beyond general awe—and a brief respite from the gloominess of the news—what can seeing all of this wildlife tell us about human-deprived spaces?

Although the media has mainly covered unexpected sightings occurring in urban settings as a result of lockdowns, there are other important, more subtle changes that are happening in the world. First, some species, including those currently living in cities and those making the occasional forays into them, are now able to use habitats and resources that they had never before been able to exploit. Second, species in cities are experiencing a less harassing environment—less active, less noisy. An unknown number of adjustments have been taking place within a short window of time, and thus we are experiencing a once-in-a-lifetime opportunity to study how humans affect other animals in our cities. Although many of us had assumed that some of these individuals avoided cities mainly due to changes in habitats, it is clear now that we humans scare many of them out.

OPPOSITE: *Master of Affligem, "Joseph Sold to Potiphar (Peacock)," c. 1490–1500*
BODE-MUSEUM, BERLIN

Fear in wildlife has been broadly addressed, and we know that a lack of fear can occur in the wild. Some of the most common examples come from the Galapagos Islands, where many of the animals on the archipelago (particularly as adults) have no natural predators. For example, adult Galapagos tortoises are fearless because they have evolved in an environment with no predators. Indeed, some research has shown that insular systems have tamer animals. Lizards from islands flee from human approach at closer distances than those from mainland habitats, a pattern that has been shown to override phylogenetic closeness

(Cooper et al., 2014). Although studies are often correlational, evidence points to the lack of insular predators as one of the main causes behind this intriguing pattern. Yet, things are not very straightforward when we add humans to the formula. Empirical findings have shown that there is substantial variation in inland species' and populations' responses to human presence. A behavioral study focused on six different Galapagos species has shown that some of them do respond with fear to human tourist activities, but some do not (González-Pérez & Cubero-Pardo 2010). Furthermore, species such as the small ground finch (*Geospiza fuliginosa*) still show fear responses to introduced predators years after predators have been exterminated from the island (Gotanda 2020), suggesting that evolutionary changes, and not just phenotypic plasticity related to wildlife behavioral responses, can occur in a relatively short time frame.

So where does this leave us in terms of understanding responses of animals to rapid increases and decreases in human presence? There are few clear patterns, and even fewer clear results, regarding some key ecological and evolutionary questions: Are the changes we see evolutionary in nature, or due just to phenotypic plasticity? How does the environment an individual is born into reflect future behavior, ecology, and fitness? Do populations adapt as quickly to the sudden absence of humans as they do to their sudden presence?

In the relatively short time period of the lockdowns, animals could make adjustments that range from the expected relaxation of alarm systems, from a behavioral approach, to changes at the evolutionary level (see Schilthuizen 2018 for amazing examples). In Southern California, we have been conducting a long-term study of several populations of an urban songbird, the dark-eyed junco (*Junco hyemalis*). Since this woodland species began to move down from the mountains into the cities of Southern California a few decades ago, these birds have adapted remarkably well to the sunny coastal lifestyle (Yeh 2004). Their population numbers are rising, and they have begun to move from college campuses to both dense city centers and leafy residential suburbs. Not surprisingly, these birds are not too scared of humans. After all, if these individuals were to fly away at every human who was approaching or passing by, there would never be enough time to eat, find mates, and feed their offspring! But the sudden quietude in which they now find themselves allows us to examine something very different: How will these birds respond to a rapid decrease in the levels of human disturbance—both the adults birds now, and the offspring born into these much calmer and more serene situations? Will they become more fearful of people or will they change their perception of us as threats? Will we see a decrease in fitness as human chaos starts up again or will our absence pass unnoticed, evolutionarily speaking?

If many species find our "urban habitat" so stressful, should we reconsider how we urbanize and how we develop cities? Wouldn't we want to generate scenarios that may attract many of the surprising species that have been sighted after lockdowns as part of our cities in our day-to-day lives?

There are obviously no answers to our questions yet. And that's why lockdowns constitute a unique and unparalleled opportunity to examine what happens to wildlife experiencing these rapid changes in their living environments. We may expect differences in the behavior of some animal populations, most likely leading to unexpected results. For instance, the newborn chick in human-deprived cities whose parents were not bold enough to use a resource in an urban area (e.g., insects from university campus lawns) may start using them now and continue to do so if they lack the perception of humans as potential threats.

But besides all of the above, the current scenario leaves us with some more philosophical questions: If many species find our "urban habitat" so stressful, should we reconsider how we urbanize and how we develop cities? Wouldn't we want to generate scenarios that may attract many of the surprising species that have been sighted after lockdowns as part of our cities in our day-to-day lives? Headline news and social media across the country are stating that "we can't go back to normal" (Baker 2020).

In some senses, we hope this is true! That is, in addition to the health, social, and economic ramifications of this pandemic, which have been fairly catastrophic (Ruiz Estrada 2020), we hope this pandemic allows us to consider how we can have more animals in our daily lives in the city, and more fully share the planet.

REFERENCES

P.C. Baker, "'We Can't Go Back to Normal': How Will Coronavirus Change the World?" *The Guardian,* March 31, 2020, https://www.theguardian.com/world/2020/mar/31/how-will-the-world-emerge-from-the-coronavirus-crisis

W.E. Cooper, R.A. Pyron, and T. Garland, 2014, "Island Tameness: Living on Islands Reduces Flight Initiation Distance," *Proceedings of the Royal Society B* 281(1777): 20133019, doi: 10.1098/rspb.2013.3019

F. González-Pérez and P. Cubero-Pardo, 2010, "Short-Term Effects of Tourism Activities on the Behavior of Representative Fauna on the Galapagos Islands, Ecuador," *Latin American Journal of Aquatic Research* 38(3): 493–500

K.M. Gotanda, 2020, "Human Influences on Antipredator Behaviour in Darwin's Finches," *Journal of Animal Ecology* 89(2): 614-622, doi: 10.1111/1365-2656.13127

M.A. Ruiz Estrada, 2020, "Economic Waves: The Effect of the Wuhan COVID-19 on the World Economy (2019-2020)," *SSRN,* https://dx.doi.org/10.2139/ssrn.3545758

M. Schilthuizen, 2018, *Darwin Comes to Town – How the Urban Jungle Drives Evolution,* New York, NY: Picador.

P. Yeh, 2004, "Rapid Evolution of a Sexually Selected Trait Following Population Establishment in a Novel Habitat," *Evolution* 58(1): 166-174, doi: 10.1111/j.0014-3820.2004.tb01583.x

REFLECTION

MAKING LEMONADE OUT OF LEMONS

DATE: *20 August 2021*

FROM: *Pamela Yeh, University of California, Los Angeles;*
Santa Fe Institute;
Ian MacGregor-Fors, University of Helsinki

Over a year ago, we discussed a new and unexpected research direction we were taking: looking at how the sudden and dramatic emptying of our cities impacts the animals that have been living there. Will the animals behave the same way? Will they change, and if so, can we make predictions about the directions of change? In particular, we were wondering how dark-eyed juncos—a fairly new urban inhabitant—might change with regards to their fearfulness of humans and in their aggression levels towards other juncos. We also concentrated on the response of bird communities to the much calmer and emptier cities resulting from the strict COVID-19 lockdowns. Additionally, both our labs worked together to keep track of wildlife species that were being recorded by the general public and the media in quarantined cities, underscoring the role that community science could have on unexpected scales. We now have some preliminary answers that we will discuss later, but the bigger surprise for us is how increasing complexity in our natural world requires increasing nimbleness on the part of researchers and educators. Until now, we had rarely thought about the role of the researcher in the research itself.

Science is often considered an impartial pursuit, where we are supposed to take the researcher out of the equation, or at least avoid biases as much as possible. This year we learned that this task is much harder to accomplish than commonly thought. During the pandemic, our labs and those of many colleagues across disciplines were decimated not just by the virus itself, although tragically many of our families were severely affected. Depression, anxiety, fear, anger, confusion, and exhaustion tore through our labs. One of the most talented and promising PhD students of the Yeh lab—a woman of color, who grew up in deep poverty, the child of a single teenage mother—dropped out of graduate school. On and off, together with students and postdocs, we were all stuck at some point; some of us were simultaneously bored by the quarantine and also completely unable to get work done. Others were completely overwhelmed by the extra work the pandemic placed on so many of us, with the academic machinery trying to keep the regular paces going even in the middle of a pandemic. We worked hard to scratch out successes, with some students graduating on time, and interesting work still being published under the pressing circumstances. But more so than at any time in our careers, we were faced with the human part of science, and that part that has proven to be beyond complex!

As we were madly pivoting in an attempt to save future careers (and our own), one of the Yeh Lab graduate students offered this saying: "We have to make lemonade out of lemons." We have all heard this saying before, but for our decimated and depleted labs, it became a mantra and we clung to it. We also realized how similar we were to our study system: Like humans, birds also had to pivot because of the pandemic. When we started making the connection of the universality of the "pivots," we had a more intimate connection with the organisms we studied, and I think it was a starting point for our lab to collectively heal.

In retrospect, it should not have been such a surprise to us that who does the research, and how we vary as individuals is a key component of research. The knowledge that who does the research is crucial has been demonstrated in our field of behavioral ecology, where the idea that there is "female-choice" in mating systems rather than "male–male combat" only started being fully understood and fleshed out in the 1970s, when women were finally allowed to join the ranks of men in field research[1].

In practice, how do these realizations change how we do our work? We are trying to understand how our own mental well-being affects the work we do, sometimes as individuals, both individually and collectively. This impacts not only our productivity but also the quality of our interactions, the questions we ask, the ability to be open to new directions, and the excitement we have about our work. We are starting to talk with psychologists, sociologists, and even artists to determine how we can measure

1 K. Borgmann, "The Forgotten Female: How a Generation of Women Scientists Changed Our View of Evolution," *All About Birds*, June 17, 2019, https://www.allaboutbirds.org/news/the-forgotten-female-how-a-generation-of-women-scientists-changed-our-view-of-evolution

all these aspects, and how we can work within our own mental, emotional, and physical capacities to produce *whole* scientists, who can in turn produce whole science—science that considers context and understands that sometimes one cannot take the *scientist* out of the science.

> Like humans, birds also had to pivot because of the pandemic. When we started making the connection of the universality of the "pivots," we had a more intimate connection with the organisms we studied and I think it was a starting point for our lab to collectively heal.

As for the birds? Well, in California dark-eyed juncos appeared to have a rough couple of years as far as breeding—although it is currently unclear how much of this has to do with the drought and how much of it has to do with a change in human activity. But the juncos pivoted substantially in their behavior, appearing to be less aggressive than the already fairly relaxed prepandemic city juncos. In Colombia, we could see how calmer cities opened an avenue for more birds, both diurnal and nocturnal, showing how our day-to-day activities act synergistically with urbanization in molding the wildlife communities we currently have. We do not yet know if these responses will have long-term effects on the behavior and ecology of urban birds, but we do know that we, as scientists, have learned and are learning more than we have ever imagined from this trying phase for humanity.

TRANSMISSION NO. 028

EXPONENTIAL GROWTH PROCESSES

DATE: *4 May 2020*

FROM: *Sidney Redner, Santa Fe Institute*

STRATEGIC INSIGHT: *Forecasting ambiguity is inevitable in exponential growth processes that underlie epidemics.*

One can't help noticing that forecasts of the COVID-19 epidemic's toll are inconsistent with each other, and a single forecast can change dramatically over a few days. Recent estimates for the US epidemic death toll have ranged from tens of thousands to millions, with the current number around 64,000. Given these uncertainties, one might conclude that none of these estimates is trustworthy. More perniciously, the wildly disparate estimates might cause a reasonable person to doubt expertise and thus rely on sources that confirm her/his own personal bias.

Our purpose is to show that huge forecasting uncertainties are an integral feature of processes that exhibit *exponential growth*, an oft-misused term in the lay literature. *Exponential growth* refers to a time-dependent process in which the size of some quantity today is its size yesterday times a factor that is greater than 1. As an example, suppose you deposit \$1 in a bank that pays 1% interest per day. How long would it take to become a millionaire? One day later, your account is worth \$1.01; two days later $\$(1.01)^2$, three days later $\$(1.01)^3$, etc. After n days your account is worth $\$(1.01)^n$. Setting $\$(1.01)^n$ equal to \$1 million leads to $n = \ln 10^6 \ln 1.01 \approx 1388$ days, or roughly 3.8 years. It's a short wait to become a millionaire at this bank. If your bank is more generous and pays 2% daily interest, you would get \$1 million in 697 days, or 1.9 years. With 5% interest, in 9.3 months; with 10% interest, just under 5 months. As an important corollary, if you initially deposited \$10

OPPOSITE: *Kimura Tangen, "Mount Fuji in the Clouds," 1753 (detail)*
KAGOSHIMA CITY MUSEUM OF ART

at 10% interest, your fortune would arise in 4 months. The initial value of the deposit hardly matters; it's the exponential growth that's crucial.

Let's now turn to epidemic forecasting. Without any mitigation, each COVID-19 patient infects an additional 2.5 people, on average, in a single incubation period. This factor 2.5 is termed the reproductive number R_0. Note that $R_0 = 2.5$ corresponds to 250% interest, for which the entire planet would be infected after only 25 incubation periods. The goal of social distancing and related interventions is to reduce the reproductive number to below 1, after which the epidemic would stop. We estimate how many people would end up getting infected when the epidemic stops.

Suppose that public health authorities detect the epidemic when N people are infected, after which mitigation strategies are imposed to reduce R_0 to less than 1. Because society is complex, it is not possible to reduce R_0 instantaneously, but rather, it happens gradually. Thus our model: each successive day R_0 is decreased by a random number r whose average value $\langle r \rangle$ is less than 1. (For simplicity of language, we use "days" rather than "incubation period" as the basic time unit.) The number of days k until R_0 reaches 1 is determined by $R_0 \langle r \rangle^k = 1$ or $k = \ln 0.4$ $\ln \langle r \rangle$. Thus, if R_0 decreases by a factor $\langle r \rangle = 0.95$ daily, the epidemic is extinguished in 18 days.

We simulate this mitigation strategy by starting with reproductive number $R_0 = 2.5$ and choosing random numbers $r_1, r_2, r_3, \ldots$, each of which is uniformly distributed between 0.9 and 1, so that $\langle r \rangle = 0.95$. The reproductive number after k days of this mitigation is the product $R_k =$

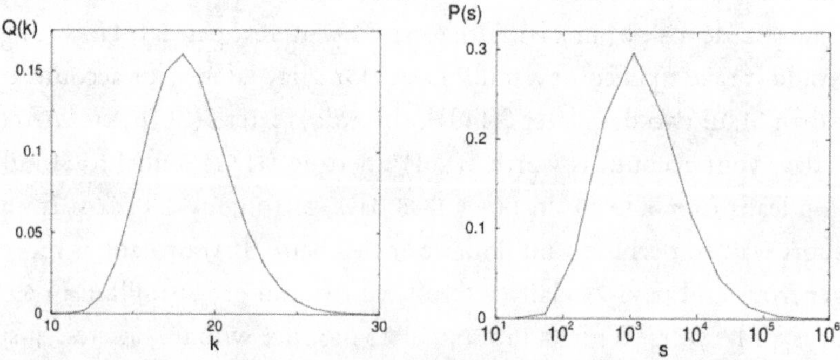

Figure 1. *(a) The probability Q(k) that the epidemic lasts k days. (b) The probability P(s) that the epidemic has increased in size by a factor s from its initial size.*

$R_0 r_1 r_2 r_3 \ldots r_k$. We measure how long it takes until R_k is reduced to 1 and then perform this same measurement for 106 different choices of the set of random numbers $r_1, r_2, \ldots, r_k$. As shown in figure 1(a), the probability $Q(k)$ that the epidemic is extinguished in k days is peaked at roughly 18 days, as we expect. If one is lucky, that is, if more of the reduction factors r_i are less than 0.95, the epidemic is extinguished in as little as 13 days. If one is unlucky (e.g., many of the r_i are close to 1), the epidemic can last more than 25 days. However, the probability of these extreme events is small and almost all instances of the epidemic last close to 18 days.

Here is the crucial point: at the end of the epidemic, the size of different instances of the outbreak can be different by several orders of magnitude, that is, by several factors of 10. As shown in figure 1(b), while the most likely epidemic size is $s = 1{,}000N$, there is an appreciable probability that the outbreak size can be as small as $100N$ or as large as $10{,}000N$! This huge disparity in outbreak size is the salient point—tiny changes in how the epidemic is mitigated can lead to huge changes in the outbreak size.

More dramatically, consider a less effective mitigation in which the reproductive number is reduced, on average, by 0.975 rather than 0.95. Now the epidemic lasts between 30 and 45 days, with an average duration of 36 days. However, the epidemic size ranges between $500{,}000N$ and $50{,}000{,}000N$, with an average value of $3{,}000{,}000N$. Although the epidemic lasts twice as long, it is typically 3,000 times larger!

It bears emphasizing that this idealized model represents a dramatic oversimplification of realistic epidemic mitigation strategies. However, the model captures the huge uncertainties that arise in predicting the final outbreak size. This unpredictability is intrinsic to epidemic dynamics and not indicative of shortcomings in modeling. In short, forecasting ambiguity is unavoidable in exponential growth processes that underlie epidemics.

TRANSMISSION NO. 029

SARS-COV-2 & LANDAUER'S BOUND

DATE: *4 May 2020*

FROM: *David Wolpert, Santa Fe Institute*

STRATEGIC INSIGHT: *The concept of the extended phenotype provides a way to circumvent Landauer's bound.*

Landauer's bound is a deep principle of physics, relating information-processing with the expenditure of energy. Loosely speaking, it says that any physical system must produce more heat if it implements more computation. Among other things, this means that even if you were to use the technology of some far-future civilization of demigods to build yourself a new laptop, that laptop would still generate heat when it runs.

While Landauer's bound cannot be avoided, it can be circumvented. To see how, note that the bound concerns physical systems that *perform a computation themselves*, and that *expend energy themselves*. But what if you get some other system to do those things on your behalf?

Getting outside systems to do your bidding this way is a well-established phenomenon in biology. One way it is often formulated is by extending the concept of a biological "self" from just your physical body to all those aspects of the surrounding environment that you can influence or control. This perspective on what the "self" is in biology, tracing back to Richard Dawkins, is known as the extended phenotype.

There are many versions of the extended phenotype. Some of them concern scenarios in which one gets a *nonliving* outside system to do one's bidding. A canonical example of this type of extended phenotype is the dirt mounds that termite colonies make and then control, in order to stay alive. Another example of this type of extended phenotype is the wearing of clothes by a human being in order to stay warm.

OPPOSITE: *John Gerrard Keulemans for* History of the Birds of New Zealand, *"Long-tailed Cuckoo Young fed by Grey Warbler," 1873 (detail)*

As Dawkins emphasized, the concept of "self" can also be extended from your physical body to include aspects of *other biological systems* that you can influence or control. An example of this second type of extended phenotype arises when crickets infected by hairworms commit suicide by drowning—behavior that is essential to the reproductive cycle of the hairworm infecting it. In this example, the cricket and its behavior are part of the extended phenotype of the hairworm.

The way this relates to Landauer's bound is that, if you happen to have a biological system around, it's often far easier to get *it* to perform computation and expend energy on your behalf rather than do those yourself. In other words, the concept of the extended phenotype provides a way to circumvent Landauer's bound.

There are many examples of this in nature. A tongue-in-cheek example is that in which the physical system "getting biological systems to do its bidding" is just a molecule: caffeine. Coffee, after all, wakes up the person drinking it. It makes that person engage in many computations that coffee cannot do itself, and makes them expend lots more energy than the coffee can expend itself. So, strictly speaking, coffee "gets biological systems to perform computation on its behalf, and to expend energy on its behalf," basically by waking someone up. In this sense, a waking human is part of the extended phenotype of coffee. (Thanks to Josh Grochow for discussion of this example.)

In more conventional examples of the extended phenotype and Landauer's bound, the controlling system is living, not just a molecule. An example of this arises in social systems. After all, a leader of a social group "gets other systems to perform computation on its behalf, and to expend energy on its behalf." Leaders behaving this way can be found in social systems ranging in size from small groups of friends to (most) large governments.

In a more sobering vein, there are other systems, lying halfway between single molecules and leaders of social systems, that also "get others to perform computation on its behalf, and to expend energy on its behalf." In particular, a virus can be viewed this way.

"Severe acute respiratory syndrome coronavirus 2 (SARS-CoV-2)," to give it its full name, is a small package of molecules: basically, just an RNA molecule, surrounded by several protein molecules. Any single such package of molecules hijacks human cells to make more such packages of molecules. By doing that, the virus gets the human cell to expend a huge amount of energy (involved in duplicating the virus many times over) and to perform an elaborate computation (namely, the biochemical computation involved in that manifold duplication). Both of those expenditures are far beyond what the virus could do itself.

Evil as SARS-CoV-2 might seem, we scientists who investigate the fundamental relationship between computation and energy expenditure can't help but look at it with awe.

So each single virus can be viewed as a system that circumvents Landauer's bound by using its extended phenotype to duplicate itself. In fact, COVID-19 also circumvents Landauer's bound on a far larger scale. Not only does a single SARS-CoV-2 virus hijack an individual human cell, getting that cell to do its bidding; a SARS-CoV-2 *population as a whole* also hijacks the entire immune system of the host, inducing that infected host to, for instance, cough and sneeze (a huge energy expenditure), thereby spreading the virus population to entirely new hosts (a process which involves all kinds of information-processing by the surrounding atmosphere, surrounding humans, etc.). So at the population level as well the individual level, SARS-CoV-2 circumvents Landauer's bound by using its extended phenotype.

Evil as SARS-CoV-2 might seem, we scientists who investigate the fundamental relationship between computation and energy expenditure can't help but look at it with awe.

THE TRANSCRIPT OF *COMPLEXITY PODCAST* TRANSMISSION SERIES EP. 6, DISCUSSING THE FOLLOWING TRANSMISSIONS

[Batch 6, released 4 May 2020]

T-025: Understanding Cities to Respond to Pandemics
Christopher P. Kempes & Geoffrey West

T-026: Mechanism Design for the Market
Eric Maskin

T-027: Studying Wildlife in Empty Cities
Pamela Yeh & Ian MacGregor-Fors

T-028: Exponential Growth Processes
Sidney Redner

T-029: SARS-CoV-2 & Landauer's Bound
David Wolpert

This podcast transcript has been abridged for length and clarity. Find the full podcast at https://complexity.simplecast.com/episodes/32

COMPLEXITY PODCAST: TRANSMISSION SERIES NO. 006

EXPONENTIALS, ECONOMICS & ECOLOGY

DATE: *11 May 2020* **HOSTED BY:** *Michael Garfield, featuring David Krakauer*

MICHAEL GARFIELD: This week we have a set of really fundamental and potent contributions to the series. Let's start again in the micro with David Wolpert's piece[1] on the Landauer bound.

DAVID KRAKAUER: This is really interesting. It's a field that I find fascinating, and of course, it's of big interest here at the Santa Fe Institute. I think of David's contribution as the "virus as a computer virus," and it's all about the energetics of computation. It introduces energetic limits to computation, how they're connected to this idea of the extended phenotype and the virus as a minimal replicator that outsources its functions to its host. And it's worth, in this case, to just do a little history:

When the great Alan Turing invented his famous thought experiment that led to the computer to prove an undecidability theorem in 1936, the computer was a completely idealized logical construct. It didn't live in the physical world. It was a mathematical thought experiment. And he devised that thought experiment to answer a question—that is, can an algorithm tell you in advance whether or not a problem can be solved? The answer, famously, was no. This construct that Turing invented, called the Turing machine, has an infinite tape and you write binary strings to the tape and you can erase these strings, but there's no friction in the Turing tape, right? The Turing computer doesn't live in the physical world; it lives in a kind of mathematical utopia.

And Rolf Landauer, a physicist working at IBM in the 1960s, became very interested in this fact that most real computers do live in the physical world. That means that they're subject to all the laws of physics, like the laws of thermodynamics, in this case, the most important of which is the second law: that entropy increases in any closed system. And he formulated the second law for computation, called the Landauer principle, and it basically says that any logical manipulation of information of the sort

---

1 See T-029 on page 367.

that Turing envisaged—writing bits, raising bits, combining bits—will generate heat, and that heat will be dissipated into the environment. This was important, because it led to limits to militarization because as we pack more and more transistors in integrated circuits, they generate more and more heat and at some point the circuit as a whole will melt.

So, a virus is a little bit like a very tiny computer. But the virus is an organic computer and its Turing tape, if you like, is its genome, the sequence of RNA bases. And what the virus computes is the function "copy." It gets into the cell and it copies its RNA genome. What David points out is the virus, like any computers, should be subject to the Landauer constraints, which would put a limit on how much information it can encode. What the virus ingeniously does is outsource almost all the computational work to the host. That's where this extended phenotype concept comes in—sometimes also called ecosystem engineering, or niche construction—where the virus basically allows us to in some sense do the bulk of the work. So, the virus doesn't encode all the ribosome complex . . . we do. We encode the Turing tape head. And, by virtue of outsourcing to its host most of the computational work, the virus gets away with being an extraordinarily efficient little bit of computer code.

That's why I alluded at the beginning to David's contribution of a virus as a computer virus. This leads to this very counterintuitive fact that most of the virus is not virus. Most of the virus is, in fact us, because we contribute most of the machinery necessary for it to complete its life cycle.

One way to think about this is like technology. We talk about iPhones as these miracles of miniaturization, but that's only partly true, because most of what the iPhone does is actually outsource to vast server farms that are cooled by river systems. So, the phone is a bit like a technological virus that heats up the world in order that you can keep this cool piece of glass in your pocket.

Timothy Morton, the philosopher of ecology, calls it a hyperobject.[2] It's an object that only partly lives in the dimension of your immediate experience, most of which is out of sight. As computing trends move toward outsourcing computation, they're becoming more and more viral.

---

2 T. Morton. *Hyperobjects: Philosophy and Ecology after the End of the World.* Minneapolis, MN: University of Minnesota Press, 2013.

There's this beautiful physics of computation link that David points out between the microbial world and the macroscopic world of technology.

> ... most of the virus is not virus. Most of the virus is, in fact us, because we contribute most of the machinery necessary for it to complete its life cycle.

M. GARFIELD: This brings us to an overarching theme in the series about the problems of bringing overly simple models to complex situations. When we had you on the show way back at its beginning, for the live recording at InterPlanetary Fest last year[3] on the Origins of Life course offered by Complexity Explorer,[4] someone in the audience asked the classic "is a virus alive" question. That's a binary question for which there is not a binary answer. You just lead-authored a piece on an information theory of individuality. I really enjoyed this piece, the way that you talk about different kinds of individuals that are scaffolded in different ways by the environment, and how we tend to think of the individual as more or less only the kind of individual that humans believe ourselves to be—that we have a sort of internal driver and informational continuity that pushes us forward in time. It's this classic nature vs. nurture question.

And then, with the Landauer bound and the extended phenotype, bringing this subtlety that you addressed in that paper: where does the spider end? The web allows the spider to be small and to outboard most of that sensitivity to the web, which is really light and comparably easier to build than an enormous spider body that would have to sit there and sense the whole thing.

D. KRAKAUER: That's absolutely right. Our perceptual apparatus operates on preferred space and timescales, and doesn't see the spider for what it really is, which is, as you say, spider plus web plus ... True individuals are these much more interesting hyperobjects which extend into

3 Complexity Podcast, episode 2, aired on October 9, 2019. https://complexity.simplecast.com/episodes/2

4 https://www.complexityexplorer.org/courses

the world in complicated ways. Human beings are their genomes, their bodies, their social networks, their homes, their cities. And the key question, of course, in that work was trying to determine where you should draw a boundary in a principled way. It has something to do with causality, which is: where does your causality, the center of causality, extend out and attenuate to the point where you have no more control?

I think that there needs to be a real revolution in how we think about what objects are, particularly adaptive objects, partly for conservation reasons. Because, if you believe that your cell phone is just that thing that you carry and can charge every night when you go to bed, and you forget that part of that phone is actually somewhere hidden away, drawing down huge quantities of energy that are radiating out heat, then there's a sense in which you're not truly responsible for your extended phenotype. So, there are ethical implications of rethinking what the individual is.

M. GARFIELD: And in that movement towards a deeper apprehension of our identity as hyperobjects, outboarding the cost of viral reproduction allows the very small to drive the very large. A lot of listeners are probably familiar with Bucky Fuller's analogy of the trim tab on the rudder of the Queen Mary, and how that tiny little thing can push a much larger craft, and steer it in a different direction.[5] That brings us to Sid Redner's piece[6] on exponentials, and how important it is to embrace the enormous range in a lot of these epidemiological models, because they're inherent in the nature of grappling with an exponential phenomenon.

D. KRAKAUER: So, the consequence of all of that efficiency that David described, whereby the virus outsources its essential computational machinery to us, is the possibility of exponentials. We pay the entropic cost for viral overpopulation, not the virus.

And exponentials are notoriously hard to get your brain around. In popular culture, they're behind the so-called "butterfly effect" in chaotic dynamics. That's where you have two arbitrarily close initial conditions, where the two associated trajectories diverge exponentially

5 See "A candid conversation with the visionary architect/inventor/philosopher R. Buckminster Fuller," *Playboy*, 1972. http://bfi.org/sites/default/files/attachments/pages/CandidConversation-Playboy.pdf

6 See T-028 on page 363

in time. So, exponentials are what we are talking about when we say extreme sensitivity to initial conditions. That's the butterfly effect. Sid gives this example from compound interest, which is the exponential that most of us know from our bank accounts. And he makes this point: you deposit a dollar with an outrageous interest rate, say 5% per day, and in about nine months you'd have a million dollars, right? When he says it's a 10% interest rate, you'd be a millionaire in just under five months. That's exponentials. The one that I like, because I think it's even more crazy to come to terms with, is the consequence of Moore's law.

Folks will remember Moore's law is this empirical observation that the number of transistors and so forth in an integrated circuit doubles approximately every two years. And this is a very general technological observation. Take this one: In the 1960s, a gigaflop of processing power—a gigaflop is about 1,000 floating point operations per second—cost $18 billion. In the 1980s that costs $20 million. In the year 2000, you get a gigaflop for about a $1,000, and iPhone X, which costs about $500 on eBay, runs at 600 gigaflops. Now here's the weird thing about exponentials. That means that if you could time travel just to the 1960s with your $500 iPhone, it would be worth hundreds, if not thousands, of *billions* of dollars. In other words, your $500 iPhone is a trillion-dollar hyperobject.

M. GARFIELD: Well, I mean, that's if you could travel back with the entire network architecture that supports it, right?

D. KRAKAUER: That's a very good point! But to be fair, the actual chip on the phone is running at 600 gigaflops. It wouldn't be very useful; you couldn't do much with it. But that little device contains, you know, by 1960s standards, trillions of dollars of technology. That's what exponentials do.

Okay, let's get back to Sid. Sid is making the point that R_0, which we've talked about a lot, is like the interest rate of the viral investment in its hosts. So, we are the virus's genome bank and the virus does not make more money—it makes more genomes. That's its profit. If you start with one infection, COVID-19 has an R_0 of about 2.5. That's a 250% interest rate. That's pretty high. At those rates you not only get huge returns, but you could potentially get huge variations in return.

And this takes us to the Cris Moore insight[7] about long tails. It's a very similar style of reasoning in this Transmission. Here's a simple mathematical model where you basically say, let's imagine that our social-distancing policies—we don't have to imagine it, it's true—are aimed at reducing R_0 below 1. Of course, you can't do that overnight. Policies aren't perfect. So, he says, let's assume that every day you can reduce from 2.5 in some range between 10% and 0%. So, on average the daily decrease would be about 5%. That means you're expected to drop beneath 1 in about twenty days. But, of course, we've already established that that's the average, between 10% and 0%, so there'll be variation, and because this is an exponential, this leads to massive differences in outbreak sizes.

> We are the virus's genome bank . . . If you start with one infection, COVID-19 has an R_0 of about 2.5. That's a 250% interest rate. . . . At those rates you not only get huge returns, but you could potentially get huge variations in return.

Even though you could reduce it in about twenty days, let's say, on average, the size of the epidemic could vary between a hundred times the initial number of infected to about 10,000 times the initial number of infected. This is, like Moore's law, one of those very counterintuitive outcomes that relates to the uncertainty. Not the uncertainty of ignorance, but the uncertainty of the butterfly effect. Exponentials build this other kind of radical uncertainty into our models, which has little to do with ignorance, but more to do with the nonlinearities that can massively amplify tiny measurement errors, or tiny differences in initial conditions.

M. GARFIELD: That calls pretty directly to Michael Hochberg's piece[8] that we discussed in the second episode of the series on the importance of timing. And then more largely, this cuts across articles we've discussed

7 See T-024 on page 309

8 See T-008 on page 121

by Lenny Smith and colleagues,[9] by Luu Hoang Duc and Jürgen Jost,[10] Simon DeDeo,[11] Melanie Mitchell,[12] and this issue we've addressed here on embracing radical uncertainty. I really liked Redner's point that, in some ways, whether you saturate the system or you saturate it five times over, is kind of beside the point. That the initial investment in some ways matters less than the exponent and *when* you start investing, So, at what point do we accept this and act on it? This is linked to Chris Kempes's and Geoff West's piece,[13] in that we live in a world now where exponentials are not only difficult for a lot of people to grasp, but also much like hyperobjects; they're a central feature, a dominant feature, of our world, because we live in cities.

D. KRAKAUER: Yes. This has been, of course, the subject of Geoffrey's work for a long time, and this is a very nice contribution from Chris and Geoff. You know, it was not until about around 2007 when for the first time in history more human beings lived in cities than in rural populations. The history of humanity by and large has been a rural history, and our institutions and our intuitions come from that world. But the point that Chris and Geoff are making is that cities are kind of time-travel machines. They speed everything up. They increase the rate parameters of these exponentials for all sorts of different processes. So, let's get to the virus. It's been observed that over two-thirds of all COVID-19 cases in the US come from about eight jurisdictions, and all of those are concentrated around cities. So, transmission spread and potential mitigation are all tied to an understanding of the dynamics of cities.

What Geoff has shown in the past about cities are very interesting regularities, the best known of which is that socioeconomic quantities—wages, patents, wealth—all scale superlinearly in population size, which means that the wealth scales as the size of the city raised to a power, and that power tends to be 1.15. So, you get a compounding effect of 15% per doubling. There's a direct connection to Sid here: with an exponential

9 See T-023 on page 305

10 See T-003 on page 37

11 See T-004 on page 57

12 See T-009 on page 129

13 See T-025 on page 337

process that's exponential in time, then the rate parameter will vary with the city size raised to this power. What that means is that a city with a million people will double the number of cases in half the time of a city of 10,000. That's like the Moore's law—there's this huge exponential effect of being larger.

The neat thing about this work that I find really intriguing and, I have to say, when I first saw it many years ago, very surprising, is that it's not just things like patents and wealth that scale to this power, but also the rate of disease transmission. We've talked earlier in this series about the connections between memetics and genetics, but cities are a very interesting machine because they impose, through mechanisms that we still don't fully understand, this extraordinary universality on the way very different transmission dynamics work. In the city, disease transmission scales the way wealth scales and those are really amazing results that we have to understand.

The reason that's important, I think, for culture, is that there's been so much emphasis in this period on understanding epidemiology, understanding virology. The Gates Foundation, that's been extraordinary in its funding of disease, has declared that it will dedicate all of its funds towards COVID-19, and that's a huge problem because it's one factor in this multidimensional hyperobject, if you like, that we call "the city." To understand these kinds of complex events means *not* disaggregating them and treating epidemics as if they're independent of the economy, which they're clearly not. The city is a super integrator of complex phenomenology where you can take nothing for granted. So, I think what Chris and Geoff are really arguing for here is that, in just the same way that we have surveillance systems and prediction systems for weather and for tsunamis and for earthquakes, we need to think very carefully about having very thoughtful, rigorous surveillance, and theories of surveillance, for the integrated set of phenomena that all adhere to the same scaling laws. Because the city, as I said before, imposes convergence on their collective dynamics.

M. GARFIELD: Last year when she was at InterPlanetary Festival, Ann Pendleton-Julian had a really lovely riff on something that you just touched on, which is that we're not just talking about cities as one network, that sort of one-dimensional thing, but it's a multidimensional thing that includes the infrastructure, the environmental resources, the complex political, economic, and social systems, the stories that we tell one another, the virtuality that we inhabit now that we're distancing from one another physically. A lot of us seem to be engaged in more social instances per day with this proliferation of video calls. Last week when we were discussing understanding this complexity in terms of identifying points of systemic intervention, Luís Bettencourt, in episode four of this show,[14] talked about his research on intervening in slums by finding ways to bring new vasculature into the slums to allow for better distribution of resources in infrastructural networks. But then, you know, Cris Moore talks about how understanding the hidden structure under the average of R_0 allows us to find points of intervention for lowering the node degree to slow the transmission of virus. Your point about surveillance needing to integrate across all of these dimensions . . . it's too simple to say, "let's grow the network to solve problems" or to say, "let's break the network to solve problems." It's understanding how to modulate that across many different dimensions at once.

> . . . cities are a very interesting machine because they impose, through mechanisms that we still don't fully understand, this extraordinary universality on the way very different transmission dynamics work.

D. KRAKAUER: I would make two points here. The first is, why have we heard so much from the epidemiologists and so little from the economists, even though for many people—for the majority of people—COVID-19 is an economic catastrophe, not a health catastrophe? The reason for that is because we have models and principles of transmission where points of intervention are understood and principles of vaccine development are somewhat understood. Whereas when we talk about economic markets, we don't have comparable insight. One of the deeper points being made

14 Complexity Podcast, episode 4, aired on October 23, 2019. https://complexity.simplecast.com/episodes/4

here is that we now know that there are these scaling laws that unify biological and cultural phenomena, and if the underlying dynamics on the fractal infrastructure of the city are what give rise to these universalities, it does suggest the possibility of a very principled understanding of economic response, not just epidemic response. So, I view that as the deeper point here, that it's time to take complexity very seriously in order to be as useful in responding to the socioeconomic catastrophe as we have been in responding to the healthcare catastrophe.

M. GARFIELD: Indeed. Eric Maskin's piece[15] is really interesting in light of this question of what happens when we use an insufficiently complex model. Like you were just saying, we look at it across only one dimension. For *Vox EU*, Wendy Carlin and Sam Bowles just published this lovely piece[16] on the revival of the civic society. Rather than thinking of things as merely just "the market" or "the state," and allowing one or the other to drive those things, but to understand how we need to fill this in so that governance is operating at multiple scales at once. I think we can get to this idea that this is a way in which economics is related to evolutionary questions, like: What forces lead to different kinds of nervous system architectures? And when is it good to have a head, or not have a head? Let's talk about what happens when the city breaks, when the market doesn't work fast enough to address the problems that we're facing.

D. KRAKAUER: That's what this crisis has shown us, that these systems are coupled. It also shows us how they all break together, and Eric's contribution is, what do we do when we don't have markets? Of course, we don't have markets—we do in some sense—but by virtue of social distancing, market efficiency is compromised and the pricing mechanism that mediates supply and demand doesn't work. Eric is one of the inventors of the solution to this problem and the solution is called mechanism design.[17] Eric won the Nobel Prize with Hurwicz and Myerson in 2007 for this work. Just a nice footnote on SFI history

15 See T-026 on page 345

16 S. Bowles and W. Carlin, "The coming battle for the COVID-19 narrative," *Vox EU*, April 10, 2020, https://voxeu.org/article/coming-battle-covid-19-narrative

17 E.S. Maskin, 2008, "Mechanism Design: How to Implement Social Goals," *The American Economic Review* 98 (3): 567–76

here, because Eric is a longstanding member of our science board, but Eric himself, as an undergraduate, stumbled into a course being taught by Ken Arrow on information economics. And Ken, of course, was one of the founders of the Santa Fe Institute[18] who had won the Nobel Prize himself in 1972, and he inspired Eric to pursue this work, particularly through his familiarity with the work of Hurwicz.

I need to explain a little bit where mechanism design comes in by explaining what game theory is. And I apologize because I'm sure most folks know this, but a little bit of background on this, because it's important. Game theory is associated with extraordinary figures like John von Neumann and Oskar Morgenstern and John Nash. The essential idea behind game theory is to find the best strategy to use in a game. What's a game? Well, the game has a set of agents—think of chess—us, the players. There are strategy options, and there are payoffs associated with playing those strategies against one another. Most importantly, there is a solution concept, that is, this choice of strategy that is the best against all other strategies, which gives you the highest payoff that allows you to win the game. And what game theory does is it looks for those solutions. It looks for their accessibility and it looks for their stability. The most famous of which, of course, is the Nash equilibrium.

Now, mechanism design turns that on its head. Sometimes mechanism design is called "reverse game theory," because it doesn't look for the solution given the strategy set and the agents. It starts with the solution. It says, "this is where we want to be," and then it asks how would we go about defining the strategies and the payoffs to ensure that that solution is reached? And if you think about a market, it's made up of strategies—strategies of buyers, strategies of sellers—and the pricing mechanism is a mechanism of the market that ensures, under very idealized conditions, of course, that at least for a short while, some fixed point—a Nash solution—is reached that satisfies the majority of suppliers and consumers... one hopes. So, if the market has gone, what do you do? How does mechanism design work for something as complicated as a market?

It sounds arcane, but most people interact with mechanism design regularly through auctions like eBay or by voting. Here, we want to

18 For more on Kenneth Arrow's role in the founding of SFI, see *Emerging Syntheses in Science* (SFI Press, 2019)

elect the preferred candidate, or we want to maximize the profit of the seller. The mechanism design in those cases is how you design the voting system or how you design the auction to achieve the goal. The example that everyone will know, that Eric often gives, is imagine, Michael, that we're both super greedy. Now of course you are and I'm not, but let's, for the sake of a thought experiment, imagine we both are.

The critical point about mechanism design is it's not despotic. It's not the command economy that says thou shalt produce *x* levels of *y*.

M. GARFIELD: I think we just failed the prisoner's dilemma . . .

D. KRAKAUER: We are given a cake and you get to cut up the cake. Now imagine that you get to cut the cake up, and you also get to divvy it up. Under those conditions, unfortunately, you'll get a very big chunk of cake and I'll get a tiny piece. That's what's called a "tragedy of the commons," because your selfish interests don't maximize the social-welfare function, the well-being of everyone, which in this case is two people. Mechanism design comes to the rescue. The mechanism in this case is one person cuts the cake, and the other person chooses. If you do that, there's actually no incentive to take more than half, so you get a fair sharing of the cake. This is the key to mechanism design: you have a tragedy-of-the-commons situation, you want somehow to reach fairness, and the way you do it is by outsourcing the utility function to others. And Eric discusses the more technical case of how we would distribute fairly ventilators to treat the acute condition of COVID-19 infection without a market.

We all saw in the news this spontaneous emergence of aberrant auctions in which state governors were bidding against each other very inefficiently to try and save lives. The point that Eric's making is we can do better than that! The mathematics of mechanism design show how. This article, while on the surface is arithmetically a little bit challenging, is actually quite simple: the idea is that buyers and sellers all report their preferred costs to the so-called mechanism—in that case, that would be a federal agency of some kind. And the mechanism creates a payment rule for both parties. The key is, just like that divide-and-choose mechanism of the cake, to align the individual incentive with the global social welfare function. And so, all mandatory transfers from the mechanism turn each of our personal utilities into social utility.

That's what the mathematics does. It provides an incentive to both buyer and seller to do the right thing. Sometimes this is called the Vickrey–Clarke–Groves mechanism, and the essential concept is always that one: you turn the individual utility-maximizing strategy into the social utility-maximizing strategy.

There are some simple examples I can give. Economists often talk about externalities. These are the things that are not typically on the spreadsheet, right? In other words, they're the hidden cost to the world of trade and production. For example, you might be manufacturing bricks, but in the process emitting a huge amount of toxic pollutants. But here's what a mechanism would do. It would say your costs to production are proportional to the damage to the environment. Well, if that was your true cost, then there would be every incentive for you to reduce the pollution, and deal with the tragedy of the commons, and achieve the fair outcome. The critical point about mechanism design is it's not despotic. It's not the command economy that says thou shalt produce x levels of y. It says you still choose, you can still make a profit, but we're going to tell you how much you're going to make and incentivize you to do the right thing.

M. GARFIELD: To draw back to a talk that you gave at UBS Financial Services last year, this links again to the information theory of individuality paper. This notion that the stable, predictable environment of the flow of nutrients that allows for the rare cases in evolution where an animal might lose its head, where sea squirts and clams and so on are capable of just fixing themselves to a rock and hanging out there and filtering. The process of evolution, as we discussed with Brian Arthur on this show back in episode 13,[19] that when you run evolutionary simulations where you iterate game-theoretical situations like the prisoner's dilemma over and over, thousands and thousands of times, you see what

19 Complexity Podcast, episode 13, aired on January 8, 2020. https://complexity.simplecast.com/episodes/13

looks kind of analogous to an evolutionary trend to grow a head, to grow this kind of mechanism, this regulatory coordination. When we think about this in light of Geoff West's and Chris Kempes's comments on cities, COVID-19 might be bringing us to a turning point where cities are seen as self-organizing, kind of headless, processes. We're seeing a kind of a historical shift towards more balance between the emergent bottom-up and the regulatory top-down that aligns the incentives of all of the various actors in the emergence of or the steering of a city.

D. KRAKAUER: This is in some sense the holy grail of thinking about decentralized architectures. There's a lot of ideology here and we talked about it in relation to federalism and anti-federalism. It's talked about a lot in terms of collectivism versus liberalism or you know, libertarianism, and so forth. I think complex systems show us that these categorical distinctions are bogus, that each of these kinds of solutions evolves out of circumstance. I think you're absolutely right. Mechanism design is not the reverse game theory in the sense of being totalitarian game theory. It's the right kind of game theory when the market mechanics don't work. There is a list of reasons for when that doesn't work, and we can go into them, they're the basis of Economics 101 class, but it's a very natural evolutionary adaptation of the economy when a certain kind of information is not present. I think having a sophisticated take of all the strategies available to us to achieve desirable outcomes is important.

We talked about that in relation to genetic regulatory architectures, and how different species according to the complexity of their genome assume the more autocratic versus more democratic topologies. Exactly the same applies here, and I think what we should all be understanding increasingly is that there isn't one best way to do things. The circumstances tell us which of the strategies we should be pursuing.

M. GARFIELD: We've got one more piece here by Pamela Yeh and Ian MacGregor-Fors[20] on animals in cities. The question they're gesturing toward in this piece is whether it's possible for us to apply this kind of reasoning to create mechanisms by which we can grow or adjust the way that we live in cities to align the incentives of humans and nonhumans.

D. KRAKAUER: I liked this a lot. It reminded me of the film *Logan's Run*. Have you ever seen that film? It was directed by Michael Anderson,

20 See T-027 on page 355

and Michael York and Jenny Agutter and Peter Ustinov were in that film, and that was a film where—I don't know when it was set, sometime in the far distant future where human civilization lives in these geodesic domes—and were all placed in these small populations run by these horrible computer systems that take care of our lives, including when we get to reproduce. And so, to reduce overpopulation, by the age of thirty we're all exterminated in this horribly weird ritual. And the reason I mentioned it is because at the end the characters that are played by Michael York and Jenny Agutter escape outside of their geodesic dome and find themselves in a Washington of the future, and it's all overgrown with beautiful, lush vegetation and ferns and so forth. It's the world that was reclaimed by nature when humans abandoned it, presumably for some folly that goes unspecified.

And in 2008, Alan Weisman wrote a beautiful book called *The World Without Us*, which was a catalog of events like that where for one reason or another buildings, various edifices, cities had been abandoned, and how long it takes for nature to reclaim its world. In that book, in fact, he interviews our colleague Doug Erwin. And so, what Pam and Ian do is report on that real-world *World Without Us* during the COVID-19 crisis, and I think they're making two points. One is, what is going on, really? And the second question is what kind of science can we do? What's going on that we should be recording during this natural experiment?

We've all seen, you know, mountain goats in the streets of Wales, and buffalo walking along highways in Delhi, and dolphins frolicking in the Bosporus. It was all of this wonderful natural history going on now in urban settings. But there have been long-term evolutionary experiments, and they cite the particular example of the Galapagos tortoise. This is an animal that's evolved on an island—because of its size, it's essentially free of predators. And Darwin, when he was there in the 1830s says, "Met an immense turpin; took little notice of me." They just don't care. They're sort of indifferent to humans. Pam and Ian point out that there is research showing that, in these very insular ecosystems, you have far tamer animals. That looks like a real effect, and you can ask, how long would it take for that to happen in our world?

There is some work that they cite from 2010 where researchers from Spain looked across six different Galapagos species that have been reported to be somewhat indifferent to humans, showing that now, about half of them respond with fear to human tourists. So, it didn't take very long for them to evolve an appropriate fear response. That's the first part, how long does evolution take? The second part is their own work, really. They've been studying dark-eyed juncos: songbirds in Southern California. These are birds that, prior to the shutdown, had been moving down from the mountains into the cities and are actually quite well adapted to human presence, and have exploited our profligacy very efficiently. I think what they're asking now is what will happen to that species now that humans are no longer there? Will they become more like the Galapagos tortoise, even less concerned about human presence? That could actually be a problem when we come back into the environment and assert our ecological dominance.

So, I think the questions they're asking are very deep and very interesting. It's also worth pointing out that there have been negative impacts on animals by people staying indoors. There are many species that have, for good or ill, become dependent on us. I remember reading an article about a macaque species outside of Bangkok that had been fed by humans around monasteries, but now the food has run dry because the humans are no longer out in the world and they're engaging in these huge conflicts over scarcity of resources. That's a frightening fact, too. It's worth bearing in mind the negative consequences of humans retreating.

The larger picture, to your point, Michael, is that what the pandemic is teaching us, among many, many other things is that we share the planet, and we share it not only with viruses, but numerous other species. And those species are sensitive to our behavior. So, you know, when we return to the world, we kind of mentally also want to return to the wild. And it's an opportunity for us as a culture to evolve. I found that a very uplifting insight into what's going on today.

M. GARFIELD: Andy Dobson has spoken both on the show[21] and elsewhere about the rate at which habitat destruction has increased the frequency of zoonotic diseases—that more and more of these diseases like COVID-19 are emerging from the wilderness that we have unsettled by settling. I think about Geoff West's piece on cities as disease incubators,

21 Complexity Podcast, episode 16, aired on January 29, 2020 (https://complexity.simplecast.com/episodes/16) and episode 23 aired on March 18, 2020 (https://complexity.simplecast.com/episodes/23).

and a piece of the conversation I had with Jennifer Dunne back when she was on the show in episodes five[22] and six,[23] about how the disruption to biodiversity is complex with cities. That empirical work that's been done over the last few years suggests that certain regions of cities promote different kinds of biodiversity. There's a change in the philosophy of conservation that is less about a retro-romantic restoration of lost biodiversity, but an active cultivation of an ongoing and open process. So, it gets us back again to this looking at the city as a multidimensional and wild process. I liked your turn, that as animals returned to these urban spaces, there's a kind of counterflow which is the rewilding of the human imagination and the understanding, through complex-systems science, that the city is itself a phenomenon of nature and so how can we design them to minimize the ecological disturbance and, counterintuitively, to improve their capacity for epidemic proofing?

D. KRAKAUER: I always thought that the great existential crisis of modernity accelerated by isolation is loneliness, to feel alone in the world. And what the theory of evolution did for me was connect me to the rest of the world, right? Trees are my cousins. Insects are my cousins. There's a sense of true connectedness that comes from understanding our common evolutionary histories. And now we're living in this time where we're fearful of a virus, and one would hope that the resolution of this kind of dialectical standoff would be a greater appreciation of our wild nature and our susceptibility to that world—that we're not isolated from it. We never have been. In learning about the microbiome, we realized that we were dominated by the genetics of a bacterium, that we're a tiny fraction of who we really think we are in terms of our individuality. It would be nice to imagine a world where we can rethink our relation to the natural world.

22 Complexity Podcast, episode 5, aired on October 30, 2019. https://complexity.simplecast.com/episodes/5

23 Complexity Podcast, episode 6, aired on November 6, 2019. https://complexity.simplecast.com/episodes/6

7

[BATCH 7, RELEASED 1 JUNE 2020]

FAILURE *to address*

in part a failure to

that people **DO NOT**

in an **ISOLATED**

BENEFIT.

mistrust in vaccines is

UNDERSTAND

REACT *to the vaccine*

equation of **RISK** *and*

—MELANIE MOSES & KATHY POWERS, Reflection 031

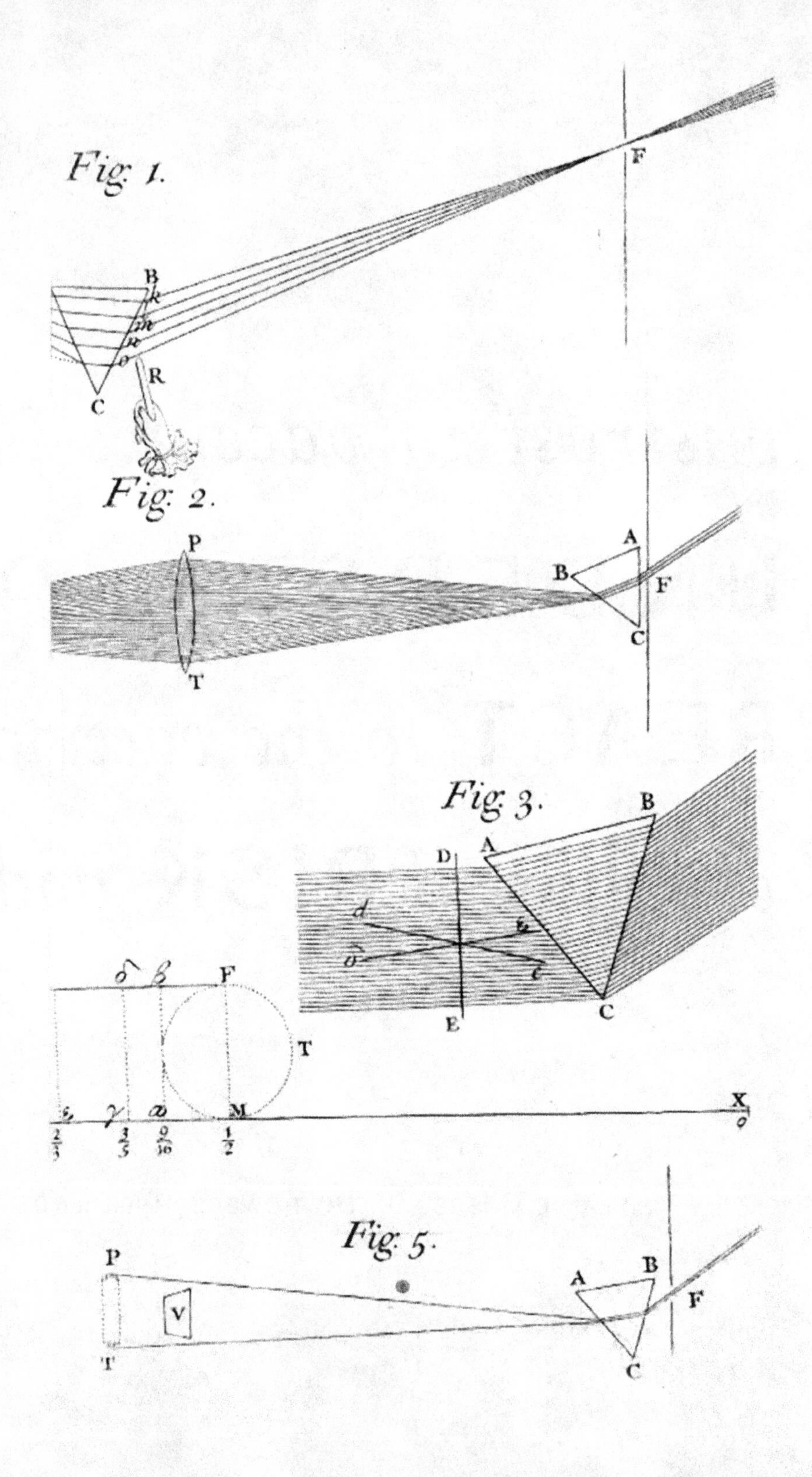

Fig. 1.
B
k
m
n
o
R
C
F
Fig. 2.
P
T
A
B
F
C
Fig. 3.
B
A
D
d
δ
E
C
δ
β
F
T
γ
M
X

TRANSMISSION NO. 030

OUT-EVOLVING COVID-19

DATE: *1 June 2020*

FROM: *David Krakauer, Santa Fe Institute*
Dan Rockmore, Dartmouth College; Santa Fe Institute

STRATEGIC INSIGHT: *The COVID-19 pandemic offers an opportunity to out-evolve the virus by evolving our own scientific ingenuity and social practices.*

If there is one thing that the coronavirus pandemic has exposed, it is that there is much that we still don't know about the world around us. Forget about the trillions—okay, more than trillions—of galaxies in the universe that we'll never explore. Just at our feet or in the air around us are cohabitants of our own world—viruses—that occupy an odd liminal space, pushing our understanding of the meaning of *life*. They exist in what is effectively a hidden world, almost a "first Earth" that is both just offstage and right in front of us, and even inside of us. It's a world teeming with activity, full of blooming, buzzing confusion, competition, and evolution. Sometimes we explore this world intentionally, but at other times we run into it by accident, most noticeably when the alarms on one of the megafauna bio-detectors—people and animals—go off. It's when these encounters happen that we remember that the space of things we don't know is truly unfathomable.

Puttering around on the edge of the known and the unknown is the standard work of science and scientists. While the twinkling of the night sky may often be an inspiration for meditations on how little we understand, it's actually what we can't see in the cosmos that is the best reminder of our limited vision. In 1933 Fritz Zwicky observed a huge discrepancy in the amount

---

OPPOSITE: *Isaac Newton, illustration for* Opticks: or a Treatise of the Reflexions, Refractions, Inflexions and Colours of Light, *1704 (detail)*

COURTESY BOSTON PUBLIC LIBRARY

of gravitational force needed to account for the rotational movement of galaxies and the amount that could be attributed to the visible matter in the galaxy. Naturally, he called this "dark matter." In 1980 Vera Rubin and Kent Ford used spectrographic data—its own form of making visible the invisible—to show definitively that galaxies contain at least six times as much dark mass as visible mass. As it turns out, Aristotle was wrong: nature *loves* a vacuum—that's where it stores most of its gravitational potential.

Countless studies and observations point to the conclusion that nearly 30% of the universe is made of dark matter. Dark matter is a big part of what holds the universe together—more stuff for creating attractive forces between things. But, as you may know, ever since the Big Bang, the universe is actually expanding. The cause of this is a different force of darkness, or, rather, something that goes by the name "dark energy."

Some forms of "infection" are deadly, but some are necessary.

Our understanding of the biological world has also been a story of the discovery of dark matter and dark energy, and our collision with the coronavirus is just the most recent reminder of that theme. Early censuses dramatically underestimated the amount of living matter, a darkness of understanding largely borne of poor optics. Our inability to see at the scale of microorganisms was a source of a good deal of pseudoscience bordering on mythology, especially when it came to disease. "Vapors" and humors were the first dark matter. It was only in the 1880s, twenty years or so after the publication of Darwin's *The Origin of Species*, that Robert Koch discovered bacteria and, in so doing, revealed a material cause for infection. Among Koch's great advances was his use of staining and culturing to make visible the agent of infection. We now know that bacteria and other micro-organisms account for most of the world's genetic diversity, not only in the environment at large, but also within our own bodies, where our inner microbial ecosystem of the microbiome turns out to be crucial to human health. Some forms of "infection" are deadly, but some are necessary.

Viruses like the coronavirus are even smaller than bacteria, and so were also dark for some time. They were brought to light in the late nineteenth century, discovered by the Dutch microbiologist Martinus Beijerinck in the course of investigating the etiology of mosaic disease in tobacco plants. Repeated efforts to culture the source of the disease failed, so it wasn't bacterial in nature—biologists were the first to understand that you need culture to live—but whatever was causing the disease was able to replicate. It was alive in some ways, but dead in others. Beijerinck called this "infectious agent" a virus.

We now know that viruses are basically nano-encapsulated genetic information. They have existed from the beginning of biological time, emergent from the proverbial primordial soup, a string of atoms, clumped into molecules, wrapped in another kind of molecular shell, a kind of biological M&M. The *raison d'être* of the virus is reproduction, which ironically leaves a fair amount of death in its wake. But really, the virus is an engine of life whose dynamics and mechanisms of existence and reproduction make it the agent of genetic expansion, a "dark life" biological force to the dark energy physical force fueling universal expansion that is dark energy. Not quite twins separated at birth, but siblings separated by several billion years, give or take.

It is now believed that the nucleated eukaryotic cell, upon which all animal and plant life is based, would not exist were it not for simple viral genes that first copied themselves into ancient host genomes. This led to selection pressures fomenting the formation of cellular membranes and cellular inclusions. It has been hypothesized that without the constant drive from highly mutable viruses there would have been no need for the evolution of recombinational sex—the kind that all animals and plants use, not to be confused with recreational sex!

In a kind of Nietzschean "that which does not kill us makes us stronger" way, any ability that we do have to fight off some diseases can also be at least partly attributed to viruses. Had jawed vertebrates (all vertebrates but lampreys and hagfish) not acquired genes of viral provenance some 500 million years ago, they would have no adaptive immune system and thus minimal means of fending off viruses.

Table 1. **VIRAL EDUCATION.** *This table represents paths from nature to therapy. It shows viral genetics (viral innovation) that have been co-opted by other forms of life (exapted adaptation) and that have been used as a part of human genetic engineering (engineering utility).*

VIRAL INNOVATION	EXAPTED ADAPTATION	ENGINEERING UTILITY
Mobile Genetic Element	Prokaryotic CRISPR	Genetic Engineering
RAGs	Adaptive Immunity	Immuno-therapy
Endonuclease	Restriction Enzyme System	Molecular cloning
Group II Introns	Eukaryogenesis	Targetrons
Oncolytic virus	Tumor antigens	Oncolytic virus therapy
Membrane proteins	Cellular tropism	Vaccine vectors

If bacteria had not absconded with viral endonucleases, they would have had no restriction enzymes to protect themselves against viral infection. And without restriction enzymes, our society would have had no science of genetics, which made possible progress based on the unique ability of these enzymes to cleave DNA. Furthermore, CRISPR, the most revolutionary genetic engineering tool in the history of biological science, is effectively the recapitulation of a bacterial antivirus defense system that kills an infiltrating virus by slicing it into genetic pieces. Current vaccine delivery techniques and other forms of biological therapies rely on a mimicry or instigation of virus insertion mechanisms. What was once dark was eventually brought to light, and once brought to light, helped to bring light—and life.

The COVID-19 crisis has made the terrifying dark energy of evolution visible and has brought us closer than is comfortable to the engines of selection

Our all-too-human tendency to focus on what is directly or instrumentally visible, or of comparable scale to ourselves, has blinded us to both the largest and smallest scales of the universe—scales where physical forces shape the elementary structure of matter. We are also blinded to those living scales invisible to the eye that have shaped the form and function of adaptive matter. The COVID-19 crisis has made the terrifying dark energy of evolution visible and has brought us closer than is comfortable to the engines of selection. We live in an invisible ocean of microbial diversity and menace, one that is insensitive to the transience of multicellular life. Maybe now is the moment for us as a culture to learn from our microbial allies in the universe of dark matter—the bacteria—from whom we acquire our symbiotic microbiome—that the best way to defeat the dark energy of the virus is to turn its entropic ingenuity against itself and out-evolve the virus by evolving our scientific ingenuity, and probably our social practices too. We'll have to adapt; what choice do we have?

REFLECTION

THE VIRUS THAT INFECTED THE WORLD

DATE: *4 April 2021*

FROM: *David Krakauer, Santa Fe Institute*
Dan Rockmore, Dartmouth College; Santa Fe Institute

In our article, "Out-Evolving COVID-19," we wanted to stress the fact that much of the matter selected and formed through evolutionary processes is effectively invisible to us—a biological manifestation of a much more general process in the sciences, where the space of "unknown unknowns" is, well, *unknown*, but present. We generally knock up against it by accident, sometimes manifest to our senses (even if requiring a technological assist), at other times a nagging error term in a model (mathematical or otherwise) that had looked like it was working, only to come up short in some new and unanticipated environment or dataset. That's the scientific process: you're right until you are wrong.

Dark matter and dark energy were inferred through a deeper and more careful understanding of the universe, identified as important parts of the engine that drives evolution on the scale of galaxies. Microbial dark matter works closer to home and has exerted a powerful influence on the evolution of life. We tabulated how many of the adaptations that we consider constitutive of being a mammal had their origin in our ancient entanglements with viral environments.

LOOKING BACK OVER THE LAST YEAR, the evolutionary force of infection speaks for itself. The virus attacked vulnerable human populations based on comorbidities precipitously terminating social life. As it transitioned from the dark to the light—and was attacked by vaccine makers—the original virus evolved and evolved again as variants spilled forth in an ongoing co-evolution with the vaccine-making process. This kind of evolutionary force—even as extended to the world of health sciences—was expected.

What was unexpected—at least to us, and thus outside of our discussion—was the way it would spill over from the sphere of natural history into that of cultural history. Not only were humans host to

SARS-CoV-2, but in a larger sense, communities, companies, transport networks, treasuries, and nations all played host to the pathogen. This terrible pandemic was a natural experiment reminding us that not only were we blissfully ignorant of phenomena at the scale of the microbe, but also that there were dark elements driving an evolution of the fundamental structures of society.

> As it transitioned from the dark to the light—and was attacked by vaccine makers—the original virus evolved and evolved again as variants spilled forth in an ongoing co-evolution with the vaccine-making process.

Mutational events and selective pressures worked through the biology of genetic elements, endonucleases, and introns. Their "epiviral" effects were now felt in travel and supply networks, the elastic markets of Main Street, the inelastic resilience of Wall Street's forecasting mechanics, and, most of all, in a fear of death that quickly transformed the basic social dynamics of modern interpersonal relationships. Spheres of friendship shrank quickly to encompass only close family. The fundamental materials of the social fabric frayed and eroded. The "triumph" of globalism—at least in the West—had elided the basic dark energy of tribalism, which needed only a push from a lethal pandemic to come (back) to light. In essence, the virus had not just leapt species, but it had leapt scales—it was zoonotic on the scale of societies.

Ever since Darwin puzzled over the social life of bees, with their altruism and sterile casts, and the gaudy displays of birds that render them conspicuous and vulnerable to predators, biologists have pondered the question of the levels of selection. What does selection see: the gene, the cell, the whole organism, or the population? And, by extension, what does a virus infect: a cell, an organism, or a community?

Richard Dawkins argued that, ultimately, it is only the gene that matters and persists, and that everything larger, from bodies to populations and species, are just forms of bookkeeping that allow us to track nature's true currency. Genes are our coinage and bodies merely organic bank accounts or blockchains that record nature's transactions.

Over the last several decades we have gained a deeper understanding into this question, largely through the use of mathematical models, and now understand better that selection can work at any level that respects a few key constraints. These include the relative rates of replication versus dispersal, and the rates of contact with one's own descendants. And the results of these models are to show that evolution operates at a multiplicity of scales. The same sort of argument might be made for a virus infecting a community or a company. As long as that community or company is densely connected, and as long as the virus depends on their internal organization to be spread—from office layouts, org charts, and water-cooler sites—then we can justifiably speak of social or corporate hosts.

Richard Dawkins argued that, ultimately, it is only the gene that matters and persists, and that everything larger, from bodies to populations and species, are just forms of bookkeeping that allow us to track nature's true currency.

And why does this matter? It matters for much the same reason that understanding immune systems is important for the control of microbial infection within the body. We are really talking about the mechanisms of robustness: How, after 2020, we should rethink the robustness of society from the ground up. And how we might use our understanding of the working of the innate and adaptive immune system within the organism to modify social life, work spaces, travel, and economic support, to minimize the effects of infection at these higher levels of organization.

So, what have we learned? We now understand that the host of a virus extends beyond the cell and body to the social networks, communities, and cities in which "life" takes place. We thought that treating a disease meant finding cures, palliatives, and vaccines on the level of the organism—that all we needed were the tools of biological sciences. What we failed to grasp was its effect at scale and the remediations required of the downstream social, economic, and even political behaviors. We can no longer see a virus as just a minimal form of life that expands by siphoning the energy of bodies. It is one of the forms of dark matter shaping all complex systems at every scale of existence.

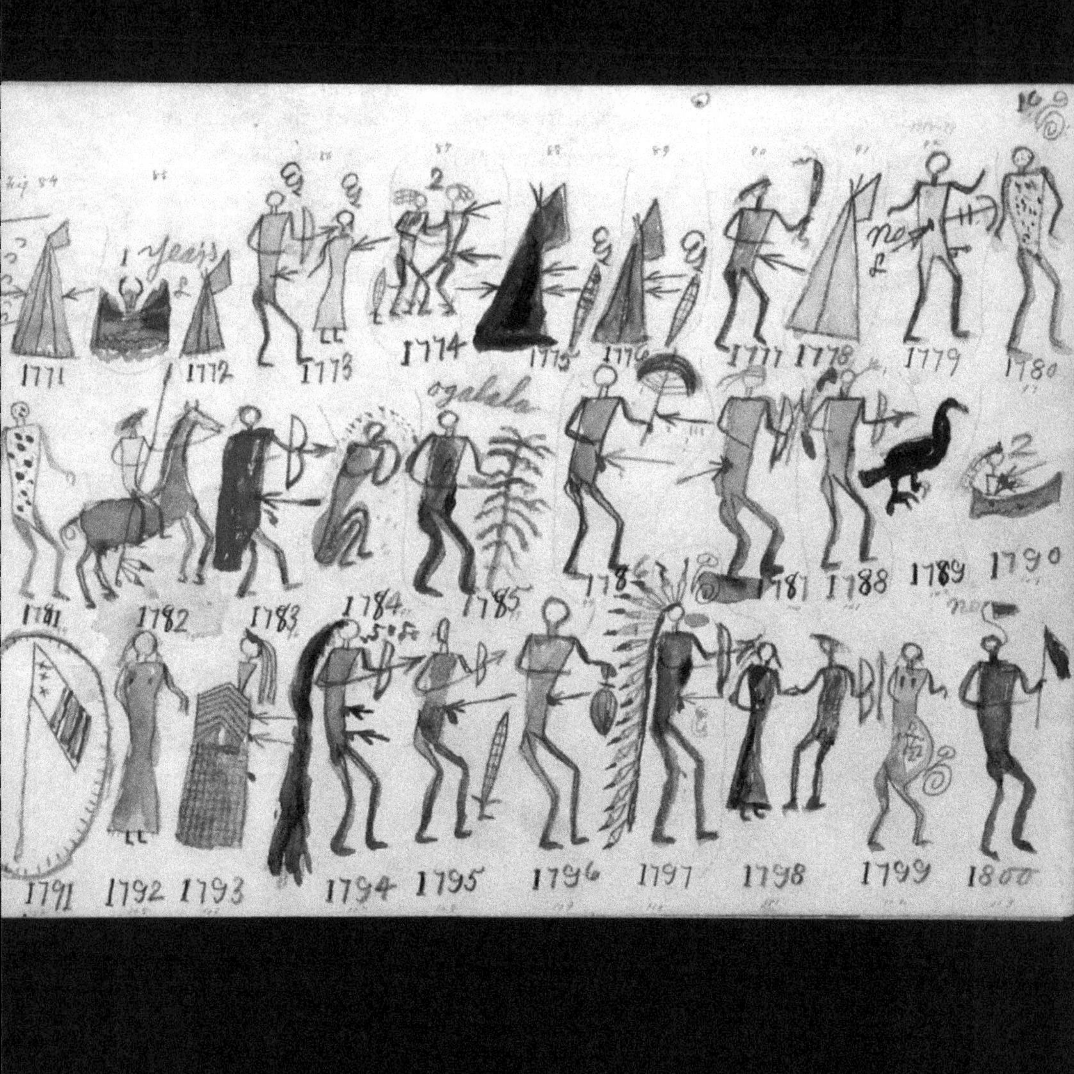
1771
1 years
1772
1773
2
1774
1775
1776
1777
1778
1779
1780
ogalala
1781
1782
1783
1784
1785
1787
1788
1789
1790
1791
1792
1793
1794
1795
1796
1797
1798
1799
1800

TRANSMISSION NO. 031

MODELS THAT PROTECT THE VULNERABLE

DATE: *1 June 2020*

FROM: *Melanie Moses, University of New Mexico; Santa Fe Institute*
Kathy Powers, University of New Mexico; Santa Fe Institute[1]

STRATEGIC INSIGHT: *Well-mixed models do not protect the vulnerable in segregated societies.*

African Americans are dying from COVID-19 at a rate two to four times higher than White Americans.[1,2] Per capita cases are higher on the Navajo reservation than in every US state.[3] What are the causes of this disparate impact of COVID-19, why did epidemic models not predict it, and what can be done to address it?

Myriad explanations for racial differences in COVID-19 exposure and mortality have been proposed, particularly focused on the work circumstances, living conditions, health status, and healthcare access among African Americans. African Americans tend to live in denser urban areas and multi-generational households, and to be essential employees who cannot work from home. They are likely to have less paid sick leave and health insurance, but more underlying medical conditions such as diabetes, cardiovascular disease, sickle cell disease, asthma, and exposure to environmental pollutants that elevate COVID-19 mortality risk. African Americans are often turned away or have the severity of their condition underestimated when they seek medical care. The history of unethical medical experimentation and exploitation (e.g., the Tuskegee syphilis experiment and Henrietta Lacks's cancer cells) has

1 Both authors are members of the SFI/UNM project on Algorithmic Justice: https://www.santafe.edu/research/projects/algorithmic-justice

---

OPPOSITE: *Battiste Good Year, "The Eruption and Pains in the Stomach and Bowels; Smallpox Used Them Up Winter," 1880*

led to community-level distrust of the medical system. When the novel coronavirus initially spread among international travelers centered in China and Europe, myths gained traction on social media that people of African descent could not contract COVID-19.

Higher COVID-19 mortality, particularly at younger ages, is consistent with other health disparities in African Americans: a fourfold greater probability of dying from complications during childbirth, 20 times greater chance of heart failure before age fifty, and a four-year-shorter lifespan.[4] The stress of living under the threat of racism appears to age Black bodies faster than White ones. The multiplication of many inequities over time results in greater disparity; Black income is 60% of White income, but Black wealth is only 10% that of Whites.

Native American COVID-19 risk is also influenced by extreme inequity in health and economic circumstances, including lack of services as basic as running water and inadequate federal funding of health care.[5] The Native population is a staggering seventeen times more likely to be diagnosed and over ten times more likely to die from COVID-19 than the White population in New Mexico[6] (one of the few states reporting sufficient data to make such comparisons). Previously isolated Native populations were decimated by diseases upon European contact and were also four times more likely than others to die in the 1918 flu pandemic.[7] COVID-19 threatens the very survival of small Pueblos like the Zia with fewer than 1,000 members. The proximate socioeconomic factors correlated with COVID vulnerabilities are themselves the manifestation of hundreds of years of structural racism that has left formerly enslaved and colonized populations with poor physical and economic health and often in spatially segregated places.

How do epidemic models incorporate these racial and socioeconomic realities? Thus far, they don't.

How do epidemic models incorporate these racial and socioeconomic realities? Thus far, they don't. Initial epidemic models, and most that still influence pandemic policies, assume "well-mixed" populations. More sophisticated network models consider different categories of people and their risk of infection and mortality,[8] but far more work is needed to incorporate systematic correlations among epidemic factors within particular groups in particular places. More often, people are modeled like identical balls bouncing randomly in a lottery machine, equally likely to contact infection, become sick, infect others, or have their number drawn as the unlucky one to die. However, people are not equally vulnerable, and America is not well mixed.

One of the greatest determinants of physical and economic health is the zip code in which you were born. Zip code is a powerful predictor of the chance that you will go to college or prison, that you will have heart disease or asthma, whether you will drive a bus or work safely at home and order your groceries online. Like wildfires that burn where the winds blow and the grass is driest, the spread of disease is determined by a template of risk and vulnerability laid out in physical space. In America, that space is largely determined by race.

Zip code has long been used as a proxy for race. Redlining was the deliberate practice of denying Black Americans access to safe neighborhoods with high property values and good schools. Now zip code is used in algorithms that determine where predictive policing is concentrated and eligibility for loans and higher credit.[9] The spatial segregations of Native and African Americans have different historical causes, but similar consequences are clearly visible (fig. 1). The zip codes, counties, and territories where African and Native Americans are concentrated are the deadliest places in America.

Complex-systems thinking developed powerful and mathematically tractable epidemic modeling approaches that mitigated unfettered exponential growth of COVID-19 in the general population. Yet, as many authors in the *Transmission* series have pointed out, the simplifying assumptions that allow for powerful broad-scale predictions are only a starting point. Models need to incorporate the systematic and structural features of societies that determine how both policies and physical space mediate disease spread.

Complex-systems thinking should emphasize that viral spread is not an idealized mathematical process taking place in a vacuum. Disease is an emergent phenomenon whose spread and severity is a consequence of the properties of the SARS-CoV-2 virus, the age, health, occupation, and socioeconomic status of individuals, and societal structures that cause disease to spread differently among different people in different places for different reasons.

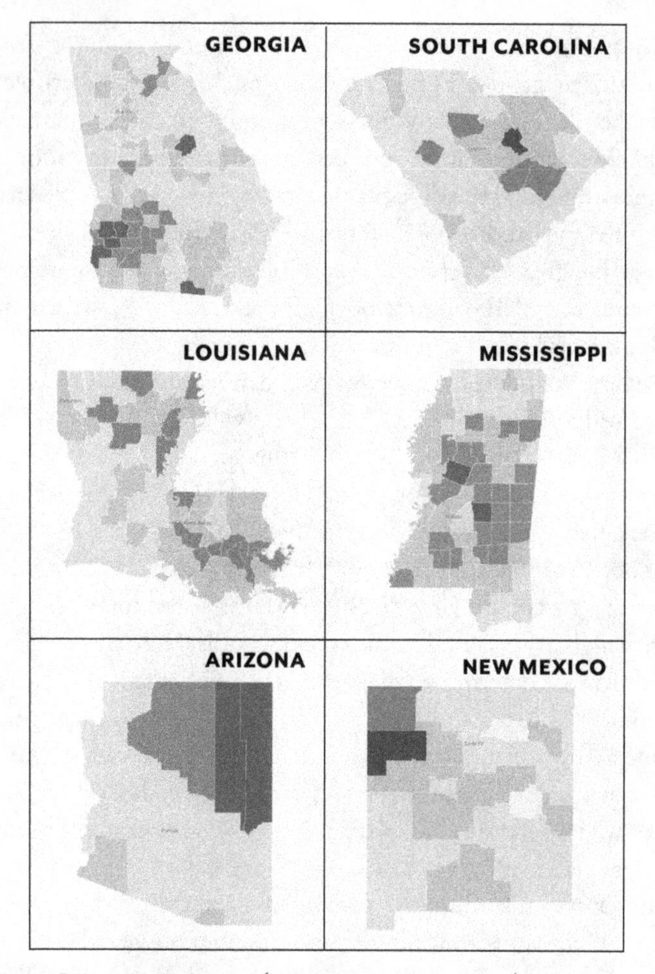

Figure 1. *Per capita positive cases (higher values in darker colors) in the four states with the highest percentages of African Americans, and in Arizona and New Mexico with large Native American populations.*

In Mississippi, Holmes County has the highest per capita death rate. It is 83% African American, the third highest proportion of the 82 counties in the state. Four counties in Georgia have among the highest COVID-19 mortality in the nation (1 death per 250–290 people). They are ranked 1st, 5th, 7th, and 20th in proportion of African Americans (49–73%) of 159 counties. In South Carolina's 46 counties, the highest mortality is in Lee and Clarendon Counties, ranked 11th and 3rd (49% and 64% Black). St. John the Baptist is the highest mortality parish in Louisiana, the 5th most African American (54%) of 64 parishes. The highest mortality counties in adjacent regions of northern New Mexico and Arizona have the highest Native American populations (27–73%).

Data and images from The New York Times, *May 26, 2020.*

Disease spread is an inherently multiscale spatial process. It is caused by the entry of a nanometer-sized virus into a cell, where the probability of that infection is influenced by the space in a crowded factory, nursing home, or prison, by the availability of healthy food or prevalence of air pollution in a neighborhood, and by a nation's social structure, which determines the level of stress, inflammation, and cardiovascular health of the body that cell inhabits. COVID-19 also disproportionally impacts ethnic minorities across the globe.[10] It will grow where the most vulnerable are concentrated in refugee camps, *favelas*, and war-ravaged cities.

Complex-systems thinking should emphasize that viral spread is not an idealized mathematical process taking place in a vacuum.

How do we find a path forward? Disparate impact matters not just for predicting who is at risk, but also for prioritizing who is tested, isolated, economically supported, treated, and vaccinated. As in civil rights law, mitigating disparate impact does not require ascribing intent or completely disentangling causal factors. How can our response to this disease increase how resilient the most vulnerable populations are to the next pandemic, the next economic shock, and the looming climate crisis?

COVID-19 racial and ethnic health disparities are the new frontier of human rights in the United States and globally. The international efforts in reparations for human rights violations provide a framework not just for redress of past injustices that have led to increased community risk from COVID-19, but also for forward-looking provision of free access to community health initiatives located specifically in the places where the most vulnerable populations live. COVID-19 maps show us where those places are (fig. 1).

A complex-systems approach to addressing past inequities would consider the historical, layered, and interacting factors that may be impossible to disentangle, but which collectively put certain populations at highest risk. The goal is not just to repair harm, but also a concerted effort to make people and places less vulnerable to the next pandemic: lowering rates of diabetes, increasing access to healthy food and caring doctors, and reducing air pollution. Such systemic changes support resiliency to survive the next outbreak in the places we already know are most vulnerable. These investments are far less expensive than blindly shutting down an entire economy. We will know we have succeeded if the next map looks different—if the location of the biggest outbreak is due to the random draw of a lottery, and not the inevitable consequence of generations of inequity.

ENDNOTES

1 C.W. Yancy, 2020, "COVID-19 and African Americans," *Journal of the American Medical Association* 323(19): 1891–1892, doi: 10.1001/jama.2020.6548

2 E.G. Price-Haywood, J. Burton, et. al., 2020, "Hospitalization and Mortality among Black Patients and White Patients with COVID-19," *The New England Journal of Medicine* 382: 2534–2543, doi: 10.1056/NEJMsa2011686

3 https://www.ndoh.navajo-nsn.gov/COVID-19

4 D.R. Williams, 2012, "Miles to Go Before We Sleep: Racial Inequities in Health." *Journal of Health and Social Behavior* 53(3): 279–295, doi: 10.1177/0022146512455804

5 E. Warren and D. Haaland, "The Federal Government Fiddles as COVID-19 Ravages Native Americans," *The Washington Post*, May 26, 2020, https://www.washingtonpost.com/opinions/2020/05/26/federal-government-fiddles-covid-19-ravages-native-americans/

6 https://cv.nmhealth.org; calculated from the 10% Native population accounting for 58% of cases and the 37% White population accounting for 12% of cases.

7 M. Kakol, D. Upson, and A. Sood, 2020, "Susceptibility of Southwestern American Indian Tribes to Coronavirus Disease 2019 (COVID-19)," *Journal of Rural Health*, doi: 10.1111/jrh.12451

8 S. V. Scarpino, A. Allard, and L. Hébert-Dufresne, 2016, "The Effect of a Prudent Adaptive Behaviour on Disease Transmission," *Nature Physics* 12: 1042–1046, doi: 10.1038/nphys3832

9 B. Lepri, N. Oliver, et al., 2018, "Fair, Transparent, and Accountable Algorithmic Decision-Making Processes," *Philosophy & Technology* 31: 611–627, doi: 10.1007/s13347-017-0279-x

10 T. Kirby, 2020, "Evidence Mounts on the Disproportionate Effect of COVID-19 on Ethnic Minorities," *The Lancet.* 8(6): 547–548, doi: 10.1016/S2213-2600(20)30228-9

REFLECTION

LEGACIES OF HARM, SOCIAL MISTRUST & POLITICAL BLAME IMPEDE A ROBUST SOCIETAL RESPONSE TO THE EVOLVING COVID-19 PANDEMIC

DATE: *21 August 2021*

FROM: *Melanie Moses, University of New Mexico; Santa Fe Institute*
Kathy Powers, University of New Mexico; Santa Fe Institute

In our Transmission of June 2020 we wrote that our scientific models and our COVID-19 mitigation strategies did not protect the most vulnerable members of society as African-American, Native American and Latinx/Hispanic Americans died at rates up to four times that of White Americans. In November 2020 we wrote that "the complexity of harms that lead to disproportionate exposure rates, infection rates, and death also limit social trust in new vaccines that might mitigate these risks in the future,"[1] and then in January 2021 that "an untrustworthy system has created many who are understandably vaccine hesitant" and lamented online disinformation campaigns that are fueled by baseless conspiracies and prey on distrust.[2]

Now, we are again facing nearly 10,000 global deaths per day as the reality sinks in that the vaccines, despite extraordinary success in protecting individuals from hospitalization and death, may not sufficiently block transmission to provide societal-level "herd immunity" from the hyper-transmissible Delta variant. Even if vaccines alone will not completely stop the virus from circulating, they are the best tool we have to protect ourselves individually from severe disease and death and to reduce transmission to others. In combination with masks, social distancing, testing, and antibody therapies, vaccines can dramatically reduce suffering from COVID-19 in the US and worldwide. Yet, less

---

1 M. Moses and K. Powers, "A Model for a just COVID-19 Vaccination Program," *Nautilus*, November 25, 2020, http://nautil.us/issue/93/forerunners/a-model-for-a-just-covid_19-vaccination-program

2 M. Moses, "How to Fix the Vaccine Rollout," *Nautilus*, January 20, 2021, https://nautil.us/issue/95/escape/how-to-fix-the-vaccine-rollout

than 2% of people on the African continent are vaccinated. That the successful, urgent production of vaccines for the US and other wealthy nations has done so little for the global poor may ultimately be considered the greatest moral failure of this pandemic, a failure that may contribute to the evolution of more dangerous variants.

That the successful, urgent production of vaccines for the US and other wealthy nations has done so little for the global poor may ultimately be considered the greatest moral failure of this pandemic, a failure that may contribute to the evolution of more dangerous variants.

In addition to the unequal global distribution of the vaccine—with most of the doses administered in just ten countries—we must also address the question of why so many refuse to accept the vaccines that are available to them and could save them from suffering and death. Scientific tools like vaccines are produced by and embedded in complex social systems. Our socio-economic, political, and biomedical systems have often engendered mistrust. Governments that have sanctioned unethical medical experimentation in the past are now devising strategies to incentivize or compel their citizens to be vaccinated today. But memories are long with respect to human-rights violations, especially when they have been repeated over time. In the Democratic Republic of the Congo and West Africa, refusal of Ebola vaccination was explained as disbelief that the virus was real, but it may instead have reflected an understandably deep distrust of external forces that brought colonialism and slavery and contemporary exploitation.[3]

3 E.T. Richardson, T. McGinnis, and R. Frankfurter, 2021, "Ebola and the Narrative of Mistrust," *BMJ Global Health*: 4:e001932, doi: 10.1136/bmjgh-2019-001932

Memory of past atrocities can stifle vaccination rates as is seen in communities of color in the US and in other populations around the world.

Restoring trust requires reckoning with contemporary inequities as well as past harms. Why should the global poor trust pharmaceutical companies and international health organizations who have provided so little access to vaccines? Why should a young Black American who is threatened by his government's criminal justice system trust his government's public health system? The ability to trust is itself a privilege. Trust can act as a ratchet that exacerbates inequity when those with more privilege have greater trust in solutions to problems that cause greater harm for the less privileged who are then blamed for their individual choices. The contribution of societal factors to those choices are not sufficiently examined.

Failure to address mistrust in vaccines is in part a failure to understand that people do not react to the vaccine in an isolated equation of risk and benefit. They also react to it as a reflection of the society that produced it and seeks to benefit from its adoption. One would logically answer *yes* to the question, "Would you take a vaccine that reduces your chance of death by over 95% if the risk of death from taking it is far less than a 1 in a million?" Yet the question some are actually answering is, "Do you trust a vaccine produced by a society that has experimented on your ancestors and will bankrupt you if you get cancer?"

Mistrust is amplified when public-health messaging changes (as it must do as the virus and our scientific understanding of it change) and by political structures that thrive on blame, scapegoating and, increasingly, disinformation, rather than cooperation. But mistrust can be overcome. Native Americans who suffered the most in the early days of the pandemic overcame mistrust of the vaccine to become the most vaccinated racial group in the US due in part to effective advocacy by trusted tribal leaders.[4] The gap in vaccination rates between African-American and Latinx populations and White populations is narrowing, and vaccination rates have increased in the places hardest hit by the wave of Delta infections. SFI researchers found that in Austria, less than half of the surveyed population trusted

4 R. Read, "Despite Obstacles, Native Americans Have the Nation's Highest COVID-19 Vaccination Rate," *Los Angeles Times*, August 12, 2021, https://www.latimes.com/world-nation/story/2021-08-12/native-american-covid-19-vaccination

their government to provide a safe vaccine,[5,6] but over 60% have now gotten their first shot. An effective and equitable vaccine rollout can help to build the trust that is needed for future public-health efforts.

The ability to trust is itself a privilege. Trust can act as a ratchet that exacerbates inequity when those with more privilege have greater trust in solutions to problems that cause greater harm for the less privileged who are then blamed for their individual choices.

Complexity science suggests that we could make further progress if we transcend the reductionist approach that attempts to isolate vaccines and other scientific tools from the societies in which they are embedded. As beautifully stated by W. Brian Arthur, people in complex economies "explore, try to make sense, react and re-react to the outcomes they together create."[7] Much of the reaction to vaccines reflects that they are seen as inseparable from unjust systems we have created. Developing a more equitable and trustworthy society is a crucial prerequisite to achieving trust in vaccines. Building that trust now will foster cooperation to address ongoing and future pandemics and the even greater challenges of climate change.

5 E. Schellhammer, J. Weitzer, et al., 2021 "Correlates of COVID-19 Vaccine Hesitancy in Austria: Trust and the Government," *Journal of Public Health*: fdab122, doi: 10.1093/pubmed/fdab122

6 J. Weitzer, M.D. Laubichler, et al., 2021, "Comment on Alley, S.J., et al., 'As the Pandemic Progresses, How Does Willingness to Vaccinate against COVID-19 Evolve?'" *International Journal of Environmental Research and Public Health* 18(6): 797, doi: 10.3390/ijerph18062809

7 https://sites.santafe.edu/~wbarthur/

Understanding and confronting the causes of social mistrust is not a magic bullet. There will still be miscreants who peddle lies and exploit ignorance and grievance in opposition to common interests. A more trustworthy society may mitigate their influence, but not likely on the time scales to address this pandemic. Regardless, we must confront past harms and current societal inequities if we wish to have robust and adaptable collective responses to increasingly complex global problems in the future.

TRANSMISSION NO. 032

THE NOISY EQUILIBRIUM OF DISEASE CONTAINMENT & ECONOMIC PAIN

DATE: *1 June 2020*

FROM: *Jon Machta, University of Massachusetts Amherst; Santa Fe Institute*

STRATEGIC INSIGHT: *The countervailing pressures of economic pain and disease containment are keeping the COVID-19 pandemic at a noisy equilibrium.*

There's a magic number in epidemic modeling. When it exceeds 1, new infections grow exponentially, spreading like wildfire. When it falls short of 1, the virus succumbs to exponential death.

That basic reproductive number, R_0, is simply the average number of new infections each existing case will spawn. If every coronavirus victim infects, on average, two new people, R_0 (pronounced "r-naught") is 2 and the virus doubles its way around the globe. If R_0 is near 0, whether due to the host's immune system or lack of contact with others, the virus dies out quickly.

R_0 equaling 1 is a critical equilibrium. Here the caseload stays constant on average but is subject to large fluctuations.[1] In this essay, we argue that R_0 for COVID-19 in the US is likely to hover near 1 or oscillate around 1 for a lengthy period. This is an example of a phenomenon known as "self-organizing criticality," which is observed in many complex systems. Another example of self-organized criticality is seen in earthquakes. Here the inexorable motion of tectonic plates is balanced by friction until some part of the fault ruptures and moves suddenly. The result is a critical equilibrium with frequent small earthquakes and rare large earthquakes.

OPPOSITE: *Vasily Vereshchagin, "The Apotheosis of War," 1871 (detail)*

The natural value of R_0 for COVID-19 in a naïve population appears to be above 2, leading to rapid spread in the absence of mitigation. Mitigation efforts have brought R_0 below 1 in many countries but at great economic cost. In some countries such as Taiwan and New Zealand, very effective mitigation has brought case numbers to a low enough value that relatively inexpensive measures now suffice.

However, in the US with its large and heterogeneous population and weak central leadership on this issue, suppressing R_0 below 1 is currently extremely costly and likely to remain so for the foreseeable

future. In this regime, where mitigation is extremely expensive but having an exponentially exploding number of cases is also unacceptable, R_0 will remain near 1. If R_0 exceeds 1, then exponential growth quickly leads to unacceptable rates of infection, hospitalization, and death, but as R_0 is suppressed by the crude tools of social distancing and business closures, the economic pain creates great pressure to open up the economy, pushing R_0 back above 1. The combination of these two forces generically and robustly keeps R_0 close to the critical value of 1. Many current models of the epidemic include both the biology of infection and policy responses in more or less detail[2] (for example, see [2]), but it is useful to understand that these models will display the general feature of self-organized criticality so long as the two strong and opposing forces are incorporated.

A simple, two-variable class of dynamical models demonstrates the idea of self-organization to $R_0 = 1$. These models cannot be used to make quantitative predictions, but they reveal important qualitative features. The two variables are $C(t)$, the number of new infections at time t, and $R_0(t)$, the (time-dependent) reproductive number. Roughly speaking, time is measured in weeks. The equation for the number of new infections, $C(t+1) = R_0(t)C(t)$, is simply the definition of R_0. The dynamics of the reproductive number is the new ingredient, and it takes the form $R_0(t+1) = R_0(t) + F[C(t), R_0(t)]$. There is wide latitude in constructing the function $F[C,R_0]$. It should be negative when the number of new infections is above an acceptable threshold, pushing the system toward mitigation. $F[C,R_0]$ should be positive when R_0 is below 1, leading to large economic costs that push the system away from mitigation. Depending on the details of the function F and the initial conditions, there are two generic behaviors. Either $R_0(t)$ will evolve to the critical value of 1 and $C(t)$ will settle to a steady state value given by solution to $F[C, 1]=0$. The result from a simulation of a specific choice of F and initial conditions is plotted in figure 1.

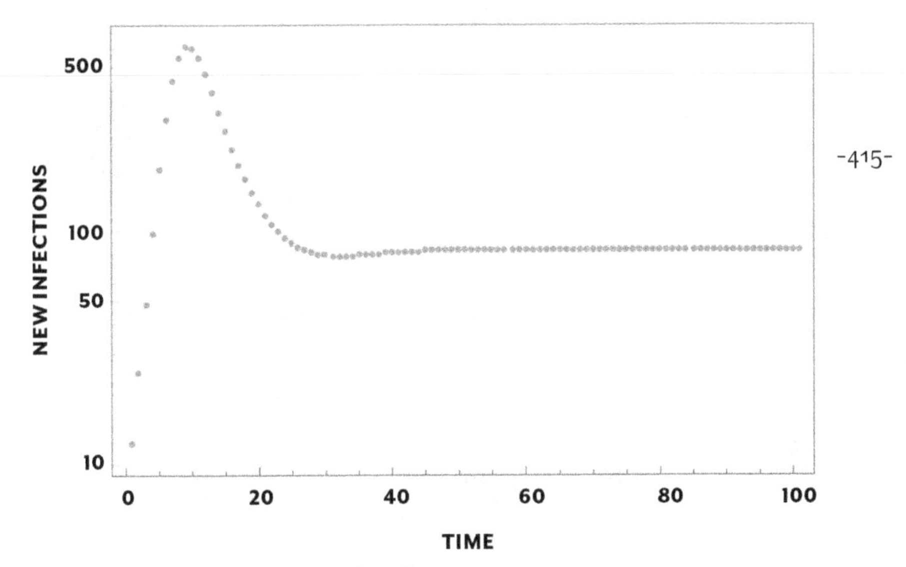

Figure 1. *New infections versus time (steady)*

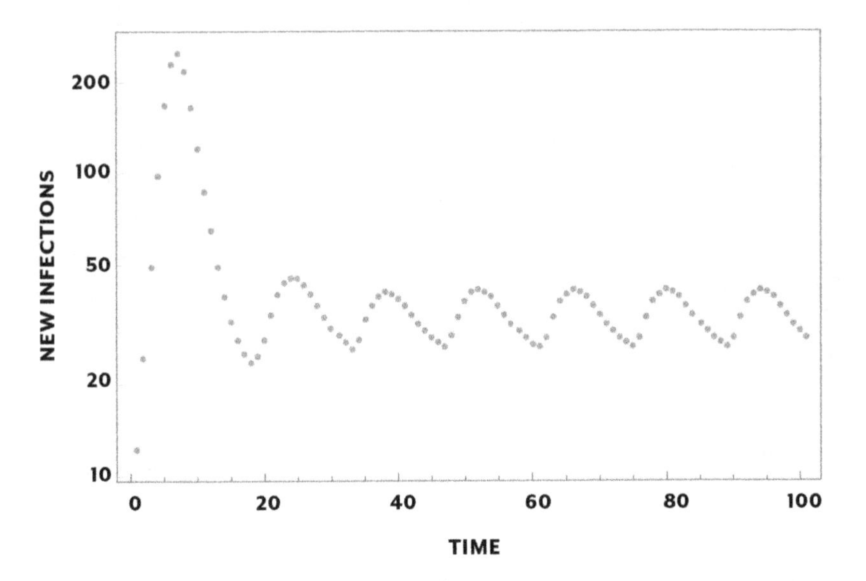

Figure 2. *New infections versus time (oscillatory)*

Alternatively, for other choices of F and initial conditions,] $R_0(t)$ oscillates around 1 and $C(t)$ oscillates as shown in figure 2. Figure 3 shows the behavior of $R_0(t)$ for the two cases of figures 1 and 2. Finally, one can add noise to either equation in various ways. When weak noise

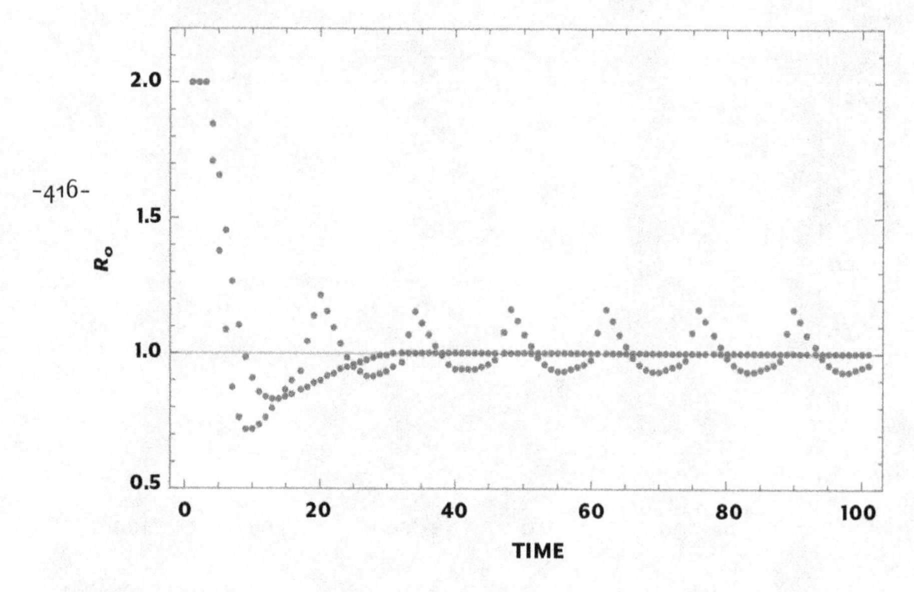

Figure 3. *Reproductive number versus time (both steady and oscillatory regimes shown)*

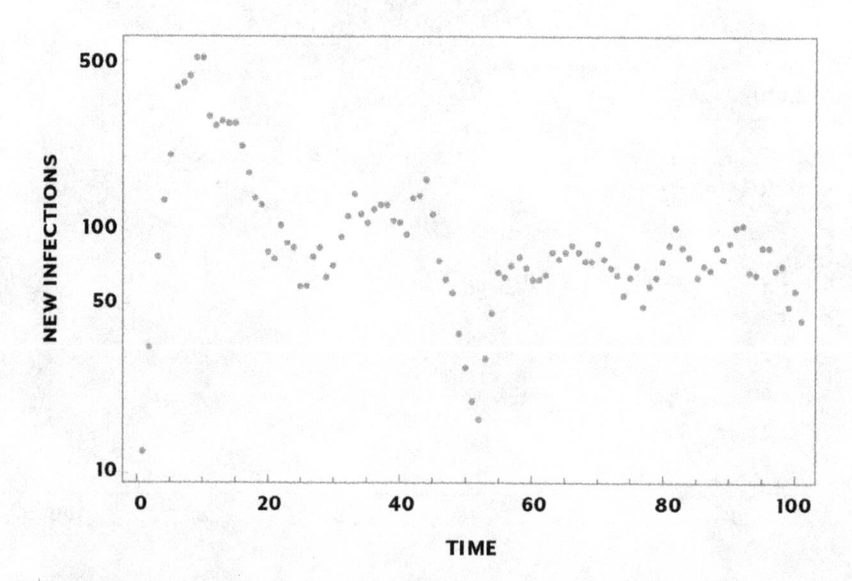

Figure 4. *New infections versus time (same as figure 1 but with noise)*

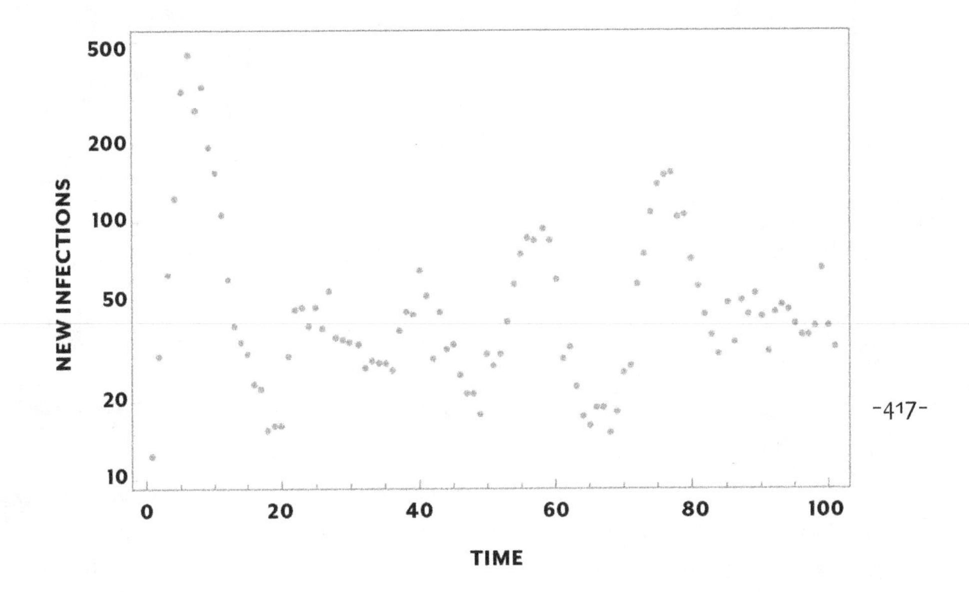

Figure 5. *New infections versus time (same as figure 2 but with noise)*

is added to the system, these qualitative behaviors persist but become quite noisy, as shown for specific simulations in figures 4 and 5.

In conclusion, one can expect R_0 to stay near 1 and the number of new cases to stay relatively high for an extended period, but with lots of unpredictable and perhaps oscillatory behavior along the way. The only way out is with an effective treatment or to dramatically lower the cost of an R_0 below 1 through natural herd immunity, a vaccine, or effective testing and contact tracing.

ENDNOTES

1 See this volume, C. Moore Transmission T-024 on "The Heavy Tail of Outbreaks" on page 309 and Redner Transmission T-028 on "Exponential Growth Processes" on page 363.

2 S.M. Kissler, C. Tedijanto, et al., 2020, "Projecting the Transmission Dynamics of SARS-CoV-2 through the Postpandemic Period," *Science* 368(6493): 860–868, doi: 10.1126/science.abb5793

REFLECTION

UNPREDICTABLE & UNSURPRISING EVENTUALITIES

DATE: *12 August 2021*

FROM: *Jon Machta, University of Massachusetts Amherst; Santa Fe Institute*

As I write this reflection, much of the world—including the US—is experiencing yet another surge in COVID-19 cases, attributed to the highly contagious Delta variant and to the relaxation of restrictions meant to limit the spread of the virus. The good news is that we now have highly effective vaccines and a substantial fraction of the population in many wealthy countries has been vaccinated. The current surge is primarily affecting unvaccinated populations. None of these eventualities could have been predicted a year ago when we were still early in the pandemic and the Transmissions articles were written, and yet none are entirely surprising. The collaboration between scientists, the pharmaceutical industry, and government produced safe and highly effective vaccines at scale in an amazingly short time. One lesson of the pandemic is that human ingenuity in the face of adversity should never be underestimated. Another lesson, however, is that the virus has also proved to be ingenious in evolving ever greater transmissibility—the Delta variant is far more infectious than the virus that originally spawned the pandemic.

My Transmission predicted that we would see a series of surges in infection based on the competing effects of the natural exponential growth in infections and the human response to the virus. High caseloads would cause individuals and governments to respond with measures to decrease the spread of the disease. These measures would work but at a high cost, both economically and socially, so that upon successfully mitigating the current surge, they would be partially or completely abandoned, setting the stage for the next surge.

Did this prediction prove to be true? The pandemic has indeed been marked by a series of surges, and the response has also followed the predicted pattern of tightening restrictions after caseloads increase and relaxing restrictions when caseloads decrease. In this sense the simple model did a great job at qualitatively predicting the trajectory of the pandemic, and I believe it is substantially the correct explanation. However, establishing causality is difficult since factors such as seasonality and the advent of more infectious variants may have played an important or even dominant role in the natural history of the disease. Academic research and common sense show that social distancing, closures, masks, and contact tracing reduce the transmission of the virus, but they do not prove that the imposition and relaxation of these measures were the dominant reason that the graph of infections looks like a series of waves.

None of these eventualities could have been predicted a year ago when we were still early in the pandemic and the Transmissions articles were written, and yet none are entirely surprising.

The model described in my Transmission is extremely simple. It is a mathematical model with two variables that incorporates the ideas described above. It produces curves of infection numbers, but these curves are only useful qualitatively. Far more elaborate models developed by a number of research teams include spatial structure and many details of the transmission process that are lumped together in the simple model. These models are able to make useful semiquantitative predictions but involve many parameters and must be frequently updated by teams of scientists. So, what's the point of a well-designed simple model such as the one described in my article? The answer is insight! The reasons that a complex system behaves qualitatively in a certain way, independently of many details, can be revealed by simple but robust models and obscured in more elaborate models with many parameters. Complex-systems science needs both types of models in both research and policy making.

TRANSMISSION NO. 033

HOW PANDEMICS RAPIDLY RESHAPE THE EVOLUTIONARY & ECOLOGICAL LANDSCAPE

DATE: *1 June 2020*

FROM: *Brian Enquist, University of Arizona; Santa Fe Institute*

STRATEGIC INSIGHT: *Pandemics rapidly reshape the evolutionary and ecological landscape and have cascading social, economic, and other system-level effects.*

In the 1950s, the planet still had isolated islands, in both geographical and cultural terms—lands of unique mysteries, societies, and resources. By the end of the 20th century, expanding numbers of people, powerful technology, and economic demands had linked Earth's formerly isolated, relatively nonindustrialized places with highly developed ones into an expansive and complex network of ideas, materials, and wealth.

—JULIANNE LUTZ WARREN AND SUSAN KIEFFER (2010)

Believe me who have tried. Thou wilt find something more in woods than in books. Trees and rocks will teach what thou canst not hear from a master.

—ST. BERNARD OF CLAIRVAUX (1841)

This oak tree and me, we're made of the same stuff. —CARL SAGAN (1980)

It is remarkable how closely the history of the apple tree is connected with that of man.

—HENRY DAVID THOREAU (1862)

AMONG THE MOST PRESSING ecological and socioeconomic imperatives of our time is that of assessing the impact of novel external perturbations to manage complex systems for resilience. Broadly defined, resilience is the capacity of a system to maintain function in response to perturbation.

OPPOSITE: *Daniel MacDonald, "An Irish Peasant Family Discovering the Blight of their Store," 1847 (detail)*

NATIONAL FOLKLORE COLLECTION, UNIVERSITY COLLEGE DUBLIN

Pandemics—the spread of epidemic disease around the world—are a signature of the Anthropocene. Before the rise of an interconnected world, epidemics were largely contained to a limited area and pandemics were rare. Pre-Anthropocene biology was defined by splendid isolation. As continents moved, mountains formed, and rivers eroded, populations, species, and clades become isolated and gene flow was limited. Indeed, the amazing biodiversity of the planet is due largely to this isolation. Within isolated regions, separate co-evolutionary dynamics between individuals and their enemies produced much of Earth's biological diversity. The co-evolution of host populations and pathogens is a stepwise reciprocal evolutionary interaction and, if transmissible, can result in epidemics. Epidemics result in intense co-evolutionary selection pressures on both host and pathogen populations, ultimately allowing long-term persistence and ecosystem stability.

Pandemics—the spread of epidemic disease around the world—are a signature of the Anthropocene.

In the new biology of the Anthropocene, however, the barriers of isolation and limited gene flow are now rapidly disappearing.[1] Global dispersal and immigration of species between regions are now increasingly frequent, entirely due to human movement, human-caused changes to the environment, and economic trade connections. The new biology is a biology of panmixia, or random mating within a population, writ large.

A pathogen is any agent that disrupts homeostasis in an individual. When a pathogen spreads among hosts, the effects on the total population can be substantial. Novel pathogen introductions can impact both host and pathogen unpredictably, often resulting in associations with no previous co-evolutionary history. When geographically isolated pathogens and their hosts are suddenly transported great distances and mixed with new hosts and pathogens, remarkable and unpredictable pathways of evolution are possible. A science of the Anthropocene necessarily would include a description of complete biological planetary interconnectedness. This is a lofty goal given the degree of global interconnectedness; indeed, the biosphere has never experienced panmixia and pandemics on the scale we see today.

August Krogh, a Danish professor at the department of zoophysiology at the University of Copenhagen from 1916 to 1945, specialized in comparative physiology, particularly areas relating to cellular respiration. In 1920 he was awarded the Nobel Prize in Physiology for his discovery of the mechanism regulating capillaries in skeletal muscle. Krogh also succinctly encapsulated the idea of comparative biology, which recognizes that parts or processes that are difficult or impossible to study in one species may be much more accessible in an alternate species.[2] To paraphrase the Krogh principle, for many problems there is another organism or organisms on which it can be most conveniently studied. Krogh's principle is the guiding foundation of comparative biology—the use of biological variation and disparity to understand the patterns and rules of life at all levels.

Given the global impact of COVID-19, the pandemic is often compared to past pandemics. For example, the news is filled with references to the Spanish influenza, smallpox, and the potential pandemic of Ebola. These past human pandemics have helped guide our current response to COVID-19. However, there is more to a viral disease pandemic than the human host and pathogens. Spread or transmission of an infectious agent relies on interactions of individuals and populations.

Stepping back to earlier concepts, we can use Krogh's principle to assess and build a comparative science that addresses a world of complete biological planetary interconnectedness. Specifically, we ask: what can other organisms reveal about how biological systems respond to the rise of a massively connected biosphere? Plants, and trees in particular, are remarkable teachers—like us, they are dominant, cover most of the earth, and disproportionately impact the functioning of the biosphere. They also prefer environments similar to those most habitable for humans, and their life cycles span time frames relatively close to our own. They, too, have experienced an increase in pandemics, some with disastrous consequences.

There are at least three immediate lessons from a comparison of pandemics in humans with pandemics in trees. These lessons point to common threads for a science of the Anthropocene as well as a focus on the need to rigorously assess the concept of resilience.

1. **Pandemics cause sudden rapid transformations across interconnected ecological and economic systems.**

The parallels between COVID-19 and the 1904 blight of chestnut trees are striking. COVID-19 (caused by the SARS-CoV-2 coronavirus) and chestnut blight (caused by the Ascomycota fungus *Cryphonectrica parasitica*) both likely arose in Southeast Asia and were transported via international travel networks that enable pathogens to circumnavigate the globe. Over the course of about a month, both spread rapidly and simultaneously across Europe and North America, causing a broad spectrum of disease, from mild to severe and even death. In the US, both pathogens first devastated the New York region. Both infections first revealed their presence by several symptoms, but ultimately the more stressed and susceptible individuals died first as the pathogen more easily blocked and "suffocated" the vascular network. The collateral damage of both pandemics was remarkable. Spreading unimpeded, the chestnut blight decimated what had been one of the most dominant species of trees. Similarly, the novel coronavirus hit the local subsistence economies, mainly in the poor areas of the continent. Spikes in unemployment and dramatic changes in the traditional ways of life left culture irreversibly altered and shifted economic models.

For years, the American chestnut tree largely defined American deciduous forests, ranging from Maine to Georgia and west to the prairies. It survived all evolutionary adversaries for 40 million years, but then, within 40 years, it effectively disappeared. The American chestnut had no evolutionary history of interaction with this new exotic fungal pathogen. First discovered in 1904 in New York City, the chestnut blight was a global pandemic hitting forests around the world. In fact, it has been called the first great ecological disaster to strike the world's forests. It is estimated that in some places, such as the Appalachians, one in every four trees was an American chestnut. Within the span of a generation the North American forests had no chestnut trees. Chestnut blight spread rapidly and caused significant tree loss on different continents. The primary plant tissues targeted by *C. parasitica* are the vascular tissue and the cambium, a layer of actively dividing cells. The fungus girdles the stem, severing the flow of nutrients and water to the vital vegetative tissues. The absence of nutrient dispersal from leaves to roots causes the main stem to die, though, ironically, the root system may survive. The American chestnut still exists, but poorly; its individuals are continuously knocked back by a resurgence of the pathogen when any one grows bigger than a sapling.

The wood of mature chestnut trees was lightweight, soft, easy to split, and resistant to decay. For three centuries, barns and homes on the East Coast were made from American chestnut. Chestnut wood literally made the new American colony—industries grew up around making posts, poles, pilings, lumber, railroad ties, and split-rail fences. In 1907, approximately 600 million board feet of chestnut were cut in the United States. The estimated modern retail value of that wood exceeds three billion dollars. But the edible nut was also a significant contributor to the rural economy. A *New York Times* article from 1892 describes how the gathering of chestnuts represented perhaps the best opportunity for a family to make money and to help enable self-sufficient agriculture. Domestic hogs and cattle could be fattened for market quickly by allowing them to forage in chestnut-dominated forests.

Pandemics offer a sober reminder that no matter how dominant the host or how grand and impressive the ecological or economic impact of a species, the exponential growth of a highly transmissible pathogen can be overwhelming.

The chestnut pandemic had a devastating economic and cultural impact on communities in the eastern United States that we are still trying to understand.[3,4,5,6] In the first half of the twentieth century, an estimated four billion chestnut trees were killed. It is clear that the chestnut blight brought about the rapid decline of American subsistence culture and expedited the rise of industrialization. Despite these major societal and cultural changes, the ecological and evolutionary impacts of the chestnut blight are surprisingly understudied. Subsequent studies of forest succession show that biodiversity equilibrium still has not been attained. Furthermore, they indicate that no single species will replace the role of the chestnut in the foreseeable future. The impacts on native wildlife and ecology are still being assessed.

Pandemics offer a sober reminder that no matter how dominant the host or how grand and impressive the ecological or economic impact of a species, the exponential growth of a highly transmissible pathogen can be overwhelming. Pandemics rapidly reshape the evolutionary and ecological landscape and have cascading social, economic, and other system-level effects. They remind us that selective pressures are not a mere abstraction: they can happen quickly and are a grim reality of mortality and adaptation. A pandemic represents the impact of novel selective events that change the rules of interactions and allow some genotypes and phenotypes to be favored at the expense of others.

2. The impact of novel pathogens on complex systems can be quantified by assessing deviations from scaling laws.

How does one measure the impact of a novel pathogen, or, for that matter, how does one define a healthy forest? How does one predict where the system will go once the perturbation has subsided? One powerful way to determine the impact of pathogens on a population is to assess age-dependent population dynamics, where the impact is measured in changes in birth and death rates.

In 1898, François de Liocourt published a manuscript in *De l'amenagement des sapinières Bulletin trimestriel, Société forestière de Franche-Comté et Belfort* called "The Management of Silver Fir Forests."[7] The pattern he discovered, now known as the law of de Liocourt, was that, in unimpacted forests, the distribution of tree sizes in different forests converged to a similar distribution or J-shaped curve. In such a forest there is an uneven age and size structure. De Liocourt argued that if disturbed forests were left alone, they would converge on distributions

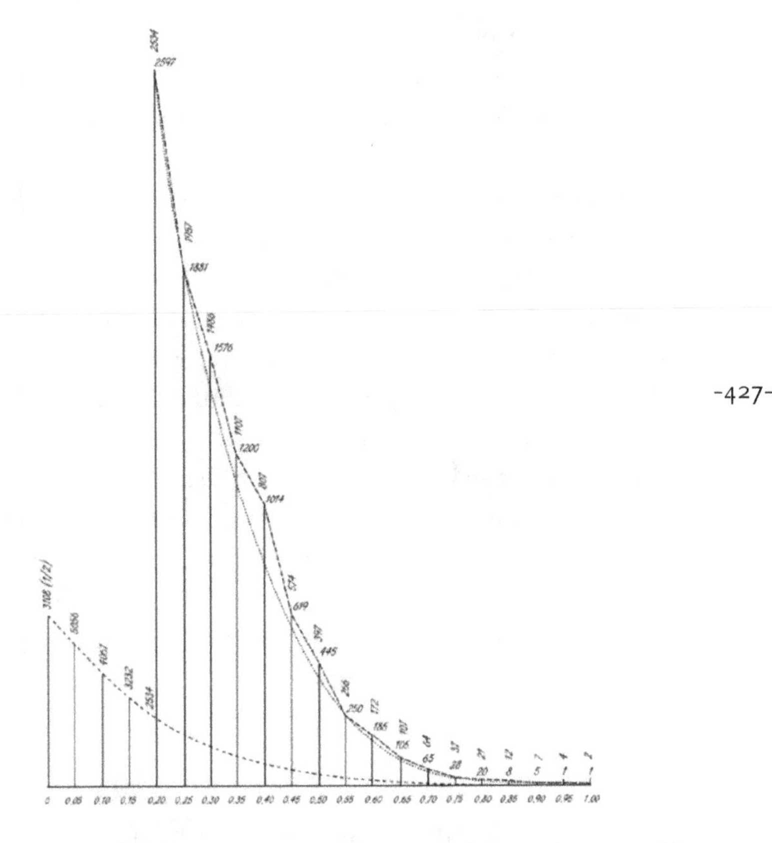

Figure 1. *From de Liocourt's 1898 paper showing the frequency distribution of tree sizes and the mean number of trees in a healthy forest (solid line and horizontal numbers) and his "fitted distribution" (dotted line and vertical numbers). The x-axis legend is diameter (cm), and the y-axis legend is frequency.*

common to other unmanaged forests, as shown in the bottom curve of figure 1.

It was not until 1931 when the roots of quantitative models of demography emerged, due in part to the insight of Soviet mathematician Andrei Kolmogorov. Kolmogorov was working on the problem of diffusion starting fr om the theory of discrete-time Markov processes.[8] Now known as the Chapman–Kolmogorov equation, he sought to develop a theory of continuous-time Markov processes by extending this equation. Kolmogorov detailed two versions of his theory for discrete-time Markov processes—a forward and backward system applied to many areas of biology. For the life sciences, the forward equation, also known as the Fokker–Planck equation, has been the most useful. Specifically, here n is the population index, with reference to the initial population, β is the birth rate, and finally $p_n(t)=\Pr(N(t)=n)p_n(t) = \Pr(N(t)=n)$ is the probability of achieving a certain total population size, N. This foundational work has been extended across biology and used to model population growth with birth, cell growth, and epidemiology. The insight from this work is that in populations, if disturbance and external mortality are low, we expect to see an age or size structure that is a balance between size-dependent growth and mortality.[9,10] The result is a characteristic age or size distribution, also called a size spectra. The implication is that an increasing impact of disturbance will increasingly shift that spectra as older or younger individuals are disproportionately impacted by the pathogen until a new population equilibrium emerges.[11] In the case of a plant pathogen that does not cause 100% mortality, we expect the new size structure to mirror a new demographic equilibrium, as was seen in the American chestnut.

Once a pathogen spreads globally, eradication becomes difficult or even impossible. Pathogens do not respect international boundaries.

Today botanists and forest ecologists often turn to de Liocourt for a useful definition of a "healthy" sustainable forest ecosystem.[12] The balance of size-dependent birth and death in the absence of a major disturbance, such as a new lethal pathogen, might indeed be characterized by a size distribution with a characteristic mathematical signature—one that maintains a stable size–structure relationship by balancing growth with mortality.

An accurate quantification of the impact caused by novel pathogens forms the basis for management strategies.[12] For fungal forest pathogens, tree mortality is the most commonly used variable to quantify losses. Since tree mortality is an inherent process of forest dynamics, defining a baseline against which mortality caused by a new pathogen can be compared presents a major challenge. Changes in the population of the host would occur if the new pathogen caused a higher mortality or changes in birth or death rate than that occurring under pre-pathogen conditions in the forest. Foresters are now obtaining the baseline mortality from a de Liocourt curve calculated from tree inventories carried out over large areas to use as a baseline to quantify the impact of a pathogen.[13] In other words, shifts in size distributions are perhaps the best general measure to link pathogen impacts on death, growth, and birth—the fundamental biological processes that apply across all living organisms. These patterns need not apply only to biology; similar inverse size spectra appear in the distribution of city and settlement size as well as the distribution of businesses, social networks, and groups.[14] De Liocourt's law and the Chapman–Kolmogorov equation offer a novel way to define resilience of a system to external impacts and quantify pandemic impacts on age and size structure. The hypothesis is that exogenous disturbance produces systematic deviations from size spectra and provides a link between the dynamics of complex systems and their age- or size- dependent scaling properties.

3. **There is no going back–our management systems are now faced with "adaptive management" of sudden rapid transformations across interconnected ecological and economic systems.**

> *That all the [chestnut] trees in the United States are doomed to destruction by a mysterious disease called chestnut blight . . . is the gloomy prediction of Dr. W.A. Murrill . . . now he asserts there is nothing to be done against it; that it must run its course like all epidemics . . . a vast loss will be entailed on the eastern forest region should this disease prove as destructive as is at present threatened.*
>
> —*The New York Times,* Sunday May 31, 1908

Like human pandemics, plant pandemics are also associated with the coming of the Anthropocene and first started appearing with the rise of an interconnected world.[15] Some of the first recorded pathogen outbreaks were associated with wheat, as recorded by the Romans

(2100-1950 BP). In fact, the Romans had a god/goddess of rust (Robigus/Robigine) because these new pathogens were so feared.[15] Feasts, processions, and sacrifices in their name were conducted in order to prevent crop destruction and stop future waves of reinfection. Over the past 200 years the number and severity of plant diseases has increased exponentially.[16] Once a pathogen spreads globally, eradication becomes difficult or even impossible. Pathogens do not respect international boundaries. Efforts to reduce the movement of pathogens across borders or by quarantines are easily frustrated by globalization, travel, and trade.

In particular, what lessons have we learned from trees that inform our ability to deal with the losses and envision a recovery from any pandemic pathogen?

Governmental plans to deal with plant pandemics—mainly associated with agriculture—now largely revolve around prevention, response, and recovery. The effort to protect the food supply and human health[17] is largely focused on limiting spread and impact. The chestnut blight was one of the first major pandemics in forests, but there are many more that impact wild ecosystems. Dutch elm disease, sudden oak death, *Phytophthora cinnamomi*, and *Armillaria* honey fungus are all pathogens in an age of pandemics with potential to alter ecology and cultures, if not entire civilizations. Human pathogens and the threat of new pandemics reveal a sobering reality. While considerable effort is now expended on learning how pathogens emerge and identifying potential pandemic pathogens, less has been done to recognize the general signatures of pathogens within and across populations. So this raises the question: what, in terms of ecological and cultural impact, will the next chestnut blight or COVID-19 bring? In particular, what lessons have we learned from trees that inform our ability to deal with the losses and envision a recovery from any pandemic pathogen?

An accurate quantification of the impacts caused by transmissible pathogens must form the basis for management strategies. Developing a general theory of ecological or system resilience is critical. Resilience is a dynamical property that depends on both the trajectory and the rate of change through time.[18] A distinct but related dynamical property is termed "engineering resilience," and is the return time of the system to a steady, or equilibrium, state. Resilience relates to the degree of perturbation that a system can tolerate. The regularities embodied by both empirically documented and theoretically derived emergent properties of systems, with and without disturbance, provide important baselines for understanding resilience and change in complex systems.11 In the context of adaptive management, such baselines can be used to formulate hypotheses and derive surrogate parameter estimates in the absence of more specific data. Furthermore, ecological scaling relationships such as de Liocourt's law, which often (but not always) take the form of power laws, may describe attractors for, or constraints on, the structure and dynamics of complex systems.[11] Deviations from scaling relationships can be used as a signature of specific underlying structuring processes—such as the impact of a pathogen—or may indicate the transient reorganization of the system. Although the focus here has been on forest communities, the concepts and research opportunities described should apply more generally to human and economic systems.

ACKNOWLEDGMENTS

I thank Lynn Enquist, Erica Newman, Robbie Burger, and Matiss Castorena Salaks for providing constructive input, editing suggestions, and encouragement on earlier drafts.

ENDNOTES

1 J.L. Warren and S. Kieffer, 2010, "Risk Management and the Wisdom of Aldo Leopold," *Risk Analysis* 30:165–174, doi: 10.1111/j.1539-6924.2009.01348.x

2 H.A. Krebs, 1975, "The August Krogh Principle: 'For Many Problems There is an Animal on which It Can Be Most Conveniently Studied,'" *Journal of Experimental Zoology* 194(1): 221–226, doi: 10.1002/jez.1401940115

3 D.E. Davis, 2006, "Historical Significance of American Chestnut to Appalachian Culture and Ecology," in *Restoration of American Chestnut to Forest Lands,*

K.C. Steiner and J.E. Carlson, eds. (Washington, DC: National Park Service): 53–60.

4 Missouri Botanical Garden, "Chestnut Blight," http://www.missouribotanicalgarden.org/gardens-gardening/your-garden/help-for-the-home-gardener/advice-tips-resources/pests-and-problems/diseases/cankers/chestnut-blight.aspx

5 F.L. Paillet, 2002, "Chestnut: History and Ecology of a Transformed Species." *Journal of Biogeography* 29: 1517–1530, doi: 10.1046/j.1365-2699.2002.00767.x

6 D. Rigling and S. Prospero, 2018, "*Cryphonectria parasitica*, the Causal Agent of Chestnut Blight: Invasion History, Population Biology and Disease Control." *Molecular Plant Pathology* 19(1): 7–20, 10.1111/mpp.12542

7 F. de Liocourt 1898, "De l'Amenagement des Sapiniers." *Bulletin — Société forestière de Franche-Comté & Belfort* 4(6): 396–409.

8 A. Kolmogorov, 1931, "On Analytical Methods in the Theory of Probability," *Mathematical Annals* 104: 415–458.

9 T. Hara, 1984, "A Stochastic Model and the Moment Dynamics of the Growth and Size Distribution in Plant Populations," *Journal of Theoretical Biology* 109(2): 173–190, doi: 10.1016/S0022-5193(84)80002-8

10 D.A. Coomes, R.P. Duncan, et al., 2003, "Disturbances Prevent Stem Size-Density Distributions in Natural Forests from Following Scaling Relationships." *Ecology Letters* 6: 980–989, doi: 10.1046/j.1461-0248.2003.00520.x

11 A.J. Kerkhoff and B.J. Enquist, 2007, "The Implications of Scaling Approaches for Understanding Resilience and Reorganization in Ecosystems," *BioScience* 57(6): 489–499, doi: 10.1641/B570606

12 O. Díaz-Yáñez, B. Mola-Yudego, et al., 2020, "The Invasive Forest Pathogen *Hymenoscyphus fraxineus* Boosts Mortality and Triggers Niche Replacement of European Ash (*Fraxinus excelsior*)," *Scientific Reports* 10: 5310, doi: 10.1038/s41598-020-61990-4

13 P.D. Manion and D.H. Griffin, 2001, "Large Landscape Scale Analysis of Tree Death in the Adirondack Park, New York." *Forest Science* 47(4): 542–549, doi: 10.1093/forestscience/47.4.542

14 M.E.J. Newman, 2005, "Power Laws, Pareto Distributions and Zipf's Law." *Contemporary Physics* 46: 323–351, doi: 10.1080/00107510500052444

15 A. Santini, A. Liebhold, et al., 2018, "Tracing the Role of Human Civilization in the Globalization of Plant Pathogens," *ISME Journal* 12(3): 647–652, doi: 10.1038/s41396-017-0013-9

16 A. Santini, L. Ghelardini, et al., 2013, "Biogeographical Patterns and Determinants of Invasion by Forest Pathogens in Europe," *New Phytologist* 197(1): 238–250, doi: 10.1111/j.1469-8137.2012.04364.x

17 J. Martin, I. Ardjosoediro, et al., 2008, *Agricultural Recovery Responses in Post-Pandemic Situations Arising from Major Animal and Plant Diseases.* United States Agency for International Development.

18 C.S. Holling, 1973, "Resilience and Stability of Ecological Systems," *Annual Review of Ecology and Systematics* 4: 1–23, doi: 10.1146/annurev.es.04.110173.000245

Batch 7 Podcast

THE TRANSCRIPT OF *COMPLEXITY PODCAST* TRANSMISSION SERIES EP. 7, DISCUSSING THE FOLLOWING TRANSMISSIONS

[Batch 7, released 1 June 2020]

T-030: Out-Evolving COVID-19
David Krakauer & Dan Rockmore

T-031: Models that Protect the Vulnerable
Melanie Moses & Kathy Powers

T-032: The Noisy Equilibrium of Disease Containment & Economic Pain
Jon Machta

T-033: How Pandemics Rapidly Reshape the Evolutionary & Ecological Landscape
Brian Enquist

This podcast transcript has been abridged for length and clarity. Find the full podcast at https://complexity.simplecast.com/episodes/34

COMPLEXITY PODCAST: TRANSMISSION SERIES NO. 007

BETTER SCIENTIFIC MODELING FOR ECOLOGICAL & SOCIAL JUSTICE

DATE: *8 June 2020* **HOSTED BY:** *Michael Garfield, featuring David Krakauer*

MICHAEL GARFIELD: We are back for the seventh in this miniseries about the pandemic. This one is also timed very well for the crisis in legitimacy and inequality and justice that we're seeing around the world right now, because this week we get to talk about the obvious realities that are really only obvious in retrospect, that have been hidden in plain sight, and what it means to bring those things into visibility.

DAVID KRAKAUER: We have to remark on the current moment and just that we are living in a very complex world, and it's more important than ever for us to face up to, as you say, the reality that has been in plain sight, but we have treated as if it were hidden. And I think that, to the best of our ability in our work, we try and address these issues.

I edited a book of essays called *Worlds Hidden in Plain Sight*, and I was very inspired in that title by a short story that many people will know by Edgar Allan Poe called *The Purloined Letter*. It's a very interesting short story, and the detective in that short story is the great Auguste Dupin. It's the absolute opposite of the standard mystery. The standard mystery is something like a Dan Brown mystery, where the secret is elaborately hidden, and it requires some kind of torturous ingenuity to solve. But Poe subverted the genre by saying, you know, the hardest crime, the one that's truly difficult to solve, is the crime that's right in front of your eyes. And of course, I'm not just editorializing by pointing out that, in the larger, more expansive sense, that's what society is dealing with now: the crime that has been there right in front of their eyes, but is extremely difficult to solve. Scientists love puzzles, you know—we like to solve the hidden mysteries. And we discover things like subatomic particles and devise ingenious machines to allow us to see them. But I think today's discussions are going to be about the kinds of ideas necessary to see the forces that have been shaping both the biological and cultural world around us. I just wanted to make that introduction.

So, the first contribution to discuss is the one that I wrote with Dan Rockmore[1]. We were just talking about this interesting analogy between how the forces that shape evolution have been hidden and the way that the forces that shaped the universe have been hidden. In the 1930s, an astrophysicist, Fritz Zwicky, pointed out that there is a discrepancy between the amount of gravitational force needed to account for the movements of the galaxies, in terms of the visible luminous matter, and the invisible matter that he called the dark matter. This was confirmed much, much later by Vera Rubin and Kent Ford, who used more detailed spectrographic data to show that, in fact, most of the universe is made up of both dark matter that accounts for gravitation, and dark energy that is the pressure that's moving everything apart. And so, we're living in this very strange world now where physics, for all its extraordinary contributions, can only account for 5% of what we see.

Dan and I got very interested in that and said, you know what we see in the living world, how much of what we see can be accounted for by what is visible to the naked eye? And an interesting observation to make is that, when Darwin wrote *On the Origin of Species* in 1859, he had no idea that there was a microbial world. If you look through that book very carefully, you might be surprised to find not one mention of a bacterium, or certainly not one mention of a virus. And that's because they weren't discovered until several decades later. In fact, for most of our history—not in the nineteenth century, but certainly in the sixteenth and seventeenth—people believed that small things like ants and flies were spontaneously generated. They didn't need an explanation beyond the fact that they just sort of mutated into existence spontaneously.

And it was in the 1880s that Koch and Beijerinck[2] discovered bacteria, and then viruses. The point that we wanted to make in this contribution is that the evolution of complex life—multicellular life, the kind of life that Darwin described—is deeply entangled with the evolution of simple life. And that the evolution of scientific ideas is deeply entangled with the plundering of microbial innovations. And this is

1 See T-030 on page 391

2 See: S.M. Blevins and M.S. Bronze, 2010, "Robert Koch and the 'Golden Age' of Bacteriology," *International Journal of Infectious Diseases*, 14 (9): e744–51, doi: 10.1016/j.ijid.2009.12.003

the interesting fact of the world, that the complex world that we now recognize would not exist if it were not for viruses, and I guess we'll go into that. Moreover, many of our technologies—of molecular biology, of genetic engineering—are essentially directly borrowing from innovations that viruses have come up with over hundreds of millions of years.

M. GARFIELD: I've been talking about this for quite a while, the discovery of CRISPR by Jennifer Doudna and her lab, and how this is such a perfect example of . . . you have a list of biotechnologies that were found in this kind of way, but it draws attention to the way that so much of human innovation is just accidental or intentional biomimicry. When we're talking about dark matter, there's something about how it links to this broader theme. We want to unpack in this conversation how the models that we make of the world are made at the timescale of the systems adapting to the world that they're modeling—which is the brain, which is the human lifespan, at the human scale, and the social environment that dominates our attention. So, it's dark matter when we're talking about cosmology, because 95% of the cosmos is irrelevant to the human timescale, to the human social world. This points me toward humility, because evolution as a distributed intelligent process has made a lot of discoveries because it's operating at a different timescale. I guess what I'm saying is that we have this way . . . like IP Fest veteran Cory Doctorow talks about the ridiculousness of "Terra Nullius," this idea that we invent things out of nothing, that we can take credit somehow for our inventions when, really, so much has to be said about being in the right place at the right time, and having our search through the space of possible innovations directed by the agency of our landscapes and the evolutionary pressures of our moment. This question comes up in the Complexity Explorers Facebook group a lot about whether emergence is an epistemological or ontological question: Does complexity really emerge, or do we just notice it? I think a lot of this conversation is about how, really, we're not inventing these things or creating them. In many cases, there's a good argument that we're simply discovering what was already there because the times call for it. And that's a shot called on Brian Enquist's piece later in this conversation, but I'll pin it there.

D. KRAKAUER: There are some very intriguing discussions to be had in the interstices of your remarks. I was very taken by Singer's Darwinian ethics. I've always believed that a deep familiarity with evolutionary thinking changes your ethical responsibilities in the living world, because we come to an understanding that we basically are primates—we're almost indistinguishable from chimpanzees and gorillas—and were very closely related to invertebrates and into the trees. That awareness of the common origin of all living things should extend deeply our sympathies and empathy to the nonhuman world. I think your argument is correct. The same argument should apply to human ingenuity. You're right; I think CRISPR is a beautiful example. I make these kinds of remarks and people think I'm nuts, but I stick by them . . . I believe that if Doudna and others were to win the Nobel Prize for CRISPR, they should share it with prokaryotes who invented it. They just tinkered with a solution that the bacteria discovered. And I think that's really true. If people understood how much of our science—certainly in the biological sciences—is essentially a repurposing of discoveries made by evolution, it would change the way they think about creativity in this networked world, right? Human intelligence is a part of a larger ecological system and it shouldn't be separated from it.

Let me just give some examples, because I think it's important that folks understand what we're talking about here. In this paper we go through the history of genetics itself; genetics would not exist as a science if it wasn't for the endonuclease enzyme. And that came directly from the restriction enzyme system of bacteria, it is the basis of most cellular biology, and certainly molecular cloning. CRISPR, as you pointed out, is based on an immune system that bacteria have to deal with viruses that might turn out to be the most important source of gene therapy ever appropriated from the natural world. Our own immune system—the adaptive immune system that is helping us fight off this horrible COVID-19 virus and other viruses—is actually based on a system called the RAG[3] system, which was essentially stolen from a virus, and placed in a human, and now is the basis of immunotherapy. You can go on and on and on. For example, modern cancer therapies—where you target, preferentially, cancer cells—are borrowing the cellular

3 Recombination activating genes

tropism of viruses in order to deliver appropriate drugs. It's extraordinary the extent to which engineering ingenuity has relied on viruses, but then you actually have to think about life itself.

Sex. There's a very interesting hypothesis that the late William Hamilton proposed, which is that the reason we have recombinational sex is to generate diversity to fend off pathogenic viruses. If it wasn't for the variability of the virus, we wouldn't have required a system to generate variability to counteract it, and so, we would be asexual. So, you know, even at the level of our own lives and the meaning we derive from our relationships, there's a deep evolutionary origin in our relationship to microbes.

... a deep familiarity with evolutionary thinking changes your ethical responsibilities in the living world ...

M. GARFIELD: That links us to the next piece by Jon Machta,[4] which is—as David Kinney talked about when he was on the show[5]—about explanatory depth and the way that different disciplines of science seem to focus at different scales. It's because the requisite explanatory depth to explore the world at that scale comes with a certain metabolic cost: the cost of fine graining things, or course graining and seeing them as big as is required. There are differing computational loads and energetic investments required at all of these different scales.

I love that you used the word "appropriation" of the products of evolutionary search because, again, to call back to the piece on Terra Nullius and intellectual property, appropriation is easy when the systems that you're looking at have been conveniently reduced to featureless points. You can talk about a network, but if you're talking about a COVID-19 transmission network, that transmission network is made out of people, and it's easy to forget if you're tuning things for convenience,

4 See T-032 on page 413

5 Complexity Podcast episode 19 aired on February 19, 2020. https://complexity.simplecast.com/episodes/19

for a parsimonious epidemic model, that there are other crucial dimensions involved in the way that that disease will spread. One is that the network is made out of people who are making economic decisions, that they have their own incentives. This has been a big piece of a huge amount of the work that's been going on at SFI with agent-based modeling, and a huge piece of the kind of models that people like Lauren Ancel Meyers and Caroline Buckee have been doing. But Jon's got an interesting take on it.

D. KRAKAUER: Yes, Jon brings us back to that enigmatic hieroglyph, R_0, this critical epidemic parameter that when greater than 1 tells us that an infection will grow exponentially, and when it's less than 1, it will peter out. The technical terms for that are supercritical and subcritical. What Jon asks is, exactly as you said, what dictates the value of what R_0? Part of what dictates the value of the R_0 is the transmissibility of the pathogen—so, biological factors. And part of what dictates it are social, behavioral, and cultural factors like social isolation and quarantine. So, R_0 captures both of these contributions, and what he goes on to point out is that the tension between these two contributions makes R_0 hover around 1. I want to make this point more broadly, because it's such a key idea in complexity science that it's worth investigating the history of this idea a little bit.

So, just consider water. Think about the ocean. The average surface temperature of the ocean is about 60 degrees Fahrenheit, and it boils at about 212 Fahrenheit. The ocean isn't poised at the critical point where water would vaporize. That would be a disaster, because if there were a tiny fluctuation in the temperature, the oceans would vaporize, and that would be the end of it all. In complex systems, though, the opposite things happen. These critical points where you observe phase transitions—like the R_0 from an epidemic to it being contained—are attractors of the dynamical system. That's naturally where the system wants to live, at this extraordinary unstable point, which is very surprising. And there's data, by the way, from physiology, twenty years of data on the study of the brain, that shows that neurons are tuning themselves to near-critical points, to a sort of order–disorder transition.

In 1987, Per Bak, Chao Tang, and Kurt Wiesenfeld wrote a paper in *Physical Review Letters*, a very famous letter, where they introduce the idea of self-organized criticality, and this is the term we use to explain why a system would be driven to the critical point.[6] Jon's explanation is the one that you described, which is that the biology of the pathogen wants R_0 to be very large because the virus spreads very widely. Human society wants to drive that down. So, what happens as we approach R_0 equals 1? Well, as R_0 falls below 1, all of the news agencies and all of the government agencies and the media say, look, R_0 is below 1, so we can now go out and socialize, and we can return to work, and restore the markets. But, as soon as we do that, the biology takes over, and it drives it above 1. And then, when it's driven above 1 culture takes over, and it drives it below 1. And it just oscillates back and forth around 1.

The ocean isn't poised at the critical point where water would vaporize. That would be a disaster, because if there were a tiny fluctuation in the temperature, the oceans would vaporize, and that would be the end of it all. In complex systems, though, the opposite things happen.

At that value—take the contributions that, say, Cris Moore made,[7] and that Sid Redner made,[8] on long tails kick in—because one of the characteristics of hovering by a critical point is that you get power law scaling of outbreaks. In fact, in the original paper of Per Bak and his colleagues, they looked at what's called the sand-pile model. That's what made this field famous. Imagine you have a level surface and you slowly pour sand onto it, it accumulates into a little sand mound, and it reaches what's called a minimally stable state with a fixed angle. That is,

6 P. Bak, C. Tang, and K. Wiesenfeld, 1987, "Self-Organized Criticality: An Explanation of the 1/*f* noise," *Physical Review Letters* 59(4): 381, doi: 10.1103/PhysRevLett.59.381

7 See T-024 on page 309

8 See T-028 on page 363

the slope, if you like, of the sand pile reaches a fixed value. And at that value, you get avalanches of sand of different scales that follow a heavy-tail distribution. That was considered to be the correct idealized model for all systems that attract to critical points. Of course, it has its limits, but I think this is a very important point for complex systems generally, because it means that the interplay of the biological and cultural factors will make it extremely difficult for this disease to ever go away. It's one of the mechanisms that will lead to endemism.

M. GARFIELD: So, we're talking about sand piles, and cascading collapses, and life poised at the edge of chaos, which is exactly what we're seeing here in the US and around the world as the coronavirus has revealed the inequities of what, for people of privilege, was an invisible environment, an invisible evolutionary landscape. We talked about this a lot in episode 29,[9] when we were talking about mass extinctions and the relationship between crisis and creative opportunity. I like your calling forth of Cris Moore's piece, because his writing on how, if you take a fine grain look at R_0, it reveals the heterogeneity of our networks and how opportunity is unequally distributed in space and time. But Brian Enquist has this really cool piece,[10] which calls back to Manfred Laubichler's piece on the evolutionary fitness landscape.[11] It reminds me of a statement I saw from William Gibson, who was quoted apocryphally as saying that the future is distributed unevenly. And, he recently revised it[12] to say that dystopia is distributed unevenly, not just in space and time, but within a single ecosystem or society for agents capable of reaching different affordances—capitalizing on available free energy, or evolutionary opportunities.

So, there's the sense in which the opportunities available to generalists after a mass extinction, the opportunities available to the virus with its high-beta-mutation strategy, and the opportunities available to us now, socially, are described by the same mathematics. And Brian's got a really beautiful example of this from landscape ecology, and from trees,

---

9 See Transmission Series 3 beginning on page 192.

10 See T-033 on page 421

11 See T-012 on page 171

12 A. Riesman, "William Gibson Has a Theory about Our Cultural Obsession with Dystopias," *Vulture*, August 1, 2017, https://www.vulture.com/2017/08/william-gibson-archangel-apocalypses-dystopias.html

and how, over the history of the human era, we have contributed to a series of radical regime changes in ecological settings that are not always as obvious to us: that this kind of mulching and boiling of possibility and creative destruction is going on in these areas.

D. KRAKAUER: Your introduction there covers both Melanie Moses's and Kathy Power's contribution, and elements of Brian's piece. Let's just jump to Brian's, because I think it fits in with this larger question of the human relationship to the natural world. Because it's not just us, and it's certainly not just animals, it's also plants... the world that sustains us. What Brian is doing is drawing a really interesting parallel between the effects of COVID-19 in our time, and the 1904 blight of chestnut trees. He makes this point that the first god, if you like, of pathogens was a Roman god of rust, a fungus that infects all sorts of wheats and ryes and apples and so forth. But I thought that the parallels that Brian drew are extremely interesting and at multiple different levels reveal a different kind of complexity.

So, this blight was a fungus that seemed to have originated in Southeast Asia. It was transported via trade networks that were global. The primary tissue that this parasite attacks is the vascular tissue, thereby reducing the respiration capability of the plant. It's interesting that the American chestnut tree defined the American forest. In his telling, it survived for 40 million years and then, in the space of forty years, it effectively disappeared. The chestnut tree, interestingly, also touched American economic life. The wood of the tree, because it's soft and it's light and it's easy to split and is very resistant to decay, was the basis for construction, and the barns and homes on the East Coast of the US were made from chestnut. The nut of the chestnut tree was a significant source of protein and very important in rural economies. So, the chestnut pandemic was not only about a biological disease that took down the chestnut tree, but ramified out into culture to have a significant economic impact on the eastern United States.

So that's the first parallel, which I thought was absolutely fascinating. It just makes it very clear, once again, that you can't really separate the impact of biology from the impact on the economy, and the impact on culture. We have to come to a much better understanding of the entangled nature of reality, and stop pretending in our departments and disciplines that everything can be modularized, because it cannot. So, that was very important to me.

He goes on to point out some more technical concepts, which touch on the power law–like results that Per Bak and Jon Machta describe. In the late nineteenth century, a Frenchman, François de Liocourt, published a paper called "The Management of Silver Fir Forests."[13] In that paper, he presents the distribution of tree sizes in the natural state. It's what we would now call—and have called in our previous episodes—null models. In other words, this is the distribution of tree sizes that you would expect if the system were healthy. It wasn't until the 1930s that the great Russian mathematician Andrey Kolmogorov wrote down the equations which describe that distribution using his diffusion equations and they become now the baseline against which we can measure deviations that are diagnostic of the disease state. I think that's also very important, and Brian points this out, that we always need to understand how far we've moved from the distribution that defines the healthy, and it's no less true for plants than it is for animals.

It's very difficult for the naked eye, and the timescales in which we live, to observe the impact on the system. We require complexity analogs of microscopes and telescopes and mass spectrometry machines to see the world properly.

M. GARFIELD: To use that piece on the distribution of tree sizes, I've read elsewhere that we have clear evidence that forests around the world are getting younger and shorter. If we think about Geoff West's work on evolutionary networks following a space-filling algorithm, a young, short forest is not filling the available space, filling the opportunity of

13 F. de Liocourt, 1898, "De l'amenagement des Sapinières," *Bulletin Trimestriel, Société Forestière de Franche-Comté et Belfort, Julliet*: 396–409.

sunlight, as efficiently as an older, mature forest. There's a way to look at this in a kind of a geopolitical sense as a disrupted regime. This calls to Andy Dobson's remarks recently, on how our rampant development around the world has led to a quadrupling of zoonotic animal origin infections over the last fifty years. That by disrupting wild ecosystems we're inviting "revolution" from agents that are ordinarily kept in check by the incumbent regimes, like the mature trophic networks.

We just issued our institutional statement on current events,[14] and we included a link to Jessica Flack's work on nonviolent power: her study of primate dominance, hierarchies, and policing. There is a way to look at violent policing as an attempt to impose a kind of order that is a system that has not effectively encoded environmental information, and sort of releases these invisible troubles in a way that is very similar to how our development and attempts to exert a technocratic management over the natural world has released all of these infections upon us.

D. KRAKAUER: Jessica's work, or Steve Lansing's work on the beautiful ritualized water irrigation system for the Subak in the Balinese Highlands, is an example of that. I think what François de Liocourt did and what, in fact, Brian has been doing recently is giving us a sense of the invisible order that we're perturbing. It's one of the problems with climate. In other words, it's very difficult for the naked eye, and the timescales in which we live, to observe the impact on the system. We require complexity analogs of microscopes and telescopes and mass spectrometry machines to see the world properly. And then we need mathematical models to tell us what the equilibrium states of that world are, and what effect we're having on them. I think it's very important to understand that it's very difficult to see the order in a forest with a naked eye. You need to do more work. One of the things that Brian is saying is that these kinds of models that de Liocourt and Kolmogorov have developed with the necessary microscopes to understand healthy states of complex reality.

14 SFI's statement in support of victims of injustice: https://www.santafe.edu/news-center/news/sfis-statement-support-victims-injustice

M. GARFIELD: This brings us to Melanie Moses's and Kathy Powers's piece,[15] because what we're really talking about here is better math and science for fairer social outcomes and justice through empiricism.

D. KRAKAUER: Yes. This is a very, very timely contribution, a very important one. I do want to relate it to exactly the point you just made, and to Brian's point, and to previous contributions, which is how we—those of us trying to understand the complex world with highly idealized mathematical and computational models—remain faithful to it, and just in our projects. I want to give a bit of history here, because I think it's a very important area to understand, so just some of the nuts and bolts: One of the pioneers in mathematical biology—that is the use of mathematics to understand biological phenomenon—was Nicolas Rashevsky. Rashevsky was born at the end of the nineteenth century in Ukraine, was educated at the University of Kyiv. He was an immigrant to the United States, where he got a job at Westinghouse Labs in Pittsburgh, and eventually went on to the University of Chicago, where he became a professor in the department of physiology.

In the 1920s and 30s, he read a book that we could have several programs about, Michael, by D'Arcy Wentworth Thompson, called *On Growth and Form*, which is one of the defining urtexts of complexity science, published in 1917. In response to that book, which presents a physical theory of the biological world, he wrote his book called *Mathematical Biophysics: Physical Mathematical Foundations of Biology*. Now, when we look back on that book, we have several very major criticisms of it. One is that it's too idealized. Another is that it's too beholden to the parsimonious dream of physics. Another is that it's a little too concerned with presenting almost a Platonic view of reality, and insufficiently concerned with the complications of reality. Out of Rashevsky's work came models, or more like chess games, inspired by reality than tools for comprehending reality. I like to use Rashevsky when I teach, because he represents a persistent challenge to the work we do, because as you move away from the sort of simple description of orbits and charges and fields—the world that physicists work on—it just gets worse and worse, and your models slowly metamorphose into

15 Transmission T-031 on page 401

metaphors. This is often true without the practitioners being aware, and you're ultimately left with these rigorous, vacuous, vaporous statements. I think at this moment in history, numerous well-intentioned people are writing down mathematical models with enormous deductive rigor and absolutely no value. What Melanie and Kathy describe is how the standard model of the epidemic—the SIR model or the SEIR model and so forth—has neglected elements of reality. It's neglected elements of reality in such a way that they are, in some endogenous sense, racist, right? And let me explain that.

. . . as you move away from the sort of simple description of orbits and charges and fields—the world that physicists work on—it just gets worse and worse, and your models slowly metamorphose into metaphors.

They point out in their article that African Americans, as we know, are dying from COVID-19 at a rate that are two or four times higher than White Americans, and that per-capita cases are higher on the Navajo reservation in our own state, Michael, than in every other US state. Why is this? They go on to point out, well, African Americans tend to live in very dense urban areas. They live in multigenerational households, and, as we know, older individuals are more susceptible to this disease. They're less likely to have paid sick leave and health insurance. And they have a larger number of preexisting medical conditions, such as diabetes and so forth. Exactly the same arguments go for the Native American population: extreme inequalities of health and economic circumstance, a lack of basic services—running water, access to healthcare—and, as a consequence, the Native population is seventeen times more likely to be diagnosed with this disease than the White population. And that's just the facts. You could argue, look, mathematics doesn't care. You can't accuse a mathematical model of being, in some sense, racist because it's just the math. But it's not really true, because when you formulate a mathematical model you make the decision about what to include and what to throw away.

The reason I mentioned Rashevsky, who is quite rightly pilloried by my community for oversimplifying the natural world, is that I think we should be criticized for not dealing with the critical factors that would allow our models to help those communities at greatest risk. I think what Melanie and Kathy ask is, how should we do this? I will say that the Santa Fe Institute, to be honest, is perhaps one of the places that's been most aware of the importance of these factors—as you pointed out, agent-based models are models that do allow us to include things like the zip code, which is a primary determinant of susceptibility, unfortunately. Network theory allows us to look at the more structured interactions amongst populations instead of treating a population as fully mixed or well-mixed. So, we have been working on formalisms that allow us to address the factors that models are designed to help us explain and treat. But I do believe this is a very important interface between the power of mathematics in helping us understand the world, and the hidden ethical suppositions or social assumptions that go into our thinking about our mathematical models.

There's no point in denying that complexity. Melanie and Kathy are right to make us respond to this debate, now more than ever, without compromising the rigorous, empirical quality of the work that we do.

M. GARFIELD: Indeed, just as a way of linking back to a super-important conversation we've had on the show before, when I had Rajiv Sethi on for episode seven,[16] talking about his work on stereotypes and criminal justice—and then, also, Mahzarin Banaji's 2012 study on the development of racial stereotypes in children:[17] these are examples of too-simple-to-be-useful models, now. The maladaptive oversimplicity of our models has been revealed by this situation, and nobody has the luxury to ignore these realities, anymore. You know, we need new norms. Again, for convenience, for parsimony, we've regarded things like health

16 Complexity Podcast episode 7 aired on November 13, 2019. https://complexity.simplecast.com/episodes/7

17 T. Ziv and M.R. Banaji, "Representations of Social Groups in the Early Years of Life," in *The Sage Handbook of Social Cognition*, ed. S.T. Fiske and C.N. Macrae (London, UK: SAGE Publications, 2012), 372–389.

as a private good, and it's become obvious that it's a public good, and the same can be said for social justice. So, really, what I hear in this is just that we can do better than three-year-olds.

D. KRAKAUER: The Institute addresses these very deep issues. I raised Rashevsky for a reason. He was so enamored with the simplicities of physical law, and their power to enable the human mind to grasp non-living reality, that he hoped that you could apply the same simplicities to humanity into the living world. Behind that was a kind of aesthetic impulse—as you pointed out, simplicity—but behind models, just as easily, there can be an ethical impulse. That's what we're describing here, that the move towards these more sophisticated models that encompass some of these essential facts of life should go into how we model and how we formalize complex systems. That doesn't in any way detract from their objectivity and their mathematical rigor. It just said that the reason we made the decision to include that variable in our models, that we then subject to all of the analysis that we would any variable in the model, is because it's important. It's part of what we're trying to explain. And that's, I think, the argument that's being made in this contribution, and it's one that we should all be very aware of.

8

[BATCH 8, RELEASED 6 JULY–25 NOV 2020]

SCIENCE *can help*

evaluate potential

but **ULTIMATELY**

to determine our

us understand and

risks and responses,

SOCIETY *will have*

path.

—MICHAEL LACHMANN & SPENCER FOX
T-034 REFLECTION

TRANSMISSION NO. 034

CAN SCHOOLS OPEN AND STAY OPEN?

DATE: *6 July 2020*

FROM: *Michael Lachmann, Santa Fe Institute;*
Spencer Fox, University of Texas at Austin

STRATEGIC INSIGHT: *When thinking about reopening schools, an important factor to consider is the rate of community transmission.*

We are all tired of the pandemic. We'd like to wake up one day and have life returned to normal. A big step towards normalcy would be allowing schools to open in the fall. Schools provide education, social environments, and meals to their students. Schools also provide a chance for parents to have free time for work or relaxation. However, schools are an optimal place for the virus to spread—a daily mass gathering. Following John Harte's great piece on mass gatherings,[1] we would like to discuss important considerations for reopening schools, and what to look for over the next couple of months to better understand the risks.

Key uncertainties still exist in thinking about school reopenings. For example, we don't fully know the role that children and young adults play in amplifying transmission, and early reports have suggested they might be both less susceptible to the virus and less infectious—with the strong caveat that, in many cases, schools were closed when these studies were conducted. We thus do not have much information about the rate at which the virus spreads in the school setting. Understanding how rearranging chairs, requiring masks, staggering classes, and other precautionary measures can reduce transmission will be key to safe reopenings in the fall. However, school-reopening risk depends both on these

1 See T-002 on page 29

OPPOSITE: *Thomas Brooks, "Der Neue Schüler," 1854 (detail)*

factors that modulate transmission potential within the school, as well as on the probability of someone coming to school infected.

Thus, we would like to ask the simple question: *What is the chance that, in a given week, one of the students comes to school infected with the virus?* This chance will depend on two things: (1) the chance that a student gets infected over the week, and (2) the size of the school.

We first look at the chance of infection. Austin, Texas, for example, is currently (July 2020) seeing around 300 new reported cases a day, which is around 2,000 a week, in a population of two million. Not accounting for underreporting, the average person in Austin therefore had a 1 in 1,000 chance of getting infected last week.

As an example, let us use this number, 1 in 1,000, as the chance for infection of each student. A group of ten students in Austin will have roughly a 1% (10/1000 or 1 in 100) chance that one of them becomes infected in a week, and the chance becomes roughly 10% (100/1000 or 1 in 10) for a group of one hundred students. Multiplying the number of students by the chance is, of course, only an approximation. A school with 2,000 students does not have a chance of 200% to have an infection! Though in this school of 2,000 students, there will be on average two

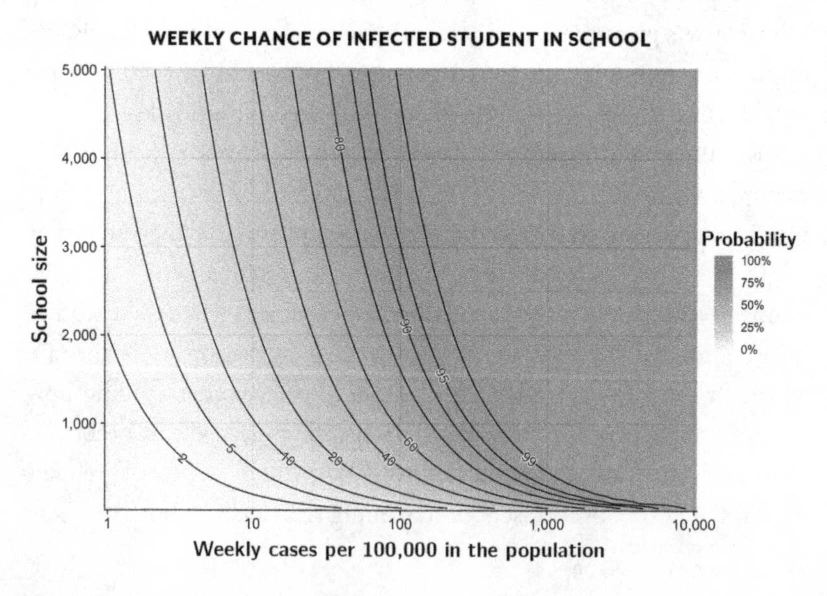

Figure 1.

infected students coming in every week. To calculate this chance exactly, we reframe the question to first ask about the chance that not a single student gets infected. The probability for a student to not get infected is 999 in 1000—a high chance. That event has to happen 2,000 times in a school of 2,000 students, giving a chance of 999/1000 to the power of 2,000, which is 14% (or about $1/e^2$). Thus the chance that at least one student gets infected is 86%. Every week, the school has an 86% chance that an infected student will come to study in class.

When thinking about reopening schools, an important factor to consider is the rate of community transmission.

Using this simple math, we get the graph above (fig. 1), which shows us, as calculated before, that in a school with 2,000 kids and a weekly infection rate of 100 in 100,000—like the current general rate in Austin—there is about an 85% weekly chance a student gets infected. If this infection rate was lowered to 10 in 100,000 (roughly the current rate in the UK), then the chance drops to 20%, whereas if we halved the school size to 1,000 students it drops to 60%, and to 40% for 500 students. It's clear that both of these factors can drastically alter a school's risks in the fall. We also see that dividing a school in two has a larger effect when the infection rate is lower. For example, halving the school size can decrease the weekly risk from 85% to 60%, from 99% to 90%, and from 99.9% to 97%. But when the rate is lower, risk is approximately halved. We go from 20% to 10.5% and from 10% to 5.1%. Though the average chance for an outbreak is not halved, the average number of outbreaks is; of course, the maximum size of each outbreak would also be halved.

There are many complexities we haven't considered here. It is clear that not everyone in a community will have the same chance of getting infected—essential workers and lower socioeconomic groups and their families may be at higher infection risk, because they can't fully isolate at home. Preventative measures like checking for fever and other symptoms could also reduce the risk, though this will likely be leaky given the high proportion of transmission occurring before or in the absence of symptoms. Widespread testing for presence of the RNA of the virus could reduce the risk further, by not relying on symptoms.

We also have not said anything about what will happen when the infected student comes to school. Will the infection spread to a sizable fraction of the students in the same class, or to the school as a whole, or will it only spread to a small number of individuals? What will the response be? Will the school close? Or maybe, given the low transmission in younger individuals, the schools will do nothing. All these are much more complex questions. We can learn from other countries. Israel, for example, decided to close each school upon a detection of a single infection after several earlier large school outbreaks. In such a situation, it would be nearly impossible to keep a school of size 2,500 open if the daily infection rate is 10 in 100,000, as 80% of the time the school will then close the week after opening. Thus, when thinking about reopening schools, an important factor to consider is the rate of community transmission, and it may be untenable in areas currently experiencing surges in cases.

REFLECTION

COMPETING WITH PANDEMICS

DATE: *30 August 2021*

FROM: *Michael Lachmann, Santa Fe Institute;*
Spencer Fox, University of Texas at Austin

We wrote our Transmission on school opening in June of 2020—it seems eons ago. At the time, communities were experiencing a major summer surge, and they were considering how to safely reopen schools for in-person education. There were questions about whether children could transmit SARS-CoV-2, questions about mitigation measures that would be most impactful in reducing school spread, and questions about how school reopening might impact community transmission and healthcare needs. In our piece we did a simple calculation, pointing out that, although one might decide to open schools, public-health guidelines such as quarantining classes and/or schools when cases are detected might not allow schools to stay open if the background COVID-19 rate in the community was high. While transmission within schools depends on intricate interaction and movement patterns on school buses and within and between classrooms, school safety cannot be disentangled from the communities in which they reside. Over a year later, with children under twelve unvaccinated and in the midst of a variant-fueled surge in much of the world, similar questions are being asked; parents and communities may commiserate with Alice and the Red Queen that it "takes all the running you can do to keep in the same place."

In the intervening year we and the COVID-19 modeling consortium at UT Austin[1] worked on questions surrounding COVID-19 and schools. Our Transmission made its way to *The New York Times*,[2] together with a dashboard that reports the chance for infections in schools in different regions of the US.[3] We also worked with schools to implement

---

1 https://covid-19.tacc.utexas.edu/

2 J. Glanz, B. Carey, and M. Conlon, "The Risk That Students Could Arrive at School with the Coronavirus," *The New York Times*, July 31, 2020, https://www.nytimes.com/interactive/2020/07/31/us/coronavirus-school-reopening-risk.html

3 https://covid-19.tacc.utexas.edu/dashboards/school-risk/

nonpharmaceutical interventions, including testing and surveillance strategies. This allowed us to conduct analyses of the outcomes in these schools showing that, with the right tools, COVID-19 transmission could be prevented and that safe school environments may even protect children from COVID-19 infection.[4] With minimal mitigating efforts in place across many school districts for the fall of 2021, community transmission dynamics may be more important than ever in determining the risks of school reopening.

> Over a year later, with children under twelve unvaccinated and in the midst of a variant-fueled surge in much of the world, similar questions are being asked; parents and communities may commiserate with Alice and the Red Queen that it "takes all the running you can do to keep in the same place."

While no two community experiences were exactly the same, we can distinguish four key pandemic stages for understanding and modeling community transmission dynamics. First, in March of 2020, growth was very predictable—simple exponential growth as the disease spread silently around the world. Differences in local dynamics arose from different movement patterns within and between communities. Next, the rise and fall of infections were governed by local policies, community behaviors, and seasonal dynamics, as regions began to enact and relax restrictions and lockdowns. The arrival of safe and efficacious vaccines in late 2020 was a beacon of hope for the world and, for a time, communities with large vaccine stockpiles saw simple decline of the virus as vaccinations slowed transmission. However, SARS-CoV-2 continued to evolve, and variants like Alpha and Delta emerged, leading to new

4 D.L. Gillespie, L.A. Meyers, et al., 2021, "The Experience of 2 Independent Schools with In-Person Learning during the COVID-19 Pandemic," *Journal of School Health* 91(5): 345–436, doi: 10.1111/josh.13008

resurgences. The immunological landscape across which the virus currently spreads has never been more complex: there is uncertainty about the fraction of individuals who have infection- or vaccine-induced immunity, the degree to which either (and the variations within—e.g., the vaccine type and doses) provides protection against infection and severe symptoms, and how immunity is distributed within and across communities. Accurate epidemiological models must now find ways to quantify and capture these complex dynamics.

In retrospect, the long-term behavior of the pandemic has little to do with the simple, uncontrolled epidemic dynamics of exponential growth followed by exponential decline that we might see in textbooks. The actual dynamics involve complicated interactions of several variants: not just RNA viruses but also of memes. The spread of commensal memes, such as staying at home, masking, and vaccinating promote protective behaviors, while parasitic memes also spread, such as purported cures like hydroxychloroquine and bleach or contrarian behaviors like avoiding masks and vaccines. What can we learn from our collective experience? Do we have to track and eliminate not just viruses and microbes but also parasitic memes? We need new models and data that can help us understand, track, and predict the co-circulation and interaction between viruses and memes, so that we can learn how to harness these processes to fight future pandemics. Over the past year we have tried to jumpstart an immune system that would work not only within the body, but within the population. Can we harness such a system to efficiently prevent or mitigate future epidemics and pandemics? Science can help us understand and evaluate potential risks and responses, but ultimately society will have to determine our path.

TRANSMISSION NO. 035

INFO-METRICS FOR MODELING & INFERENCE WITH COMPLEX & UNCERTAIN PANDEMIC INFORMATION

DATE: *6 July 2020*

FROM: *Amos Golan, American University; Santa Fe Institute; Pembroke College, University of Oxford*

STRATEGIC INSIGHT: *We must use a modeling approach to COVID-19 data that will yield the least biased inference and prediction.*

As the world faces the possibility of recurring waves of the current novel coronavirus pandemic, it is critical to identify patterns and dynamics that could be leveraged to decrease future transmission, infection, and death rates. At this stage in the pandemic, data on disease patterns and dynamics are emerging from almost all countries in the world. Variations across countries with respect to coronavirus infection rates, public health policies, social structure, norms, health conditions, environmental policy, climate, and other factors provide us with the data to investigate the impact of different underlying factors and governmental policies on COVID-19 transmission, infection, and death rates.

Despite the fact that millions have been infected and hundreds of thousands have died from COVID-19, the available information is still insufficient for reaching precise inferences and predictions. This is because the available data on each patient are very limited, the variables of interest are highly correlated, and great uncertainty surrounds the underlying process. In addition, though the death rate from COVID-19 is high relative to other infectious diseases, from an inferential point of view, it is still very small since the number of deaths relative to those who did not die is extremely small. As a result, the observations are in the tail of the survival probability distribution. In short, the available data for analysis of COVID-19 are complex, constantly

---

OPPOSITE: *Edvard Munch, "The Sick Child I," 1896 (detail)*

COURTESY BERGEN KUNSTMUSEUM

evolving, and ill-behaved. Inferring and modeling with such data results in a continuum of explanations and predictions. We need to use a modeling and inferential approach that will yield the least biased inference and prediction. Unfortunately, traditional approaches impose strong assumptions and structures—most of which are incorrect or cannot be verified—leading to biased, unstable, and misguided inferences and predictions. Information theory offers a solution. It provides a rational inference framework for dealing with mathematically underdetermined problems, allowing us to achieve the least biased inferences.

Despite the fact that millions have been infected and hundreds of thousands have died from COVID-19, the available information is still insufficient for reaching precise inferences and predictions.

An information-theoretic approach—specifically, info-metrics—is situated at the intersection of information theory, statistical inference, decision-making under uncertainty, and modeling. In this framework, all information enters as constraints plus added uncertainty within a constrained optimization setup, and the decision function is an information-theoretic one. That decision function is defined simultaneously as the entities of interest—say, patients' survival probabilities—and the uncertainty surrounding the constraints. That framework extends the maximum entropy principle of Jaynes,[1] which uses Shannon's entropy[2] as the decision function for problems that are surrounded with much uncertainty.[3] Info-metrics has clear parallels with more traditional approaches, where the joint choice of the information used (within the optimization setting) and a particular decision function will determine a likelihood function. The encompassing role of constrained optimization ensures that the info-metrics framework is suitable for constructing and validating new theories and models, using all types of information. It also enables us to test hypotheses about competing theories or causal mechanisms. For certain problems, the traditional maximum likelihood is a special case of info-metrics.

The info-metrics approach is well suited to dealing with the complex and uncertain cross-country COVID-19 pandemic data, specifically the relatively small sample size of detailed data, high correlations in the data, and the observations in the tail of the distribution. For this analysis,

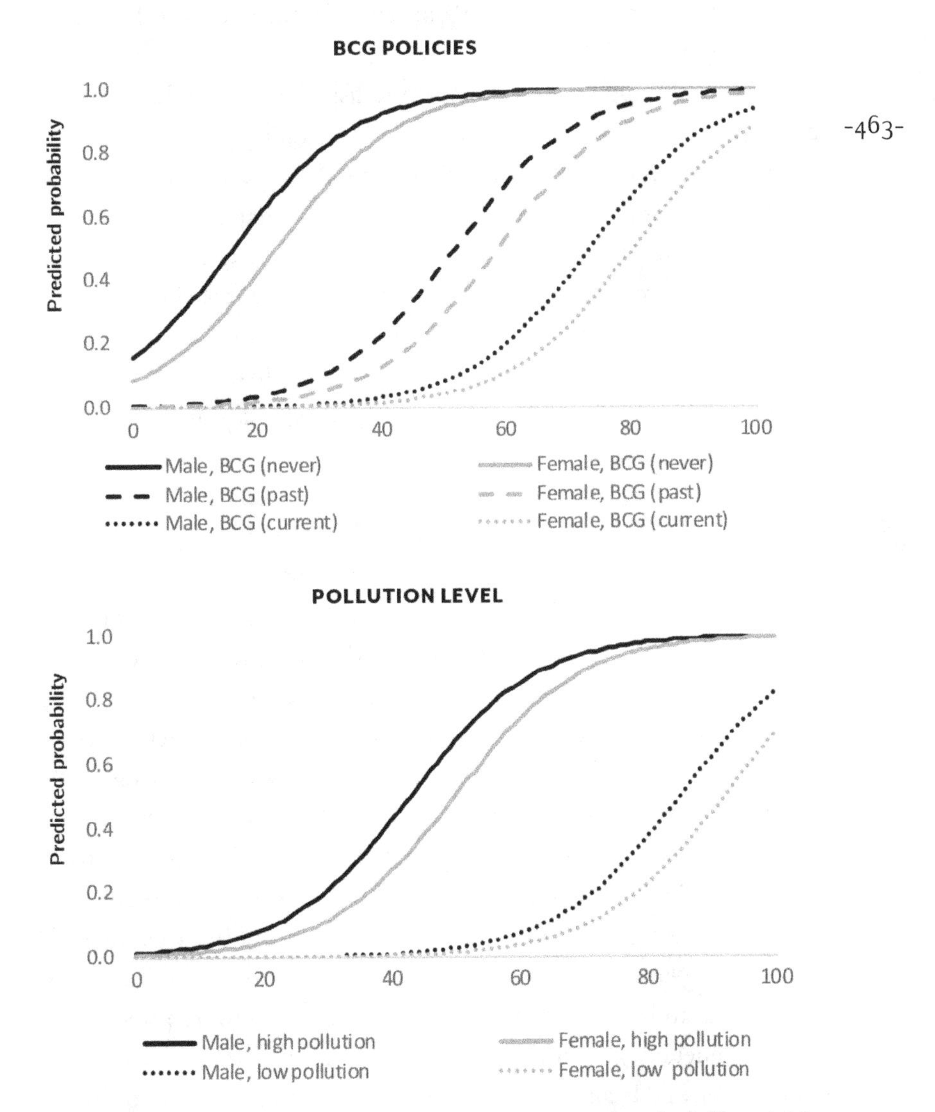

Figure 1. *The probability of dying conditional on BCG vaccination policies (top). The probability of dying conditional on pollution (bottom) showing the death rate in the 10th percentile (dots) vs. those at the 90th percentile (continuous). The x-axis is the age of the patients.*

we developed a discrete-choice, binary (recovered/died) model to infer the association between the underlying country-level factors and death. The model controls for age, sex, and whether the country had universal vaccination for measles and Hepatitis B. This information-theoretic approach also allows us to complement existing data with priors constructed from the death frequency (by age and sex) of individuals who were infected with Severe Acute Respiratory Syndrome (SARS). For the detailed study, see Golan et al.[4]

Using data from twenty countries published on the public server on April 24, 2020, our study found a number of country-level factors with a significant impact on the survival rate of COVID-19. One of these is a country's past or present universal TB (BCG) vaccination. Another one is the air-pollution death rate in the country. Some quantified results (by age—the *x*-axis—and sex) are presented in the figure below. The left panel shows the predicted death probability conditional on a universal BCG vaccination. There are three universal vaccination possibilities: countries that never had it (say, the United States), that currently have it (say, the Philippines), or that had it in the past (say, Australia). The huge impact on survival rates, across ages, of a universal BCG vaccination, is clear. The right panel demonstrates the probability of dying conditional on air-pollution death—the number of deaths attributable to the joint effects of household and ambient air pollution in a year per 100,000 population. The continuous line reflects the 90th percentile of pollution. The dashed line reflects the 10th percentile of pollution.

The same framework can be used for modeling all other pandemic-related problems, even under much uncertainty and evolving, complex data. Examples include conditional Markov processes, dynamical systems, and systems that evolve simultaneously. The info-metrics framework allows us to construct theories and models and to perform consistent inferences and predictions with all types of information and uncertainty. Naturally, each problem is different and demands its own information and structure, but the info-metrics framework provides us with the general logical foundations and tools for approaching all inferential problems. It also allows us to incorporate priors and guides us toward a correct specification of the constraints—the information we have and use—which is a nontrivial problem.

So, should we always use info-metrics? To answer this, it is necessary to compare info-metrics with other methods used for policy analysis and causal inference. All inferential methods force choices, impose structures, and require assumptions. With complex and ill-behaved pandemic data, more assumptions are needed. Together with the data used, these imposed assumptions determine the inferred solutions. The assumptions and structures include the likelihood function, the decision function, and other parametric (or even nonparametric) assumptions on the functional form or constraints used. The reason for that is, without this additional information, all problems are under-determined. A logical way to compare different inferential approaches (classical and Bayesian), especially in relation to complex and ill-behaved pandemic data, is within a constrained optimization setup. That way, the comparison is on a fair basis as we can account for the information used in each approach.[3] But such a detailed comparison, including other approaches like agent-based models (ABM), deserves its own paper and is outside of the scope of this essay.

Naturally, each problem is different and demands its own information and structure, but the info-metrics framework provides us with the general logical foundations and tools for approaching all inferential problems.

Here, I point toward two basic choices we need to make when using the info-metrics approach. First, the choice of the constraints; the constraints are chosen based on the symmetry conditions or the theory we know (or hypothesize) about the problem. They capture the rules that govern the system we study. Mathematically, they must be satisfied within the optimization. Statistically, if specified correctly, they are sufficient statistics. In the more classical and Bayesian approaches, the constraints are directly related to the parametric functional form used (say, linear, nonlinear, etc.). But specifying the constraints within info-metrics, or the functional forms in other approaches, is far from trivial and affects the inferred solution. Info-metrics provides us with a way to falsify the constraints and points us in the direction of improving them. That choice, together with the decision function used, determines the exact functional form of the solution, or the inference.

> Whenever we deal with insufficient and uncertain information, [info-metrics] allows us to account for all types of uncertainties and to handle ill-behaved data. It provides us with a way to make inferences even under much uncertainty and ill-behaved data.

The second choice we make in the info-metrics framework is constructing the constraints as stochastic. This is different than the classical maximum-entropy approach where the constraints must be perfectly satisfied. This is also different than classical approaches where the likelihood and functional forms must be perfectly specified. But there is no free lunch. To achieve this more generalized framework, which allows us to model and infer a larger class of problems, we must bear the cost of specifying the bounds on the uncertainty. These bounds are theoretically or empirically derived. But, regardless of that derivation, it implies that what we give up is the assurance that our solution is first-best; rather, it may be a second-best solution, a solution describing an approximate theory, or the evolution of a complex theory derived from a mix of different underlying elements and distributions. The benefit is that whenever we deal with insufficient and uncertain information, it allows us to account for all types of uncertainties and to handle ill-behaved data. It provides us with a way to make inferences even under much uncertainty and ill-behaved data. Out of all possible methods, it is the one that uses the least amount of information and therefore tends to produce the least biased inference.

Whether it is more convenient or appropriate to choose a likelihood function or to determine the structure of the constraints from symmetry conditions and other information is a decision faced by each researcher. When approaching this decision, we should keep in mind that the constraints are only one part of the decision. The choice, however, of what method to use, depends on the problem we try to solve, the information we have, and the researcher's preference.

ENDNOTES

1 E.T. Jaynes, 1957, "Information Theory and Statistical Mechanics," *Physical Review* 106(4): 620–630, doi: 10.1103/PhysRev.106.620

2 C.E. Shannon, 1948, "A Mathematical Theory of Communication," *Bell System Technical Journal* 27: 379–423, doi: 10.1002/j.1538-7305.1948.tb01338.x

3 A. Golan, 2018, *Foundations of Info-Metrics: Modeling, Inference, and Imperfect Information*, Oxford, UK: Oxford University Press, http://info-metrics.org

4 A. Golan, T. Mumladze, et al., 2020, "Effect of Universal TB Vaccination and Other Policy-Relevant Factors on the Probability of Patient Death from COVID-19," Working Paper 2020-041, Human Capital and Economic Opportunity Working Group (University of Chicago), https://hceconomics.uchicago.edu/research/working-paper/effect-universal-tb-vaccination-and-other-policy-relevant-factors-probability

REFLECTION

FINDING WAYS TO MODEL COMPLEX DATA AMID THE ABSOLUTELY UNEXPECTED

DATE: *23 July 2021*

FROM: *Amos Golan, American University; Santa Fe Institute; Pembroke College, University of Oxford*

In my Transmission, "Info-Metrics for Modeling and Inference with Complex and Uncertain Pandemic Information," I discussed the problem of modeling and trying to understand complex pandemic data under insufficient and imperfect information. I stressed the fact that such problems have multiple solutions due to model and theory ambiguity, a high level of uncertainty in the information, and the ill-behaved nature of the observed information. I argued in favor of the information-theoretic approach for modeling and inference, called info-metrics, as the least biased and most efficient approach for studying such complex problems.

Looking at the available worldwide data early in the pandemic period—when the COVID-19 disease and its exact transmission were still mysterious—I used the info-metrics approach to probe potential policies and interventions that could be applied to increase the survival rate of infected individuals. The universal Bacillus Calmette–Guérin (BCG) tuberculosis (TB) vaccination was the most pronounced factor, followed by pollution rates.

Looking back over the pandemic to date, the COVID-19 infection rate grew exponentially, more than 190 million were infected and about 4.1 million infected individuals died as of July 23, 2021. This is about a 0.021 probability of dying (or 2.1% of the infected) conditional only on being infected. We now have much more data and understanding of how to treat the disease. Related to my discussion on modeling and inference of complex systems, this means that we can get improved results and predictions. My original study controlled for age, sex, and other basic countrywide indicators and explored the possible effects of existing factors. The dataset I had was extremely small and the observed number of patients who died was much above the 2.1% mentioned above. Now we have a vaccine, but it is still important to study the potential implications for COVID-19 infection of different health, environmental, and economic policies.

A year later, I see two key lessons. First, though my original study was done with very little information and data, the basic results seem to be quite robust, even months later. Recent evidence confirms the higher survival rate for patients from countries with a TB vaccination. Similarly, using newly available data, the conditional probabilities by age and sex shown in the original study are somewhat lower but are qualitatively unchanged. I expected the model to work, but I didn't expect it to be so robust when millions of new observations became available. Second, it is important to keep in mind that regardless of the amount of information (and data points) we have, there will always remain uncertainty and model ambiguity when modeling complex systems. Furthermore, such data are often highly correlated and, as discussed in my Transmission, the interest lies in the tail of the distribution (conditional death rate of infected individuals). This means that to understand such systems, and regardless of the amount of data, one must accommodate these complications. The information-theoretic and least-biased approach seems like a good candidate for the task. That method allows us to determine the simplest possible (approximate) theory with the information we have, but with minimal assumptions that cannot be validated. It also provides a natural way to falsify the theory and directs us toward a better one. But to reemphasize, each problem is unique and specifying the information (constraints) is far from trivial. When specifying the information, it is essential to accommodate for possible uncertainty around that information. Overall, the framework I described allows us to model and better understand very complex problems. If done correctly, it provides us with the best possible approximate theory given the uncertain and insufficient information we have.

Looking back, I also see that, though I have no knowledge in the medical sciences, my own work helped me to characterize a close connection between basic vaccinations and the individual relative (and conditional on their own characteristics and other country factors) immunity for the coronavirus. I was also able to understand the direct impact of the environment on the probability of survival. More generally, it helped me realize that certain basic policies affecting health and wellness, some of which are very simple, should be implemented (or reimplemented) in places where they are currently unavailable.

. . . it is important to keep in mind that regardless of the amount of information (and data points) we have, there will always remain uncertainty and model ambiguity when modeling complex systems.

On a more fundamental level, looking back, it is also clear—and expected—that we can never accurately predict future pandemics and events originating from complex systems. But this is not only because these are complex systems and data. Rather, from our observers' point of view, these complex systems are surrounded by much uncertainty and ignorance (defined below). This means that we must accommodate for model and theory ambiguity. Furthermore, these systems are constantly evolving, so we need to allow our theories to evolve as well, but this is practically impossible. Therefore, we have to keep testing our theories and adjust them according to the new information, but they will always remain "approximate theories."

What have we learned? Modeling should be not just about prediction or forecasting, but rather about causal relationships. These relationships allow us to understand the direct and indirect impacts of different policies on the transmission and survival rates conditional on patients' characteristics and other environmental and economic conditions. With that said, I believe the current theories are incomplete and cannot yet explain the pandemic, its sources, and the conditional transmission and survival rates (at least for unvaccinated individuals).

Are current models and theories satisfactory? Are they sufficient to reduce the risk of future pandemics or other unexpected natural or human-made disasters? All systems, including complex ones, evolve constantly. Society and nature coevolve simultaneously. The models and theories constructed to describe social outcomes must constantly change. Modeling the dynamics of the change is very difficult (or practically impossible). And this is not just due to uncertainty and insufficient information. This is also due to ignorance about new pathogens such as SARS-CoV-2 and our resulting difficulties in responding to them. It is not quantifiable uncertainty, which we can handle, but rather the absolute unexpected. So, yes, we have learned much, and we have improved our understanding, but we need to always update and learn from the new events and information we observe.

Mathematically, the basic problems described above also hold for modeling recent social unrest and different types of discrimination in society and the marketplace. The information is insufficient and uncertain and, regardless of the amount of data, the problems are ill-behaved. Like the pandemic data, the solution is at the tail of the distribution. The info-metrics framework summarized very briefly in my earlier Transmission, and commented on here, provides us with a basic framework for studying such problems.

The pandemic revealed to us how much we still do not understand and cannot predict. But it also taught us that science endowed us with tools that allow us to partially understand complex data even when the available information remains insufficient and complex. If these tools are used correctly, they provide us with ways to make causal inferences despite the complexities discussed above.

TRANSMISSION NO. 036

BUILDING TRUST IN COVID-19 VACCINES[1]

DATE: *14 December 2020*

FROM: *Ramanan Laxminarayan, Center for Disease Dynamics, Economics & Policy, Princeton University*
Susan Fitzpatrick, James S. McDonnell Foundation; Santa Fe Institute
Simon Levin, Princeton University; Santa Fe Institute

STRATEGIC INSIGHT: *How to build trust in COVID-19 vaccines: Why people distrust vaccines and how they can be convinced otherwise.*

Safe, effective, and available vaccines are the best long-term solution to the coronavirus pandemic.[1] So it's welcome news that two vaccines are poised for distribution, and others will soon be on their way. Preliminary clinical trial data from these two vaccines, made by Pfizer and Moderna, indicate the vaccines are effective at stimulating a strong and long-lasting response against the virus responsible for COVID-19. A recent FDA briefing[2] confirmed the efficacy and safety profile of Pfizer's vaccine, reiterating that the shot was 95% effective at preventing COVID-19 after two doses with no serious safety concerns.

This is all good news, but it's still unlikely that a vaccine will be widely distributed until mid-2021. Supply constraints could cause additional delays.[3] In the meantime, health professionals must continue to develop, trial, and improve the therapeutic management of COVID-19, and people must continue to attenuate transmission through social distancing and adherence to recommended public health practices such as wearing a mask and avoiding crowded indoor activities.

---

1 This Transmission originally appeared in *Nautilus*: https://nautil.us/issue/93/forerunners/how-to-build-trust-in-covid_19-vaccines

---

OPPOSITE: *Joaquín Sorolla, "Triste Herencia," 1899 (detail)*

Looming over our best efforts, however, is a factor that could derail the campaign to end the pandemic: distrust of vaccines. In a survey of 1,000 individuals in New York City in April 2020, at the height of the epidemic's first wave, only 59% of respondents said they would get a vaccine, and only 53% would have their children vaccinated.[4] The large public and private investments in developing vaccines need to be matched by investments in laying the groundwork for vaccine acceptance.

Much is at stake. If the vaccination rollout is to succeed, it behooves us to be sensitive to general perceptions about vaccination. COVID-19 vaccination has to be viewed within the broader context of attitudes toward routine immunization. An effective and safe vaccine that stops the pandemic and restores our ability to resume social, educational, and economic activity could improve public attitudes toward science and enhance vaccine acceptance more generally. Issues around public acceptance of COVID-19 vaccination could influence immunization campaigns for years to come.

Vaccine hesitancy in the United States now ranges from narrow concerns about specific vaccines to broad coalitions of "anti-vaxxers" who reject all vaccines. The roots of these sentiments, most without scientific basis, run deep. The continual expansion of vaccines for children (from seven vaccines forty years ago to fifteen today) is mistakenly thought to burden young immune systems. Other factors include suspicion about the financial motives of the vaccine industry, mistrust of scientific evidence and government institutions,[5,6] the effect of vaccine vehicles on neurological disorders, and the distrust in vaccines deliberately sown by Russian-linked social media accounts.[7] The result: even though annual flu shots are available at most neighborhood pharmacies, fewer than half of Americans now receive a flu shot each year.[8]

Reluctance to get vaccinated for COVID-19 is widespread. Only two-thirds of crowdsourced participants (68.2%) in a survey of Canadians reported being very likely to voluntarily get vaccinated.[9] In Israel, people who had received influenza vaccinations in regular years were more likely to report that they would accept a COVID-19 vaccine, but others were hesitant.[10] A nationally representative study in Ireland and the United Kingdom reported vaccine hesitancy for 26% and 25% of respondents, and resistance for 9% and 6%, respectively.[11] A survey of 450 Americans found that 67% would accept a COVID-19 vaccine if one were recommended by the government. Males (72%), adults over fifty-five (78%), Asians (81%), and college and/or graduate degree holders (75%) were more likely to accept a vaccine.[12] Other surveys have indicated that 35% of Americans would not accept a COVID-19 vaccine even if one were available today.[13]

Vaccine hesitancy may not be uniform for all age groups. Parents are more reluctant to vaccinate their children than themselves,[14] but children have been found to be key spreaders of infection and should arguably be among the priority groups to be vaccinated, especially if the antibody response is weak in the elderly population.

If the vaccination rollout is to succeed, it behooves us to be sensitive to general perceptions about vaccination. . . . Issues around public acceptance of COVID-19 vaccination could influence immunization campaigns for years to come.

Attitudes toward a COVID-19 vaccine are still evolving. People's preferences are a function of the perceived safety and efficacy of a vaccine, and the health burden and outcomes related to the disease—both of which could change. The virus may mutate and become less (or more) virulent. An effective therapeutic may depress the perceived need to be vaccinated, since individuals in good health are often willing to risk getting ill and receive treatment rather than get a vaccine. Risk-benefit calculations may be shaped by new information on the long-term respiratory and neurological effects of the virus. People worldwide may develop pandemic fatigue and desire only a return to normalcy.

The perceived risk posed by the infectious agent is a major motivator for vaccination,[15] and this works in favor of acceptance of a COVID-19 vaccine. Perceptions of personal vulnerability[16] and anticipated regret[17] are also correlated with the choice to be vaccinated, and both are likely to be high for this disease. The more effective the COVID-19 vaccine, the more likely people will choose to be vaccinated. Paradoxically, however, this could reduce vaccine uptake over time if the incidence of COVID-19 declines.[18] Fear of a disease does not always translate into high vaccination rates, as a study of tetanus in the 1960s showed.[19]

Widespread vaccination against COVID-19 requires that scientists and healthcare workers go beyond just providing information: the issue is not simply a matter of educating people about the importance of vaccination. A more sophisticated approach is needed to address convenience, affordability, and trust in public institutions.

The first hurdle is engendering confidence in the vaccine development process. The speed at which the first vaccines have been developed could itself prove a barrier to vaccine acceptance. The language of "Warp Speed," intended to communicate urgency, unintentionally signals a prioritization of urgency over safety. In this context, the commitment of pharma CEOs to "only submit for approval or emergency use authorization after demonstrating safety and efficacy through a phase 3 clinical study that is designed and conducted to meet requirements of expert regulatory authorities such as FDA" inspires confidence.[20] The advantages of getting a vaccine to market quickly could be squandered if too many people believe it has been rushed to market without adequate safety and efficacy testing. Time is wasted if they need to be convinced otherwise. Communicating the vaccine's safety and efficacy is critical. Without this, even healthcare workers may be reluctant to be vaccinated.[10]

Second, trust in public authorities is likely to be a significant factor.[10] People with diminished trust in the country's leaders are far less likely to accept a COVID-19 vaccine. Public officials, community leaders, celebrities, and their families should themselves get vaccinated and be transparent in a nonlegalistic, commonsensical manner about the risks and benefits. Trusted voices must be used to improve public understanding. The enthusiastic declaration by three former US presidents that they want to be among those receiving the vaccine will hopefully inspire other influencers to step forward.

Third, it is important to find local solutions to vaccine hesitancy. Demographic and geographic disparities in vaccine acceptance could lead to creating clusters of unvaccinated individuals who could then sustain repeated outbreaks of COVID-19. Achieving a high national average vaccination coverage rate would then not be sufficient. A targeted approach to messaging and improving coverage could expedite our exit from the pandemic.

Fourth, mandatory vaccination policies should be avoided because they could backfire. More acceptable would be tying vaccination status to travel or access to public places. Such policies preserve individual agency to reject the vaccine while also protecting public health, much like regulations that prohibit smoking in public places: individuals can do as they like in private.

Finally, social media should be used to help shape preferences around COVID-19 vaccination and counter those who oppose vaccines under any circumstances. Researchers found that 54% of anti-vaccination ads on Facebook came from just two organizations: the World Mercury Project and Stop Mandatory Vaccinations.[21] Facebook's recent pledge to remove discredited claims about COVID-19 vaccinations[22] could be a significant step toward countering misinformation from these groups, despite the inherent difficulties around detecting, discerning, and removing misinformation at scale.[23]

Another study[24] found that anti-vaccine groups were small and had few followers but were more likely to be linked into social media of decision-making bodies, such as parent associations at schools. Emotive and fear-based online messages are seeding doubt about vaccines and gaining disproportionate strength over neutral, scientific postings.

There is little time to lose in the campaign to end the pandemic with a vaccine. At least 70% of the population must be covered to reach herd immunity, with higher or lower levels in some communities. This is achievable. Mississippi, with strict laws against nonmedical vaccine exemptions for school attendance, has vaccination rates for measles and diphtheria–tetanus–pertussis exceeding 99%, and California, Maine, and West Virginia have followed the Mississippi example. Successful vaccination programs are possible in all states, regardless of the political leanings of their residents.

Fresh thinking and behavioral research are needed to build trust and complement authoritative and data-oriented communications. A lot has been written on the importance of known, trusted messengers of information, including doctors and nurses, community thought leaders like ministers, or heads of important community groups. However, our existing knowledge base is in the context of childhood vaccines. Trusted voices and strategies to support a potential COVID-19 vaccine need to be identified. Discussions with family and friends can help address people's concerns, yet this is not part of the current strategy to encourage vaccination.[25] There is evidence that traditionally trusted sources like the US Centers for Disease Control and Prevention are seen as less reliable in the context of a COVID-19 vaccine than are doctors and nurses.[26] Given the importance of community opinions and social norms, vaccine messaging and engagement should address specific communities known for low vaccination compliance.[27] We have a window of opportunity to course-correct to ensure that we have a vaccine that is not just safe, effective, and available but is also trusted. That work should begin immediately.

ENDNOTES

1 K. Leung, J.T. Wu, et al., 2020, "First-Wave COVID-19 Transmissibility and Severity in China outside Hubei after Control Measures, and Second-Wave Scenario Planning: A Modeling Impact Assessment." *Lancet* 395: 1382–1393.

2 Pfizer-BioNTech COVID-19 Vaccine (BNT162, PF-07302048), 2020, "Vaccines and Related Biological Products Advisory Committee Briefing Document—FDA.gov." https://www.fda.gov/media/144246/download

3 K. Sheikh, "Find a Vaccine. Next: Produce 300 Million Vials of It," *The New York Times*, May 1, 2020, https://www.nytimes.com/2020/05/01/health/coronavirus-vaccine-supplies.html

4 CUNY Graduate School of Public Health, 2020, "COVID-19 Tracking Survey," https://sph.cuny.edu/research/covid-19-tracking-survey/

5 D.A. Salmon, M.Z. Dudley, et al., 2015, "Vaccine Hesitancy: Causes, Consequences, and a Call to Action," *Vaccine* 33: D66–D71.

6 H.J. Larson, L.Z. Cooper, et al., 2011, "Addressing the Vaccine Confidence Gap," *Lancet* 378: 526–535.

7 D. Walter, Y. Ophir, and K.H. Jamieson, 2020, "Russian Twitter Accounts and the Partisan Polarization of Vaccine Discourse," *American Journal of Public Health* 110: 715-724.

8 Centers for Disease Control and Prevention, 2019, "Flu Vaccination Coverage, United States, 2018–19 Influenza Season," https://www.cdc.gov/flu/fluvaxview/coverage-1819estimates.htm#summary

9 Leger, 2020, "Concerns about COVID-19—A view from Canada—Coronavirus (Covid-19) researches," https://researchchoices.org/covid19/findings/report/41/concerns-about-covid-19-april-13-2020

10 A. Dror, N. Eisenbach, et al., 2020, "Vaccine Hesitancy: The Next Challenge in the Fight against COVID-19," *European Journal of Epidemiology* 35: 775–779, doi: 10.1007/s10654-020-00671-y

11 J. Murphy, F. Vallières, et al., 2020, "Preparing for a COVID-19 Vaccine: Identifying and Psychologically Profiling Those Who are Vaccine Hesitant or Resistant in Two General Population Samples," *PsyArXiv.*

12 A.A. Malik, S.M. McFadden, et al., 2020, "Determinants of COVID-19 Vaccine Acceptance in the US," *medRxiv.*

13 S. Mullen O'Keefe, 2020, "One in Three Americans Would Not Get COVID-19 Vaccine," https://news.gallup.com/poll/317018/one-three-americans-not-covid-vaccine.aspx.

14 G. Lima, H. Hwang, et al., 2020, "Public Willingness to Get Vaccinated against COVID-19: How AI-Developed Vaccines Can Affect Acceptance," *arXiv.*

15 N.T. Brewer, G.B. Chapman, et al., 2007, "Meta-Analysis of the Relationship between Risk Perception and Health Behavior: The Example of Vaccination," *Health Psychology* 26: 136–145.

16 N.T. Brewer, J.T. DeFrank, and M.B. Gilkey, 2016 "Anticipated Regret and Health Behavior: A Meta-Analysis," *Health Psychology* 35: 1264–1275.

17 N.T. Brewer, G.B. Chapman, et al., 2017, "Increasing Vaccination: Putting Psychological Science into Action," *Psychological Science in the Public Interest* 18: 149–207.

18 H. Leventhal, R. Singer, and S. Jones, 1965, "Effects of Fear and Specificity of Recommendation upon Attitudes and Behavior," *Journal of Personality and Social Psychology* 2: 20–29.

19 K. Frank and R. Arim, "Canadians' Willingness to Get a COVID-19 Vaccine When One Becomes Available: What Role Does Trust Play?" Statistics Canada, July 7, 2020, www150.statcan.gc.ca.

20 N.P. Taylor, 2020, "COVID-19 Vaccine CEOs Vow to Wait for Phase 3 Data before Filing for Approval," *Fierce Biotech*, September 8, 2020.

21 A.M. Jamison, D.A. Broniatowski, et al., 2020, "Vaccine-Related Advertising in the Facebook Ad Archive," *Vaccine* 38: 512–520.

22 M. Isaac, 2020, "Facebook Says It Will Remove Coronavirus Misinformation," *The New York Times.*

23 N. Bliss, E. Bradley, et al., 2020, "An Agenda for Disinformation Research. Computing Community Consortium (CCC) Quadrennial Paper," https://arxiv.org/pdf/2012.08572.pdf

24 N.F. Johnson, N. Velásquez, et al., "The Online Competition between Pro- and Anti-Vaccination Views," *Nature* 582: 230–233.

25 M.S. Chan, K.H. Jamieson, and D. Albarracin, 2020, "Prospective Associations of Regional Social Media Messages with Attitudes and Actual Vaccination: A Big Data and Survey Study of the Influenza Vaccine in the United States," *Vaccine* 38: 6236–6247.

26 T. Reed, 2020, "Americans See Hospitals as More Trustworthy than FDA or CDC on COVID-19 Vaccine Information, Poll Finds," *Fierce Healthcare.*

27 S.C. Quinn, K.M. Hilyard, et al., 2017, "The Influence of Social Norms on Flu Vaccination among African American and White Adults," *Health Education Research* 32(6): 473–486, doi: 10.1093/her/cyx070

REFLECTION

THE NON-COVID VACCINATED: REACHING THE RELUCTANT

DATE: *23 September 2021*

FROM: *Ramanan Laxminarayan, Center for Disease Dynamics, Economics & Policy, Princeton University*
Susan Fitzpatrick, James S. McDonnell Foundation; Santa Fe Institute
Simon Levin, Princeton University; Santa Fe Institute

Much has been written about vaccine hesitancy since our piece "How to Build Trust in COVID-19 Vaccines" discussed the need for multiple communication strategies if society were to reach the all-important goal of vaccinating 70% or greater of the population. Unfortunately, during 2021, as vaccines became safely produced and made widely available, we have witnessed a hardening of the vaccine/anti-vaccine lines that could have been anticipated and mitigated to some extent as outreach to different communities occurred. For many in the scientific, public health, and medical communities it can be hard to fathom that, with the pandemic dragging on and curtailing daily life, and the emergence of the Delta variant threatening lives and livelihoods, a significant percentage of the US population (45% as of September 23, 2021) remains unvaccinated. The proportions of the unvaccinated around the globe is far greater, but that is largely because of shortage of vaccine supply. For a wealthy country like the United States where vaccines are conveniently available in virtually every neighborhood, the shortfall in vaccination coverage is difficult to explain. Vaccination coverage (fully vaccinated) in the US hovers around 50%;[1] rates are higher in France and Germany (64%) and the United Kingdom (67%). Canada is currently reporting that 75% or more of the eligible population is fully vaccinated.

Vaccination coverage in the US is highly skewed, with the greatest uncovered populations in the South and Midwest. Who are the unvaccinated? There is the core set of committed anti-vaxxers, who are not

1 "See How Vaccinations are Going in Your County and State," *The New York Times*, https://www.nytimes.com/interactive/2020/us/covid-19-vaccine-doses.html

satisfied with rejecting vaccines for themselves but also actively voice their objections to all vaccines and disseminate false or incomplete information to stop others from obtaining recommended vaccines. This group, although small in number, is not receptive to any messaging or debate regarding the benefits of getting a COVID-19 vaccine.

The larger "vaccine-resistant" population is partially a direct product of the political polarization the US and other countries are experiencing. A percentage of the hesitant are skeptical of COVID as a disease and of the need for a COVID vaccine as a remedy. For some, resisting "coercion" from the government or the dictates of coastal elites appears to be a point of pride.

Our bet is that reaching the resistant or the hesitant requires a small-scale, personal, and hyperlocal "carrot"-based approach. Mandates and harsh sticks are likely to fail to be persuasive.

There is a third category of the "vaccine hesitant," who represent a variety of individuals who are not yet vaccinated because they do not believe they are at risk for the disease, doubt the new vaccine technologies are safe, fear possible side-effects could derail them from home and work responsibilities, or must overcome the inconvenience of seeking vaccination, especially if they live in sparsely populated rural areas. For the hesitant, the right message delivered by the right messengers (trusted medical care or other professionals, employees and coworkers, family, or friends), could sway some individuals. Our bet is that reaching the resistant or the hesitant requires a small-scale, personal, and hyperlocal "carrot"-based approach. Mandates and harsh sticks are likely to fail to be persuasive.

Within the unvaccinated population, there is also a group of people who have not received much attention and could be characterized as the "reluctant." One of us (SF), on a routine trip to a local drugstore, observed another such customer awaiting her vaccination against

COVID-19. She talked rapidly and incessantly, as the anxious tend to do. Her stream of consciousness monologue centered on why she had waited so long. It was early September 2021; she could have been vaccinated months ago. When the vaccinator told her she was done, the customer's surprised relief that it was over and painless was palpable, revealing a possible reason for her "hesitancy": a dread of needles.

Like that drugstore customer, many unvaccinated individuals are not committed anti-vaxxers; they neither mistrust vaccines nor disbelieve COVID is causing serious illness and death. Instead, the reluctant are those who fear needles, dislike engaging with the medical system, or procrastinate when it comes to other routine medical interventions such as annual exams or health screens. Lack of adherence to medical advice is a major issue the medical community continually grapples to solve.[2]

The current situation is both an opportunity to rebuild in a broader sense and an essential barrier to bringing the country back from the depths of the pandemic.

The vaccine reluctant offer the greatest opportunity for increases in coverage essential for the United States to exit the pandemic. Nudges could be helpful, and many have been proposed, including employer mandates or the granting of vaccine passports for highly valued activities. Some individuals could be reached by bringing vaccinators to workplaces, neighborhood stores, shopping centers, social activities (convenient vaccine booths at community events such as fairs, festivals, etc.) or by campaigns to get friends and family members to act as vaccine coaches. Approaches such as vaccine lotteries have had limited effect, most likely because they do not directly address the sources of the reluctance.

---

2 "Medication Adherence in America: A National Report Card," National Community Pharmacists Association, http://www.ncpa.co/adherence/AdherenceReportCard_Full.pdf

The silver lining in the COVID pandemic has been the rapid response of science in helping develop and deploy vaccines at a pace that is unprecedented in history. However, the lack of progress on vaccination holds back everyone, including those who have been vaccinated. Paradoxically, this monumental success of science, public health, and clinical care is somehow deepening a political divide that predates COVID-19.

Unless we take serious measures to close the vaccination gap, conditions for viral mutations and evolution will continue to pose a threat for everyone. We must begin by acknowledging that the lack of confidence in political and other institutions has been building for a while. The current situation is both an opportunity to rebuild in a broader sense and an essential barrier to bringing the country back from the depths of the pandemic. Frustration notwithstanding, treating one another with humanity is still the right way to go.

Part Two: Interviews & Essays

INTERVIEW NO. I-001

HALTING THE SPREAD OF COVID-19

DATE: *25 March 2020*

FROM: *Laurent Hébert-Dufresne, Vermont Complex Systems Center*

CREDIT: *This interview was released as Ep. 24 of SFI's Complexity Podcast and has been abridged for length and clarity. Find the full podcast at https://complexity.simplecast.com/episodes/24*

MICHAEL GARFIELD: Laurent, it is a pleasure to have you here on Complexity—in spite of the fact that you're on the show because we're in a pandemic. It's one of those mixed blessings that you get to step forward and speak with some expertise and authority here. You've been looking at epidemic transmission, social reinforcement, contagion networks, and so on for years. You've given us a couple papers to look at. You did a paper with Sam Scarpino and Antoine Allard, "The Effect of a Prudent Adaptive Behavior on Disease Transmission," from November 2016.[1] That seems like the simplest place to start.

LAURENT HÉBERT-DUFRESNE: This paper started during the Ebola epidemic in West Africa in 2014. The news was focused on how different the Western African societies were than our own. There were a lot of conversations and new stories about traditional burial practice in West Africa and how that contributed to the spread of Ebola: funeral practices that involve touching the body a lot and that, of course, would spread Ebola, and those people are obviously going to be at risk. One of the first insights of network science and epidemiology is that you have individuals with just a lot more social connections, whether that's

1 S. V. Scarpino A. Allard, and L. Hébert-Dufresne, 2016, "The Effect of a Prudent Adaptive Behaviour on Disease Transmission," *Nature Physics* 12:1042–1046, doi: 10.1038/nphys3832

OPPOSITE: *William Roberts, sketch for "A Shell Dump, France," 1918 (detail)*
COURTESY OF IMPERIAL WAR MUSEUMS

school teachers or funeral workers in West Africa, and those are going to be high-risk individuals. If the disease is going to reach someone first, it's probably going to be them. They have so many connections.

Now, healthcare workers and funeral workers are in this weird position where they're really at risk, but they're also essential. If a funeral worker gets sick from Ebola because of their job, chances are they get sick, they stay home for the next few days, someone else is going to take care of the funerals. We're replacing high-risk individuals with new healthy individuals.

The same thing happens in any outbreak, really. Right now, we're talking about the capacity of our healthcare system. If nurses and doctors who are at risk get sick, we've got to bring other people in, ask other people to put themselves at higher risk. It's the idea that you have those high-risk essential individuals that you keep replacing, and it seems like a good idea because you want to treat people and you want to take care of the burials, but that can accelerate the spread of the disease because you have this turnover of hubs in your social network.

Sometimes we make things worse for the collective by having a locally prudent behavior.

M. GARFIELD: Talk about the model you created for this, which you then compared to empirical evidence from a number of different outbreaks.

L. HÉBERT-DUFRESNE: The model is fairly simple. We use the standard network models for disease transmission, which is simply, you track individuals based on their current state—whether they're healthy, infectious, dead, or recovered—but you also track how many interactions occur between healthy people or one healthy person and one infectious individual. Just having this network perspective gives you the opportunity to model a social mechanism. In our case, it's the idea that people don't like being in contact with infectious individuals. The classic example I always give in talks is, you go to the butcher to buy a steak for dinner and you see the person behind the counter sneezing all over the meat. You're just going to eat something else that day. You're just rewiring that potential edge that you had with the butcher to whoever, the cheese maker, because you're going to buy cheese instead.

Having a network perspective allows us to account for local rewiring mechanisms that we all do, like that social distancing. That's us cutting links. It's all very local network mechanisms. We had a simple model of that with the idea that some connections have to be preserved because they're essential. If my nurse gets sick, I'm just going to get rewired to a different nurse that's healthy, and now that nurse is being put at risk. That was a simple model, and it gave us this idea of super-exponential spread. The classic picture is, I infect two people, they each infect two people, these four infect two people, to get this exponential transmission tree for a disease. With this mechanism, because you keep re-feeding new, highly connected, high-risk individuals into your population, it can get this super-exponential spread—so, faster than exponential.

We got the theoretical results and were like, "Well, has there ever been a disease where we think that there is a faster than exponential spread?" It turns out if you go in almost any flu season in the US in almost any state, you're going to get this regime just before the peak of accelerating spread. That can be social patterns—what I'm describing now. It could be related to weather—the flu virus likes dry weather—that could be driving that. It could be it's flu season—we think about it more, we go to the doctor more. It could be a lot of things, but it seems like one of the most potent drivers of disease dynamics has always been social mechanisms and social behavior. So it seems likely that this could be one of them: sometimes we make things worse for the collective by having a locally prudent behavior.

M. GARFIELD: There are some implications that you discuss in this paper. How do these results fit in with COVID-19? Obviously, it's a lot more difficult to identify who has been infected, because there's a latency with symptoms. How does that complicate this situation?

L. HÉBERT-DUFRESNE: When we published that paper in 2016, we got some media attention, not all of which understood the subtlety of the model and of the results. There were a few headlines that read, "Don't Stay Home Sick," as if we were arguing against people staying home when they're sick. Of course, that's not the message. The message is, stay home sick, and if your job needs someone to take over, then they should take some preventive measures. With coronavirus, if you are sick, absolutely stay home sick. If you can do your job remotely, do that. If you have to bring someone in, it seems like it's probably a high-risk job because you got sick, and so that person should take additional measures to prevent their infection.

The idea is, if the teacher gets sick, we're going to bring in a substitute if the school is still open, but that substitute should be aware that the virus might be going around the school, and, therefore, the way they teach and the way they interact with people should change accordingly, which is something we're not used to doing in normal seasonal flu outbreaks.

○ ○ ○ ○ ○ ○

M. GARFIELD: So, there's another paper you coauthored on school closures, event cancellations, and the "Mesoscopic Localization of Epidemics on Networks" with higher-order structure.[2]

L. HÉBERT-DUFRESNE: I love that title so much! It was meant to reflect that sometimes very technical results like the mesoscopic localization of epidemics in higher-order networks can inform us on why and how useful it is to do very concrete things like close school and cancel events.

M. GARFIELD: As you've already mentioned, one of the main challenges here is the failures of the surveillance system in this particular case. We're pretty deep into the outbreak already. I want to spend a little time on some questions that have been coming up in our social media audience. One was about how models show how containment strategies at different scales work. When is this about the distance that you're taking from another person, or about a school or event closure, or about closing border crossings? How did you model it in this particular paper?

L. HÉBERT-DUFRESNE: Oh, this is such a good question. I want to first answer what we know about the different scales and why I felt like we needed this mesoscopic scale. Microscopic is the individual—so

2 G. St-Onge, V. Thibeault, et al., 2021, "Social Confinement and Mesoscopic Localization of Epidemics on Networks," *Physical Review Letters* 126(9): 098301, doi:10.1103/PhysRevLett.126.098301

that's me choosing to avoid a friend for a night. Mesoscopic is anything that involves multiple individuals—so that's school closures or events. And then the macroscopic would be mass quarantines at the country level. We know from history that mass quarantines don't work well, especially late in an outbreak. People find a way to avoid them. People need to travel and it's easier to make it easy for people to travel but actually test them and track them as they do. So, I checked my visa status and requirements recently because I've been asked to comment about some of the decisions in the intervention from the US. Apparently I'm allowed to criticize my host country—I'm Canadian.

But that's why the intervention has been shit: because it's focused a little bit too much on the macro scale and hasn't given much guideline to the meso scale. I've been organizing conferences and it's up to me to decide whether I cancel my conferences or not. It was up to SFI to decide whether to shut down or not. And some places can't make those decisions because they have to weigh the collective good when facing the outbreak with the fact that their business maybe can't afford to close down for a month or two. And that's not a decision that anyone should be asked to make, to choose between their own business and livelihood or the health of their community.

We know from history that mass quarantines don't work well, especially late in an outbreak. People find a way to avoid them. People need to travel and it's easier to make it easy for people to travel but actually test them and track them as they do.

And we don't have enough information or even knowledge, really, at the individual level to make those decisions. So the intervention has been messy in that regard. And, the kind of information that we would need is also not available. So, we cancel a conference and we get emails from speakers who are like, "Oh, I think you're overreacting." At the time of the cancellation there were only nine cases in Quebec, which is where I'm from. I'm sure other people you will be talking too will bring up the fact that this is an exponential curve. So, early on, you don't want to use the number at $T = 0$ to try and predict what is going to be in two months, because it's an exponential spread and might be gigantic.

But also it's not true that those cases are uniformly distributed in the population. We tend to think of, "Oh, it's nine cases, every person has 9 over n chances of being infected in this population," which is just not the case. Some people are more at risk—that's what network epidemiology has been telling us for decades—and some structures are more at risk. That's why the meso scale is really important. You're much more likely to find cases in schools or hospitals or big events just because that's where most of the social connections are out there, and therefore that's where most of the infections are going to occur. So if the disease lives at this meso scale—which not all diseases do, but some do, and I think this might be the case for coronavirus—our intervention is much more powerful if it operates on that same scale.

And what it does mean is that, by closing a school, you remove, a lot of those social connections that might spread the disease, but you also reduce the coupling across different schools that might still be open or across other structures. So it's really like an intervention that works in two ways by both protecting the individuals in that school, but also reducing the overall mesoscopic spread of the outbreak.

The important thing is, try and think about, What are the key mechanisms for the spread of this disease? And individual interventions are super useful—washing your hands, social distancing—but you can spread to someone very inadvertently by just touching something you didn't even realized you touched and ten minutes later they touch the same surface and touch their face. So unless you do it extremely well, those microscopic interventions are not going to be super efficient.

They'll work well for sexually transmitted infections, not so well for something where you don't even know if you're infected. We don't clearly know that the mechanisms with coronavirus. In that case, mesoscopic works much better, in my opinion.

M. GARFIELD: So, in the model, this is just about pruning at different levels. The question I see coming up from people is why 500 people going to a church service is the number that people are picking. It seems relatively arbitrary, but you talk about the efficacy of interventions at different levels. How do you actually model these interventions, and how it is that you and other epidemiologists are coming to these numbers and making these specific suggestions?

L. HÉBERT-DUFRESNE: When we wrote the paper, I remember France had a ban on all gatherings of above a thousand people. California at another number, maybe 5,500, and now my university, University of Vermont, we can still have seminars on campus, but nothing with a crowd of more than twenty-five people. So the question is, how do you pick twenty-five? How do you pick a thousand? I think we have a good idea of just broad distribution. So we know those distributions are heterogeneous. We know there's hockey games in Montreal. That's what, 21,000 people? There's not that many big events that occur weekly in Quebec and there are a lot of events with fifty people. We know how many classes occur and all that. So we have an idea that those distributions are very heterogeneous, and we have an idea of how heterogeneous they are. Similarly, we have an idea of how many social groups and events a given individual will participate in.

So we have broad ways to parameterize these types of networks that are created not by simple pairwise interactions but really by higher-order-structure events and groups and classes. And once you have that, you can calculate the coupling between different structures. So if I start an outbreak in this hospital, maybe it's going to stay in the hospital. But if the coupling is strong enough, maybe it's going to be able to jump to other hospitals or schools. So then the idea is, how much do I have to play with this distributions of participants per event to decrease the coupling? It's almost like bringing the reproduction number down below 1, not on the individual level but at the level of events and structures. So it's just a coupling that you decrease by making a less and less big event and eventually you get this critical threshold, which is 1 in the case of R_0, something else in the case of this mesoscopic transmission.

But then you stop the spread at the mesoscopic level. And that's where those interventions can be super beneficial.

Actually, you end up cutting fewer social interactions by doing it at that level than if you were just randomly cutting pairwise interactions or social interactions through social distancing. So it is a powerful metric, but it is very hard to calculate or to estimate the critical size that you need because it's specific to different populations. We still don't know just how transmissible the virus even is. So it is tricky. In the paper, twenty-seven . . . but that's for one set of parameters and a bunch of assumptions. But the idea is that you want to be safe, so this idea that one thousand might be too much—that's already a really big event and twenty-five might be too low because then it's hard to operate. Not everything can move online. So in my mind it's this intermediate thing of numbers like fifty and one hundred sort of make sense—they're in the right ballpark. Twenty-five is very safe. One thousand is too much. It's hard to explain what it is, but we have a good understanding of what they should be.

And of course, the safer we are now, the better it's going to be in the long run. And my new philosophy is that almost no event is really critical. Nothing is that important. I could cancel my class right now. My students might miss some of my good jokes and a little bit of good material, but it's not that important. At the end of the day, it's not that critical. So canceling events is really not that big of a deal.

M. GARFIELD: So, let's talk about your 2015 paper in PNAS with Ben Althouse, "Complex Dynamics of Synergistic Coinfections on Realistically Clustered Networks."[3] Coronavirus isn't happening in isolation and, for a lot of people, this has been a particularly vicious flu season. You've done some research into how modeling networks of different structures changes our expectations when we're looking at diseases, not in isolation, but in co-infected individuals.

L. HÉBERT-DUFRESNE: Yeah, this is an interesting line of work because we often or almost always model disease in a vacuum. If you

---

3 L. Hébert-Dufresne and B.M. Althouse, 2015, "Complex Dynamics of Synergistic Coinfections on Realistically Clustered Networks," *Proceedings of the National Academy of Sciences* 112(33): 10551–10556, doi: 10.1073/pnas.1507820112

look at any models, it's rare that they're going to take more than one disease into account. And yet we know that those things interact, right? The classic examples are more sexually transmitted infection, so if you have syphilis and HIV, your transmission rate of HIV goes up by maybe a factor of forty, right? Syphilis gives you surface wounds which are going to just create more bodily fluid exchange and you're going to transmit HIV a lot more than you would if you only had HIV and not syphilis.

We know that for some diseases it's incredibly important to take interactions into account. Interactions can be mechanistic, like the one I just described. It could just be that you have the flu. It makes you weaker. It makes your immune system weaker and then you get colonized by pneumonia or something else. That's almost an indirect interaction, but it can also be that you declare a state of emergency to fight Ebola in Congo. Well then you're sending resources to fight Ebola and those resources—whether it's healthcare workers, or hospital beds, or actual dollars—have to come from somewhere else. They often come from somewhere else in public-health interventions. That's why Congo is now seeing more deaths because of malaria or cholera than they did before Ebola. That's an indirect interaction that goes through political and social systems.

We know those things: they all interact in one way or another. We were just curious about, how different would those models be? Of course they become very hard to track because you need to keep track of my state regarding flu, my state regarding pneumonia, my state regarding something else. The number of variables and mechanisms just blows up. It is harder, but it's doable. When you do it well, you realize that they also have completely different dynamical patterns. The big one for me that was a big surprise back then, in 2015, was the fact that classic models almost always give us this monotonous relationship between the expected epidemic size—so, how many people are going to get sick—and the transmission rate or R_0 of a disease. If R_0 is below 1, nothing happens. When R_0 goes above 1, there's this monotonous relationship: the larger it is, the bigger the outbreak.

When diseases interact, it's not quite like that because the idea is maybe syphilis can spread alone and HIV can't. If syphilis just gets a little stronger, then you can get this discontinuous jump where you go from almost no HIV epidemic to a huge HIV epidemic. I don't want to use HIV and syphilis as an example too much because I don't know that much about sexually transmitted infections, but the idea is that when things interact, you can get this latent heat where one disease would be like, "I would really blow up. If you spread a little more I'll blow up." The other one could say the same thing, so if one of them gets a random mutation, just gets a little more transmissible, then they both blow up. Right? You get these discontinuous jumps in epidemic size, which are just unheard of in classic models.

We know that dynamics can be very different. The super-exponential spread that we get with some social behavior we can get as well with interacting diseases so that the range of possible behavior is just incredible. It means that if we want to do robust forecasts, we need to take more than one thing into account and we can't model them in a vacuum, really. I've been talking about different diseases, but what the coronavirus outbreak is showing us is that those contagions can also, or maybe more easily, be social and not so much other infectious diseases, right? You get spread of news information of social media . . . that's going to help coronavirus spread. If you want to get a good model to forecast the spread of coronavirus or to forecast the effectiveness of interventions, you have to take those contagions of misinformation into account because they're just as critical to the public health questions you're trying to ask as the virus itself.

M. GARFIELD: We know that it's not just the size of the events, but it's the network structure within them. When we're talking about limiting events to, say, fifty to one hundred, but some of those events are school closures, some of them are sporting events and the contact networks in those different kinds of events look different. How does randomizing a network affect the spread of co-infections, and what does that mean for people who might have to make difficult choices about the priority and structure of events of similar sizes?

L. HÉBERT-DUFRESNE: When you have multiple things that are spreading on a network, but interact together in a synergistic way—so news information and coronavirus, or syphilis and HIV—they help each other spread. Well, these synergistic contagions, as I like to call them, also benefit from this mesoscopic clustering of many people being part of the same structure. That's also a little surprising because the classic models tell us that clustering your social connections—social isolation—is great, right? If the disease spreads from me to my sister and my mom, then it can go from my mom to my sister. There's a triangle, so that triangle ends up being a wasted link for the disease. Clustering of connections is good in classic models.

For synergistic interactions, not so much. The diseases or the different contagions like to be kept together, right? They want to be kept together because these spread better if they're together. Then, having these dense groups actually helps their spread and makes those groups a hotspot for the spread of contagions. That's just another way that the dynamics of interacting contagions is very different from classic models. They interact differently with the mesoscopic scale. I think if we went back to that paper where we use a really simple model to show this phenomenon of mesoscopic localization, if we did it with some form of interacting contagions that benefit from mesoscopic clustering, I think the results would be even stronger and would make even a better case for closing certain structures.

The problem is we don't have good data about co-infection almost ever, right? For privacy reasons, you don't share the identity of the cases. When we do studies on flu, we have flu incidence data. When we do a study for coronavirus, we have coronavirus incidence data. We rarely know who had both. The one thing we know for coronavirus now is that it seems like most cases of deaths are due to people with preexisting conditions. That could be asthma. That could be another respiratory infection. It could be cancer. That's one way that there seems to be an interaction, but just in terms of spread or incidence, we rarely have coincidence data so it's very, very hard to be able to do these models in the first place.

Coronavirus is showing us how important the social contagion aspect around diseases can be. That's one place where we could have access to data. If you wanted to parameterize a good model for coronavirus, you would want to know what fraction of that population thinks coronavirus is just like another flu, or is a hoax, or that social distancing doesn't work, or that everything is dumb, right? You want to know what kind of misinformation is spreading and where, so that your model can take that into account. That data exists out there. I think the idea that public-health messaging, whether from officials or just individuals on social media, should probably be treated as public-health data, and therefore should be public in some anonymized way, but that would be incredibly useful for modelers and public-health officials alike.

> If you have ten friends telling you to go see the new Wonder Woman movie, that has more impact than one friend telling you ten times in a row to go see the movie, right? In the realm of infectious disease, if you're healthy, whether it's one friend sneezing on you ten times or ten friends sneezing on you once each, it's roughly the same exposure.

M. GARFIELD: You've led right into this last paper to discuss "Macroscopic Patterns of Interacting Contagions are Indistinguishable from Social Reinforcement." This is one you wrote with Sam Scarpino and Jean-Gabriel Young.[4] This paper, and what you've just been talking about, suggests that if we treat beliefs as something that we can be infected by, then this is an instance where it becomes especially obvious that there are benefits to associating online with people who have different beliefs than you do. That's sort of a speculation. We can just table that for a moment. But that does seem like it would sort of help randomize a contact network in the meme structure of our society. I'd love to hear you talk about how you modeled and what insights you drew from those models about co-infection of a biological contagion and a memetic contagion and how that might change behaviors.

4 L. Hébert-Dufresne, S.V. Scarpino, and J.-G. Young, 2020, "Macroscopic Patterns of Interacting Contagions are Indistinguishable from Social Reinforcement," *Nature Physics* 16(4): 426–431, doi: 10.1038/s41567-020-0791-2

L. HÉBERT-DUFRESNE: Before I get into paper, that's again a very interesting point you bring up. It's one thing that we haven't studied enough. We tend to assume one or multiple network structures, but the connection between the physical, social, or contact network on which a virus might spread, and the sort of virtual nature or delocalized nature of information networks on social media, is something that we don't study enough. We know they overlap, so we assume that to some extent they're very much correlated, but it would be interesting to look at the importance of how different are your physical networks and your information networks that mostly live online? The reason why that'd be important is in part because of this paper.

In 2015, when Ben Althouse and I published our first paper on interacting contagions, a lot of the dynamical features—so the idea that expected outbreak size is non-monotonous with respect to R_0, or that you can have super-exponential spread, or that these contagions can interact and benefit from social isolation in clustering—all of that was brand new in public-health modeling, but old news in social modeling.

In social sciences they've looked at contagions for a long time as well. There, contagions tend to be different. One way that I like to talk about this difference is that the number of exposures and the identity of people who expose you to a contagion matters a lot in the social realm, not so much in the infectious disease realm. The dumb example that I use all the time is if you have ten friends telling you to go see the new Wonder Woman movie, that has more impact than one friend telling you ten times in a row to go see the movie, right? It matters who signals to you, who sends you an exposure to that idea. In the realm of infectious disease, if you're healthy, whether it's one friend sneezing on you ten times or ten friends sneezing on you once each, it's roughly the same exposure. And maybe there's a genetic diversity in the different sneezes, but really, you're going to get sick or you're not, and you can sum up those exposures in a linear way. Right? So, models of infectious disease tend to be linear in terms of exposure, and that's not quite true in the social sciences.

Social sciences had all those features for exponential spread, discontinuity, the importance of clustering, so I was wondering (and that was in 2015) if you have data about the incidence of a very infectious disease through time, how can you tell whether it's a classic case of a classic model from public health—interacting contagions like the ones I described—or this type of social contagion that spreads through social reinforcement? How can you tell these things apart?

It took years to get a negative result, which is, well, actually, you're probably never going to be able to tell a social-contagion time series apart from an interacting-contagion time series. Especially if you don't know what might be interacting. So that was a negative result. The answer to my question was, no, you can't do it. But that turned out to be just as interesting, because I did mention that models of interacting contagions get really, really complicated, right? It's just, there are so many states I can be in. So if I'm looking at n contagions I need to follow like, "Do I have this one? Do I have this one? That one?" And I have so many assumptions to make about how they interact with each other.

But if in some way I can tell what comes out of those models or that dynamical system apart from social models, then why not use social models in the public-health space? They're much simpler, and while they don't make sense—like this idea of social reinforcement doesn't quite make sense from a disease perspective, but if it's useful as a forecasting tool or as a modeling tool—then maybe we should embrace that. The idea is that this thing seems to spread like you'd expect social contagion to spread, and that helps us understand who can leverage the work that has been done for decades in the social sciences to hopefully better understand interactions with contagions in the future.

> Messaging around infectious diseases, as we're seeing now in the United States and pretty much everywhere in the world, is just very, very tricky.

M. GARFIELD: Would you have gotten the same results fifty years ago? Or is what we're really looking at here a transition in the structure of society into this sort of more virtualized network structure?

L. HÉBERT-DUFRESNE: Yeah, I think we just have a lot more data. We've been able to track social contagions like never before because of social media. There have always been people refusing to take social-distancing measures but we just never knew that, and we didn't quite know what messaging they were using to talk to each other. But now, it's all mostly out in the open and on social media, so we do know a lot more and we're able to look at the interaction of social contagions with infectious disease like never before.

Before coronavirus... I think this might be my new favorite example, although it's a terrible situation, but from a dynamical-system perspective, it's just so rich. My previous example that I was using before to talk about this was the measles outbreak in the Philippines. You might have heard me talk about this before because I do talk about it a lot, but it's just a fascinating system. The Philippines also have quite a bit of dengue? And dengue has four strains—dengue 1, 2, 3, 4.

And it can interact in some ways because you can get a phenomenon by which, if you've had dengue 2 and then six months later you see dengue 3, your immune response might still be good enough so that your antibodies are typed to the virus, but not strong enough to neutralize the virus. And really, what you end up doing is that you're just providing sort of a genetic material disguise to the virus and helping it invade you. So by having had dengue 2 six months ago, you're making yourself more susceptible now to other strains of dengue, potentially.

So what this meant was that in the Philippines, they introduced... I think they were one of the first countries to introduce a tetravalent dengue vaccine, meaning a vaccine that works for all four strains. But it didn't work equally well for all four strains. A lot of especially young kids who got the vaccine ended up getting a reaction by which they were more susceptible to certain strains of dengue than if they didn't get the vaccine at all. So, the government, as far as I know, did the right thing of pulling back on the vaccine and messaging publicly saying, "This vaccine puts certain populations at risk. You're only going to get this vaccine if you're above a certain age or if you've had experienced dengue before."

Messaging around infectious diseases, as we're seeing now in the United States and pretty much everywhere in the world, is just very, very tricky. So, what people heard was not, "Take the vaccine only under these conditions." Really, people heard, "See, the vaccine is dangerous. We knew all along," and it did spark a huge anti-vaccination movement in the Philippines, which is why there's now a big measles outbreak. I like this story because it's a terrible story whose origins are very well-understood interactions between dengue strains. Everyone involved was meaning well. And then this interaction between dengue strains interacted with the existing anti-vaccination movement, which then fed back into measles, which spreads in many countries now.

. . . we're seeing so many patterns of how social viral messaging online and online media are interacting with the spread of coronavirus that I think all of this together is going to be a big wake-up call that we need to treat messaging around public-health data as public health in and of itself.

It used to be my favorite example, but now, especially in China and the US, we're seeing so many patterns of how social viral messaging online and online media are interacting with the spread of coronavirus that I think all of this together is going to be a big wake-up call that we need to treat messaging around public-health data as public health in and of itself. Right? How people talk about diseases should be considered public health and should be included in our models.

REFLECTION

ESCAPING THE MADNESS OF CROWDS

DATE: *24 September 2021*

FROM: *Laurent Hébert-Dufresne, Vermont Complex Systems Center*

In reading the history of nations, we find that, like individuals, they have their whims and their peculiarities; their seasons of excitement and recklessness, when they care not what they do. We find that whole communities suddenly fix their minds upon one object, and go mad in its pursuit; that millions of people become simultaneously impressed with one delusion, and run after it, till their attention is caught by some new folly more captivating than the first. We see one nation suddenly seized, from its highest to its lowest members, with a fierce desire of military glory; another as suddenly becoming crazed upon a religious scruple; and neither of them recovering its senses until it has shed rivers of blood and sowed a harvest of groans and tears, to be reaped by its posterity. . . . Men, it has been well said, think in herds; it will be seen that they go mad in herds, while they only recover their senses slowly, and one by one.

CHARLES MACKAY
Memoirs of Extraordinary Popular Delusions & The Madness of Crowds

FLOATING A WITCH.

Scottish poet Charles Mackay's 1841 three-volume work, excerpted in the quote above, tells various stories of social constructs that become group illnesses. *Volume I: National Delusions* covers economic bubbles; *Volume II: Peculiar Follies* recounts the crusades and the witch trials; and *Volume III: Philosophical Delusions* describes alchemy and fortune telling. Recent years have shown us that natural processes can also give rise to a *Madness of Crowds*. Perhaps a *Volume IV* would define how natural crises, like epidemics, affect groups.

Traveling, even a little, during the temporary respite from COVID-19 in the summer of 2021 was especially telling. Crossing from one town to the next, one would see sharp differences in gathering restrictions, mask wearing, and no doubt in vaccination rates. The truth is that these things also spread, but, unlike SARS-CoV-2, which jumps from one individual to the next, these norms jump from one *group* to the next.

Our current view of complex-systems modeling tends to focus on the microscopic mechanisms and therefore on the state and behavior of individual agents. Yet, to shape the dynamics of collective problems like pandemics and climate change, the actions of groups of agents, including institutions, matter just as much as individual behaviors. To extend our work on group size and connectivity from the beginning of the pandemic,[1] the next generation of epidemic models will no doubt integrate social factors like institutions, group selection, cultural adaptation, and peer pressure.[2] These mechanisms drive the collective cost–benefit analysis that determines why one town in Vermont with no cases might still be fully masked while a similar town in Colorado might not. One may feel frustrated to be the lone masked person in a crowded store, just as one may feel out of place as the lone unmasked person in a crowd.

We have seen many dangerous behaviors and pieces of misinformation about the pandemic spread through groups. Again, while information can spread from the mouth (or fingers) of one person to the ears (or eyes) of another, that process is affected by many group-level mechanisms.

---

1 G. St-Onge, V. Thibeault, et al., 2021, "Social Confinement and Mesoscopic Localization of Epidemics on Networks," *Physical Review Letters*, 126(9): 098301, doi: 10.1103/PhysRevLett.126.098301

2 J. Bedson, L.A. Skrip, et al., 2021, "A Review and Agenda for Integrated Disease Models including Social and Behavioural Factors," *Nature Human Behaviour* 5: 834–846, doi: 10.1038/s41562-021-01136-2

Local culture, diversity of media sources, peer pressure, presence of public health institutions, diversity of one's social network—all of these effects impact the spread of information. When only one group adopts dangerous behaviors (unmasking or epidemic denial), they are unlikely to pay a price if they are surrounded by more careful groups. This phenomenon of institutional free-riding will, either intentionally or unintentionally, further promote the spread of dangerous behavior to other groups.[3] Eventually, this will cause a localized epidemic resurgence.

To shape the dynamics of collective problems like pandemics and climate change, the actions of groups of agents, including institutions, matter just as much as individual behaviors.

Integrated disease models of the future will be able to account for group culture and local institutions to capture the localized phenomena often ignored by classic disease models.[4] The goal, simply stated, is to model not only how the pathogen might spread between heterogeneous individuals (we're not all the same!) but how groups also heterogeneously adapt and react (our groups are not all the same!), both critical in shaping the pandemic.

Ultimately, some of what we see as a *Madness of Crowds* could be viewed as failure of dialogue in populations. Journalism and communication can help to better share adaptations that work, counteract misinformation, and provide a more accurate picture of the pandemic's collective burden. Our social networks no doubt contribute to epidemics by spreading diseases, but they will also be part of the solution.

3 L. Hébert-Dufresne, T.M. Waring, et al., 2021, "Source-Sink Cooperation Dynamics Constrain Institutional Evolution in a Group-Structured Society," arXiv:2109.08106 [physics.soc-ph].

4 Bedson, Skrip, et al. (2021).

INTERVIEW NO. I-002

MODELING DISEASE TRANSMISSION & INTERVENTIONS

DATE: *1 April 2020*

FROM: *Samuel V. Scarpino, The Rockefeller Foundation; Santa Fe Institute; Vermont Complex Systems Center; Northeastern University*

CREDIT: *This interview was released as Ep. 25 of SFI's Complexity Podcast and has been abridged for length and clarity. Find the full podcast at https://complexity.simplecast.com/episodes/25*

MICHAEL GARFIELD: Sam Scarpino, thank you for joining us. This is a moment where folks with your expertise get to rise to the occasion. I interviewed your coauthor, Laurent Hébert-Dufresne, earlier this week. I want to briefly cover a couple of topics that I covered with him. One of those is the paper you wrote with him and Jean-Gabriel Young, "Macroscopic Patterns of Interacting Contagions are Indistinguishable from Social Reinforcement."[1] Can you talk about the thoughts behind this particular study, what drove you to do this, and how the results differ from traditional thinking about this?

SAM SCARPINO: Absolutely. The reason that we started thinking about this project is that for years—probably at least a decade or more now—there's been an observation that the mathematics behind how we think certain kinds of social contagions spread—so you could think of a meme or a hashtag on Twitter or anything that we think is spreading like a contagion but is not a biological pathogen—the mathematics of

1 L. Hébert-Dufresne, S.V. Scarpino, and J.-G. Young, 2020, "Macroscopic Patterns of Interacting Contagions are Indistinguishable from Social Reinforcement," *Nature Physics*, 16: 426–431, doi: 10.1038/s41567-020-0791-2

OPPOSITE: *Ferdinand Hodler, "Die Lebensmüden," 1892*

how some of those move from person to person look very similar to how some biological pathogens may be interacting with each other as they spread through a population of hosts.

In particular, you get an increase in the importance of the nonlinear dynamics associated with the disease spread, similar to what happens when you have certain kinds of social contagions. And this led us to try to answer the question, is the similarity kind of superficial—doesn't mean that it's not interesting—or is it something that actually is indicative of a much deeper relationship between the way in which we study social contagion and the way in which we study biological contagion?

M. GARFIELD: One take-home is that beliefs that engender specific kinds of behaviors act as a sort of co-infection with actual biological pathogens, and that certain ideas spread in a way that primes a population for an epidemic.

S. SCARPINO: There are sort of two main conclusions of the paper. The first—the answer to the question that we were posing—is, yes, there is a deep relationship. In particular, you can show mathematically that these models have interacting pathogens. So, two viruses spreading through the population can be mapped onto a model of social contagion.

What this means is that there are lots of ways in which two social contagions could interact with each other. Two biological contagions could interact with each other and, as you were pointing out, a social contagion could be interacting with a biological contagion. A couple of examples of that from the COVID outbreak: one of the first cases of COVID that we detected in the United States was in Boston. This was a college student who had come back recently from China and had pretty mild symptoms—mild enough that, almost certainly, we would not have picked up on this case, except for the fact that everybody was hyper-attuned to this new emerging disease outbreak and as a result, we were identifying earlier cases. Then these individuals can self-isolate, as we're doing now, and try to prevent ongoing transmission.

Not only does the social contagion of fear or concern around COVID increase the chance that we detect some of these spaces—so, it affects our datasets that affect our public health understanding—it actually can feed back on the transmission process, because once we identify that individual, they self-isolate until the symptoms clear, and it prevents ongoing transmission.

M. GARFIELD: In your discussion, you say, "Interacting simple contagions are mathematically equivalent to complex contagions if we assume well-mixed populations." That seems like a segue into a *medRxiv* preprint, "The Effect of Human Mobility and Control Measures on the COVID-19 Epidemic in China,"[2] where you're using real-time mobility data from Wuhan and other detailed case data, including travel history. You're looking at how social-isolation measures were able to interrupt this there.

S. SCARPINO: To your point about mass action mixing . . . what we mean is that individuals mix randomly following kind of an ideal gas law for how they interact. It's clearly not a realistic assumption. In fact, it's interesting that we have that in our paper, because one of the things that we spend a lot of time working on, both in my group and in collaboration with Laurent and others, is really relaxing those assumptions. So this is kind of unusual territory for us to be in. However, we were able to show that in some limited network cases you get the same results. One of the things that happens with networks—and this is why we were interested in the mobility for COVID—is that lots of different kinds of social-network structures can also drive the sorts of nonlinear dynamics that we see in these interacting pathogens.

So, if you have an arbitrarily complex social network, you can get arbitrarily anything with respect to the dynamics. We need to place some constraints on what we're studying in order to get a clean solution mathematically. That then takes us to thinking about the mobility stuff for COVID-19—we know that social networks, mobility networks, are what's moving these pathogens around. And there were a number of things that were unprecedented with respect to the initial Chinese response. Wuhan is over ten million people. It's one of the largest cities in the world. Cordoning off a city of that size is completely unprecedented in the history of humanity. However, there were also a bunch of other measures, some of which we're seeing now in the US and South Korea and Italy—shelter-in-place, reduced social gatherings, remote

2 Published now in *Science:* M.U.G. Kraemer, C.-H. Yang, et al., 2020, "The Effect of Human Mobility and Control Measures on the COVID-19 Epidemic in China," *Science* 368(6490): 493–497, doi: 10.1126/science.abb4218

work, those kinds of things. We wanted to understand the relative importance of those different measures and also incorporate other things that we're all grappling with, like the availability of molecular testing, etc. So that was really the focus of this study: given the importance we know exists for social networks and mobility and all of these things that happened in China, what can we learn about what works, and what the complications were?

M. GARFIELD: In figure 4 in this paper, which talks about shifting age and sex distributions over time, it looks as though you see a demographic shift in age cohorts and in male–female sex ratios during the course of this intervention. What is going on with that?

As soon as you give up a little bit of ground on privacy, it can be hard to get that ground back.

S. SCARPINO: We think that the situation there is differences—that we still don't fully understand—in the early part of the outbreak in Wuhan caused there to be a biased sex ratio in terms of the number of cases. So, more men than women. For a respiratory pathogen, approximately 50% male, 50% female in the population, we would not necessarily expect there to be a statistically significant bias.

And then also with the age . . . we see it in Wuhan . . . you have a different age distribution than in the rest of China. And with the mobility, what we're able to see is you actually get the kind of shifting of these age distributions in such a way that it is strongly indicative of the importance of the mobility in moving the pathogen around. So, we took the fact that you have this unusual age distribution to then try and assess the role of mobility, or at least provide an additional line of evidence that mobility was really key for moving this pathogen around.

M. GARFIELD: And their combination of interventions was successful.

S. SCARPINO: That's right. Wuhan has basically had multiple days strung together of almost no cases, and all over China they're coming back to work. You know, manufacturing is coming back online and so it really does seem to have worked. Similar measures worked in South Korea, although they didn't end up implementing the strict mobility cordons that we saw in China. Of course, South Korea caught it a little bit earlier, which makes it easier to control with less severe measures.

M. GARFIELD: You end this by making an important point, which is, "More analyses will be required to determine how to optimally balance expected positive effects on health with negative impacts on individual liberties, the economy, and society at large." On social media for this mini series, one of the big questions is: how can we possibly begin to understand something as complex and multidimensional as, "Where do we draw the line and who's holding the knife?" Do you have any thoughts on the longer-term implications of the virus and its containment?

S. SCARPINO: I had a conversation this morning with a journalist who had very similar concerns around, "How do we balance the high value that we place on privacy with the need to respond effectively to this outbreak?" That's something that we're going to have to have a conversation about very quickly as a society. And I do think that there's a role here for technology, because there should be a way in which we can get some of the critical information that we need to respond without necessarily the wholesale sacrifice of some of our individual privacy. My hope is that, for example, we would be interested in finding out the fraction of people that have symptoms, and how many social contacts they might've had in a particular window of time around getting symptoms.

Those are the kinds of things that we could have, even if it's collected at the individual level, reported out in aggregate such that you're protecting privacy, and would still get most of the benefits from capturing that information. So, this is a situation where we're going to have to balance carefully the costs and benefits. There are of course some longer-term consequences. As soon as you give up a little bit of ground on privacy, it can be hard to get that ground back. It's not just that we're looking at this outbreak; we're looking at what kind of precedents, legal and otherwise, does this set for privacy concerns going forward?

M. GARFIELD: There was a *New York Times* article recently on Kinsa Smart Thermometers.[3] They were talking about how they've been using these internet-connected thermometers to predict the spread of the flu and that they believe they are tracking the coronavirus in real time.

Millions of people have these smart thermometers. I think this dovetails nicely into another paper that you lead-authored on *arXiv* right now, "Socioeconomic Bias and Influenza Surveillance,"[4] and how we're currently deploying different datasets and how we might come up with more even-handed and humane ways of tracking these things. This piece was about influenza specifically, but how is the US currently tracking epidemics, what kind of datasets are they using, and what kind of biases in the data did you find in this study?

. . . the data that we have is biased against the individuals who are most at risk precisely *because* they don't have access to the healthcare system.

S. SCARPINO: It's something I'm really interested in and I think has also really got to be top of mind for us with respect to COVID-19. Some of us are fortunate enough to have jobs that, at least seemingly, will continue through the epidemic, have enough money to buy food and keep our families fed and healthy. Large percentages of the population don't. I think the number is something like 40 to 50% of the US doesn't have over $400 in emergency money that they can tap into for any reason.

That's one of the things we need to be very cautious and thoughtful about with respect to our COVID response—individuals who are in either at-risk or marginalized communities. You close the schools . . . well, schools also are an important source of lunch and before- and

3 D.G. McNeil, Jr.,"Can Smart Thermometers Track the Spread of the Coronavirus?" *The New York Times*, March 18, 2020, https://www.nytimes.com/2020/03/18/health/coronavirus-fever-thermometers.html

4 S.V. Scarpino, J.G. Scott, et al., 2018, "Socioeconomic Bias in Influenza Surveillance," *arXiv*:1804.00327 [stat.AP], https://arxiv.org/abs/1804.00327

after-work care for large percentages of the population. Making sure we can continue to provide these kinds of services is really critical.

So what we show in this paper that you mentioned is, the longest-running results in epidemiology is that individuals who are in lower socioeconomic groups have higher health burdens. And oftentimes that's because they are forced to live in more environmentally marginal parts of our communities. They have lower nutrition. More recently, especially in countries like the US with our healthcare system, they don't have access to the same kinds of health care, which causes an increase in health burden. That's known. We show that quite clearly in the state of Texas: three times the population-controlled number of hospitalizations due to influenza in the lowest 25% of individuals by income.

However, what we also show is that our ability to forecast the hospital demand in those most-at-risk populations—keeping in mind they have the largest burden of hospitalization—is much less accurate in terms of our ability to forecast and in the individuals in the upper three-fourths of the income distribution. That is precisely because all of the data that we have, for the most part, comes from healthcare systems, and these individuals are sicker because they don't have access to the healthcare system.

So that's why we refer to this as kind of a blind spot; the data that we have is biased against the individuals who are most at risk precisely *because* they don't have access to the healthcare system. This also, at least in our minds, fits into this broader narrative around biases when it comes to data-driven decision-making, algorithmic decision-making, machine learning, artificial intelligence . . . We don't have any data on these populations and as a result it is essentially impossible to generate actionable forecasts on their demand.

M. GARFIELD: You make a point here that it's important to improve the timeliness, the accuracy, of situational-awareness forecasting, but it's sort of a garbage-in–garbage-out scenario unless you actually are getting a complete, or a reasonably complete, dataset.

One of the paper's coauthors, Lauren Ancel Meyers, an SFI External Professor, gave her Ulam Lecture last year on "Preventing the Next

Pandemic,"[5] and she talked about some of these new methods of data collection. It's been canceled, but Google Flu Trends correlated search data—people looking up signs, symptoms, and treatment. What are some other new ways of harvesting large datasets that you think might help patch up this sort of under-reported quartile of the population?

S. SCARPINO: That's something we looked at a little bit in the paper. We show that Google Flu Trends does not ameliorate any of the data bias. And one of the things, again, that's sort of a blind spot, in our minds, is we think of everybody having access to high-speed internet and smartphones. Even though a large percentage of the population does, there's still a sizable percentage that does not. Not surprisingly, that overlaps with the percentage of the population that is most at risk for influenza—individuals in lower socioeconomic groups and the elderly. And so we think that's a reason that there's a gap there.

I think we're probably going to be in a situation where "normal" is going to be very different certainly for a period of time into the future

In this case it's really tough, because if they don't have access to any of these healthcare providers' systems, how do you actually get information on what's going on in these populations? It's something that lots of states and governments are constantly wrestling with. I think the tech solutions that we're all excited about and can provide lots of really valuable information about other parts of the population are still largely going to be blind to these individuals.

M. GARFIELD: What do you find to be the most salient metrics right now in updating your understanding and modeling of this outbreak?

S. SCARPINO: I think the most important data points to track are things like ventilator demand, number of individuals in ICUs, those pieces, because that's what we're watching very carefully for any risks of hospitals

5 https://www.youtube.com/watch?v=vwVDJVbw10k&t=3s

being overwhelmed, but also because they're going to be much less sensitive to reporting issues. They're going to be biased in their own way, but the test rates don't really affect the number of people that need ventilators. If you need a ventilator, you need a ventilator, even if you haven't gotten the COVID test yet. So those kinds of information tell us what the risks are for how close we're getting to a hospital reaching capacity. But they also help us benchmark the test positivity rates, the case reports, all the other pieces of information that we know are biased because of under-reporting. We start to leverage these different datasets against each other to get a more complete picture of how many cases are out there.

M. GARFIELD: When I spoke to Andy Dobson earlier this week,[6] he made a good point about how herd immunity requires the majority of the population to develop an immunity. And right now China reported, what was it, 100,000 confirmed cases or something like that? It was so extremely below the number of people that would have to have contracted it for the population to have developed herd immunity. And so it looks like we're sort of lurching back into business-as-usual or at least the first few countries to be infected are . . . and Andy was kind of convinced that this is going to become an endemic infection, that it's something that we're going to be dealing with for a long time. I saw that *MIT Technology Review* put up a thing about modeling intermittent social distancing over the next eighteen-plus months. It seems like a lot of people are expecting us to be able to get back to normal, but normal is no longer in view anymore.

S. SCARPINO: I think we're probably going to be in a situation where "normal" is going to be very different certainly for a period of time into the future. The point around how we're going to deal with herd immunity is a really complex one because, as you pointed out, China didn't get anywhere near herd immunity. Obviously they put in these fairly incredible measures, and I'm not saying "incredible" in terms of whether they're good or bad, I'm saying "incredible" in that they're totally unprecedented. But still, maybe somewhere in the range of a 0.5% to 2 or 3% of the population got infected. That's not even anywhere

6 Complexity Podcast, episode 23 aired on March 18, 2020. https://complexity.simplecast.com/episodes/23

near close to the herd-immunity threshold. The other piece of it is we don't know about how long-lived the immunity is for this pathogen.

There's good evidence that individuals will be immune for a period of time. We don't know what that period of time is. And the herd immunity also presupposes that individuals will be immune for a year or two years. They might not be. It also assumes that people are not going to take measures into their own hands. We could see all over the United States that people were reducing social contacts well before any mandatory orders were put into place because the social contagions are spreading as well as the biological contagion, and we're starting to change our behavior. And so what you're really risking with that herd-immunity strategy is really just killing a bunch of people and not actually getting the outcome you want. It's not really an ethical question, for example, because it was clear the UK had already made that ethical decision. I think the real question is, it's very damn unlikely to work.

The real problem is the ethical problem. We should not have a strategy that involves killing a sizable percentage of the population. But, even if you were going to get over that ethical hurdle, it still isn't going to work. And I think that's the situation that was really unfortunate about those conversations—although, I think it's very clear that the scientific community, especially in UK, were able to inform the government and help them think through that scenario.

The real problem is the ethical problem. We should not have a strategy that involves killing a sizable percentage of the population.

One of the really bright spots, such as there can be—aside from penguins roaming around the Shedd Aquarium during this outbreak—has been the tight interface between public health, academic, scientific research, private sector, everybody, government, kind of coalescing NGOs around this initiative: sharing data, sharing resources, building models, helping to inform public policy, supporting each other. That has been really exciting to see and I think is a big part of the reason why we're starting to move in the right direction with respect to the response.

M. GARFIELD: You sort of preempted my last question to you, which is, what is the good news? And it sounds like: collaboration, learning an effective real-time response to crisis.

S. SCARPINO: Collaboration is good news. It's good news that we're taking this thing seriously. It's super important. We're ramping up the testing, we're going to need that testing when we try to come back to normal. The strategies that they have in South Korea that are working really well. Singapore, Hong Kong, Taiwan, China, mainland China now, involve this sort of high rate of testing, coupled with case isolation when you identify them, to help bring things back to normal. And so this test volume that we have now is going to be super important for our current response, but it's also going to be incredibly important for our ability to come a little bit more back to normal a little bit earlier because we can engage in this kind of "test, isolate, measure" once the cases start to come back down, that will allow us to try to control the outbreak as best as possible.

REFLECTION

COMPLEX-SYSTEMS THINKING: A NECESSARY TOOL FOR PANDEMIC PREVENTION

DATE: *3 October 2021*

FROM: *Sam Scarpino, The Rockefeller Foundation; Santa Fe Institute; Vermont Complex Systems Center; Northeastern University*

COVID-19 became a pandemic because the world does not understand complex systems. I use these words to start nearly all my seminars on COVID-19. And I believe them. But what do they mean?

To me, much of what went wrong (and right) over the past two years should have been anticipated. That is not to say we could have intervened with the models and data we currently have, but we should have at least been aware of the risks ahead of us. The role of human behavior in deciding to flee cities and countries before restrictions went into effect, the politicization of nonpharmaceutical interventions like mask wearing, the arrival of more fit variants, the anti-vaccine movement, etc., did not surprise complex-systems scientists. But they seemed to surprise just about everyone else.

On the other hand, complex-systems scientists learned hard lessons about the importance—and difficulty—of clear, actionable science communication. We learned that accurate projections are not enough to drive policy changes. We witnessed data systems collapse firsthand and millions of lives and livelihoods lost because the world failed to act. We also saw the challenges inherent in the all-too-common dichotomy of prediction in complex systems: we often know what is going to happen, just not when or where.

Although my training as a quantitative epidemiologist goes back to graduate school, it was my time as an Omidyar Fellow at the Santa Fe Institute that structured much of my current thinking on pandemics. In response to the Ebola outbreak that began in 2013, I leveraged SFI's flexible funding model to pivot my research closer towards using models from complex systems to support public health decision-making. Working with teams at Yale University and the University of Texas at

Austin, I built models—and integrated genomic, epidemiological, and social-network data—to demonstrate the importance of social clustering on the pace and tempo of cases in Sierra Leone.[1] I also documented[2] the organizing effects that racism and poverty had on Ebola in West Africa, a theme common to all public health crises.

A central challenge during the Ebola epidemic was our inability to collect and share data. As a result of what I learned in 2014 about the importance of good data, by the end of January 2020 I was embedded in a volunteer group collecting and sharing anonymized, individual-level case records of COVID-19 (although back then we still called it nCoV19). This project now houses over fifty million records and is called Global.health.[3] After a year working on the Global.health project, in July, 2021, I stepped away from full-time academia and joined The Rockefeller Foundation's effort to establish an interdisciplinary institute committed to preventing pandemics. Our goal at this newly formed Pandemic Prevention Institute is to leverage data, science, and technology to contain outbreaks before they become established.

One of the questions we are asking at the Pandemic Prevention Institute is, what are the gaps in our theory of epidemics? In spring 2020, our first academic semester dealing with COVID-19, I taught a graduate course in the physics department at Northeastern University entitled "Contagion on Networks." The question I asked students about every model we learned is whether it is just a good analogy or whether we really think the mechanisms are correct. It turns out that answering this question is no easy task (and in fact may be unanswerable for epidemics[4]). Nevertheless, determining mechanisms is central to our ability to contain outbreaks with math and data.

---

1 S. V. Scarpino, A. Iamarino, et al., 2015, "Epidemiological and Viral Genomic Sequence Analysis of the 2014 Ebola Outbreak Reveals Clustered Transmission," *Clinical Infectious Diseases* 60(7): 1079–1082, doi: 10.1093/cid/ciu1131

2 S.V. Scarpino, "3 Graphs that Help Show Why Ebola Goes Viral or Dies Out," *Nautilus*, January 12, 2015, https://nautil.us/blog/3-graphs-that-help-show-why-ebola-goes-viral-or-dies-out

3 A. Maxmen, "Massive Google-Funded COVID Database Will Track Variants and Immunity," *Nature*, February 24, 2021, https://www.nature.com/articles/d41586-021-00490-5

4 S.V. Scarpino and G. Petri, 2019, "On the Predictability of Infectious Disease Outbreaks," *Nature Communications* 10(1): 1–8, doi: 10.1038/s41467-019-08616-0

Just as COVID-19 was spreading out of Wuhan, I published a manuscript in *Nature Physics*[5] with Laurent Hébert-Dufresne and Jean-Gabriel Young on the mathematical equivalence of models used to describe memes spreading through a social network and multiple biological pathogens infecting the same population of hosts. In plain English, our results show that you literally cannot tell the models apart. I spoke to *Axios*[6] about our new publication, and, interestingly, on that same day in early March 2020, I got a call from Helen Branswell at *Stat News*[7] about a cluster of COVID-19 deaths at a nursing home in Seattle. Our lives were really never the same after that.

What we showed in our paper on the equivalence of social and biological contagion is how models we know to be wrong can recreate many observable properties of epidemics. Similar results exist for quantum mechanics[8] and genetics.[9] If incorrect models can recreate the data, how then can we determine whether theories need to be updated to prevent another pandemic? If our policies rely on a mechanistic understanding of epidemics, how can we inform decision-making if we cannot validate our models? Not surprisingly, I think the answer lies in asking the right questions.

Predicting the next pandemic feels to me like earthquake forecasting. It may be that we can say something meaningful about the expected distribution of epidemic sizes over the next twenty, fifty, or hundred years, but we cannot predict precisely when and where they will occur. Nor can we say which pathogen will cause them. That is to say, focusing our

---

5 L. Hébert-Dufresne, S.V. Scarpino, and J.G. Young, 2020, "Macroscopic Patterns of Interacting Contagions are Indistinguishable from Social Reinforcement," *Nature Physics* 16(4): 426–431, doi: 10.1038/s41567-020-0791-2

6 N. Rothschild and S. Fischer, "Coronavirus Panic Sells as Alarmist Information Flies on Social Media," *Axios*, March 6, 2020, https://www.axios.com/coronavirus-social-media-b56326b6-ab16-4c8a-bc86-e29b06e5ab2b.html

7 H. Branswell, "Washington State Risks Seeing Explosion in Coronavirus Cases without Dramatic Action, New Analysis Says," *Stat*, March 3, 2020, https://www.statnews.com/2020/03/03/washington-state-risks-seeing-explosion-in-coronavirus-without-dramatic-action-new-analysis-says/

8 E. Schrödinger, 1968, "Quantization as an Eigenvalue Problem, Four Communications," in G. Ludwig, ed., *Wave Mechanics*, Oxford, UK: Pergamon Press, pp. 94–157.

9 W. Huang and T.F. Mackay, 2016, "The Genetic Architecture of Quantitative Traits Cannot be Inferred from Variance Component Analysis," *PLoS Genetics* 12(11): e1006421, doi: 10.1371/journal.pgen.1006421

efforts on spillovers may not be the best use of finite resources. Instead, we need data systems and models capable of quickly and accurately calculating the risk that a small number of observed cases will grow into a large number in the absence of interventions. Ideally these models will also speak to the necessary interventions. I wrote about the importance of focusing on what we can predict about epidemics in *Nature*[10] with Caitlin Rivers in 2018.

So, where do we go from here? Perhaps the biggest opportunity for theoretical advancement is around how we model the evolution of pathogen phenotypes. Is there a way to integrate immunity, social networks, mobility, etc., into a common framework and at least say something actionable about the risk that new variants will emerge and become established? What are the limits of predictability for viral phenotypic evolution? I believe these are areas where SFI-style thinking can make rapid advancements over the coming years and help us prepare for both the future of COVID-19 and also Pathogen *X*. From models of genotype–phenotype maps and evolvability to the evolutionary importance of metapopulation structure and mobility, work done at SFI has built much of the necessary foundation for actionable models of pathogen evolution.

At the Pandemic Prevention Institute, our goal is simple: contain outbreaks before they become established. As a complex-systems scientist who studies epidemics, I know how to mathematize both the word "containment" and "established," which means I can design surveillance and response systems to achieve quantifiable objectives. However, I also know that there will be trade-offs, the rules of the game will change, and that there is "no free lunch."[11] Only by embracing complexity can we hope to end this and prevent future pandemics.

10 C.M. Rivers and S.V. Scarpino, 2018, "Modelling the Trajectory of Disease Outbreaks Works," *Nature* 559(7715): 477, doi: 10.1038/d41586-018-05798-3

11 D.H. Wolpert and W.G. Macready, 1997, "No Free Lunch Theorems for Optimization," *IEEE Transactions on Evolutionary Computation* 1(1): 67–82, doi: 10.1109/4235.585893

INTERVIEW NO. I-003

IMPROVING COVID-19 SURVEILLANCE & RESPONSE

DATE: *16 April 2020*

FROM: *Caroline Buckee, T.H. Chan School of Public Health, Harvard University*

CREDIT: *This interview was released as Ep. 28 of SFI's Complexity Podcast and has been abridged for length and clarity. Find the full podcast at https://complexity.simplecast.com/episodes/28*

MICHAEL GARFIELD: This week's guest is Caroline Buckee, formerly an SFI Omidyar Fellow, one of *MIT Technology Review*'s "35 Innovators Under 35,"[1] and named in the "CNN 10: Thinkers,"[2] now associate director of the Center for Communicable Disease Dynamics at the Harvard School of Public Health. In this episode, we discuss the myriad challenges involved in monitoring and preventing the spread of epidemics like COVID-19, from the ethical concerns of high-resolution mobility data to an academic research ecosystem ill-equipped for rapid response, to the uneven distribution of international science funding.[3]

There are two issues I'd like to discuss from two papers. One is a letter that you coauthored for *Science*, "Aggregated Mobility Data Could Help Fight COVID-19."[4] The other is a letter that you wrote for

---

1 T. Maher, "Innovator: Caroline Buckee," *MIT Technology Review*, August 21, 2013, https://www.technologyreview.com/innovator/caroline-buckee/

2 B. Griggs, 2013, "The CNN 10: Thinkers — Caroline Buckee," http://www.cnn.com/interactive/2013/10/tech/cnn10-thinkers/

3 Note that this episode was recorded on April 8, 2020, and our understanding of this pandemic evolves by the hour. We believe this information to be up-to-date as of publication, but the findings discussed in this episode could soon be refined.

4 C.O. Buckee, S. Balsari, et al., 2020, "Aggregated Mobility Data Could Help Fight COVID-19," Science 368(6487): 145–146, doi: 10.1126/science.abb8021

OPPOSITE: *J.J. Grandville for* Un Autre Monde, *"Venus at the Opera," 1844*

The Lancet, "Improving Epidemic Surveillance and Response: Big Data is Dead, Long Live Big Data."[5]

CAROLINE BUCKEE: I actually wrote that before this outbreak, and subsequently modified it to include reference to COVID because it's so pertinent right now.

M. GARFIELD: Let's start with the state of epidemic surveillance as it is now. And, before we get into the specific problems that you bring up in the *Science* piece, talk about how we were actually doing disease surveillance at the end of 2019 and the beginning of 2020. What are the sources and how are they being employed?

C. BUCKEE: There's a big difference between routine surveillance for infections that we already have diagnostic capabilities and programs for versus surveillance for an emerging infection that we don't know anything about. Surveillance systems are in place for many different kinds of infectious diseases and other diseases. And those have kind of different types of data-collection methods and different diagnostics and different reporting mechanisms. For an *emerging* infectious disease like COVID-19, what starts out happening is a handful or cluster of weird-looking disease in a particular place.

In terms of surveillance, what that looks like—and what it looked like in this context—was strange pneumonia that started to happen in association with a particular time and place (in this case, a wet market in Wuhan). And what "unusual" is—unusual-looking pneumonia in a lot of patients—will depend on your baseline and how well you can detect an uptick in unusual symptoms.

Our ability to detect it is also going to depend on how much the symptoms from a new disease look like other symptoms that we already understand. In this case, a fever and a cough, for example, is fairly non-specific. And for surveillance systems in general, this is often an issue. For many, many diseases, we won't have a diagnostic test; we'll just have patterns of, for example, influenza-like illness. And we'll try to *infer* what's happening with transmission of a particular disease, from patterns of fever or other types of symptoms rather than confirmed cases

---

5 C. Buckee, 2020, "Improving Epidemic Surveillance and Response: Big Data is Dead, Long Live Big Data," *The Lancet* 2(5): E218–E220, doi: 10.1016/S2589-7500(20)30059-5

that are circulating in different places over time. So the first thing to say is that the state of surveillance for an emerging disease is, by definition, going to rely on symptoms and being able to tell the difference between an unusual cluster of patients with particular symptoms versus a baseline. In global terms, that's kind of how we're going to first detect something unusual happening.

Our ability to track and monitor the spread of an epidemic depends on how well we're measuring cases, and that has a whole bunch of different features to it. In this context, fairly quickly it was identified that this was a coronavirus, but in other contexts, the etiology—which is what the actual underlying pathogen is that's causing a disease—may not be known initially. So that's the first thing: trying to figure out what's causing the disease. We knew fairly early on that it was a coronavirus of some kind. Being able to track it, then, depends on how well we're able to capture cases within a surveillance system. So the whole debacle with testing really emphasizes the importance of this piece, of being able to track the outbreak.

Because testing has been slow to ramp up—and that's understandable for a new disease, because you need to develop a quick diagnostic—that means that we're not sure how many cases are occurring in different places in different times. Already that makes it quite difficult to figure out how quickly it might spread, you know, estimate the basic parameters of the disease, like the reproduction number and so on. In the context of this outbreak—and this is true for quite a few different diseases—we also think that there's a huge number of mild cases and cases that don't have any symptoms at all. And so, of course, in a normal surveillance system, you're not going to capture those people, because they won't be tested and they won't be showing up at the clinic or the hospital. So again, you have different types of biases both from a testing-capacity standpoint and from the epidemiological standpoint where you're just not capturing a lot of cases because people are either home sick but they're not going to hospital or they may not have symptoms. That's why the surveillance aspect is quite tricky at the beginning and it remains a big challenge in this epidemic.

What you do with that data on how many cases there are, and where, is then going to depend on the kind of infectious disease it is, what kinds of interventions you have available for an emerging infectious disease. Nonpharmaceutical interventions are often our only available tools before we have treatments and vaccines, which require, of course, a huge scale-up of pharmaceutical capacity and randomized controlled trials to establish what's going to work. That's why social distancing is one of the only tools we have, at the beginning of a scary epidemic like this, to shut down transmission sufficiently to give us some time to figure out how to combat the disease itself.

M. GARFIELD: Just this morning in the SFI inhouse emails, Geoffrey West shared this thing that some of the listeners may have seen going around suggesting that this novel coronavirus is like Schrödinger's virus because you both have to act as though you have it and that you don't have it. This is a profound uncertainty that we're all acting under here.

I think you brought us right up to the concern around when we really only have network interventions, when we really can't do effective contact tracing, when we really don't know who has already had [the virus] and recovered at the resolution that we would like.

> ... we live in a global community and, depending on the epidemiology of a disease, we are not immune to global outbreaks.

C. BUCKEE: We *have* contact tracing. In fact, places like South Korea and Singapore did very effective contact-tracing programs. We wrote a paper[6] where we showed that, in combination with social distancing, contact tracing—aggressive case finding—can be very effective. It's not that we can't do contact tracing, but we certainly don't have pharmaceutical interventions just now. Contact tracing is one in a suite of tools that will be helpful in combination with social distancing more broadly.

M. GARFIELD: A lot of the conversations I've been having about COVID-19 are about how, in a sense, it didn't just infect us as individuals;

---

6 C.M. Peak, R. Kahn, et al., 2020, "Individual Quarantine versus Active Monitoring of Contacts for the Mitigation of COVID-19: A Modelling Study," *The Lancet Infectious Diseases* 20: 1025–1033, doi: 10.1016/S1473-3099(20)30361-3

it revealed to us, by breaking all of these other global networks—supply chains and so on—it showed us the vulnerabilities in those networks. One of the vulnerabilities that you address in both of these writings—although at different angles, in places where we don't have sort of modularity built into the structure—is how we aggregate data. I'm thinking about Albert Kao's work on modular decision-making and how local aggregations coarse-grain things.[7] So in this case, it's about anonymizing data at the level of individual people, but still providing useful information to people at different points at county-, state-, and national-level decision structures.

Thinking about disease surveillance brings up all kinds of issues with privacy. I'd love to hear how you and your colleagues have been thinking through this.

C. BUCKEE: Just to back up to your first point about global connectivity and supply chains: I think absolutely this has brought home to people what epidemiologists already knew, which is that we live in a global community and, depending on the epidemiology of a disease, we are not immune to global outbreaks. I think the supply chain issue is very interesting, and one of the things that we're seeing that's worrying me is the disruption to the humanitarian aid supply. That's going to have a huge impact on low- and middle-income countries that rely on, for example, a distribution of food and other humanitarian aid, and how they're going to be able to manage that. It's an interesting scenario: In Vanuatu right now, there's a tropical cyclone coming and they're going to probably need international aid. But there's a possibility that those international actors will bring COVID-19 to Vanuatu with them.

What's more, when you have a natural disaster, everybody shelters in place and they're all crowded together. That promotes the spread of the disease. So there's all kinds of interconnectedness that happens, not just in terms of the initial outbreak and people traveling around, but also, as you point out, with supply chains—and then this issue of international aid and humanitarian interventions that we need to continue.

We really need to think through some of the factors for low-income settings where the reality is that the elderly populations there are lower

7 A.B. Kao and I.D. Couzin, 2019, "Modular Structure within Groups Causes Information Loss but Can Improve Decision Accuracy," *Philosophical Transactions of the Royal Society B*, 374: 20180378, doi: 10.1098/rstb.2018.0378

than they are in the Western world. From a disease standpoint, the trade-offs that they're having to make with respect to humanitarian issues—food, routine vaccination—they may have to make different types of decisions about what they want to do with respect to COVID-19 because of the way that their societies are set up and their demographic distributions and their reliance on global supply of a different kind.

With respect to privacy: I think the most important thing to emphasize here is that there's a very important distinction between apps designed for contact tracing, which are individual-level data and are designed specifically to try and look at chains of transmission to aid contact-tracing programs. In contrast, the type of data that we've been working with is aggregated to the extent that it's no longer human-subjects research. We've done this work for quite a long time. The principle behind the aggregation that happens is that you want the lowest resolution—so the coarsest spatial scale—at which you can still say something useful. And if it's the case that the spatial scale needs to be on a smaller spatial scale or higher resolution for you to be able to say anything sensible, then we don't do that.

So, under the DUAs[8] and the privacy protocols that we have in place, re-identification is a big deal and we really take that seriously. The aggregated data itself tells you something much more general than these contact-tracing apps. What it tells you is generally what's happening: how far are people going when they're traveling around, how much movement is there? And that's roughly related to the contact rate within an epidemiological model. We don't really know yet exactly how those two things are linked mechanistically, but once we have better data on COVID transmission in different places, coupled with specific policies being put in place, we should be able to start to disentangle what these aggregated mobility metrics mean for social distancing. Down the road we're going to need evidence if we're going to make decisions about how to relax social distancing on the other side of this epidemic. Without having a way to monitor social distancing interventions and understand what that will do in terms of transmission, we won't be able to relax them based on data; we'll be guessing. So again, that's going to be really key, especially if we're going to be in a scenario where we have to go

8 Data Use Agreement

into lockdown multiple times, which is one possible scenario for the future. So we really need to start measuring this in a sensible way.

M. GARFIELD: The episode that came out just before this was with David Krakauer discussing the first few submissions for the SFI Transmission article series.[9] The theme linking all of the first five articles in that series had to do with this issue of the correct resolution of the model and deciding at what point you're making the trade-off. Where are the most effective trade-offs made between an honest account of the spread of probabilities to the actionability of the knowledge itself. So that links this work on aggregating mobility data to this other piece that you wrote for *The Lancet*, which addresses this issue in a lot of ways,[10] one of which is that right now, epidemiology is a crisis discipline where you're working these things out in real time. And much as Rajiv Sethi mentioned and an early episode of this show,[11] talking about the way that we run models of one another in the criminal justice context, you know, if you meet someone in a dark alley, you don't have a lot of time to make a decision about them. And that's where all of these implicit biases come out.

There's a similar thing going on right now, when we meet the coronavirus in a dark alley. We have to figure out what degree we are clustering versus treating these cases as unique cases. I think it's good to start where you started, on the urgency of better surveillance systems, but you also bring in these three other points about—three other very crucial challenges to this.

C. BUCKEE: For several years now there have been discussions around the use of different kinds of data to inform surveillance and to react when epidemics happen. Data from phones is one of them—the one that I know the best—but there's lots of different kinds of discussions around how corporate-owned data could be useful for forecasting. So in that article I really

---

9 See Transmission Podcast Episode 1 beginning on page 63

10 N. Kishore, M.V. Kiang, et al., 2020, "Measuring Mobility to Monitor Travel and Physical Distancing Interventions: A Common Framework for Mobile Phone Data Analysis,"*The Lancet Digital Health* 2(11): E622-E628, doi: 10.1016/S2589-7500(20)30193-X

11 Complexity Podcast, episode 7, November 13, 2019.https://complexity.simplecast.com/episodes/7

discussed some of the barriers, over and above the privacy concerns, which I think are very real and are being addressed currently.

I think the incentive structures are all misaligned for this to be implemented in a routine way, although I've been really surprised and pleased by this response. I think we've started to see these collaborative networks being built and hopefully they will continue to be incentivized in a way that's sustainable into the future. But academic incentive structures are not really great for translational work. And then corporate data sharing—there's all kinds of issues with that. And then of course,

for governments, it's a risky game to try and do a massive data sharing. There are lots of different types of priorities that they have. Everyone's got their own agenda and that's challenging when you're trying to build out these analytics pipelines.

The other point I made in the article was that at the moment, there's very much a separation between the methodological and high-tech world in which some of these methods are being developed, and the realities for the most vulnerable communities and the populations who are actually dealing with the implementation of surveillance on the ground and are going to be the frontline when it comes to first detecting and then responding to epidemics.

Really, I was thinking about Ebola in the article because I think that highlights it well. There are a lot of scientists in high-income settings who think through problems very far away from the people who are going to have to implement them. And so, in the article, I argue for a shift in the focus of intellectual and methodological work down the kind of translational pipeline and to the geographies where many of these challenges are being dealt with.

The last thing, that relates to what you mentioned about the previous podcast: There's a sort of divide in the modeling community that reflects attention between simple mechanistic models and very granular, detailed agent-based models, and what their utility is for different types of scenario planning versus making quick decisions during an epidemic.

I argue in *The Lancet* letter that for emerging epidemics, often the simplest models are the best or most useful because they're transparent and can be quick and easily translated. For this outbreak, one thing that modelers always have to grapple with is whether the uncertainty in your model parameters and model structure outweigh the utility of having a very detailed agent-based model. If you have a very detailed agent-based model, but you have huge amounts of uncertainty in the basic epidemiological parameters of the disease, it's not clear to me how useful that will be, not to mention it's highly computationally intensive.

There are a lot of scientists in high-income settings who think through problems very far away from the people who are going to have to implement them.

In contrast, a very simple model is limited in what it can manage to tell you because, by definition, it's simple. On the other hand, you can clearly explain the major uncertainties in your parameters and in the model structure, and you can see very quickly the types of broad qualitative impacts some of those uncertainties are going to have on different types of intervention. While the spectrum of model complexity may be useful for different kinds of responses and different kinds of research playing into different policy decisions, right now there's a huge amount of uncertainty in the basic parameters. I think one of the biggest ones is how many people are asymptomatic, and by that I mean never symptomatic. It could be a substantial fraction, and that's an uncertainty in the model that's very hard to account for unless you have serological data on how many people have antibodies to the virus. Some of these other parameters have become more clear as the epidemic has progressed, but still, there are big questions about that in particular that are making detailed predictions difficult.

The other thing that's very uncertain is what impact all these social-distancing interventions are having on the contact rate. And that's something that we will hopefully start to be able to parameterize a bit more as we move further into this epidemic. But right now, again, it's quite hard to parameterize; If you imagine an agent-based model with people moving around, it's quite hard to know what's happening now in terms of the contact rate. So linking that to a prediction or a forecast is extremely challenging. Again, I think that the simple models are pretty good for general scenario planning—getting a quick idea for what's going on and highlighting the major uncertainties in some of the parameters—whereas the agent-based models can obviously give you a lot more granular resolution on particular kinds of question. But I think there's a huge amount of uncertainty associated with that right now.

M. GARFIELD: Let's wheel this back for a moment to the first challenge that you laid out, which is the misalignment of incentives in public–private partnerships with respect to this kind of urgent situation. I've actually been really impressed; it's been inspiring to see academics and corporations and governments working together as well as they have. But it does call into question not just policies around corporate and governmental data sharing, but also the timescale at which academic research is normally conducted. I'm curious how you see these things shifting in an adaptation to the crisis and what kind of hopeful developments you've observed on that front.

C. BUCKEE: The timescale issue is a big one. If you look at the NIH grants that many of us rely on, the turnaround time is long, and for good reasons: because we want peer review and so forth. But I think it's fairly extreme if you think about epidemics and how we respond to them. Also, the incentives in academia are misaligned in the sense that we are still incentivized in terms of promotion based on first and last authorship and these types of archaic metrics, which don't reflect how science works these days. So now if you look at the papers that are coming out on COVID-19 and you look at the people actually doing the science, these are big team efforts. These are consortia of modelers and virologists and clinicians and they're working together to come up with a solution. And I think there has to be a shift in both how we allocate credit to scientists in those big teams and how we think about funding. So, single PI-led grants and things like that, I think, are just not going to be conducive to being able to push out this kind of work.

The other thing is that, for example, the MIDAS Network[12] was funding centers of modelers and other collaborators with the express purpose of developing methods that could then be deployed during an epidemic. And I think what you've seen in the US response is that

12 Models of Infections Disease Agent Study: https://midasnetwork.us

network has been absolutely central—like *really* central—to responding to this challenge and doing the science and rolling out models and everything. Those centers are about to lose funding and I hope that this epidemic will prove to the powers that be that, actually, we need to invest in those kinds of centers. We need to give academics sustainable funding to do the slow methodological work, and then to have the flexibility to respond when a pandemic hits. So that's one thing and I hope that we will be able to use this to showcase how science works now and get it funded in a more appropriate way.

As far as the corporate side, I think there's still some evidence that there are kind of competitive forces at work that aren't necessarily helpful for data sharing more broadly. We also have seen a number of corporate and other actors generating models and stepping in to be the interface with state governments on their response, and I think this pandemic shouldn't be seen as an opportunity to monetize. I feel that very strongly and I worry that, given the current economic setup, we're still seeing evidence that those types of competitive and profit-driven motives are evident in this response. And I think it's really important that, as a society, we think about that moving forward, because that strikes me as potentially quite problematic.

M. GARFIELD: This seems like it ties into your third challenge, the methodological challenge with respect to uncertainty here. There's something about the way that this pandemic is unfolding that is distinctly information age, right? When we had Laurent Hébert-Dufresne[13] and Sam Scarpino[14] on the show, both of them were talking about complex contagions that involve both biological as well as informational social components, in terms of how people are understanding it, making sense of it, behaving in regard to that. All of this seems somewhat structurally similar to the Thirty Years War. The way that a lot of the structures for how we would normally verify information are themselves going through a crisis. We're seeing a lot of competition at the level of people without what we would think of as normal credentials stepping forward. In terms of the printing press, the analogy is like, their

13 Complexity Podcast Episode 24, March 25, 2020. See also page 487

14 Complexity Podcast Episode 25, April 1, 2020. See also page 507

own version of Christianity, their own version of epidemiology. And on the one hand, this is really inspiring. You're seeing people step up en masse. But in another sense, you make the case in this *Lancet* article that it's really dangerous because what's happening now is that we have a lot of people who, when the barrier to entry to participation in modelling and sharing this information has been lowered like it has, there's a narrative collapse. It becomes very difficult to coordinate action. People are buying models from people that aren't really authorized or they lack the expertise. Disinformation and misinformation are rampant and it's a tricky thing, right? Because on the one hand, we do need to be able to move fast enough to do this, but it's unclear how to move fast enough and still be able to apply the brakes when necessary, or steer it.

C. BUCKEE: The specific point I was making the article was that you can't get away from needing solid epidemiological data and solid epidemiological analysis. In any model, you need to know how many cases you have. So, there's this feeling in this world of AI and deep learning and big data that, somehow, we're going to be able to make up for a lack of solid epidemiological data with all these other datasets. And while I agree that if you have *nothing*, it's helpful to have some other sources to inform your estimates, but you can't replace it and you can't ultimately make any sensible statement without these key, very basic, epidemiological pieces of information.

I think people conflate forecasting epidemiological models with forecasting the weather. But epidemiology involves human behavior and feedback and you change the situation when human behavior changes. If you say it's going to rain tomorrow and everyone carries an umbrella, that doesn't change the fact that it's going to rain. If you say there's going to be an outbreak tomorrow and everyone stays home from work, there's no longer an outbreak, your prediction is wrong, and everybody loses trust in your model. So these things are not the same. And I think it's very easy to underestimate the difficulties and the amount of expertise that you might need to make a sensible model that's actually useful.

I'm all for democratizing science and I think that that's great. The problem is whether you're using that to inform policy and how it's being communicated to the public. And *there* you really have to be very clear, and we've seen that with this COVID-19 outbreak where there's a lot of confusion in the media and among different levels of government about what these models really say and what they can say, what they can't say. You know, we've had the press say, "Oh, the model said this last week, and now it's saying that," when there's no inconsistency in the modeling framework, it's just that the model under scenario *X* is not the same as the model under scenario *Y*, for example. So I think that there's a huge communication problem here that has the potential to be quite dangerous, and the rise of so-called armchair epidemiologist is part of that general attempt. People want answers, right? And so people step up.

I think people conflate forecasting epidemiological models with forecasting the weather.

You know, there was an article yesterday in *The New York Times* about the new heroes of the coronavirus outbreak and all six of them were white men.[15] I do think that there are gender differences in who is stepping up to have an opinion. And not just that, but there's also a very big divide in terms of who's being portrayed in the media, who is coming out on TV and talking. So we talk about disinformation; I think there's also a misrepresentation and a problem there. If you look at the people who are coming forward to have strong opinions based on potentially not that much expertise, it definitely has a racial and gender bias.

M. GARFIELD: So that's contributing to a problem that you identify and prescribe with a reallocation of resources—money research, etc.—into those populations that are at risk, those populations that are under-surveilled, those populations that are actually the stakeholders locally, the recipients of these interventions. And this is something that touches on all three points in this piece—the misalignment in the analysis pipeline, the gap between the innovation and the implementation, and then

15 M. Stevis-Gridneff, "The Rising Heroes of the Coronavirus Era? Nations' Top Scientists," *The New York Times*, April 5, 2020, https://www.nytimes.com/2020/04/05/world/europe/scientists-coronavirus-heroes.html

the inherent uncertainty of it. It would be better if, rather than trying to make a global theory of everything for the coronavirus, we looked at local solutions, and then how those fit into a broader understanding and solution. It would make sense if we designed these interventions in the context of where they would actually be deployed.

But, when I had Sam Scarpino on the show, he made the point that often the populations that are most in need are the ones about which we have the least information. There's a misrepresentation among the heroes as well as the sort of victims in this situation. And so how could we align our incentives and reinvest when we're basically shooting blind?

C. BUCKEE: It's true that there has been a kind of neglect in terms of allocation of resources for science globally. That has a very distinct geographic flavor. So I think a lot of these issues—not necessarily for COVID, although for COVID as well . . . being able to tailor a response in a way that is not just reflective of the situation in the US and Europe, for example, but reflects the different realities around the world—different demographics, different comorbidities, different issues for routine medical care. What you need is to invest in scientific groups in the Global South, for example. We should be building centers, supporting excellent scientists—who exist all across Africa and India and Bangladesh and so on—and we should be making sure that we fund centers of excellence and the researchers there who are in a much better position to be able to interface between policymakers and scientists and public health in the context that is going to be relevant for their response in a way that will ultimately protect everyone better.

Given that we live in a global world, we need to think about adjusting resources for the scientific community so that they can respond where epidemics start. And that's going to be agnostic to many different factors, but right now the focus of expertise and money and funding for science for this kind of thing is in the US and Europe and Canada and places like that. We need to shift it. We need to recognize that we live in a globally connected world, and that we have systematically neglected funding in the Global South. And then we need to address that directly.

M. GARFIELD: How do you see it as working within the United States? In Europe, for example, the lowest quartile in income are the ones that both have less access to healthcare, and then also in a weird way, a kind of privacy through their disclusion from the system that would provide data on them in the first place. So how do you imagine creating flexible and distributed teams that are able to serve the poor in that regard, even within healthy nations?[16]

Given that we live in a global world, we need to think about adjusting resources for the scientific community so that they can respond where epidemics start. . . . We need to recognize that we live in a globally connected world and we have systematically neglected funding in the Global South. And then we need to address that directly.

C. BUCKEE: A lot of these problems stem from economic and political injustice against particular groups of people and the poor. The best thing that the government could do for those groups is to make sure that they're supported economically. Unemployment's just skyrocketing. We need to support those populations and make sure that they are at least economically looked after, as well as provided with excellent medical care and access. But I think that this data issue is really key. So one thing that we have talked about with this aggregated mobility data is the reason that we provide very coarse-scale data is because we really want to avoid punitive targeting of particular groups of people based on their ability or inability to social distance. And with these contact tracing apps and personal data, one of the critical things here is that it's not used in a way that exacerbates inequality and is punitive. So that's something that I think this whole field is going to have to really think hard about in any context. But particularly in the US, I think it's going to be an issue. And then we should be thinking creatively about how to

16 See, for example, P. Turchin, "The Double Helix of Inequality and Well-Being," *Cliodynamica*, February 8, 2013, https://peterturchin.com/cliodynamica/the-double-helix-of-inequality-and-well-being/

re-employ a lot of the newly unemployed populations to help us fight this thing. Let's try and put people to work to distribute different kinds of PPE or help with contact tracing protocols, while making sure that we're not putting them at excess risk, but there should be creative ways that we can manage this threat and simultaneously think about the economic hardship that the impoverished communities here are going to have to suffer through.

M. GARFIELD: Right now, what do you find, for you personally, are the most interesting or the most salient points that you're tracking? What do you think are the channels or the types of data that are most relevant to a randomly selected audience member right now?

C. BUCKEE: There are two things. First, there is a race right now to develop serological tests, and that's going to be absolutely critical. So that's a test that looks for antibodies to this virus. It's a marker of infection. Once we have serological tests, we're going to need to roll them out to make sure that we are sending healthy people who have some level of immunity—and we still don't know whether people are immune, but hopefully they are. So first of all, the serological tests are going to be critical, and that will help us estimate how many people never had symptoms, the epidemic size, parameterize a lot of the models, think through policy scenarios, as well as get people back to work and back to their normal lives as quickly as possible. The second thing I think we really need to do is monitor and measure what social distancing interventions are doing, whether they're working, how much reduction in mobility is needed to reduce transmission by a certain amount.

That's going to be critical as we move towards a situation where we're going to relax some social distancing interventions and not others, right? Is it okay to open schools? Is it okay to open workplaces? We're going to need to make those decisions in an evidence-based way. So again, linking some of this mobility data to actual COVID is going to be critical. For that, we need testing. So like everyone, I would call for more testing, randomized testing, good study design. And then moving forward, for me this is a probably once in a lifetime—I hope once in a lifetime—event. So, I hope that it leads to a radically restructured society and a much more inclusive society that recognizes how interconnected we are globally and within our communities. And I think a positive outcome from all of this could be that we start to be much more community-minded and we start to think through how we provide healthcare to people in a way that reflects their needs.

The good outcome could be that we are more equitable on the other side of this, and we've restructured supply chains so our dependencies are more local and so on, while not giving up the really vital international aid and cooperation that we need for our global world.

My fear is that that won't happen, and like with other pandemics that have come and gone, we will come through this, we'll scrape through—it will be a disaster for many people but we'll scrape through—and we will have learned nothing. My biggest fear is that we'll end up on the other side of this with a weakened society that has suffered a lot, both economically and socially, and that we will not have taken the opportunity to restructure our society in ways that will benefit everyone. So, I don't know. I think people need to really think through what their values are and how we move forward to make political decisions that reflect those values. That's my hope.

ESSAY NO. E-001

THE DAMAGE WE'RE NOT ATTENDING TO

DATE: *8 July 2020*

FROM: *David Krakauer, Santa Fe Institute;*
Geoffrey West, Santa Fe Institute

CREDIT: *This essay originally appeared in Nautilus:*
nautil.us/issue/87/risk/the-damage-were-not-attending-to

World War II bomber planes returned from their missions riddled with bullet holes. The first response was, not surprisingly, to add armor to those areas most heavily damaged. However, the statistician Abraham Wald made what seemed like the counterintuitive recommendation to add armor to those parts with no damage. Wald had uniquely understood that the planes that had been shot where no bullet holes were seen were the planes that never made it back. That's, of course, where the real problem was. Armor was added to the seemingly undamaged places, and losses decreased dramatically.

The visible bullet holes of this pandemic are the virus and its transmission. Understandably, a near-universal response to the COVID-19 pandemic has been to double down on those disciplines where we already possess deep and powerful knowledge: immunology and epidemiology. Massive resources have been directed at combating the virus by providing fast grants for disciplinary work on vaccines. Federal agencies have called for even more rapid response from the scientific community. This is a natural reaction to the immediate short-term crisis.

The damage we are *not* attending to is the deeper nature of the crisis—the collapse of multiple coupled complex systems.

Societies the world over are experiencing what might be called the first complexity crisis in history. We should not have been surprised that

---

OPPOSITE: *A printed triangular bandage of the kind developed by Friedrich von Esmarch*
COURTESY NATIONAL LIBRARY OF MEDICINE,
NATIONAL INSTITUTES OF HEALTH, HEALTH & HUMAN SERVICES

a random mutation of a virus in a far-off city in China could lead in just a few short months to the crash of financial markets worldwide, the end of football in Spain, a shortage of flour in the United Kingdom, the bankruptcy of Hertz and Niemann–Marcus in the United States, the collapse of travel, and so much more.

Societies the world over are experiencing what might be called the first complexity crisis in history.

As scientists who study complex systems, we conceive of a complexity crisis as a twofold event. First, it is the failure of multiple coupled systems—our physical bodies, cities, societies, economies, and ecosystems. Second, it involves solutions, such as social distancing, that involve unavoidable trade-offs, some of which amplify the primary failures. In other words, the way we respond to failing systems can accelerate their decline.

We and our colleagues in the Santa Fe Institute Transmission Project believe there are some non-obvious insights and solutions to this crisis that can be gleaned from studying complex systems and their universal properties. One useful way to think about a complexity crisis is in terms of the strategic trade-offs that need to be managed and the complex mechanisms that these trade-offs involve. These mechanisms include ideas of contagion, epidemic cycles, superspreading events, critical phenomena, scaling, and path dependence.

THE TRADE-OFF BETWEEN BIOLOGICAL & SOCIAL CONTAGION

Once an outbreak has occurred, contagion in biology is dominated by contact networks—the direct physical interactions among infected and susceptible individuals. The crucial variable describing disease contagion is R_0, the number of secondary infections traceable to a primary infection. Social contagion is different and can depend on both a longer history of prior contact and a multitude of non-physical interactions. This means that disease versus ideas about disease spread in fundamentally different ways and at different rates. Whereas the R_0 value for COVID-19 has been estimated at around 2.5, a cultural R_0 describing the spread of ideas about the COVID-19 span around 3.5 for YouTube and 5 for Twitter.

The implication of these disparities is that we are infected by the idea of a disease long before the disease itself. This establishes the possibility of forecasting and adaptive behavioral change, exactly as is the case with weather and earthquake forecasting. In both of these cases, large-scale changes in habit and activity are seen as necessary and logical adjustments to mitigate far greater damage. Highly profitable industries and technological infrastructures have emerged to manage the trade-off between the differential spread of information and physical phenomena. Weather forecasting alone is around a $6 billion industry.

THE TRADE-OFF BETWEEN AVERAGE & HEAVY-TAILED SPREADING

The spread of a disease varies from event to event, with some contacts spreading to only a few people as given by the average R_0 value (biological or cultural) and some spreading to far more—so-called superspreader events—that can come to dominate the numbers of infectious individuals and the eventual size of an epidemic. This means that the variability of contagion is potentially more informative than the average contagion.

This is important because efforts at mitigation that involve finite resources are best concentrated in those parts of a network that are likely to generate superspreading events, reducing the need to deploy systemic and draconian nationwide strategies, in favor of well-informed, local, and targeted interventions. The opportunity for efficiencies are significant and already heavily exploited in adjacent superspreading domains such as the use of influencers (taste superspreaders) in social media campaigns.

THE TRADE-OFF BETWEEN BIOLOGICAL & SOCIAL NEEDS

It's the interaction of culture and biology that influences R_0, and the ultimate value established in a population is derived from feedback between the two. We have seen this sociobiological feedback loop played out in the past few weeks. As isolation has reduced contact and reduced the contagion of disease, the biological R_0 has dropped in some places to below 1. The social contagion of this fact has led some states to reopen and physical contact to increase, driving the R_0 above 1. This is analogous to a thermostat that seeks to maintain a constant temperature, but in this case the thermostat is the news, causing R_0 to hover around 1. In complex systems, this is called self-organized criticality, and can lead to the long-term persistence of a virus. Introducing delays that allow one side of the dynamic to dominate, such as the use of longer-term averages of R_0 values, could allow for a more sustained reduction in contagion.

THE TRADE-OFF BETWEEN TROPHIC (FOOD NETWORKS) & DISEASE NETWORKS

The origin of human infectious diseases and pathogens can often be traced to nonhuman species (zoonoses) that have very different life cycles to humans, often much faster birth–death processes. The COVID-19 virus is likely to have originated in bat populations that are an essential source of protein in human diets. Bats live around twenty years and bat epidemics cycle more quickly than those in humans. The implication of frequent zoonotic interactions is that the primary force driving the evolution of a pathogen is the shorter-lived host, which means that humans need to adapt to infection at a rate proportional to the evolution of the virus in a bat. It is the shorter-lived species that calls the shots on the progress of a disease.

> It is well known that it's easier to continue with an existing technology than adopt a newer and better one . . .

We have discovered cultural means of resolving the trade-off between trophic networks and health in the past. Perhaps the most notable in our species was the switch to cooking meats and tubers just over a million years ago. In our own time, the majority of humans cannot tolerate milk-sugars into adulthood and seek alternative energy sources. The costs of illness outweigh the caloric benefits of lactose leading to global and large-scale diet switching across large swathes of the globe. Providing access to alternative sources of protein for populations at risk of zoonosis would benefit from the way we have managed trophic trade-offs with coconut milk, almond milk, and soy milk.

THE TRADE-OFF BETWEEN ECONOMIC & DISEASE SCALING

Cities are machines we evolved to accelerate sociobiological interactions. The larger the city, the more the average individual interacts with other people in a multiplicative positive feedback process. It has recently been observed that high-density urban settings are associated with the super-linear scaling of a large number of biological and social variables. This means that a doubling in city size leads to more than twofold increase in these variables which include economic productivity, rates of innovation, crime, and disease. As a result, not only are there systematically more disease cases in larger cities but, equally importantly, their growth rate, like all socioeconomic urban phenomena, increases systematically faster. If the number of cases increases exponentially with time, then the rate parameter is predicted to systematically increase with city size. Consequently, a city of a million people will double the number of cases in approximately half the time as a city of 10,000. Cities are where trade-offs are most acutely felt and where the costs of trade-offs are most extreme. A means of addressing this trade-off is to invest heavily in urban infrastructure that can compensate for reduced physical contact.

THE TRADE-OFF BETWEEN THE PAST & THE FUTURE

The processes of contagion, superspreading, self-organized criticality, and urban scaling are all nonlinear and tend to result in multiple different outcomes or equilibria. One of the hallmarks of systems with multiple equilibria is path-dependence, meaning it is far easier to move in one direction than another. This often leads to "lock-in." For example, it is well known that it's easier to continue with an existing technology than adopt a newer and better one, such as the preference for silicon transistors over the superior metal oxide transistors in the early 1960s. And the same goes for social habits such as the continued preference for the QWERTY keyboard once it was widely adopted over all alternatives. And vastly more worrying is the lock-in around racial and gender-based hiring preferences, which perpetuate an historical precedent rather than rewarding ability.

The social habits we tend to see as either the fabric of society or unintended corollaries of social life—gathering at high-density, shaking hands as a greeting, traveling, and interacting when infectious—have become established as social norms. Path-dependence tells us that far more energy needs to be invested in campaigns to eliminate these habits than is required to perpetuate the habits.

Just as biological systems pay a cost for robustness and evolvability foregoing efficiency for long-term persistence, so too should we demand this of our institutions.

THE TRADE-OFF BETWEEN ROBUSTNESS & EVOLVABILITY

A key insight from complexity science is that complex systems function by the continuous trade-off between robustness and evolvability. Robustness describes the ability of a system to withstand a critical perturbation without a significant loss of function. For example, individuals are able to sustain high levels of damage to the brain, such as severing the two hemispheres, while continuing to function. Evolvability describes a mechanism that allows for the efficient exploration of adjacent novelties, whereby small changes to a mechanism or structure can engender new functions. Many enzymes can be mutated into functioning alternatives once the space of functional molecules has been mapped. Evolution has discovered the means of balancing the competing demands of these two principles and together they account for both the long-run stability of lineages and the diversity of life. Contrast this complex, evolved reality to our modern, social–technical world. A typical strategy of companies and corporations is to eliminate redundancies and degeneracies in the name of minimizing costs. This is the major reason why almost all companies have great difficulty adapting to change, and eventually disappear. Just as biological systems pay a cost for robustness and evolvability foregoing efficiency for long-term persistence, so too should we demand this of our institutions.

COMPLEXITY THINKING

It is high time we attended to issues resulting from the long-term consequences of this crisis and its interconnection with all socio-economic life across the planet. We have modest understanding in these areas, and we are in desperate need of support and new ideas. In the future, we need to be thinking more about the threats of a full complexity crisis with all their attendant trade-offs rather than the means of mitigating a single threat. The challenge for all healthy societies is deciding where to place the fulcrum that balances competing priorities and not treating priorities as if they were independent concerns.

ESSAY NO. E-002

UNCERTAIN TIMES

DATE: *21 August 2020*

FROM: *Jessica C. Flack, Santa Fe Institute*
Melanie Mitchell, Portland State University; Santa Fe Institute

CREDIT: *This essay, edited by Sally Davies, originally appeared in Aeon and can be found at aeon.co/essays/complex-systems-science-allows-us-to-see-new-paths-forward*

We're at a unique moment in the 200,000 years or so that *Homo sapiens* have walked the Earth. For the first time in that long history, humans are capable of coordinating on a global scale, using fine-grained data on individual behavior, to design robust and adaptable social systems. The pandemic of 2019–20 has brought home this potential. Never before has there been a collective, empirically informed response of the magnitude that COVID-19 has demanded. Yes, the response has been ambivalent, uneven, and chaotic—we are fumbling in low light, but it's the low light of dawn.

At this historical juncture, we should acknowledge and exploit the fact we live in a complex system—a system with many interacting agents, whose collective behavior is usually hard to predict. Understanding the key properties of complex systems can help us clarify and deal with many new and existing global challenges, from pandemics to poverty and ecological collapse.

In complex systems, the last thing that happened is almost never informative about what's coming next. The world is always changing—partly due to factors outside our control and partly due to our own interventions. In the final pages of his novel *One Hundred Years of Solitude* (1967), Gabriel García Márquez highlights the paradox of how human agency at

---

OPPOSITE: *John Tenniel, "The Red Queen and Alice Running," 1871*

once enables and interferes with our capacity to predict the future, when he describes one of the characters translating a significant manuscript:

> *Before reaching the final line, however, he had already understood that he would never leave that room, for it was foreseen that the city of mirrors (or mirages) would be wiped out by the wind and exiled from the memory of men at the precise moment when Aureliano Babilonia would finish deciphering the parchments.*

-550-

Our world is not so different from the vertiginous fantasies of Márquez—and the linear thinking of simple cause—effect reasoning, to which the human mind can default, is not a good policy tool. Instead, living in a complex system requires us to embrace and even harness uncertainty. Instead of attempting to narrowly forecast and control outcomes, we need to design systems that are robust and adaptable enough to weather a wide range of possible futures.

Think of hundreds of fireflies flashing together on a summer's evening. How does that happen? A firefly's decision to flash is thought to depend on the flashing of its neighbors. Depending on the copying rule they're using, this coordination causes the group to synchronize in either a "bursty" or "snappy" fashion. In her book *Patterns of Culture* (1934), the anthropologist Ruth Benedict argued that each part of a social system depends on its other parts in circuitous ways. Not only are such systems nonlinear—the whole is more than the sum of the parts—but the behavior of the parts themselves depends on the behavior of the whole.

Like swarms of fireflies, all human societies are *collective* and *coupled*. Collective, meaning it is our combined behavior that gives rise to society-wide effects. Coupled, in that our perceptions and behavior depend on the perceptions and behavior of others, and on the social and economic structures we collectively build. As consumers, we note a shortage of toilet paper at the supermarket, so we hoard it, and then milk, eggs, and flour, too. We see our neighbors wearing masks, so put on a mask as well. Traders in markets panic upon perceiving a downward trend, follow the herd and, to echo Márquez, end up causing the precipitous drop they fear.

These examples capture how the collective results of our actions feed back, in both virtuous and vicious circles, to affect the system in its entirety—reinforcing or changing the patterns we initially perceived, often in nonobvious ways. For instance, some coronavirus contact-tracing apps can inform users of the locations of infected persons so they can be avoided. This kind of coupling between local behavior and society-wide information is appealing because it seems to simplify decision-making for busy individuals. Yet we know from many years of work on swarming and synchronicity—think of the flashing fireflies—that the dynamics of coupled systems can be surprising.

A recent study in *Nature Physics* found transitions to orderly states such as schooling in fish (all fish swimming in the same direction) can be caused, paradoxically, by randomness, or "noise" feeding back on itself. That is, a misalignment among the fish causes further misalignment, eventually inducing a transition to schooling. Most of us wouldn't guess noise can produce predictable behavior. The result invites us to consider how technology such as contact-tracing apps, although informing us locally, might negatively impact our collective movement. If each of us changes our behavior to avoid the infected, we might generate a collective pattern we had aimed to avoid: higher levels of interaction between the infected and susceptible, or high levels of interaction among the asymptomatic.

Complex systems also suffer from a special vulnerability to events that don't follow a normal distribution or "bell curve." When events are distributed normally, most outcomes are familiar and don't seem particularly striking. Height is a good example: it's pretty unusual for a man to be over seven feet tall; most adults are between five and six feet, and there is no known person over nine feet tall. But in collective settings where contagion shapes behavior—a run on the banks, a scramble to buy toilet paper—the probability distributions for possible events are often heavy-tailed. There is a much higher probability of extreme events, such as a stock market crash or a massive surge in infections. These events are still unlikely, but they occur more frequently and are larger than would be expected under normal distributions.

What's more, once a rare but hugely significant "tail" event takes place, this raises the probability of further tail events. We might call them *second-order tail events*; they include stock market gyrations after a big fall, and earthquake aftershocks. The initial probability of second-order tail events is so tiny it's almost impossible to calculate—but once a first-order tail event occurs, the rules change, and the probability of a second-order tail event increases.

Learning changes an agent's behavior. This in turn changes the behavior of the system.

The dynamics of tail events are complicated by the fact they result from cascades of other unlikely events. When COVID-19 first struck, the stock market suffered stunning losses followed by an equally stunning recovery. Some of these dynamics are potentially attributable to former sports bettors, with no sports to bet on, entering the market as speculators rather than investors. The arrival of these new players might have increased inefficiencies, and allowed savvy long-term investors to gain an edge over bettors with different goals. In a different context, we might eventually see the explosive growth of Black Lives Matter protests in 2020 as an example of a third-order tail event: a "black swan," precipitated by the killing of George Floyd, but primed by a virus that disproportionately affected the Black community in the United States, a recession, a lockdown, and widespread frustration with a void of political leadership. The statistician and former financier Nassim Nicholas Taleb has argued that black swans can have a disproportionate role in how history plays out—perhaps in part because of their magnitude, and in part because their improbability means we are rarely prepared to handle them.

One reason a first-order tail event can induce further tail events is that it changes the perceived costs of our actions, and change the rules that we play by. This game-change is an example of another key complex-systems concept: *nonstationarity*. A second, canonical example of nonstationarity is adaptation, as illustrated by the arms race involved in the co-evolution of hosts and parasites. Like the Red Queen and

Alice in *Alice in Wonderland*, parasite and host each has to "run" faster, just to keep up with the novel solutions the other one presents as they battle it out in evolutionary time.

Learning changes an agent's behavior, which in turn changes the behavior of the system. Take a firm that fudges its numbers on quarterly earnings reports, or a high-school student who spends all her time studying specifically for a college-entrance exam rather than developing the analytical skills the test is supposed to be measuring. In these examples, a metric is introduced as a proxy for ability. Individuals in the system come to associate performance on these metrics with shareholder happiness or getting into college. As this happens, the metric becomes a target to be gamed, and as such ceases to be an objective measure of what it is purporting to assess. This is known as Goodhart's Law, summarized by the business adage: "The worst thing that can happen to you is to meet your targets."

Another type of nonstationarity relates to a concept we call *information flux*. The system might not be changing, but the amount of information we have about it is. While learning concerns the way we use the information available, information flux relates to the quality of the data we use to learn. At the beginning of the pandemic, for example, there was a dramatic range of estimates of the asymptomatic transmission rate. This variation partly came from learning how to make a good model of the COVID-19 contagion, but it was *also* due to information flux caused by the fact that viruses spread, and so early on only a small number of people are infected. This makes for sparse data on the numbers of asymptomatic and symptomatic individuals, not to mention the number of people exposed. Early on, noise in the data tends to overwhelm the signal, making learning very difficult indeed.

These forms of nonstationarity mean biological and social systems will be "out of equilibrium," as it's known in the physics and complex-systems literature. One of the biggest hazards of living in an out-of-equilibrium system is that even interventions informed by data and modeling can have unintended consequences. Consider government efforts to enforce social distancing to flatten the COVID-19 infection curve. Although social distancing has been crucial in slowing the infection rate and helping to avoid overwhelming hospitals, the strategy has created a slew of second- and third-order biological, sociological, and economic effects. Among them are massive unemployment, lost profit, market instability, mental health issues, increase in domestic violence, social shaming, neglect of other urgent problems such as climate change, and, perhaps most importantly, second-order interventions such as the reserve banks injecting liquidity into the markets, governments passing massive stimulus bills to shore up economies, and possible changes to privacy laws to accommodate the need to enforce social distancing and perform contact-tracing.

-554- Do the properties of complex systems mean prediction and control are hopeless enterprises? They certainly make prediction hard, and favor scenario planning for multiple eventualities instead of forecasting the most likely ones. But an inability to predict the future doesn't preclude the possibility of security and quality of life. Nature, after all, is full of collective, coupled systems with the same properties of nonlinearity and nonstationarity. We should therefore look to the way biological systems cope, adapt, and even thrive under such conditions.

Before we turn to nature, a few remarks about human engineering. Our species has been attempting to engineer social and ecological outcomes since the onset of cultural history. That can work well when the engineering is iterative, "bottom up," and takes place over a long time. But many such interventions have been impotent or, worse, disastrous, as discussed by the anthropologist Steve Lansing in his book *Priests and Programmers: Technologies of Power in the Engineered Landscape of Bali* (2007). In one section, Lansing compares the effective, 1,000-year-old local water distribution system in Bali with the one imposed by central government engineers during the twentieth-century green revolution. This top-down approach disrupted the fragile island and its shoreline ecosystems, and undermined collective governance.

Fiascos happen when we use crude data to make qualitative decisions. Other reasons include facile understandings of cause and effect, and the assumption that the past contains the best information about the future. This kind of "backward-looking" prediction, with a narrow focus on the last "bad" event, leaves us vulnerable to perceptual blindness. Take how the US responded to the terrorist attacks of September 11, 2001 by investing heavily in terrorism prevention, at the expense of other problems such as health care, education, and global poverty. Likewise, during the COVID-19 crisis, a deluge of commentators has stressed investment in health care as the key issue. Health care is neglected and important, as the pandemic has made clear—but to put it at the center of our efforts is to again be controlled by the past.

Fans of *The Lord of the Rings* might remember the character Aragorn's plan to draw Sauron's eye to the Black Gate, so that the protagonists Frodo and Sam could slip into Sauron's realm via another route (the lair of a terrifying spider-like monster). The plan relied on Sauron's fear of the past, when Aragorn's ancestor cut the powerful ring at the center of the story from Sauron's finger. The point is that narrow, emotionally laden focus effectively prevents us from perceiving other problems even when they are developing right under our noses. In complex systems, it is critical to build safeguards against this tendency—which, on a light-hearted note, we name Sauron's bias.

There are better ways to make consequential, society-wide decisions. As the mathematician John Allen Paulos remarked about complex systems: "Uncertainty is the only certainty there is. And knowing how to live with insecurity is the only security." Instead of prioritizing outcomes based on the last bad thing that happened—applying laser focus to terrorism or inequality, or putting vast resources into healthcare—we might take inspiration from complex systems in nature and design processes that foster adaptability and robustness for a range of scenarios that could come to pass.

This approach has been called *emergent engineering*. It's profoundly different from traditional engineering, which is dominated by forecasting, trying to control the behavior of a system, and designing it to achieve specific outcomes. By contrast, emergent engineering embraces uncertainty as a fact of life that's potentially constructive.

When applied to society-wide challenges, emergent engineering yields a different kind of problem-solving. Under a policy of constructive uncertainty, for example, individuals might be guaranteed a high minimum quality of life, but wouldn't be guaranteed social structures or institutions in any particular form. Instead, economic, social, and other systems would be designed so that they can switch states fluidly, as context demands. This would require a careful balancing act between questions of what's good and right on the one hand—fairness, equality, equal opportunity—and a commitment to robustness and adaptability on the other. It is a provocative proposal, and experimenting with it, even on a relatively small scale as in healthcare or financial market design, will require wading through a quagmire of philosophical, ethical, and technical issues. Yet nature's success suggests it has potential.

> The complex structure of the human heart is thought to confer robustness . . . by beating to a rhythm that's neither chaotic nor periodic but has a fractal structure.

Consider that the human body is remarkably functional given all that *could* go wrong with its approximately 30 trillion cells (and 38 trillion bacterial cells in the body's microbiome). Nature keeps things working with two broad classes of strategy. The first ensures that a system will continue to function in the face of disturbances or "perturbations;" the second enables a system to reduce uncertainty but allow for change, by letting processes proceed at different timescales.

The first strategy relies on what are known as *robustness mechanisms*. They allow systems to continue to operate smoothly even when perturbations damage key components. For example, gene expression patterns are said to be robust if they do not vary in the face of environmental or genetic perturbations such as mutations. There are many mechanisms that make this invariance possible, and much debate about how they work, but we can simplify here to give the basic idea. One example is *shadow enhancers*: partially redundant DNA sequences that regulate genes and work together to keep gene expression stable when a mutation occurs. Another example is *gene duplication*, in which genes have a backup copy with partial functional overlap. This redundancy can allow the duplicate to compensate if the original gene is damaged.

Robustness mechanisms can be challenging to build in both natural and engineered systems, because their utility isn't obvious until something goes wrong. They require anticipating the character of rare but damaging perturbations. Nature nonetheless has discovered a rich repertoire of robustness mechanisms. Reconciliation—making up after fights and restoring relationships to a preconflict baseline—isn't just a human invention. It's common throughout the animal kingdom and has been observed in many different species. In a different context, the complex structure of the human heart is thought to confer robustness to perturbations at a wide range of scales by beating to a rhythm that is neither chaotic nor periodic but has a fractal structure. Robust design, in contrast to typical approaches in engineering, focuses on discovering mechanisms that maintain functionality under changing or uncertain environments.

Nature has another set of tricks up her sleeve. The timescales on which a system's processes run have critical consequences for its ability to predict and adapt to the future. Prediction is easier when things change slowly—but if things change too slowly, it becomes hard to innovate and respond to change. To solve this paradox, nature builds systems that operate on multiple timescales. Genes change relatively slowly but gene expression is fast. The outcomes of fights in a monkey group change daily but their power structure takes months or years to change. Fast timescales—monkey fights—have more uncertainty, and consequently provide a mechanism for social mobility. Meanwhile, slow timescales—power structures—provide consistency and predictability, allowing individuals to figure out the regularities and develop appropriate strategies.

The degree of timescale separation between fast and slow dynamics matters, too. If there's a big separation and the power structure changes very slowly, no amount of fight-winning will get a young monkey to the top—even if that monkey, as it gained experience, became a really gifted fighter. A big separation means it will take a long time for "real" information at the individual level—e.g., that the young monkey has become a good fighter—to be reflected in the power structure. Hence, if the power structure changes too slowly, although it might guard against meaningless changes at the individual level, it won't be informative about regularities—about who can actually successfully use force when things, such as the ability of our young monkey, really do change.

Furthermore, sometimes the environment requires the system as a whole to innovate, but sometimes it demands quiescence. That means there's a benefit to being able to adjust the *degree* of timescale separation between the fast and slow processes, depending on whether it's useful for a change at the "bottom" to be felt at the "top." These points circle us back to our earlier remarks about nonstationarity—the degree of timescale separation is a way of balancing trade-offs caused by different types of nonstationarity in the system.

The detailed mechanisms by which nature accomplishes timescale separation are still largely unknown and an active area of scientific investigation. However, humans can still take inspiration from the timescale-separation idea. When we design systems of the future, we could build in mechanisms that enable users—such as market engineers and policymakers—to tune the degree of timescale separation or coupling between individual behavior on the one hand, and institutions or aggregate variables such as stock returns or time in elected office on the other. We have crude versions of this already. Financial markets are vulnerable to crashes because of an inherent lack of timescale separation between trading and stock market indices, such that it's possible in periods of panic-selling for an index to lose substantial value in a matter of hours. In recognition of this property, market engineers introduced what's called a "circuit breaker"—a rule for pausing trading when signs of a massive drop are detected. The circuit breaker doesn't really tune the separation between trades and index performance, though. It simply halts trading when a crash seems likely. A more explicit tuning approach would be to slow down trading during dangerous periods by limiting the magnitude or frequency of trades in a given window, and to allow trading to proceed at will when the environment is more predictable. There are many possible alternative tuning mechanisms; which is best suited to markets is ultimately an empirical question.

Stock market crashes are a bridge to another of nature's fascinating properties: the presence of tipping points or *critical points*, as they're called in physics. When a system "sits" near a critical point, a small shock can cause a big shift. Sometimes, this means a shift into a new state—a group of fish shoaling (weakly aligned) detects a shark (the shock) and switches to a school formation (highly aligned), which is good for speedy swimming and confusing the predator. These tipping points are often presented in popular articles as something to avoid, for example, when it comes to climate change. But, in fact, as the shark example illustrates, sitting near a critical point can allow a system to adapt appropriately if the environment changes.

As with timescale separation, tipping points can be useful design features—if distance from them can be modulated. For example, in a recent study of a large, captive monkey society it was found the social system was near a critical point such that a small rise in agitation—perhaps caused by a hot afternoon—could set off a cascade of aggression that would nudge the group from a peaceful state into one in which everyone is fighting. In this group there happened to be powerful individuals who policed conflict, breaking up fights impartially. By increasing or decreasing their frequency of intervention, these individuals could be tuning the group's sensitivity to perturbations—how far the aggression cascades travel—and thereby tuning distance from the critical point.

We still don't know how widespread this sort of tuning is in biological systems. But, like degree of timescale separation, it's something we can build into human systems to make them more fluid and adaptive—and therefore better able to respond to volatility and shocks. In the case of health care, that might mean having the financial and technological capacity to build and dismantle temporary treatment facilities at a moment's notice, perhaps using 3D-printed equipment and biodegradable or reusable materials. In the economy, market corrections that burst bubbles before they get too large serve this function to some extent—they dissipate energy that has built up within the system, but keep the cascade small enough that the market isn't forced into a crash.

Climate-change activists warning about tipping points are right to worry. Problems arise when the distance from the critical point can't be tuned, when individuals make errors (such as incorrectly thinking a shark is present), and when there's no resilience in the system—that is, no way back to an adaptive state after a system has been disturbed. Irreversible perturbations can lead to complete reconfigurations or total system failure. Reconfiguration might be necessary if the environment has changed, but it will likely involve a costly transition, in that the system will need time and resources to find satisfactory solutions to the new environment. When the world is moderately or very noisy—filled with random, uninformative events—sensitivity to perturbations is dangerous. But it's useful when a strategic shift is warranted (e.g., a predator appears) or when the environment is fundamentally changing and the old tactics simply won't do.

One of the many challenges in designing systems that flourish under uncertainty is how to improve the quality of information available in the system. We are not perfect information processors. We make mistakes and have a partial, incomplete view of the world. The same is true of markets, as the investor Bill Miller has pointed out. This lack of individual omniscience can have positive and negative effects. From the system's point of view, many windows on the world affords multiple independent (or semi-independent) assessments of the environment that provide a form of "collective intelligence." However, each individual would also like a complete view, and so is motivated to copy, share and steal information from others. Copying and observation can facilitate individual learning, but at the same time tends to reduce the independence and diversity that's valuable for the group as a whole. A commonly cited example is the so-called herd mentality of traders who, in seeing others sell, panic and sell their own shares.

For emergent engineering to succeed, we need to develop a better understanding of what makes a group intelligent. What we do know is there seem to be two phases or parts of the process—the accumulation phase, in which individuals collect information about how the world works, and the aggregation phase, in which that information is pooled. We also know that if individuals are bad at collecting good information—if they misinterpret data due to their own biases or are overconfident in their assessments—an aggregation mechanism can compensate.

One example of an aggregation mechanism is the PageRank algorithm used early on in Google searches. PageRank worked by giving more weight to those pages that have many incoming connections from other webpages. Another kind of aggregation mechanism might discount votes of individuals who are prone to come to the same conclusion because they use the same reasoning process, thereby undermining diversity. Or take the US electoral college, which was originally conceived to "correct" the popular vote so that population-dense areas didn't entirely control election outcomes. If, on the other hand, implementing or identifying good aggregation mechanisms is hard—there are, for example, many good arguments against the electoral college—it might be possible to compensate by investing in improving the information-accumulation capacity of individuals. That way, common cognitive biases such as overconfidence, anchoring, and loss-aversion are less likely at first instance. That said, in thinking through how to design aggregation algorithms that optimize for collective intelligence, ethical issues concerning privacy and fairness also present themselves.

We are not perfect information processors. We make mistakes. The same is true of markets.

Rather than attempt to precisely predict the future, we have tried to make the case for designing systems that favor robustness and adaptability—systems that can be creative and responsive when faced with an array of possible scenarios. The COVID-19 pandemic provides an unprecedented opportunity to begin to think through how we might harness collective behavior and uncertainty to shape a better future for us all. The most important term in this essay is not "chaotic," "complex," "black swan," "nonequilibrium," or "second-order effect." It's: "dawn."

Part Three: The Complexity of Crisis

SYMPOSIUM INTRODUCTION

APPLIED COMPLEXITY IN THE TURBULENT DISEQUILIBRIUM

DATE: *20 September 2021*

FROM: *William Tracy, Santa Fe Institute*
Casey Cox, Santa Fe Institute

In 2020, nearly 200 members of SFI's research and practitioner communities gathered for the 27th Annual Applied Complexity Network (ACtioN) Symposium. The topic of this year's conference was "The Complexity of Crisis," and the meeting explored the nature and interdependence of many concomitant crises afflicting humanity. The intellectual underpinning of the meeting was inspired by an article by David Krakauer and Geoffrey West, which had recently been published in *Nautilus* on "The Damage We're Not Attending To."[1] Due to the pandemic, this year's symposium was conducted virtually on November 13 and 14. This section is a continuation of our ongoing *Dialogues of the Applied Complexity Network* series, which documents some of discussions between members of SFI's scholarly and practitioner communities.[2] This includes the symposium's four talks, one panel, and subsequent full-group discussions. Some of the conversations are followed by reflections written by speakers and panelists with the benefit of six months' hindsight.

1 See E-001 on page 541.

2 Through the Applied Complexity Network (ACtioN) firms, governments, and non-profits partner with SFI to effectively explore cutting-edge applications of complexity science. This network provides member organizations with new insights for addressing their challenges, and provides SFI scientists with valuable observations and anecdotes from the world of practice.

OPPOSITE: *Vincent van Gogh, "Coal Shoveler," 1879*

Early in 2021, W. Brian Arthur spoke at a different ACtioN meeting about his recently published review of complexity economics.[3] Brian's comments provide a useful lens for understanding the 2020 symposium. When describing the results from evolutionary computational models of game-theoretic interactions, Brian noted that these models oscillate between "periods of stasis followed by ones of turbulent change." In this context, "stasis" is relative; Brian broadly emphasized the constant dynamism of complex economies. However, the extent of change during these periods of near stasis paled in comparison to the turbulence of the disequilibrium.

The transition from a relatively static state to a turbulent disequilibrium aptly characterizes humanity's shared experiences in 2020. In the computational models, this transition is often sparked by the chance mutation of a strategy. To produce the disequilibrium, that chance mutation exploits large pockets of strategic drift, which occurred over a longer period of the time but previously remained phenotypically dormant.[4] Similarly, the COVID-19 crisis sparked 2020's initial transition, but the ensuing turbulent disequilibrium ignited knock-on crises by exciting heretofore dormant social pressures that had been building for some time. Manifestations of these knock-on crises include: growing social unrest, surging hate crimes, surging murder rates, surging domestic violence, diverging beliefs about basic ground truths, broken supply chains, and an abrupt transition from high unemployment to labor shortages. The network-based processes by which crises can be transmitted between interdependent systems are an active area of complexity research. Raissa D'Souza's 2015 *Interface* paper provides an excellent example of this work.[5]

The pandemic was the most obvious initial crisis for the 2020 ACtioN Symposium; it had already ignited multiple knock-on crises in interdependent complex systems. Lauren Ancel Meyers thus presented

---

3 W.B. Arthur, 2021, "Foundations of Complexity Economics," *Nature Reviews Physics* 3(2): 136–145.

4 W.M. Tracy, 2014, "Paradox Lost: The Evolution of Strategies in Selten's Chain Store Game," *Computational Economics* 43(1): 83–103.

5 C.D. Brummitt, G. Barnett, and R. M. D'Souza, 2015, "Coupled Catastrophes: Sudden Shifts Cascade and Hop among Interdependent Systems," *Journal of The Royal Society Interface* 12(112): 20150712.

her work modeling the spread of SARS-CoV-2. Lauren began pioneering the use of networks to model the spread of communicable diseases when she was an SFI postdoctoral fellow.[6] She played a leading role as that field matured, and co-founded the UT COVID-19 Modeling Consortium[7] at the start of the pandemic. Her consortium has become a crucial voice in helping governments understand and respond to the pandemic. As such, Lauren's symposium talk provided unique insight into how disease-transmission models can inform policy trade-offs.

> The transition from a relatively static state to a turbulent disequilibrium aptly characterizes humanity's shared experiences in 2020.

Climate change is another crisis capable of triggering knock-on crises. In the run-up to the 2020 symposium, the climate crisis seemed less acute than COVID. That distinction is no longer so clear cut. In the ten months since symposium, unprecedented wild fires, droughts, and floods have applied increasing pressure to coupled social and ecological systems. Marten Scheffer's symposium presentation offered a compelling, albeit frightening, perspective on mechanisms by which climate change might trigger other social crises.

As the pandemic and climate change continued to stress human systems, the misinformation crisis has worsened. Misinformation also feeds back to exacerbate the crises of the pandemic and climate change, via mechanisms such as vaccine skepticism and climate-change denial, respectively. In her symposium talk, journalist Gillian Tett brought an anthropological perspective to social mechanisms that incentivize and operationalize opacity in some in some professions (such as the finance industry), and other professions (such as the media) ignore this

6 L.A. Meyers, M. E. J. Newman, et al., 2003, "Applying Network Theory to Epidemics: Control Measures for Mycoplasma Pneumoniae Outbreaks," *Emerging Infectious Diseases* 9(2): 204.

7 https://covid-19.tacc.utexas.edu/

opacity. In concert, these mechanisms can produce widespread blind spots within professional and social communities. This analysis is complementary to the growing body of work by SFI researchers on belief dynamics. An article that Mirta Galesic and several SFI collaborators recently published in the *Journal of the Royal Society Interface* provides a good example of contemporary work in this area, which blends ideas from psychology, anthropology, math, computer science, political science, and other fields to understand how beliefs spread.[8]

Changing institutions is hard. Predicting how actions taken during this turbulent disequilibrium will ultimately impact humanity's next relatively stable state is even harder. Insights from complexity science can help.

Vladimir Ilyich Lenin once quipped, "there are decades when nothing happens; and there are weeks when decades happen." A key insight of complexity economics is that events and decisions made in the turbulent disequilibrium impact the state around which the system next achieves a degree of stasis. More bluntly, *these* are weeks in which decades can happen. It was in this spirit of optimism that the final symposium speaker, Suresh Naidu, was asked to explore how shocks and crises can push institutions out of relatively stable states and lead to something new. Specific examples Suresh examined included the establishment of democracy, the Wagner Act, and the 1927 flooding of the Mississippi River.

Changing institutions is hard. Predicting how actions taken during this turbulent disequilibrium will ultimately impact humanity's next

8 M. Galesic, H. Olsson, et al., 2021, "Integrating Social and Cognitive Aspects of Belief Dynamics: Towards a Unifying Framework," *Journal of the Royal Society Interface* 18(176): 20200857.

relatively stable state is even harder. Insights from complexity science can help. At a minimum, complexity can disabuse practitioners and decision-makers of overconfidences born from linear mental models. Yet there is also scope for complexity to make prescriptive contributions to discussions of what might emerge from this current disequilibrium of concomitant crises.

SYMPOSIUM TALK NO. 001

FROM TUNNEL VISION TO LATERAL VISION—IN THE MEDIA AND ELSEWHERE

DATE: *13 Nov. 2020* **FROM:** *Gillian Tett, Financial Times*

WILLIAM TRACY: Gillian Tett is chair of the editorial board and editor-at-large, US, of the *Financial Times*. Her writing often brings a complexity perspective to a host of real-world problems. Gillian previously spoke in an Applied Complexity Network (ACtioN) meeting for us in October 2016, at which time we were discussing digital tribes holding nonoverlapping truths, and how that might impact the ability of pollsters to collect accurate data.

Gillian said, and this is an actual quote, "My hunch is that more people are going to vote for Trump than are willing to admit it." And remember, this was October of 2016. Gillian's comments were one of the first things I thought of when I heard the 2016 election results. While preparing for this intro, I also read a 2018 *Financial Times* article by Gillian, in which she argued that the risk of pandemic was far greater than most people realized.

Now, I'm not claiming that Gillian can predict the future, but Gillian is uniquely insightful in a very complex-systems way. And so, let me turn it over to you, Gillian.

GILLIAN TETT: Thank you very much indeed. And I should start by saying that it's a great honor to be asked to talk to you all because I am these days just working as a humble journalist scribbling the first draft of history with all the inevitable mistakes that that entails. However, I should also explain, for those of you who don't know, I was actually trained as a cultural anthropologist.

OPPOSITE: *Uwe Kils, "Iceberg" photomontage, (2005)*

I have a PhD in cultural anthropology based on fieldwork around Afghanistan back in the late 1980s/early 1990s, which informs everything I do. Because although my PhD topic was looking technically at marriage rituals and Islamic identity and communism, the skills I learned there absolutely inform the way I look at the financial system today, the political system, and so much else about our political economy.

I'm actually in the middle of finishing off a book right now about the power of anthropology to try and explain the world[1]. I think the introduction for the draft version of the book is actually in one of the things you are being sent right now, and I'm very hungry to have any feedback or ideas from people.

One way to understand the rise of the sustainability movement—the environmental, social, and governance movement—is the fact that many people instinctively recognize they need to start looking laterally, not through tunnels, and that really is part of the shift from shareholder value to stakeholder value.

I have three key messages to impart today before I go into what I'm going to talk about narrative-wise, which really play into this question of, to be honest, Why did I foresee the rise of Donald Trump? I wrote a number of columns in 2016 saying that Trump was very likely to win. I did write about pandemics in 2018. I also had a sense of the financial crisis brewing before 2008. And I think that's got very little to do with my personal ability to see things or not see things—I've got plenty of things wrong in my life as well—but it really reflects a very simple point. That is a top point I want to make, which is that we need to move from

1 This book is now published: G. Tett, 2021, *Anthro-Vision: A New Way to See in Business and Life*, New York, NY: Avid Reader Press.

a world of tunnel vision to what I call lateral vision. And I happen to think that anthropology is a truly brilliant training to do that for anybody, because it teaches you to try and look at the world in a holistic way, and above all else to look at things which we don't talk about as much as we do.

The second point arising from that is that in order to try and move from tunnel vision to lateral vision, I think we need to think much more clearly about our information flows and how we present and package information, not just right now in the world of epidemiology but also in computing, finance, etc. And that plays directly into my day job that I now hold that I've been working in for the last twenty-five years—which makes me sound truly ancient, as my teenage daughters keep telling me—which is the media. And I'm going to talk about that in just a moment.

And the last point that I want to convey is that, in some ways, a growing number of people are starting to recognize this point about a need for lateral vision and the limitations of tunnel vision and the way that tunnel vision is creating all kinds of risks, particularly in that infamous world of volatility, uncertainty, complexity, and ambiguity—VUCA, to use that military analogy.

One way to understand the rise of the sustainability movement—the environmental, social, and governance movement—is the fact that many people instinctively recognize they need to start looking laterally, not through tunnels, and that really is part of the shift from shareholder value to stakeholder value. Another way of saying it is that stakeholder-isms, stakeholder value, stakeholder debate is another way of getting an anthropological vision in action, just to talk about my own book.

Let me give you a quick background, a bit of history about these issues about information flows. I'm going to beg your forgiveness for a second by going back in what now feels like ancient history in a world where we're living from moment to moment, which is 2004. In those days I was running something called the Lex column at the *Financial Times*. That's a section of the *FT* that comments on corporate finance matters.

One day my then-editor said, "Can you write a memo explaining what you think that Lex, the newspaper, should be covering in finance?"

So I went off and did the usual memo, which was basically five pieces a week on banking, three pieces a week on industry, whatever.

And then I finally stopped and thought, well, hang on a sec, what would happen if I looked at the City of London—I was based in London at the time—as an anthropologist, and tried to go around the financial village and worked out what was really important for us in the media to cover and what was not. I did that exercise for a few months, very much using exactly the same skills I used in fieldwork in Tajikistan: I walked around, I talked to people endlessly where I could, and I tried to listen to both what they said and what they did not say. Social silences, to use a concept pioneered by the French intellectual Pierre Bourdieu, is absolutely critical in understanding how the world works.

I've actually come to the conclusion that what we had in finance in 2004, and in fact all the way up until the financial crisis, was essentially an iceberg pattern. You had a tiny bit of the financial system, which was above the waters, which was obsessively discussed, and that was the equity markets. And then you had this vast shadowy underbelly, which was incredibly important and growing very fast, which was almost completely ignored, which was the debt derivatives and credit world.

So I went back and said to my colleagues, "I think we are missing the trick. We've got to invert that iceberg as far as the *FT* is concerned." Not because, I should stress, I had any brilliant hunch that the financial crisis was looming at all, but for rather more nakedly ambitious reasons within the *FT*: I thought that the coverage of the equity market was so commoditized by then, we as a newspaper couldn't offer much of value. We should be focusing on the areas that no one else was covering, because I thought that was a way to suck more readers into the *FT*.

One thing led to another, and I ended up getting put in charge of the capital markets team. And I spent the period from 2005 to 2007 trying to build a team to cover this murky aspect of the iceberg. Again, I can't stress this strongly enough: I didn't do that because I thought the world was about to collapse; I wanted to provide a travel guide to the innovation that was happening there because I thought no one else was covering it, and this was a big area of social silence.

I subsequently became convinced that in fact there was huge danger in that and wrote about it, warning about a crisis brewing, but that really came out of my curiosity, to be honest. But the question which really continued to haunt me, and which is incredibly relevant to thinking about where we are now, was why did that financial iceberg exist? Why were people obsessively covering equities and ignoring so much else?

I later wrote a piece for the Banque de France, the French Central Bank, trying to sketch out the issues here, which I think is also in the things that have been sent over. And it's very out of date, but it's very relevant to today, extremely relevant to today.

Because after the financial crisis, there was an overwhelming presumption amongst most journalists that the reason why people hadn't seen it coming and hadn't understood the dangerous present in derivatives, and the complexity of that—to cut into your key theme today—was because everything was being hidden deliberately by those evil bankers. And it was all a great sort of James Bond–style plot to try and conceal the key elements of information.

And I came out of the crisis saying, well, yes, the bankers were not angels. There were certainly people who realized that opacity made their margins fatter and who didn't want to have a lot of attention drawn to it. But actually I think there were other things going on structurally, which was both to do with the cultural dynamics of finance, but also the cultural dynamics of the media. The latter point is the important issue I'll comment on in a moment.

The cultural dynamics of finance in some ways were fairly easy to look at. What you had, using my anthropology background and training, was a group of people who spoke a language, which was utter gobbledygook to everybody else in the planet. This was called the language of CDO [collateralized debt obligations], CDS [credit default swaps], etc.

I remember going to a banking conference in 2005, sitting in a dark conference hall and thinking, "This is as weird as a Tajik wedding." I went to a lot of Tajik wedding rituals when I was an anthropologist. And, by the way, investment banking conferences play exactly the same functional role for a financial community as a wedding in Tajikistan.

It's a way of getting together a scattered tribe, pulling them together through rituals and all kinds of symbols to not merely, in the case of a wedding, help pave the way to physical reproduction, i.e., babies, but social and cultural reproduction. Because the sheer act of getting together and engaging in rituals reinforces social ties and invokes all kinds of symbols, which actually reflect the worldview of financiers.

And the world view of financiers that I saw was very much about this elite tribe who had mastery of this gobbledygook jargon language that no one else understood of extreme complexity. They use complexity as a weapon to basically make themselves feel like an elite tribe cut apart from others who nobody else was expected to peer into and have a look at.

And they had a very distinctive creation mythology, which is all about the theory of liquification: that these complex financial instruments were going to create the perfect liquid market that would benefit everyone, which today looks like a completely insane thing to argue given the financial crisis. At the time, I can't stress this strongly enough, they told themselves that they believed it. They may not have completely believed it, but it was very convenient to think that they believed it. So the issue is partly that the bankers were very elite, tribal, fragmented. Most people inside banks had no idea what they were up to, which is the ever-present problem of technological complexity.

But the problem was also that the media and politicians thought that precisely because they'd wrapped their craft in technological complexity and jargon, they didn't really have much incentive to look into it either. And the reason is very simple: If you want to hide something in broad daylight today, then the easiest way to do it is not to create a James Bond–style plot and bury it in the ground somewhere; it's to simply wrap it up in acronyms and jargon and mathematics, and then it will be hidden in plain sight by virtue of the fact that everyone else will look away because we culturally label that as boring, geeky, and dull. That is absolutely what was going on in the financial system before 2007.

Within the rubric or the cultural traditions of the journalistic guild, if you go to journalism school, you are trained that what defines a story that you put on the front page of a newspaper or anything else is something which has a person; a sudden dramatic event—if it's a bad event, that's even better, because bad news sells better than good news; you have tangible facts and figures—things you can quote; and ideally you have images and metaphors that you can quickly use to connect to an audience. And, although people don't actually say this, you definitely don't want to have jargon, maths, acronyms, algorithms, or anything that smacks of extreme complexity and a special technological language.

> If you want to hide something in broad daylight today, then the easiest way to do it is not to create a James Bond–style plot and bury it in the ground somewhere; it's to simply wrap it up in acronyms and jargon and mathematics, and then it will be hidden in plain sight . . .

In retrospect, the reason why the equity markets were being covered a lot before 2007 and then debt and derivatives were being almost completely ignored was because almost all of the elements you need to make a story were present in equities in the sense that you had these colorful CEOs—who, by the way, before 2007 often were rock-star CEOs. These days they're much more boring. But you had colorful rock-star CEOs who were forever doing things and saying things that you could quote, and the share prices moved in a way so that you could see the numbers and they tended to be dramatic step-change events. What was happening in the debt and derivatives world, by contrast, were these very slow-moving elliptical changes, which were incredibly hard to get any numbers attached to because much of it was murky and opaque.

The people who were involved in it tended to be very, very media shy. So there weren't a lot of faces anywhere in the picture. And although the whole point of finance was supposedly to serve real people, at the end of the day, one of the other things that happened to bankers as they became completely enraptured by models and computers and algorithms is that they engaged in what some anthropologists call moral disengagement, which is that they kind of forgot that there were human beings at the other end of the chain. They never talked about humans. There were no faces in the world of debt and derivatives at all in terms of who the end users were. So there weren't many humans in the story of debt and derivatives either. And then of course there was this kiss of death, which was algorithms and acronyms.

I remember very clearly going to my editors at the *FT* because I was, for two years on a kind of one-band brigade desperately trying to get this stuff onto the front page of the *FT* and trying to do everything I could to make it sound sexy and exciting and interesting and grab attention. And I was told over and over again that any story which came labeled CPDO or CDS or CDO cubed wouldn't get on the front page.

I wasn't the only person to see this. And I will come to what this means in medicine and computing in a second. I wasn't the only person to see this. Paul Tucker, who was the deputy governor of the Bank of England at the time, was also, like me, trying to ring alarm bells. And he, like me, faced a hellish job in trying to do this, because guess what? Inside the Bank of England, like other central banks, there was a similar kind of sense of status.

I should say that the capital markets team within the *FT* was very low status compared to the high-status economics team. The economics team sat next to the editor's office overlooking the river. I sat in the basement overlooking the trash cans, and we almost never actually interacted. And it was exactly the same in the Bank of England. The people looking at the financial weeds were lower status, and they sat very much out of public sight and public eye.

So Paul Tucker used to try to invent all these phrases and words to make this stuff sound exciting. I mean, we literally sat there and we tossed ideas around, and he gave a speech at one point about saying, "Let's call this stuff or these securitized products the Russian doll financing, or let's call it vehicular finance."

It didn't work at all. None of those phrases caught on, and so you had this communication gap and no one really started to pay attention to it until two things happened. First, there was obviously and visibly the financial crisis, and everybody almost overnight went from having no interest to having a kind of yikes moment where suddenly they woke up and realized what they'd been missing, what they thought the bankers had been hiding, but it would actually have been hidden in plain sight.

The other very important thing is that in August 2007, an executive from PIMCO stood up on a stage at Jackson Hole and used a phrase for the first time, which no one had ever heard of but which immediately caught on, which was "shadow banking." And at a stroke, the world had a way to visualize and imagine what was happening in the financial system. It wasn't acronyms anymore. It was called shadow banking.

They had what was basically a Copernican mental revolution. They went from thinking that all this stuff that was outside the banking system was somehow a derivative of finance and could be chucked into the margin, to realizing it was actually in the center of finance and driving the financial crisis. Shadow banking was essentially not so much the tail wagging the dog, it was the dog at the time. And suddenly the debate changed, and you began to have the modicum of a sensible regulatory conversation and a sensible public policy debate. Still a long way to go. I'm not going to talk about that now.

I'll illustrate that to explain to you how the way we imagine and conceive things, and the way the media as a cultural entity reflects those patterns, matters enormously to what risks we perceive, how we actually understand the world around us, but also how complexity as both an issue and a reflection of what's going on actually plays into where we are today.

That's the history. Ironically, after the crisis, the financial crisis, I wrote a book capturing some of this stuff. In fact, I wrote two books about this stuff. I was asked a lot in audiences what I thought the next areas were of extreme complexity where these patterns could play out next. And I sort of said two areas.

One, I thought, was in the whole computing sphere—where, again, you have extreme information asymmetries which really matter—originally around ad tech. I was obsessed with ad tech for a while because I could see how a lot of the old derivatives traders I'd known were starting to move into ad tech, which was another area of extreme information, asymmetries, complexity, where, frankly, until the Cambridge Analytica scandal exploded, everybody was averting their eyes away from ad tech too.

There were very few journalistic articles before Cambridge Analytica looking at what was happening in ad tech; far less asking questions about what that meant for democracy and politics going forward. We now realize that was a mistake as well. I thought that cybercrime, cybersecurity was another area which was going to be very important.

But the other area that concerned me was the world of medicine and pharmaceuticals, because I could see once again that you've got these extreme information asymmetries. You've got a small group of people who understand what's going on. You have, once again, profit-seeking companies like banks, i.e., pharmaceutical companies, who are in control of that information and a world where everyone depends on them but are only too willing to avert their eyes from the complexity of it all. And then of course you have the problems of globalization and the fact that, not only do regulators find it very hard to track what's happening across borders and government officials, but also journalists do as well, let alone ordinary mortals.

The thing that in retrospect is shocking is that scientists and epidemiologists have been warning for a very long time that we were careering toward some kind of zoonotic infectious disease crisis.

So, I certainly didn't predict that this kind of COVID-19 crisis was coming, but I've been concerned and I am still concerned about the question of how you cover it. Because when the crisis first started to happen, when the COVID issue started to crop up early in 2020, two or three things happened. First, there was an inevitable response of assuming that what was happening in Wuhan was a question of something happening way out there, kind of a whole bunch of slightly weird people who we didn't understand doing stuff that we could kind of ignore.

I spent a lot of time talking to people who were involved in the Ebola crisis—partly because it's quite an important part of my book—and it's fascinating because in many ways that was the original response to the Ebola crisis in West Africa: weird people in what Donald Trump has infamously called shithole countries doing stuff that we can't really relate to or empathize with, but, hey, we're kind of going to pay attention to it but not really, and ignore it.

On top of that, of course you had this problem of complexity and the fact that people are hardwired to avert their eyes from things with algorithms, scientific equations, or anything else. The thing that in retrospect is shocking is that scientists and epidemiologists have been warning for a very long time that we were careering toward some kind of zoonotic infectious disease crisis.

I mean, that is all over the literature, whether or not you actually looked at what had happened with Ebola, and Ebola was a very, very near miss. And people like me have been going to Davos [World Economic Forum] and seeing these sessions on infectious diseases and sometimes going to some of them. I wrote a column about it in 2018 because I was so struck by what I heard, but also knowing that these sessions were incredibly thinly attended and had almost no traction in the wider media at all.

Yet in spite of having gone through the financial crisis and learned about the risks of ignoring geeks then, we saw the whole pattern essentially playing out all over again with medicine. And before that with the issue of ad tech too. So that raises the question, What can we, or could we, or should we do? Is there any way to change this? Any way to start having more intelligent, grown-up debates around information flows?

Because when I look to the future, I see a world where there are all manner of other issues coming down the tracks, which raise even more alarming issues to do with complexity potentially. I mean, right now I am particularly obsessed with the role of AI—and not just machine learning but deep learning—in finance, which I think has barely got any attention whatsoever.

Ant Financial[2] has just revealed the tip of a massive iceberg that's coming down the tracks to hit us. Which once again is swathed in complexity, algorithms, acronyms are being ignored. So what can we do? Well, a couple of things to mention. On the positive side, as I said earlier, I actually think many people I speak to are instinctively aware of this problem of tunnels and this problem that the media is not well placed to try and break down tunnels and silos of information. In fact, they tend to make it worse.

---

2 Ant Group, formerly known as Ant Financial and Alipay, is an affiliate company of the Chinese Alibaba Group.

I think that many people have realized that they need to basically take a wider view, not just of information but of the actions and of consequences. And about two or three years ago, I was very struck by the degree to which I was talking to corporate executives and financiers and they kept talking to me about ESG, environmental, social, and governance, and sustainability.

Like most journalists, I initially assumed that this was complete greenwashing, whitewashing, woke-washing baloney PR spin. And so I, like most journalists, completely ignored it for the first year. In fact, I used to delete all these emails almost on autopilot because I was so distracted by Donald Trump, and because journalists are hardwired to be cynical and assume that this is just all PR BS.

Then one day I stopped and said to myself, "Hang on a sec. I'm just ignoring this stuff that people want to talk about. Once again, I'm being a journalist and forgetting my anthropology, and I'm not listening to social silences. I'm basically imposing my own worldview on everyone else. I should at least listen to why people are talking so much about ESG and sustainability." And I did.

I came to the conclusion that what was going on was really a shift in mindset amongst, not just corporate boards and C-suites but also financiers and policymakers . . . In the corporate world, it was about moving toward a sense of purpose for companies. In the financial world, it was about moving toward a world where you thought about more than just financial returns, where you thought about the consequences of how you were investing.

I mean, "impact investing" in some ways is a misnomer because all investing has an impact. The question is, Is it a good impact or a bad impact? And do you think about the impact or not? And people who weren't thinking before, they started to do so. And in the policymaking world, they're running out of money, and they needed to basically corral private-sector capital and resources to get development projects done.

So you had these three changes under way, which were colliding, around the world of environmental, social, and governance. But the critical thing was that ESG and sustainability were driven really by three motives. And we were only focusing on one: a tiny minority of people were getting involved in sustainability because they wanted to change the world actively.

That was the impact investors. They were the Danish pension funds and the nuns and a few wealthy trust-fund families. Another group of people were getting involved because they wanted to do no harm to the world, but a bigger group was trying to do no harm to themselves. Because they'd woken up and realized that in a world of growing volatility, uncertainty, and complexity, having seen what happened in the 2008 financial crisis, having lost faith in the idea that you had the end of history, it was actually a mistake to ignore everything that economists used to call externalities, i.e., the stuff they couldn't put in their neat models, like the environment, like income inequality, like supply chains.

And people were waking up and realizing—and this is a key point—that ESG and sustainability was not just about activism. In fact, it mostly wasn't about activism, it was actually about risk management. How do I manage all the risks that are not in the model that I've created with my twentieth-century tunnel vision?

In many ways, ESG is about saying that if we don't recognize lateral vision and the need for it, then we will actually end up damaging ourselves, whether it's through reputation, whether it's through our balance sheet, because we've invested in things like fossil fuels, etc. So I would argue that one way to see the sustainability movement is a societal instinctive recognition that actually we need lateral vision. Hooray. That's a good thing.

The bad side is that obviously it is incredibly difficult to get lateral vision in place for all kinds of reasons. And going back to my day job, the media, in a world where the media is very short of resources, in a world where they, like financiers who are under pressure to provide short-term returns, quarterly earnings, are under pressure to provide short-term hits, to provide catchy news that basically grabs people's attention in a very crowded media marketplace, where everything is about attention grabbing, particularly if you have someone like Donald Trump tweeting. In a world where you basically have information systems incredibly fragmented and tribal. It's very, very hard.

In fact, in many ways, it's becoming harder to either have the luxury of time or the resources to roam and to try and join up the dots and try and dig into complexity and acronyms in a meaningful way. Yes, if there's anyone watching who has a spare billion or two to invest in the media, fabulous, we would love you.

There are areas where you can still be very encouraged. There are sites like *The Conversation*, which are trying to get academic research into the mainstream media world. I love *The Conversation*, but that's only a start. There needs to be a lot more of that. We need to actually change the way that our information flows are constructed as well.

Complexity has put a premium on having a joined-up view of the world and having joined-up policy debates. We desperately need that. Tragically, though, complexity wrapped up in acronyms and algorithms and jargon that journalists and politicians are trained to avert their eyes from makes it harder to have that kind of joined-up intelligent debate.

I hope there's a way through that. I, for my part am desperately trying to find one, the *Financial Times* as a group is trying to find one, but if anyone's worried about what extreme complexity means and the cost of that, if anyone's worried about the implications we've seen from the COVID-19 crisis and what that means about other structural shifts in the planet, they need to think about information flows along with all the other flows that are connecting us in this modern complex world.

So thanks for listening. Very keen to take any questions on anything at all, even Donald Trump. And I'm also very happy to have any feedback on anything I've sent over, because I'm desperately learning and keen to learn from anybody who's willing to give me time to help me understand the world. So thank you.

[CANNED APPLAUSE]

W. TRACY: That is our applause for you. Thank you very much, Gillian.

G. TETT: What is that? Is that canned laughter?

[GILLIAN LAUGHS]

W. TRACY: We are going to roll into Q&A, as Gillian said. Let me start with you, Esther Dyson, if you have a question for Gillian.

ESTHER DYSON:[3] There are all these markets . . . you call them shadow banking, derivatives. To me right now, the influencer market is

3 Wellville

the newest phenomenon, but it reflects the old joke about advertising: "Half of it works and half of it doesn't, but I don't know which half."

It seems like influencer marketing is the epitome of a derivatives market in the sense that so many of these influencers are paid much more than they're ever going to return in profits to whoever's paying them, but each individual payer/buyer is hoping to win. Has anybody actually looked at those numbers?

G. TETT: No, and that's a great point. And I think there is an element of Ponzi scheme, which again echoes a lot of the derivatives world, where because no one really quite understands it and they're a bit nervous if you are actually buying these services, these products.

If you were an asset manager buying a CDO back in 2006, you didn't really want to admit you hadn't got much idea of what you were buying, but you knew you kind of had to buy it because everyone else was doing it, so you didn't question too much. So whoever was selling that product had a license to engage in all kinds of things, and no one could see whether it actually made sense or not.

I think the same thing's true for a lot of the ad tech world. And the reality is there's been very little probing. There's a lot of opacity. Few people have the resources in the media world to actually look hard at it. And there's also a problem with silos because the people who have historically covered advertising as journalists, our so-called marketing teams, they're used to talking about exciting things like Saatchi & Saatchi and the clever images they use or the next Netflix drama. They don't come at it with a background in science and tech and algorithms and all of that. So you have a problem about different buckets of knowledge within the journalism world as well.

KATHERINE COLLINS:[4] Thank you for this. I love how you tie this idea of taking a lateral view to sustainability and then to ESG. As a practitioner in this field, one of my biggest concerns is that we see the same dynamic you described in other areas is starting to take root in sustainable investing.

The ESG machine of acronyms and algorithms is growing perhaps even faster than the substance of sustainability. And so, I'm curious if

4 Putnam Investments & SFI Trustee

you have any advice from the other case studies that you've witnessed and researched as to how to best manage that tension, to keep the machine from taking over the substance.

G. TETT: Well, I used to joke, when I was involved in writing about growth derivatives, that I couldn't imagine a sector that would have more acronyms. And that was before I got into covering sustainability, because sustainability is absolutely drowning in acronyms today, whether it's TCFD, SASB, GRI,[5] you name it. Even ESG.

And I say that because the combination of me trying to take the sector seriously and recognizing that it was an area of great fast growth, was that about a year and a half ago I worked with colleagues at the *FT* to create the first platform in the mainstream media world to cover it called Moral Money. And I called it Moral Money because I like the idea of Adam Smith's theory of moral sentiments, but also because I was desperate to find a name which didn't have an acronym and which will be memorable enough to be catchy, a bit like shadow banking.

It's very imperfect, the word *moral money*. We've been trying to cover it, and as we've tried to cover it, we can see that, like any new era of financial innovation, the sector is marked by opacity, label confusion, fragmentation, rapid innovation, regulators scrambling to catch up, an awful lot of hype. All of which is potentially quite dangerous in terms of creating risks that things will get oversold, etc.

So we're trying to sort of shed light on that and expose it even while recognizing that there are tangible reasons why you need to take a lateral vision of the world, not tunnel vision, like genuine climate change, like rising inequality, which could have explosive connotations. And these are all real reasons why finance and business needs to rethink its modus operandi.

We've been trying to indicate that there's a lot happening, which actually is good. And also show that just as other sectors of finance have "grown up" over time to become, not a cottage industry that's prone to all this kind of potential opacity and fragmentation, but a more grown-up part of finance, a more mature market. That could and should happen

5 TCFD: Task force on Climate-related Financial Disclosures; SASB: Sustainability Accounting Standards Board; GRI: Global Reporting Initiative

with ESG. So that's what we're trying to expose and sort of illuminate. What I find fascinating is that, because the culture of journalism—again, looking as an anthropologist—is that we're trained to be cynical, to look for bad news, to look for step-change events, drama, blah, blah, blah, and ESG isn't that.

But eventually we got the funding, a tiny bit of funding. We have turned into the most successful new product the *FT* has launched for a very long time, and the metrics are off the charts. Not because of anything we'd done pretty well—I mean, I'm full of self-criticism about the moral money platform still—but because there is this explosive demand in the marketplace out there.

The last time I saw anything like this was, I have to say, writing about the securitization world, where actually a lot of people are suddenly saying, "Yikes, I need to understand this. How do I get information?" So that's been an interesting journey, to put it mildly.

I'm always very happy to talk to anyone with my day-job hat on about that because that's my kind of day job. My anthropology stuff is my kind of non-day job, which is what I'm writing my book about.

IAN MCKINNON:[6] I was really struck by your notion of the need to move from tunnel vision to lateral vision. But there are two great progenitors of tunnel vision in society. One of course is the university system, which is supposed to turn out our most brilliant scholars who have very narrowly defined fields in silos and zealously protect them.

And then once in society, all these acronyms you speak of, all the maths and whatnot, the very professionals in those silos use those precisely to keep people out and drive up rents they're able to capture. So you have this interplay, which really does reinforce this notion of tunnel vision. How do you break through those two to get to a more broad view?

G. TETT: It's very difficult. I think there are two options. In terms of the universities, there clearly needs to be a way to get much more interdisciplinary dialogue going. And that sounds like a calling from motherhood and apple pie, it's so obvious. But the structures of the whole tenure-track system and things go in precisely the opposite direction.

---

6 Sandia Holdings, LLC & SFI Trustee

Groups like the Santa Fe Institute are obviously trying to do that, and I salute what you're doing. That needs to happen a lot more. And I think students in particular need to be encouraged when they're at the undergraduate level to have a much more broadly based curriculum and vision of the world. Again, it's obvious, it's very hard to happen.

What can be done . . . ? Well, two or three things. One is to encourage the geeks—the priests—to be forced to periodically translate what they're doing into words that "commoners" might understand.

In terms of how do you get over the fact that it is entirely human nature for elites in any society to use whatever they can to entrench their position, control of language is inevitably part of that. The medieval Catholic church priest used Latin, and the congregation sat there meekly listening to a service they did not understand, and they were happy to accept that because they appeared to be getting the blessings flowing down to them from the priests.

And no one ever challenges the way that priests of any sort use their specialized language until something goes wrong, and then they do. While things are good, the priests can keep using Latin until the cows come home. And that's essentially what's happened in the medical world, and it's happening in Silicon Valley today.

What can be done about that? Well, two or three things. One is to encourage the geeks—the priests—to be forced to periodically translate what they're doing into words that "commoners" might understand. I mean the meaning of "common sense," which was so missing in finance before 2008, is simply making sure that "common people" outside the elite class can actually look in and understand what's happening. So, asking a geek to say in words with no acronyms and no maths and no specialist language what exactly they're doing is a very good mental and cultural exercise to start with.

Another thing people can do is think about what I call the domino effect of communication. And by that I don't mean the idea of dominoes stacked up against each other where you knock one over and the entire lot fall over. I mean, think about a domino and the fact that you have two numbers on it. The domino is played by matching one of those numbers against someone else's number, but the other number is different. I used to run the editorial operations of the *FT* in America, and something I often told people was, when you're writing news stories, think about giving the reader one half of the domino that they might already have that will match up and will make them want to read what you're writing. Because people won't grab on to information unless there's something that clicks in their mind that they recognize, okay? And then in the rest of the article, give them the other half of the domino and take them somewhere else. So if you're talking about epidemiology, by all means pander first into their stereotypes or their fears or what they know about it, and then take them someplace else in the rest of the article.

I think we should be trying to find and train techies, experts, to think about not just translating what they're doing into common language but finding ways to communicate it to the wider world and say things like, "You think *x*. Well, yes, that's kind of true, but actually let me tell you, it's actually *y* and *z* as well." And that helps to expand the pool of common knowledge.

Again, I think platforms like *The Conversation* are absolutely brilliant. I really do. If any of you have not read *The Conversation*, it's a website platform which essentially takes pieces of academic work and tries to popularize it in a way that even journalists can understand in the hope that the journalists will then pick it up and not just quote other journalists or government officials on important topics but actually quote academics too.

I think that kind of initiative should be supported to the max. There are lots of things I would do to try and get it even more in the public stream, but that's the kind of encouraging thing that's going on if you want to understand.

And the last thing I'd say, there's innovation in the media landscape right now. Two other examples I'd highlight before anyone gets too depressed about media. *The Conversation* is a brilliant innovation, and then again if you've got a few spare million, by all means chuck it at *The Conversation*. I see someone said that *Aeon* is another public platform. Great, wonderful. I didn't know about that one. It's clearly not doing its marketing very well, but that's great.

ProPublica. Mainstream publications are running out of money to do investigations, but rather than simply wave your hands and say, "It's hopeless. There's no more investigations going on," there are now new operations springing up that can help do investigations in a nonprofit format. *ProPublica* is an example of a group of journalists who are diving into difficult topics to try and expose them properly.

And the third area of innovation, which again shows how the need for intelligent conversation remains there, but it may be finding new formats, is the revival of podcasts. I mean, who were the guests twenty years ago that voice would be a key form of disseminating brilliant ideas. Radio seems so retro, but actually, against the odds, podcasts have emerged as the ultimate anti-tweet. They're the place where you have intelligent, thoughtful, long-term discussion about key issues, and that's interesting. You're meeting a need there.

DAVID KRAKAUER: Thanks, Gillian. I was going to start my question with the Tridentine Mass—the ecclesiastical Latin mass—but you mentioned it. You used the word *geek* and abbreviations and acronyms as if this is one monolithic group, but it's not. It's highly divided.

So one thing, just a bit of nuance I want to introduce, is the huge distinction between what we do at SFI and other scientific institutes versus people working in machine learning. And so we actually believe that the purpose of science is to understand the world, right?

Unfortunately, narrative is very limited, so we have to deal with that fact. Nevertheless, there are ways of lucidly presenting difficult ideas. Now, machine learning doesn't try to do that at all, right? In other words, it just makes predictions. So I wanted you to comment on that. I would say that the allure of the algorithmic world is not the arcana, right, but something else, something much more immediate, is utility. It provides instant gratification in the terms of prediction, whereas science, real science—that is, mechanistic science—does not necessarily do that, not in the complex domain. So how do we deal with that dichotomy?

G. TETT: I think that's a very good way of putting it. And, of course, the great flaw of AI is that it builds its models of the future to a certain degree on what's happened in the past, and that both embeds patterns and potential prejudices into the future but also runs a risk of potentially being upended by a sudden step change in conditions. Which of course AI is supposed to be able to react to, but that's one of the questions in any kind of model construction.

I think the interesting question is, How do we as human beings have an intelligent debate about uncertainty and about the fact that prediction is based around probabilities, not about certainty? And the issue there, I think, is captured quite well by the Bank of England's fantail charts, the fan charts. The Bank of England tried a few years ago to move from giving precise inflation forecast, which give absolute numbers, to basically giving a vision of the future with sort of trailing fans like the fan of an aircraft exhaust, which showed different probabilities and potential outcomes, which is incredibly obvious and normal to anyone involved in statistics or modeling or predictions and stuff like that. But they were trying to communicate that to the wider public about the levels of uncertainty, the error bar.

How do you tell stories about complexity? How do you tell stories about numbers? And if you don't manage to tell those stories effectively, what's going to happen to us?

That's quite a clever visual device, which could and should be used in many forecasts. Certainly when you've looked at people standing up on television recently with predictions about infection rates, etc., a bit more of those types of fantail charts could be very useful looking forward.

But I agree with you. One of the attractions of AI is, you seem to have this magical divining machine—which is not that different from all the other divination tools that people have used in societies around the world—which gives you a prediction presented to the nearest decimal point.

MICHAEL SCHWARTZ:[7] Hi, Gillian. Thanks so much for your talk. It was fascinating. And I've been hoping I could hear you live sometime. I've been following your writing.

I couldn't help but think as I was listening to you about different levels or different orders of skepticism and how journalists accommodate the sorts of issues that you described. And, in particular, what I mean by that is the sort of cynicism and skepticism that you ascribe to journalists seems to me often causes them to challenge a prevailing trend in the first order. And I think the ESG is a good example where there's a sort of instinctive reaction to our faith in the market and then we realized that there are externalities, and we think, "Hey, we've got to pay attention to these." That's the first-order reaction.

But the second-order reaction is that doesn't mean that everything in the initial system is bad. And, in fact, sometimes we have unintended consequences from taking action against something that we think is bad and we don't realize that it's a little more complicated than that, and that kicks in a second order of skepticism.

My sense is that often we get the first-order reaction right, but then we don't move to the second-order reaction and say, "How do we actually accomplish what we're now concerned about?" I think that's one of the contributions that the Santa Fe Institute makes, is that we say, "Ah, these things are not so simple." And, in fact, our unassisted human minds are not always so acute at analyzing the problem and getting it right before we figure out what the solution is. All right, I'll stop.

G. TETT: Well, I'd say I agree.

M. SCHWARTZ: [*laughs*] That's a real problem, though, because it means that even if we can instinctively identify problems to solve, we don't then mobilize our resources to figure out solutions very well.

G. TETT: Well, I think the last year would show that very clearly in every sense. The last two years. But yeah. No, I agree. And I think one of the other interesting questions is that if you think about what separates man or the human brain from AI and machine learning, it's partly, you can say it's culture, that AI systems don't have culture . . . They will

7 Deloitte Center for the Edge

capture biases from the past and embed that. They don't necessarily have sort of the future-looking biases.

Another way of saying it is as far as I know—and people who are AI experts, I'd love you to correct me if I'm wrong—that AI-enabled devices don't tell stories. And we as human beings are hardwired to communicate with stories and narratives. That's what creates the symbolic patterns, and they're often embedded with rituals that allow us to basically retain information, pass it on, but also shape our own and reflect our own cultural biases.

And we are storytelling animals, as someone says. Well, as far as I can tell, AI machines, for the most part aren't storytelling animals. Maybe they will be in the future, but they don't seem to be right now. So in some ways that helps. I say that as a self-serving statement because it helps to validate my craft, but I do think it's true as well. And the question really is at the core of it: How do you tell stories about complexity? How do you tell stories about numbers? And if you don't manage to tell those stories effectively, what's going to happen to us?

W. TRACY: Thank you again, Gillian. I know there are a lot more questions in this conversation. We could go on for a lot longer, but in the interest of time, we do need to move on to our second speaker. I do think, though, that yours was perhaps the perfect talk to open up our "Complexity of Crisis" symposium.

SYMPOSIUM TALK NO. 002

LOST ATTRACTION: HOW CIVILIZATIONS TRANSFORM

DATE: *13 Nov. 2020* **FROM:** *Marten Scheffer, Wageningen University; SFI*

WILLIAM TRACY: We now move on to Marten Scheffer. Marten is an SFI External Professor and a distinguished professor at Wageningen University. He is also cofounder of the South American Institute for Resilience and Sustainability (SARAS). In many ways, Marten is a world leader on the intersection of global sustainability and resilient societies. His talk today, based on an upcoming book, examines the conditions underlying the large systemic shifts that have occurred in human societies and explores what the possibilities could be for an upcoming shift. Marten, with that, let me turn it over to you.

MARTEN SCHEFFER: Thanks for allowing me to pick your brain, because I have questions about the role of science in Santa Fe, all this kind of thing that we were talking about. I am trying to wrap my brain around it, and you're a bunch of smart people. So let me share some questions with you.

I think we all feel that we kind of live in a special time. I feel, certainly, that change is in the air, and it's not just because of the COVID pandemic. I think change was in the air before that already. Globally, you see a number of street protests that's unsurpassed since the 1960s. And also all kinds of movements that are thriving—new people become vegan, all kinds of things happening. At the same time, I would say that change is needed. I think transformation is inevitable. Just because of the trends we are in, we live in the best of times. If you read Steven Pinker's book *Enlightenment Now*, you easily get convinced that no

---

OPPOSITE: *Jean-Pierre Saint Ours, "Le tremblement de terre," c. 1790 (detail)*

previous generation had it better than we have, in terms of our wealth, how long we live. Inequality is less than previously. We are healthier.

But we got here by eating up the Earth in about one or two generations, of course. And that's a problem. If you saw the movie of David Attenborough, his latest,[1] I think it's a very nice example of storytelling. Things are changing rapidly. We just brought out a paper from a Santa Fe workshop called "Future of the Human Climate Niche."[2] That paper, in the month of May, was among the five papers that got the most media attention in the world. The other four were on COVID. This was the only non-COVID one. What we showed is that humans have, for thousands of years, remained within a certain limited temperature range. And that temperature range will change its position on the Earth. If you look at this picture,[3] you see the dark spots. Those are spots that are now the hottest spots on the Earth.

> ... civilizations age, and when they age, they become fragile. And inevitably, at some point, the thing falls apart.

Fifty years from now, the shaded area will have that same temperature—near-unlivable conditions—and about one-third of humanity lives in this range. Thus we project that about one-third of humanity may need to move to a different place. We're talking about one to three billion people possibly having to move within the lifetimes of our children. So it's one of the reasons why I think that transformation is inevitable. We're in some trends that will make it necessary to kind of reinvent society. I find it interesting, and also I think it's the responsibility of scientists to help think how we can make the transformation, because I think we can make a transformation as humanity. And I think we can do it gracefully. But it doesn't happen automatically.

---

1 *A Life on Our Planet*, directed by A. Fothergill, J. Hughes, and K. Scholey and featuring D. Attenborough (2020).

2 C. Xu, T.A. Kohler, et al., 2020, "Future of the Human Climate Niche," *PNAS* 117(21): 11350–11355, doi: 10.1073/pnas.1910114117

3 See Figure 3 in Xu, Kohler, et al. (2020).

The new book I am working on is basically trying to get bits and pieces of information together to help think about how we can transform gracefully. And one of the sources of information is looking to the past: How did past civilizations transform? When you look at the literature about that, you see that there is a lot of adaptation. Most of the time change is in the form of gradual adaptation—you find new crops that you can use, you find new ways of getting organized in your villages—but every once in a while, there is something that, if you read the literature, people say, "Well, this is something really different." We call this transformation. That's when, basically, the thing falls apart. And that allows a reassembly of a new civilization. (By the way, civilization isn't necessarily something good. It has that connotation, but civilization is just the fact that people get organized in larger groups. And with that comes hierarchy, comes inequality, and comes a whole bunch of stuff, always.)

We have a whole set of manuscripts on transformations of civilizations in progress with, from the Santa Fe Institute, Tim Kohler,[4] and other coworkers. One of the things we're looking at is the dynamics of past civilizations. . . . Each line is a civilization, and you see when they're born, and when they die.[5]

And you see there is a bit of a pattern in there. If you plot that, you see there's actually waves of change. And those waves correspond more or less to ups and downs in climate, but not really. There is a lot of that that you cannot explain from climate variation. Getting into all those different case studies in more and more detail, basically, we started thinking what many people think, and that is that civilizations age, and when they age, they become fragile. And inevitably, at some point, the thing falls apart. So the basic idea here is there is a loss of resilience with the aging of civilizations.

That's a nice idea, and it's not a new idea; it's an idea that has been coming back over and over again. The question is, Can you actually prove that? Because basically, archaeology is a lot of storytelling from little bits and pieces, and it's difficult to really find quantitative evidence.

4 SFI External Professor

5 See L. Kemp, "The Lifespans of Ancient Civilisations," *BBC*, February 19, 2019, https://www.bbc.com/future/article/20190218-the-lifespans-of-ancient-civilisations-compared

This brings me to another line of work we're doing: it is trying to find indicators of resilience, indicators that show you when a system becomes brittle, even if you don't know that system. The whole idea is based on the phenomenon that, if a system becomes more and more labile or more and more fragile—for the mathematicians, if it is approaching zero eigenvalue bifurcation—then you expect that when you perturb it a little bit, the recovery rate becomes slower and slower, and when the recovery rate approaches zero, then you're close to the criticality. So this is basically the idea.[6] If you have a resilient system, and you perturb it, it's recovering very quickly. If it's not resilient anymore, the slope around the equilibrium is not so steep; when you perturb it, it's slower to come back. If you can't experimentally perturb the system, you can still follow it in time to follow it's natural behavior. All systems are always perturbed. I'm perturbing the room around me, I'm perturbing your brains by this talk, the climate is perturbed by volcano outbreaks and all kinds of things. All systems are subject to a regime of perturbations. When that regime of perturbation stays the same, but the system changes, you'll see that systems that get close to a tipping point start responding to their perturbations differently. You see more variance and you see slowness, you see increased memory in the system, increased temporal autocorrelation.

Well, these are just fundamental mathematical principles of systems close to tipping points (so zero eigenvalue bifurcations). But we also looked for this in different real systems, and surprisingly the predictions work quite well. So if you look, for instance, to the climate system, every once in a while the climate goes through a very spectacular transition. For instance, this is the end of the greenhouse Earth.[7] Suddenly you got the big ice caps. Now, suppose you were a scientist, and you didn't know that, and you would have to predict what would happen. Maybe you would say, "Well, this time series it's getting down and down, there is a little bit of a fluctuation there but the trend will likely continue". You wouldn't predict this dramatic sudden surge. But maybe if you had

6 See Figure 1 in M. Scheffer, J. Bascompte, et al., 2009, "Early-Warning Signals for Critical Transitions," *Nature* 461: 53–59, doi: 10.1038/nature08227

7 See panel A from Figure 1 in V. Dakos, M. Scheffer, et al., 2008, "Slowing Down as an Early Warning Signal for Abrupt Climate Change," *PNAS* 105(38): 14308–14312, doi: 10.1073/pnas.0802430105

measured those dynamic indicators of resilience, you would have seen that some instability was building up over this period, signaled by rising autocorrelation, an indicator of critical slowing down. The same for the end of this glaciation—would you expect that the ice would be gone suddenly? Well, maybe if you had studied this indicator of dwindling resilience of the ice caps.

> All systems are always perturbed. I'm perturbing the room around me, I'm perturbing your brains by this talk, the climate is perturbed by volcano outbreaks and all kinds of things. All systems are subject to a regime of perturbations.

We showed it also in the lab, for living systems with tipping points.[8] We showed it for forests where you can, from remote sensing, show differences in the resilience of tropical forests.[9] And we even showed that it works for the human mood.[10] People who have mood fluctuations with a larger temporal autocorrelation are more likely to flip into depression later in life. It's remarkable that you have this universality. And it's that kind of universality . . . that you can capture in mathematical theory that the Santa Fe Institute is interested in.

And not only the Santa Fe Institute—somebody who was very interested in it was Salvador Dalí. This is Dalí,[11] and it is the last picture

8 A.J. Veraart, E.J. Faassen, et al., 2012, "Recovery Rates Reflect Distance to a Tipping Point in a Living System," *Nature* 481(7381): 357–359, doi: 10.1038/nature10723

9 J. Verbesselt, N. Umlauf, et al., 2016, "Remotely Sensed Resilience of Tropical Forests," *Nature Climate Change* 6: 1028–1031, doi: 10.1038/nclimate3108

10 I.A. van de Leemput, M. Wichers, et al., 2014, "Critical Slowing Down as Early Warning for the Onset and Termination of Depression," *PNAS* 111(1): 87–92, doi: 10.1073/pnas.1312114110

11 Speaker shows an image of Salvador Dalí sitting beside his painting "The Swallow's Tail," his last painting.

of him—he wanted to dress up one more time, because he felt he was becoming frail. He was reaching his tipping point. And he posed in front of a picture of a painting he made, the last painting he made in his life. It's a painting about the theory of tipping points, of catastrophic shifts, that was already, in those days, of quite a lot of interest and discussion. So those deep universalities make you feel like, yeah . . . I think that's the kind of excitement we have around the Santa Fe Institute also.

Now, the question getting back to the past civilizations: Could we use these kinds of universal indicators to see if those civilizations crashed -600- because they were becoming fragile, because they reached a tipping point?

Why do we see all the turbulence if Steven Pinker is right, and we live in the best of times?

Well, it's very difficult, because usually you don't have very good time series. But there is one example, from close to the Santa Fe Institute. It's the ancient Pueblo cultures for which the archaeologists have reconstructed, on an annual basis, their building activity using tree rings from pieces of wood in those buildings. And also, from the same tree rings, they could reconstruct the local climate conditions, and that's for a period of 800 years. It's quite remarkable that you have a quantitative index of social activity, and so far in the past, for such a long period.[12]

What we could do with that information is look at each period in this culture. What is special about these ancient Pueblo cultures is that they would build up, and then this building activity would crash, and they started doing things differently. They got a different culture, a different way of making the potteries, of doing the architecture, different rituals, different beliefs. And then it would start again: they would build and build and build, and it would crash. So those are different cultural periods before the moment of crash. And you see that over this period

12 See Figure 2 in M. Scheffer, E.H. van Nes, et al, 2021, "Loss of Resilience Preceded Transformations of Pre-Hispanic Pueblo Societies," *PNAS* 118(18): e2024397118, doi: 10.1073/pnas.2024397118

before the crash, the variance and autocorrelation build up as a sign of increasing social instability. It became fragile.

Archaeologists also find more broken skulls, more signs of violence. So it's consistent with the other evidence we have that gradually those civilizations became more and more fragile, and then they had to reassemble.

So, indeed, we have evidence for the systematic increase in frailty, just as in people that age, and we have actually had some nice workshops at Santa Fe about frailty in aging peoples. Well, here you see frailty in aging civilizations. And if you look at the details of the archaeological reconstructions, it's more or less this story. The civilization is a success—it grows. Inevitably, then it runs into shortages. Inevitably, also, you see more and more inequality, and this leads to discontent, which leads to violence, which leads to more discontent, and eventually, people don't believe in the old way of doing things. They don't buy it anymore. The whole set of rituals, beliefs, the credibility of the elite is gone. They abandon the whole thing, and then find new ways, start anew, and the successful experiments grow, and the thing starts again.

So, thinking that we live in a really interesting time, the best of times, but also a time of a lot of turbulence where change seems inevitable, an obvious question for me is, Could this happen again? Somehow, of course, it can never happen the same because it's an entirely different world, but would some elements be the same? Well, we do seem to see discontent. So this discontent, is it strong enough to reduce attraction, to make the social attractor that keeps us all together weak enough to allow a big transformation?

In other words, do we see this kind of situation, even though, of course, it's very different from the old times? Most people actually have enough to eat, and we don't have to fear for our lives. The kind of discontent, in ancient civilizations, that was arising from really very serious conditions, we don't see that. Most of us live in the best of times, but still we see a lot of discontent. So that's interesting, right? Why do we see all the turbulence if Steven Pinker is right, and we live in the best of times?

I think what is very interesting is that we live, since perhaps a decade, in a time in which perceptions have changed a lot. We have the internet and we have social media, and you could argue that they boost discontent, they show you very clearly how others seem to be better off than you. And they show you all kinds of problems that are really serious, something needs to be done about it—the behavior of the police agents, of soldiers. There's all kinds of conspiracy theories. None of that is new, but it's kind of amplified.

At the same time, you see that social media allows us to self-organize in new ways. It's easy to find like-mindeds; if you would be interested in becoming a vegan, or you had another idea that was novel but rare, it used to be very difficult to find like-minded people, but now it's much easier. Also, we now have a new kind of emergent leadership; you don't have to be elected. You can start writing beautiful blogs and become a celebrity on blogs and on the internet. You can co-craft views with others. I would think that this combination of generic discontent related to different perceptions of the world and the new self-organization allow, in a very new way, fighting the old order and finding new ways.

We are easily locked into things we think are important and we think we want. The question is, How can it change?

But where is it all leading? I would say it's quite fragmented. To me it looks like this: the thing is maybe becoming unstable, but it's going in all kinds of directions.[13] And it's not necessarily clear if discontent pushing you to vote for Trump would really move the world in a way that would solve the climate crisis or another big crisis. The big question, I would say, is, How can we move to a just and sustainable world? Is that an alternative attractor? Is it possible, if you think of utopias, to shift to a world that is actually sustainable and just? I think we can, and I think our main challenge is now to find feasible ways to get out of this turbulent phase and catalyze change into a direction that provides a better future for a lot of the people on Earth. It's not obvious, of course, how that should be done. And after all, why as a scientist would you want to work on a wicked problem. I started out my work studying

13 See Figure 1 in S.R. Carpenter, C. Folke, et al., 2019, "Dancing on the Volcano: Social Exploration in Times of Discontent," *Ecology and Society* 24(1): 23, doi: 10.5751/ES-10839-240123

plankton and lakes, and I was quite happy with that. Then we started looking at chaos in plankton, then tipping points in lakes, and then we started working on forests. And more and more, I liked to think about big problems, big issues that we're in. Santa Fe is a great place to do that.

So for me, we're part of a complex adaptive system, so to say. What we would really need to change if we want to change the path of humanity toward this just and sustainable goal is the worldviews. It would be nice if we wouldn't be striving for material goods, if we would be striving for satisfying lives in ways that have a smaller ecological footprint. But getting there is really a very complex thing, of course. The way we see the world, the things we believe that are important, the things that we value, the social norms may seem very hard to alter. Yet those things can change. For instance, the norms around smoking in public places have suddenly flipped. There was some enforcement but now enforcement is not needed because if I take a cigarette into university, I get such angry views that I immediately put it out. So it has become a social norm, and there is no need to enforce it anymore. Norms may look like they're set in stone, but they can change. Understanding how that works will require understanding the relationship between norms and all our institutions.

It's a bit unclear what institutions are—people sometimes call a social norm also an institution—but take for example education. Of course, what people learn at school, what children learn at school, is shaped by the worldviews we have, but also it shapes worldviews. So this creates hysteresis, inertia in the system. In the same way, the media play this role of responding to our worldviews but also affecting our worldviews. Science, in a way, too, and languages, of course. A good example: Our language is shaped by our history, by our past, and it allows us to express the worldviews that were important in our past. If you think of, for instance, a gender-neutral language and how difficult it is to make that happen, you see that language, in addition to being an instrument for narratives, for creativity, is also a prison. And you could say the same for arts: the novels that are written, the movies that are made, the pop songs that are written. All follow from our beliefs, norms, values, worldviews, but also feedback on them.

So if you think about this complex adaptive system, you can see why there is a lot of inertia there. We are easily locked into things we think are important and we think we want. The question is, How can it change? Well, of course, we have science. We can observe. This whole system, together with the business, everything we do, causes climate change, causes the loss of biodiversity. It also causes changes in social worlds—this is what drives inequality, corruption—and we observe that. And when we observe that, we interpret it, we tell a story about it. And eventually that can affect the way we look at it—our norms, our beliefs, our failures, our worldviews—in a way that helps us mitigate some of the changes like climate change or inequality that we would not like to see.

So that's the kind of the complex system that I think is really cool to study. And we have new tools, of course, to study that. We have big data; we can sense the stock markets; we can analyze the millions of tweets we have every day, look at their sentiment content, see how it correlates to certain topics, see if the sentiment around certain aspects is changing. We can do sensing of the climate and of the land cover, of the greenness of the forest. We have this tremendous wealth of data. And we have this hindsight of studying how transformations happened in the past.

I think it's a great moment to be a complexity scientist and try to think about how we can help make a graceful transition, because I'm convinced we need to make a transition as humanity. And I'm also convinced that change is the air, just the changes going all directions. And it would be really cool if we could help make a narrative that helps catalyze worldviews in ways that allow a transition to a world that is a better place. So that's the set of ideas I wanted to share with you.

W. TRACY: Outstanding. Thank you so much, Marten. I actually wanted to ask you the first question myself, Marten, if I could.

In order to get out of the discontentment trap that you highlighted—this idea that most of us live in the best of times yet we still see a lot of discontentment—it seems that we need not incrementally change our norms and worldviews but really have a transformation, as you said. Do you know of examples from history where we're able to have such a radical transformation in our worldviews and norms without first having almost a complete collapse of society? Is that possible, do you think?

M. SCHEFFER: Actually, I think all the examples from the past, they go with a violent collapse. And so I think it'll be kind of cool if we could do the next one without that. So that's, I think, the big challenge. And so if you think that three billion people will be on the move in fifty years, of course you can say, "Well, we have a great place, we'll just lock the borders and make it a well-defended fort." I think that's not going to work. I think we have to think about ways to accommodate change. And I think it is possible.

I think there are, though, examples of change. Well, if you think of the two world wars: they were not a complete collapse of the system, [but] they were very violent, they did completely change the way the world was run in a way, certainly in Europe, and reduced inequality a lot. They were a kind of a reset.

I think it's a great moment to be a complexity scientist and try to think about how we can help make a graceful transition, because I'm convinced we need to make a transition as humanity. And I'm also convinced that change is the air . . .

W. TRACY: But they were violent collapse.

M. SCHEFFER: Yeah, not complete collapse but violent. So I think that's basically the story. In the past, we saw the transitions, but they were really unpleasant, the transition themselves, even though they usually resulted in something better. It could be immediately a next civilization, so to say, and next system of hierarchy, or what they called Dark Ages . . . now the idea is a bit increasing that the Dark Ages were actually mostly the best for the people. They didn't suffer from being low in the hierarchy.

In the past, the transformations were mostly preceded by collapse that was pretty violent. And I think we should be able to do better.

JOHN CHISHOLM:[14] Marten, thank you. So often public policy confers substantial benefits on a small, visible group but also imposes smaller, mostly invisible costs on much larger groups. For example, steel tariffs benefit a few tens of thousands of steel workers in the US but impose smaller costs on hundreds of millions of Americans who buy anything made of steel. The Affordable Care Act (ACA) confers significant benefits to those with pre-existing conditions but imposes smaller costs on millions of healthier Americans. Might declining societal resilience be attributable to the growing aggregate burden of all these smaller, often invisible costs?

M. SCHEFFER: Sure. I think that definitely is something that plays a role. Those are the externalities, so to say, that small groups can put on very large groups of society that find it difficult to get organized well enough to actually push back on those costs. And that has to do, in a way, with power—power to get your act together, so power to get organized. And in our world, of course, one of the things that pushes discontent is this. I think if you look now in the world, most of the protests and discontent don't have to do with climate change or the laws of nature. Most of it has to do with the feeling that things are unfair, that others are getting rich and you're not, and that others are preventing you from expressing yourself.

But I definitely think that it's a very fundamental thing, this spillover cost thing that you mentioned. Economists would call it market imperfections: if you just would be able to tax all those things, everything would be fine. The problem is, we never get rid of the market imperfections, of course.

ESTHER DYSON: To continue this, I think there's a big difference between externalities of short-term impact, whether it's just the cost of doing something that would need to be paid by someone else, and what I would call externalities across time. So, for example, investing in public

14 John Chisholm Ventures & SFI Trustee

health actually has a high ROI to society as a whole. Everyone's taxes will go down later if we pay now to prevent illness for all these poor people who are going to get sick without current health-maintaining services—what we call "prevention." In short, there is another part of the externalities to consider—it's not just across people or across entities now, but also paying now to avoid paying later, which is called investing.

M. SCHEFFER: Sure, sure. Yeah. No, that's a great point. You can also make that argument for inequality. It's inequality between people now, but it can also be inequality between generations. So the shit will be on the next generations. So that's what economists try to handle in discount rates, discounting the future. If you don't discount the future, then the costs of things you do to the Earth sum up to infinity, if time goes on. So to keep your equations running nicely, you need to do something about it, and that's discounting, but discounting isn't fair, of course. So that's a great point.

DAVID KRAKAUER: Thank you, Marten. Thank you very much. Let me ask a sort of fairly obvious question. It has to do with issues around universality. I remember years ago, Ed Wilson, rightly, getting in trouble for extrapolating from ants to human societies in his book Sociobiology, [which] had a very sort of reductive flavor. And the question for you I have is: You talk about plankton, and now you're talking about people, and you have a set of methods that seem to apply equally to both. Where does human will, human agency, feature in the theory?

M. SCHEFFER: Do you mean that human agency would imply that there are no tipping points in society, for instance?

D. KRAKAUER: No, not at all. It's just that, as you point out, you know, Marten Scheffer has now got involved, right? So the theorist of the tipping point is now in the system, and preaching to the system. I'm just wondering how that kind of awareness manifests in the dynamics?

M. SCHEFFER: Ah yes. That sets humans apart. One of the famous examples, of course, is finance—a very, very interesting system, because it's fundamentally, in principle, unpredictable if you assume that humans are very smart and try to predict things, and that they will use anything they find, to squeeze out most of the predictability. That makes finance a really interesting system, yet there is still the phenomenon of bubbles. So still you see this kind of inevitable herd behavior. What makes societies in a sense different from ecosystems is that, in part, they learn from the past, and history never repeats. But, of course, some elements of history still do repeat. And you get back to the famous thing that it never repeats, but it rhymes. There are things that do get back.

But turning it into something positive, I would say that we have come now to a point in which we have become so good at reconstructing what happened to us in the past and understanding what is happening now and predicting what's happening in the future, that we should be able to take that human agency to the next level and actually craft our own future quite different from the plankton that I started out studying.

W. TRACY: Marten, I'd like to tie this back to some of the things that Gillian Tett said earlier in the morning,[15] particularly this idea of managing risks outside of the model, which is what I thought of, for your last comment to David about our new ability to learn from history, predict the future, and craft our future. And I apologize if I can't let go of the sort of doom and gloom of a painful transition. But it seems to me that for a number of technological reasons, as well as environmental reasons, we're much closer to wiping out the species now than we've ever been in human history. So you used the world wars at the beginning of the last century as an example of these transformations. If everyone had nuclear weapons at the beginning of that period, would humanity have survived? And I certainly don't mean to take your talk from an optimistic to a more pessimistic one. But do you look at the system as a real risk that we could wipe ourselves out, or do you think that there's good reason to believe that we are going to be able to learn our way out of this?

M. SCHEFFER: No, I think the default path would be to wipe ourselves out, and so the challenge, sad to say, is to make that not happen. Certainly there are good, nice books about the systemic risk we run. There is the Center for the Study of Existential Risk, in Cambridge and people giving very, very good thoughts to it. Yeah, there is a risk. And so I think the default will be gloom and doom, but I don't like that. And I believe that we can actually do better. I would say I force myself toward

15 See S-001 on page 571

that path. Because the gloom and doom is not a very stimulating thing for actually inviting action, so to say. There is lots of really good psychological research about that. There is this beautiful work, for instance, showing that if you let people read a piece about climate change that would be catastrophic for humans, and you ask them before and after if they believe that climate change is actually true, after reading the story, more of them think that climate change is actually not true at all. So it has this negative effect on the willingness to face it. But if you add one paragraph saying, "But if we do this and this, everything is under control," then it does not have this effect. So I think it's really important, if we want to communicate our thinking about this, that we don't stop with the gloom and doom but also at least make a beginning of the next step. Otherwise, we're not playing a very constructive role in society, I'm afraid.

W. TRACY: That is the perfect setup for the rest of the talks and panels that we're going to be doing this weekend for symposium. So thank you for that, Marten.

REFLECTION

Q&A WITH MARTEN SCHEFFER

DATE: *20 August, 2021*

FROM: *Marten Scheffer, Wageningen University; SFI*

Q: How has the pandemic changed your thinking about complex systems?

It did not prompt me to revise any basic theory, but the pandemic has made me more optimistic about the capacity of our global society to act in concert toward a common goal.

Q: What in your own work helped you to understand the events of 2020? Looking back, what would you now change or enlarge in your work to address what you saw?

None of the pandemic dynamics have really been a surprise from the complexity point of view. However, the pandemic has nudged my work more toward attempting to understand the role of worldviews, trust, institutions, and the social contract.

Q: Are the theories that we now possess adequate to explain what happened in 2020?

We have fantastic theories about lots of things already. The main challenge is tying those to what really happens in the world. What may have changed is the support for certain lines of inquiry. For instance, understanding economy and finance was always a big thing. We might now see more support for work on the complex dynamics of attitudes and worldviews.

Q: What do we need to change in our models and theories, or possibly in society at large, to minimize the risk of a repeat event or worse?

We should not shy away from addressing scary possible scenarios. There was little interest in the possible effects of a pandemic till it happened. It seemed so alarmist. The events now should give us courage to think of ways we may help to change fundamental mechanisms that could make societies more resilient.

Q: Does your research help us better understand the entangled social protest movements, political uncertainties, and market behaviors of 2020?

My research suggests a crucial role for social tension in preparing societies for fundamental change. We just showed that this was the rule for some ancient societies, too. My impression is that social media plays a fundamental role in fueling instability. I am directing some of my research in that direction.

SYMPOSIUM TALK NO. 003

LESSONS LEARNED & NEXT STEPS: AN EPIDEMIOLOGICAL VIEW OF THE COVID CRISIS & ITS COMPLEXITIES

DATE: *13 Nov. 2020* **FROM:** *Lauren Ancel Meyers, UT Austin; Santa Fe Institute*

WILLIAM TRACY: Lauren Ancel Meyers is external faculty here at SFI. She also holds the Denton A. Cooley Centennial Professorship of Zoology at the University of Texas at Austin. Her academic accomplishments are manifold, but I think her recent work on COVID is most germane for our conversation today. Lauren is the founding director of the COVID-19 Modeling Consortium. She has advised the government at all levels about the spread of COVID and has done crucial work on the impact of social distancing on the pandemic spread. With that as an introduction, let me turn it over to you, Lauren.

LAUREN ANCEL MEYERS: Thank you so much, Will. I kind of don't want you to turn off the videos. I like seeing all these familiar faces—hello to all my old friends. It's great to be here.

All right, let me share my screen and get started. Will just mentioned that I am the founding director of UT COVID-19 Modeling Consortium. I'm going to start by telling you a little bit about what that is, and then spend most of the time talking about some of our recent activities in COVID. From day one, when we first heard about COVID, my lab got right to work, like all the other labs in my field, just trying to get our hands on whatever data were available, to answer basic questions. And very quickly that escalated into requests from the federal government, the state government in Texas, who we've worked with for years and years, and local authorities asking us questions: What should they do? What's going on?

OPPOSITE: *Frank Leslie's Illustrated Newspaper, "Dallas health officers stop a train, attempting to quarantine it for fear of yellow fever," 1873 (detail)*

It really very quickly got to be much too much for us to handle. And so with the support of my university and with a generous gift from Tito's Handmade Vodka—a special form of liquid courage— and everything that I learned from being affiliated with SFI for over twenty years, I had the resources, I had the confidence in the model to set up what's now the UT COVID-19 Modeling Consortium. As described here, it's an interdisciplinary network of researchers and health professionals, building models to detect, project, and combat COVID-19. Really what it was was I just put out a call to all my colleagues at UT, those I knew, those I didn't, people in my network of collaborators, and said, "Hey, let's start talking. Let's meet every morning on Zoom. There's a lot of requests. Let's see how many of these we can field." It's very self-organized.

We've tried to tackle as many things as we can if we have the right kinds of tools. If you go to our website[1], you can see several dashboards, a bunch of publications. I'm sure this[2] is not a complete list, but these are many of the people responsible for the work that I'm going to be talking about today, people I'm working with on a daily basis, including Michael Lachmann, who's on the faculty at SFI and who's really been a key leader in this effort and contributed to a lot of the work I'm going to talk about today.

So with that, we've done a lot of different things, as have modelers all around the world, and I think that most of what we've done as a community can be categorized according to three different goals: We model to help us understand the threat; we model to help us forecast the threat; and we model to come up with good strategies and to support decision-makers in containing the threat. And I'm going to organize the examples I provide today according to these categories.

We also, in the modeling consortium, have kind of explicitly been modeling for three different audiences. The first audience is the policymakers, and this is one of my favorite emails I got in this whole thing—actually maybe my least favorite—but it was an urgent modeling request from the White House Coronavirus Task Force a couple of days after Ambassador Birx had been appointed: "This is a time-sensitive and urgent request. Please see below the series of parameters and

1 https://covid-19.tacc.utexas.edu/

2 See Acknowledgments on page 641

outcomes we would like you to model. We will need whatever results you can achieve by the close of business, Wednesday"—I think we got this Wednesday morning or something—"East Coast time, or opening of business Thursday. Your results in that timeframe will inform US policies." And we spun our models quickly. We provided results. There's no evidence that they impacted policy whatsoever.

> We model to help us understand the threat; we model to help us forecast the threat; and we model to come up with good strategies and to support decision-makers in containing the threat.

We also model for the public, and this is something sort of new to us. But we really are proactive when we feel like we have insights; when we feel like we have dashboards that can provide useful public awareness, inform behavior that's aligned with the interest of public health, we've tried to get out there and talk about it.

And then, of course, for scientific audiences . . . While this is a published paper[3], many of us are trying to get our work out by *medRxiv* as quickly as possible. I think that this very rapid sharing has really accelerated the pace of science because we're learning from each other. We're building on each other. We're not reinventing the wheel every time we try to model something different. If somebody comes with a request—for example, how to deploy testing to open a workplace or something like that—instead of building that model, I've seen that model and I can refer them to somebody else.

Now I'm going to give you a bit of a whirlwind tour of some of the studies, some of the tools that we've developed since the beginning of the pandemic, again, organized around these three goals. First, understanding the threat. A couple of the studies that we published, we did

3 D. Duque, D.P. Morton, et al., 2020, "Timing Social Distancing to Avert Unmanageable COVID-19 Hospital Surges," *PNAS* 117(33): 19873–19878, doi: 10.1073/pnas.2009033117

within that first month/month and a half of becoming aware of COVID in early January. Very early on, we didn't know what was going on in Wuhan. The first reports were that there was this anomalous virus and it was spreading from animals to humans, but it wasn't human-to-human transmissible.

Then maybe a few days later, "Oh, there's some evidence that it's actually spreading in households, but limited." Then before maybe a week or so was out, we knew that this thing had the capacity to spread from person to person in the community. We didn't know how many people were infected in Wuhan. We didn't know how fast it was spreading. We didn't know where else it might already be emerging in China or around the globe, but we wanted to answer those questions as quickly as possible. And we didn't really trust the data coming out of China. We didn't know if they were being completely transparent, but even if they were, we weren't testing for this virus yet, there was just really no way to know what the true prevalence was.

So what data could we trust at the time? Well, we felt like we had some faith in the first nineteen cases that were reported out of China. We knew the locations—Seattle, Taipei, Tokyo, the very first case was in Bangkok—and the dates of those. And we also knew—like many modelers in my field, we already had at our disposal—international daily airline mobility between all pairs of cities around the world. We also very quickly got our hands on cell phone mobility from within China, for hundreds of millions of people, that reflected daily air, rail, and ground travel patterns between 370 different cities in China. So we used those first nineteen cases to triangulate how fast the virus was spreading in Wuhan.

We imagined a very simple model. The virus is growing at some exponential rate. We know how often people are coming in and out. It's sort of like sampling from the city. Based on when known cases first arrived around the globe, we could say something about how fast the virus was probably spreading within Wuhan. And what we estimated at that time—very early on, in fact, our first estimates we did before Wuhan locked down on January 23—was that the virus had spread far faster and far farther than the data were telling us. By that January 23 lockdown, when China was reporting a total of 425 cases, we estimated that there were probably somewhere over ten thousand cases already in the city—and that by the time of that January 23 lockdown, it had probably already spread to well over a hundred other cities in China, where it was very likely there were already small outbreaks brewing. In fact, within a few days, or within a few weeks, there were data from all over China that corroborated this, and many subsequent studies have also shown that there were probably thousands and thousands of cases during that early period in Wuhan.

The second study I'm going to mention was by far the simplest mathematically but the most jaw-dropping from the perspective of just understanding, characterizing, this threat.[4] This was a study that I think we started collecting the data from this, I want to say late January. No, it was probably early February. We worked with a team of Chinese students who went to the websites of eighteen different public health agencies from different provinces in China and looked for case reports that provided a few pieces of information: A person was infected. It also reported the day they first felt symptoms. It also reported who most likely infected them—and most case reports aren't able to specify this, but there'll be a few cases where a person was infected, there was no one else yet infected in the city, and their spouse had recently come back from Wuhan or something like that. So they could connect the dots.

It also had to have a likely infector and the date when that infector first started exhibiting symptoms. We simply used what ended up being 450 case reports to estimate the time between the infector developing symptoms and the infectee developing symptoms. That's a quantity known as the *serial interval*, and it tells us something about the pace of transmission. . . . We almost simply need a histogram of those 450 serial intervals. Before I show you this, I'm just going to tell you that at this point in time, very early on, much of the public health community and the researchers really had no idea, but we were building models where this virus looks something like flu, but we were also building models where this virus looks something like SARS from 2003.

The thing about SARS in 2003 was that it was a slow and visible spreader. It had a serial interval of about eight days. So there's a good

4 Z. Du, X. Xu, et al., 2020 "Serial Interval of COVID-19 among Publicly Reported Confirmed Cases," *Emerging Infectious Diseases* 26(6): 1341–1343, doi: 10.3201/eid2606.200357

week between somebody feeling sick and a person they infect feeling sick, and people weren't contagious until they had symptoms. This one study made us realize that, in a very upsetting way, this virus was not like that first SARS virus.

So this was the histogram (fig. 1), with the serial interval along the *x*-axis.[5] Three really interesting things: Number one, this virus was spreading at about twice the speed of SARS, with the serial interval closer to four days than eight days. This was our first evidence that there was asymptomatic transmission—people actually spreading before they felt symptoms. And then on the long end, there was evidence that we would have to isolate people for quite a while to ensure against longer-term chains of transmission.

When we shared this histogram in its roughest form with our colleagues at the CDC, they looked at this and said, "You must've made a mistake. Go back and check your work." And in fact, we did make mistakes, but not that mistake. Pretty soon after, within a few days, there was another study using different data from China that corroborated that this thing was spreading quickly, and that there was evidence of presymptomatic transmission.

The third and final study I'm going to mention under this category of characterizing the threat is something closer to home in Austin.[6] Fast-forward now to March. Austin leadership—the mayor and the judge of the largest county in the Austin area—enacted our "Stay Home, Work Safe" order that started on March 24 and went through the reopening of America on May 1.

At the time when they enacted this order, they basically said, "Everybody's got to stay home except for essential workforce." And they put some pretty clear restrictions on construction work; only the most essential projects could go forward. Within a couple of days, the governor of Texas overruled that and said, "You guys can do your

5 See Figure in Z. Du, X. Xu, et al., 2020, "Serial Interval of COVID-19 among Publicly Reported Confirmed Cases," *Emerging Infectious Diseases* 26(6): 1341–1343, doi: 10.3201/eid2606.200357

6 J.F. Pasco, S.J. Fox, et al., 2020 "Estimated Association of Construction Work With Risks of COVID-19 Infection and Hospitalization in Texas," *JAMA Network Open* 3(10): e2026373, doi: 10.1001/jamanetworkopen.2020.26373

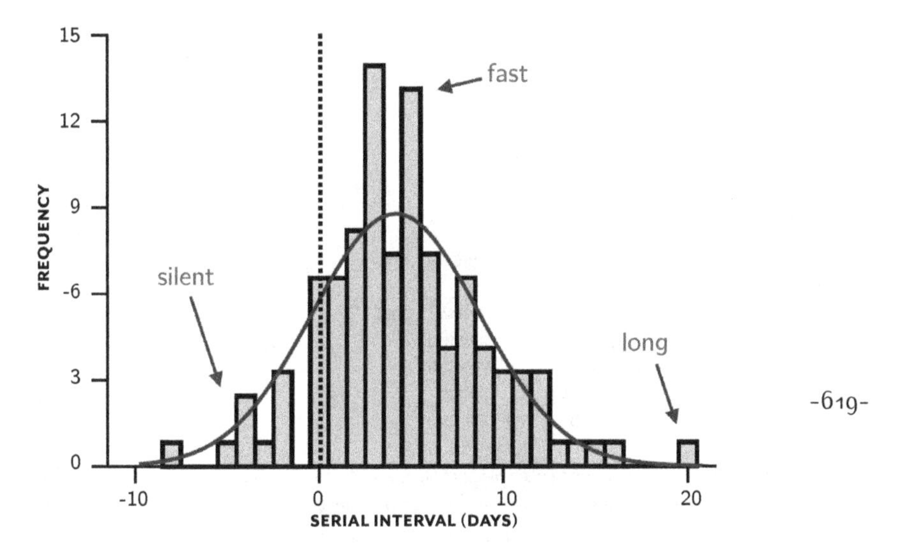

Figure 1: *Estimated serial interval distribution for coronavirus disease (COVID-19) based on 468 reported transmission events, China, January 21–February 8, 2020. Gray bars indicate the number of infection events with specified serial interval, and solid line indicates fitted normal distributions.*

stay-at-home orders, but all construction work has to be allowed to proceed." So the city was concerned about that, and they asked us if we could adapt our models that we were already using to project COVID in Austin to estimate what the impact might be in terms of increased risk to the construction workforce and the possibility that transmission at work sites could spill over and make things worse for all of us.

So we made these kinds of projections early on. This shows rate of hospitalizations on the *y*-axis, two different scenarios on the *x*-axis (fig. 2).[7] "Baseline risk" means just according to our models, what we think would be the risk of transmission on construction sites; "50% risk" means mitigation efforts are being put in place at work sites to reduce the chance of transmission.

These are just pairs of bars, comparing projections for what the hospitalization rate is going to be among construction workers compared to other people in the same age group who aren't in that same workforce. You can see at the baseline we've projected that construction workers

7 Excerpt from Figure 3 in R.F. Pasco, S.J. Fox, et al., 2020, "Estimated Association of Construction Work With Risks of COVID-19 Infection and Hospitalization in Texas," *JAMA Network Open* 3(10): e2026373, doi: 10.1001/jamanetworkopen.2020.26373

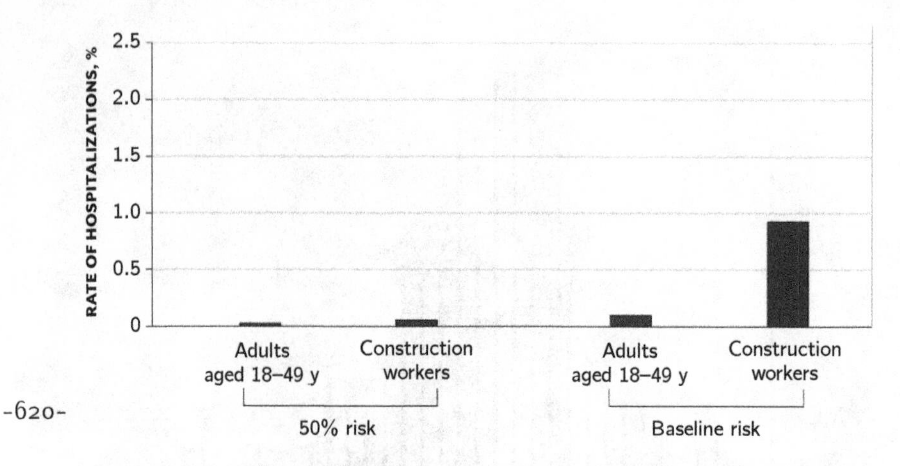

Figure 2: *This graph represents a proportion of the construction workforce being active. A bar for construction workers that is larger than the corresponding bar for the general population indicates a scenario where construction workers are at disproportionate risk for COVID-19 hospitalization.*

would probably experience four to eight full times higher COVID hospitalization rates compared to their counterparts.

When we kept an eye on things because of this, the city of Austin started asking for occupational data from COVID-19 patients in hospitals. We were tracking that, and by the time we got to early May, we looked at it, and what you're seeing here (fig. 3)[8] is a fitted model projection in that line, and all the bars are actual hospitalization data from Austin-area hospitals. The total heights of the bars is total number of hospitalizations for COVID in Austin. The black shade at the bottom are people who identified as construction workers. The medium gray represents people who gave their occupations but weren't construction workers, and the light gray is anyone who didn't give their occupation or wasn't in the age group we were comparing.

What we found when we looked at the data was that the actual excess burden in the construction workforce was predicted almost perfectly by our model. The relative risk of COVID-19 hospitalization among construction workers compared to same age group, different occupations, was almost exactly five, so they're at much higher risk.

So where is this coming from? Well, this sort of illuminates disparities we've been hearing about from all over the country, different studies, different ways of slicing this, but this is one example of the

8 See Figure 1 in Pasco, Fox, et al. (2020).

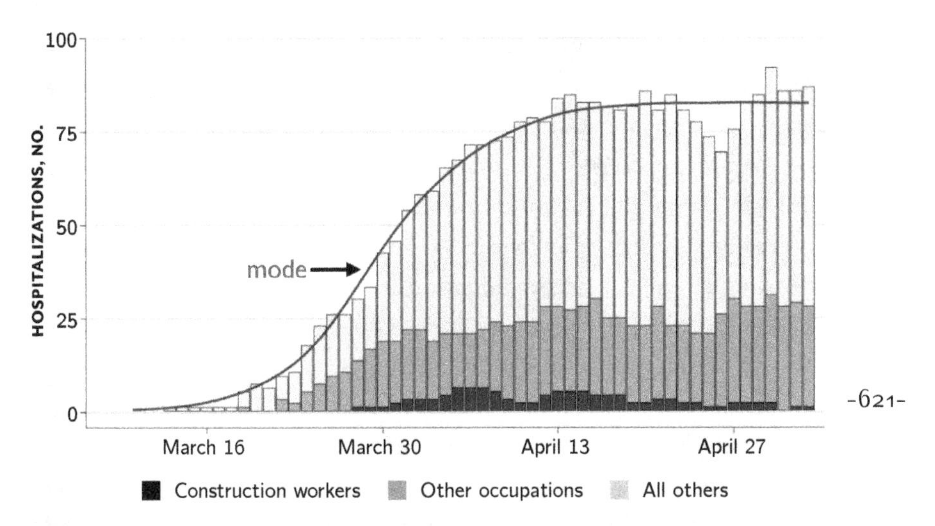

Figure 3: *Reported hospitalizations (bars) from March 10 to May 3, 2020, compared with estimates (line) based on an SEIR compartmental model that considers age-specific high-risk groups and contact rates for the general population of 2.17 million and a construction workforce of 50,000 workers.*

disproportionate burden that is felt by our racial and ethnic minorities, groups that have overlapping vulnerabilities in Austin: 66% of our construction force is Latinx; it's about 30% Latinx across the United States. Among Latinx construction workers, roughly 48% lack health insurance. This means that population has had less access to preventative care. Because of this, they have higher rates of comorbidities that put them at risk for severe outcomes. And they also, because of a lack of health insurance, are more likely to delay or not at all seek treatment for COVID-19, which can lead to worse outcomes.

In addition, the construction workforce has higher rates of smoking and may be exposed to hazardous materials that would also compromise lungs, making it more likely they'll suffer severe outcomes. And this is a community that has to work on a daily basis to put food on the table. So even if they feel symptoms and are told to stay home, if they have symptoms, they still might go to work out of necessity and thus expose their coworkers. The Latinx community in Austin—and many parts of the country—has larger-than-average household sizes, which also contributes to the community spillover, what happens on the work side.

So, this is an example of how we can use modeling to incorporate these different complexities and early on make projections to identify high-risk groups. But more importantly, I think this is just another additional piece of data telling us that we've got to do something; we have to stop throwing our essential workforce under the bus and really take steps to protect them. That includes making sure that there are basic mitigation standards in workplaces. That means incentivizing workers to stay home if they or anyone in their household is symptomatic, and providing alternative accommodations to people who are at high risk, have comorbidities if they do get infected or exposed.

So now I'm going to turn to the second category of modeling, which is modeling to forecast the threat. The very first request that we got to forecast something was from the White House Coronavirus Task Force. And the question was, *How many people are going to die from COVID by the end of the year?* I think this request came in March or February. And our answer was, There is no way we can forecast that because it absolutely depends on the policies that you put in place and the decisions that individuals make. But we'll give you three plausible scenarios (fig. 4). The scenario where we do nothing to mitigate this—so there's no control—we would anticipate that there could well be a million or more deaths in the United States by the end of 2020. If we all stay sheltered in place indefinitely, we could potentially cap the potential deaths under 50,000.

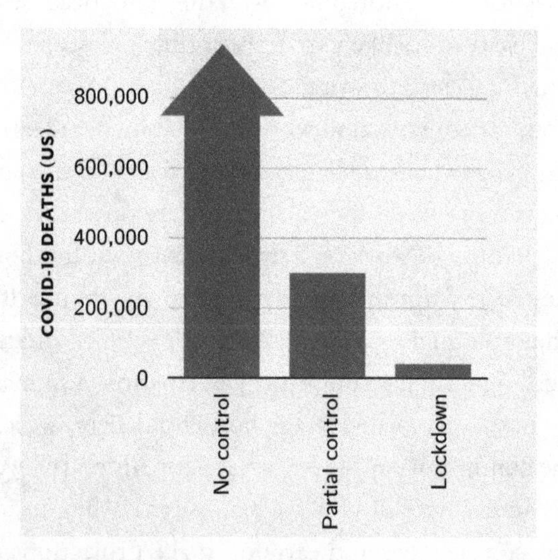

Figure 4: Forecast the Threat.

If we do something in between, which is essentially the roller coaster that we've been on, we anticipate somewhere between 200,000 and 400,000 deaths by the end of the year. This is pretty simple forecasting, but since then we have really gotten involved in much more sophisticated, more detailed, more action-oriented forecasting. Our first real endeavor into forecasting was inspired by the IHME model,[9] which many of you may be familiar with. It is I think the best-known model by the public and policymakers out of Washington. This was a group that wasn't part of the core community of modelers that I've been working with over the last decades. And they came in with an alternative approach to modeling, which is known as curve fitting or forecasting. The idea was that they assumed that deaths—and this was just the original model; they've changed it since then—but in the original models, they assume that deaths would follow a sort of Gaussian-shaped curve, a bell curve, and that the steepness and when it peaks and all those things would be governed by when governors enacted shelter-in-place orders that would lead the curve to start slowing, and that the overall shape of the curve would resemble the curves that we had just observed in China and in Europe. This model got a lot of flak, a lot of pushback—even though it was widely adopted by the press and by policymakers, a lot of the scientists didn't like it partly because it was a little bit, I think, foreign and didn't seem at face value like it should work. . . . The opposite of curve fitting is a much more mechanistic model that has some of the complex dynamics of virus transmission built into the model. I think there was an assumption that this probably wouldn't work because it was way too simplistic. And then it was also criticized because it actually didn't hold up when we looked at the backcast.[10]

We compared the projections of their model for Spain, assessing their 95% confidence interval with what actually happened. Their projections didn't quite capture reality. So we decided to dig in a little more, instead of just throwing the baby out with the bathwater. We were intrigued by the curve-fitting method. We thought there were some issues with how

---

9 http://www.healthdata.org/covid

10 https://www.statnews.com/2020/04/17/influential-covid-19-model-uses-flawed-methods-shouldnt-guide-policies-critics-say/

they approached it. And so the first forecasting model that we ever built was really just a variation on their model.

The two things we did that were really different from theirs—one is instead of making assumptions that the shape of the curve should look like China and Europe, and instead of assuming that people's behavior automatically followed what governors ordered, this was probably one of the very first models that actually incorporated cell phone mobility data as a way to directly estimate activity on the ground that might reflect pace of transmission. And so we adapted the model so that it could assimilate that kind of data. In addition, we corrected what we thought was a statistical error in how they were quantifying uncertainty. I think our first model was a substantial improvement on IHME, but what we realized—and what I'm sure IHME realized as well—is that once we exited that first nice-shaped wave and we started going in all sorts of directions, the curve no longer fit.

We replaced that model with much more mechanistic-type models, which have sort of compartmental SEIR frameworks under the hood; we've adapted those not only to make national-level forecasts of mortality, but we've used them more on the ground to . . . Oh, I meant to mention that our mortality models, in addition to models of dozens of other research groups now around the world, all feed into the CDC's ensemble. So they're looking at a lot of different forecasts to try to get the most robust assessment they can regarding what's to come.

We've also been building models on a more local scale. This is something that Michael Lachmann has spearheaded for our consortium, building models that tell us what's going on in Austin.[11] On a daily basis, we provide estimates for the reproduction number. I'll say a little bit more about that in a minute, but it tells us something about how fast the virus is spreading, the probability that the epidemic is in a growth phase, the estimated fourteen-day change. For example, as of November 12, 2020, we estimated that there were 44% more infections in Austin than there were two weeks ago.

And this model that's making forecasts—we have a bunch of forecasts of what's going to happen in hospitals and in ICUs in Austin, and that's the real purpose of this dashboard—but this model is driven by

11 See https://covid-19.tacc.utexas.edu/dashboards/austin/

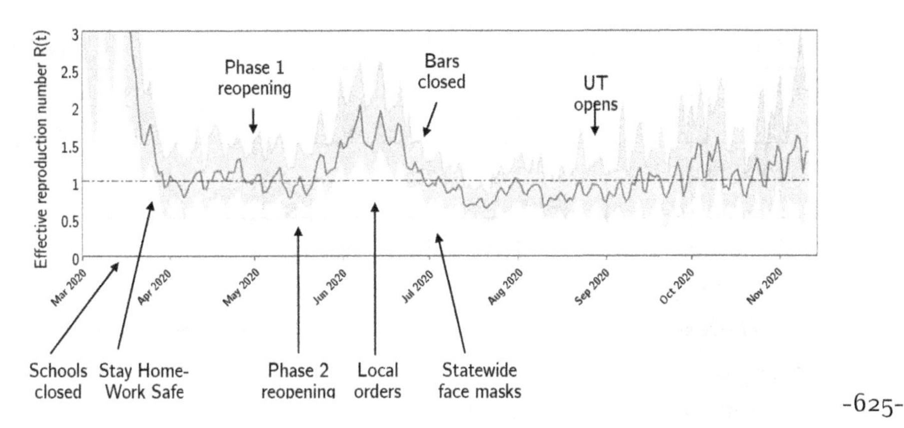

Figure 5: Looking Back.

three things: the daily number of COVID-19 hospital admissions in Austin, the daily number of COVID-19 hospital discharges in Austin, and the cell phone mobility data from SafeGraph, which gives us a sense of what fraction of the day people are staying at home and how often they're going to venues like restaurants and schools and grocery stores.

We also, on our web page, provide the historical time series of our estimated reproduction number. For those of you who aren't familiar, the reproduction number tells us something about how fast the virus is spreading. A value of 1 means that it's basically flatlining. And one way to think about it is like every person who gets infected, on average, is infecting one other person, and so every infection sort of replaces itself. Values above 1 indicate that the epidemic is going to be spreading exponentially, and values below 1 indicate that the epidemic should be dying out.

What we see in Austin going back to early March (fig. 5), prior to closing schools and enacting our stay-at-home order—before we started mitigating this threat—it probably had a reproduction number somewhere between 2 and 4. So the average number of secondary infections was quite a few. We enacted these orders, and they had the desired effect. The reproduction number came down somewhere close to 1, possibly a little bit below. And then we started reopening, and Texas was one of the first to enthusiastically embrace the reopening of America plan in early May, mid-May. And we saw the consequences of that by June, in many cities across Texas, including Austin: we saw our hospitalizations, our ICUs, going up at an alarming pace. That led the governor to allow local authorities to impose some restrictions, including some face masking, but not comprehensive face masking. We took advantage of that as soon as we could. Then the governor closed the bars, then there were statewide face mask mandates.

And now fast-forwarding, UT opened. Students came back to Austin around August 18. We've been holding our breath to see what's going to happen. There were some clusters of cases. And UT is usually 50,000 students. We probably have at least 30,000 back in town, hanging out at frats, etc. There's certainly been clusters of cases at UT. And I think just now we're beginning to see the effects of slow but certain spillover from that and other activities around the city, and we're seeing our hospitalizations start to creep up again in alarming way.

We also have a dashboard showing these kinds of projections for all of Texas.[12] Here we can look at all of Texas or each of the twenty-two regions of Texas individually. And you can see what you're hearing in the news. If you look for El Paso and view ICU capacities, you can see El Paso doesn't have enough, they can't take care of their patients, they're coming to Austin and they're taking our healthcare workers. Abilene, Amarillo—you can see that much of Texas, though not all of Texas, is either past the brink or already on the brink of a really catastrophic situation in terms of the integrity of the healthcare system and consequent loss of life.

Okay. Finally, on topic three, modeling to really support decision-making, modeling to really think about, given all this uncertainty, what kind of decisions can we still make that make sense, that are robust? This is a screenshot (fig. 6) from one of Austin's two dashboards. They call this the Key Indicators for Staging dashboard. The different shades correspond to different levels of restrictions; where darker is stay at home, lighter is mostly closed restaurants, no bars, and I don't know we're going to do about schools, etc. You probably have one of these in your own city. So what you see here is the city tracking, the seven-day rolling average of COVID-19 hospital admissions, and with a clear indication to the public that when we cross these certain thresholds, we're going to change our policies. We worked very closely with the city to

12 https://covid-19.tacc.utexas.edu/dashboards/texas/

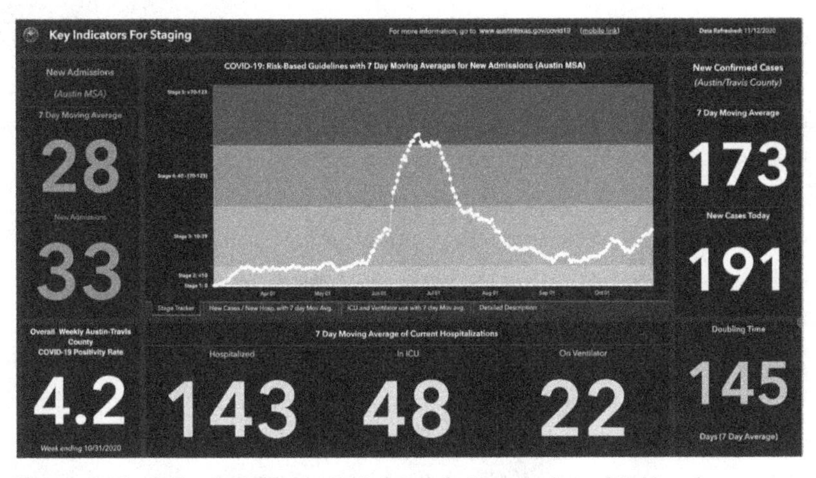

Figure 6: Contain the Threat. *Key Indicators for Staging in Austin, TX, November 12, 2020*

not only formulate this approach but to actually derive the values that separate the different shades.

Okay. So how did we come around to this? Well, first, way back when we were thinking about this policy in April, the question was, Okay, what happens if we open up America on May 1, we tell people to be careful, they're a little bit careful, but they're not very careful, and we just see what happens? Then we projected we'd have something catastrophic. In Austin, we would probably have surges that overwhelmed our hospital capacity by the end of the summer, lots of people would die. Nobody wanted that.

So the question was, What should we do given that we want to avoid loss of life, we want to avoid loss of livelihood, we have to obey the laws of Texas, which are pretty strict—strict on the side of not allowing local authorities to close things down? So what kind of policy could we put in place that is sort of allowable and aligned with what we hope to do? The first step, and arguably one of the most important steps, in building a policy is to explicitly state what your goals are. What we arrived at that satisfied those needs was that we want a policy that will ensure that we keep hospitalizations in ICUs under our local capacity, and if at all possible let's avoid stay-home orders. Let's spend as much of our time open as we possibly can.

What kind of policy options do we have to achieve those goals? Well, it's basically all the different rules under the different shades—different combinations of face masks, social distancing, gathering, restaurants, bars, etc. So those are the levers we have to achieve those goals. I'm not going to go into the details, but I'm happy to talk about it more if you'd like to. Very briefly, by building a detailed compartmental model of COVID transmission that was fit to data from Austin, and then implementing different thresholds for those different levels and using numerical stochastic optimization to figure out which thresholds were going to allow us to best achieve those goals: ensure that we keep those hospitalizations under capacity while absolutely minimizing the time we have to be under the most restrictive rules. So the solutions we came up with for the city were that we needed to track daily COVID-19 hospital admissions to give a real-time sense of how it was spreading, and then tap on the brake at the key thresholds that we derived.

So again, I'm not going to go to the details, and we don't know how it's going to play out—how it's going to play out depends on whether people comply with the policies, whether the city actually enacts the policies—but we were able to extrapolate a plausible future based on our model of the policy and fit to data up till recently. [13]

The thing we're most worried about is the integrity of our ICUs, our ability to have beds and care for ICU patients. So that fixed capacity is what we hope never to exceed. So what we expect under our current projections, if people behave and the policymakers behave as prescribed, is that we're going to have to be under pretty severe lockdowns to prevent overwhelming ICUs over the next few months. And we may even have to go into some short-lived full-blown lockdowns in order to ensure the integrity of our healthcare system.

There are so many different alert schemes, so many cities, so many countries have developed them, and they may all be based on rigorous analysis or good analysis of whatever kind. But it's hard to know when you go to these websites, you look at these dashboards, they rarely provide any information or rationale for why they're tracking the data they're tracking, why they chose the thresholds they're choosing. So hopefully they're decent. However, when you look at the state of the world today

---

13 See Figure 1 in H. Yang, Ö. Sürer, et al., 2021, "Design of COVID-19 Staged Alert Systems to Ensure Healthcare Capacity with Minimal Closures," *Nature Communications* 12: 3767, doi: 10.1038/s41467-021-23989-x

and how many communities are actually facing overwhelming numbers of hospitalizations and are really desperate, something's not working, at least in some places.

In an attempt to start to think about these more intuition-based systems—are they going to work?—we have done a comparison between different alert stages.[14] We implemented other people's alert systems in Austin. The goals are to remain under the ICU capacity and spend as few days as possible on lockdown. We managed to keep the needed days in lockdown just to about a couple of weeks.

I'm also going to mention France's policy. They have multiple stages, but the one we've heard about a lot in the news is that when 30% of their ICUs are occupied by COVID cases, that is when they lock down their cities. When we implement their whole stage process, what we find is that that trigger is too late. It's very possible that if they wait until 30%-full ICU to really lock things down, even if they have those intermediate steps beforehand, they will overshoot their ICU capacity, and at the same time, require even more time in lockdown than the Austin strategy. That is partly because they waited too long. The longer you wait, the longer you have to deal with a really high level of healthcare demand, and you end up having to stay locked down longer on the back end.

The Harvard Global Health Institute published a widely cited set of recommendations for when it's safe to reopen schools and communities. It's based on incidence, number of cases per 100,000 in the community. Their criteria for locking down is when the seven-day average of incident cases exceeds 25 per 100,000. And then they have other thresholds for partial lockdown, etc. What we estimate when we plug that into our model is that it's erring on the side of caution. It is very likely if you implement those that you will preserve the integrity of your healthcare system; however, you may end up spending more time in lockdown than needed.

So, in closing, these are some lessons learned that I thought of as I was thinking about this talk this morning. I'm sure there are many other lessons learned, but let me just share a couple.

---

14 See Figure 2 in Yang, Sürer, et al. (2021)

One is that what we have seen throughout this pandemic is an utter failure of preparation. We built a pandemic simulation app for the state of Texas[15] after the 2009 H1N1 pandemic so that they could run pandemic-preparedness exercises and simulate all different kinds of pandemic threats, see what happens if they do different things with vaccines, antivirals, social-distancing measures, even see it through the lens of a noisy surveillance system. We presented this model to representatives from CDC, from NIH, and from our peer institutions in the early teens.

I mention this because the State of Texas probably paid half a million dollars for this and some other tools that we built for them. And they promptly shelved them and totally forgot they were there, never did a single exercise. And that was because the person whose vision it was in the state agency went and got a higher-paying job in industry. And so this arguably got the closest a state agency ever got to having models inhouse to support decision-making and planning and thinking through different intervention strategies, and that never really played out. This is one of many examples, and this is what happens when we have an impoverished public health system and no good bridges between research and practice at the frontline and in public health.

Moreover, I think this is an utter failure of imagination. Even though we and others have done all this great work in the past, we all thought the next pandemic was going to be flu. None of us really thought about shutting down businesses and schools for such a long time. We really have a hard time thinking about anything beyond slight variations on what we've experienced in the past.

I think these are the two most important lessons learned, at a very high level. We also have learned, as we have been really working hard and other brilliant people around the world have been working hard to try to forecast this pandemic, is that we have a painfully short forecasting horizon. Even though we know a whole lot about this virus now—about its pace of transmission, asymptomatic, symptomatic—there are still things we don't know about age-related susceptibility and infectivity, but there's a lot we do understand.

What we absolutely don't have a crystal ball for is human behavior, human decision-making, political choices, innovations. Like, when are

15 https://flu.tacc.utexas.edu

those vaccines coming out? How effective are they going to be? For how long? Right. So we don't have a crystal ball for that or exogenous events. When the hurricanes came through Texas, we suddenly had tens of thousands of people going from Galveston to Austin, staying downtown, right? These things can impact the transmission dynamics of a pandemic. And we haven't figured out how to predict them or, even if we can't fully predict them, really incorporate them in a coherent way into our forecasting models.

. . . we all thought the next pandemic was going to be flu. None of us really thought about shutting down businesses and schools for such a long time.

Getting more down into the details of what I talked about today, some of these alert policies are better than others, probably few are validated or really given any kind of quantitative push. Timing is absolutely critical for our interventions: a week's delay in implementing lockdowns or stricter measures can really make the difference between having your hospitals overwhelmed or not.

It turns out . . . ICU personnel are absolutely the limiting resource in most communities around the United States.

Hospital admissions are a good data source to track to get good information on what's happening in your community. It turns out, I didn't really talk about this, but ICU personnel are absolutely the limiting resource in most communities around the United States. And then there are all these other things, which I didn't think I'd talk about, but probably resonate with a lot of you. It seems like in the United States, politics often trump science. Masks turned out to be useful. Contact tracing turns out to be less useful, yet we are holding onto it so much in the United States. It's not useful in the sense that if we're not using some of these new digital technologies and we're letting this virus spread at such high levels, we just don't have the resources to make a dent in transmission with contact tracing. Testing turns out to be a promising path to opening. We have seen catastrophic overlapping vulnerabilities causing extreme burdens in some of our racial and ethnic minority groups. And this thing that I just can't get over, I mean, time and time again, is that in the United States it feels like opening bars is so much more important than opening schools.

Finally, next steps: things we're working on in the consortium, that a lot of people in our world are working on: safe opening of schools and colleges, how do we use surveillance testing to do that even more so, and working with the state and locals to figure out how we're going to roll out that vaccine as safely, as effectively, as fairly as possible?

Finally, taking a step back, I think next steps for understanding the threat of pandemics, preparing our globe for the threat of pandemics—this is where SFI really can step up. I mean, it already does in so many ways.

So, not really in the failure of preparation, but the failure of imagination, understanding not just what kind of viruses or pathogens might come our way, but what is that complex cascade of things that happen because we have a pathogen in all different dimensions—in social and economic and political and cultural, and you name it, in technological . . . how do those feed back? Imagining what threats look like and then imagining the universe of solutions, figuring out, even though there's so much uncertainty, we don't know what's going to come our way. There must be things that will be at least robust paths forward, robust ways of collecting data, robust ways of making decisions, etc. And then I also think on a related note that SFI can help in tackling this challenge of expanding our forecasting horizon by really thinking about these complex interdependencies and folding them into our models of pandemics.

With that, I want to, again, acknowledge all of the people who have contributed to this work and am happy to have a discussion.

W. TRACY: Awesome. Thank you so much, Lauren. That was a great talk and extremely timely. As a few people have already noted, our governor ordered a two-week shelter in place here in New Mexico earlier this afternoon. So, yeah, that was wonderful. I do want to open it up for Q&A. David, do you want to kick us off?

DAVID KRAKAUER:[16] Yes. Thanks, Lauren. That was great. Yeah, so very early on in the pandemic, people were talking about endemism, right—would it become a permanent fixture? And Jon Machta made this really nice point about tuning to the critical point, where the critical point here is R_0 equals 1. And the problem as he saw it was that what happens of course is we implement a policy like mask use or social distancing, [and] we reduce transmission . . . You have a dashboard. It says, we're now in some safe zone with respect to our thresholds, and now we go out again and you open businesses again, and you become imprudent again, and it increases. And then of course we move into another zone, and we go down. So there's self-organizing criticality. How do you avoid that dilemma? It seems to me almost inevitable.

L. ANCEL MEYERS: One way you avoid it as you get safe and effective vaccines, and you vaccinate everybody and then you achieve true herd immunity, right? Herd immunity is such a complicated concept. But this idea of herd immunity: some fraction of the population has to be immunized either through vaccines or through direct exposure to the virus. And once we get to a certain point, the reproduction number will come below 1 and the epidemic should fade out on its own. However, if everybody's wearing face masks, then that herd immunity threshold gets lower. If only 20% instead of 50% of people are infected or vaccinated, then under a world of face masks, we achieve sort of eradication. But if then we stop wearing face masks—and this is what you're talking about David, right? So if we have a safe and effective vaccine, and we vaccinate up to the true naive herd immunity threshold, that's one way out. And until then, we really are stuck in this world where we may have to live with precautions that are themselves bringing the reproduction number down below 1.

---

16 SFI

JOHN GEANAKOPLOS:[17] Hi. Thanks, Lauren. That was terrific. I have a very simple, practical question. I'm at a university, like most people in this Zoom are, with a great hospital. I am amazed that Yale isn't doing any testing, farming the testing out to somebody else in Cambridge, actually. So my question is, if Yale, say, could test every student every morning and get the results back in fifteen minutes every day, could Yale operate as normal, with students going to Glee Club rehearsals and Football practices? Does your model answer questions like that?

Would that degree of testing make any difference? Nobody seems to try to do more than twice a week. Would it make a difference to do it every day? And especially if you have one of these fast tests, let's say it had a high accuracy rate, would that allow businesses to open and act normally? Or is testing more often not really worth it? Apparently, it's not, because nobody's doing it, but I'm surprised.

L. ANCEL MEYERS: Yeah. So, there's been some really nice work done on that, including by Michael Mina, who I know has spoken to SFI before. And Michael Lachmann at SFI has actually developed a nice little app for looking at, as you change your frequency of testing, how does that change the speed with which you would catch clusters or whether you prevent clusters altogether. So there are definitely models out there that address that. Twice per week or every three days has become this rule of thumb where you can almost go back to normal. If you can achieve that level of testing—everybody every three days—you're almost guaranteed that nothing's going to slip through, especially if people wear face masks, right. Then you've got that extra guard.

I don't actually know off the top of my head whether, if you go to every single day testing—given [that] with these rapid tests, they're antigen tests and they're less accurate than the PCR tests—if you factor in the lower accuracy of the antigen test and you go to everyday testing, what kind of assurance it gives you. But certainly it gets you even closer to going back to life as normal. And if everybody's going to wear a face mask, then it may be overkill.

J. GEANAKOPLOS: So that suggests that things really aren't that bad. Many businesses could maybe afford to do it. It's not that expensive.

17 Yale University & SFI

L. ANCEL MEYERS: Yes. I mean, I do think that testing does provide us a way to open things up even more. But I think we also have to think about . . . it might make it easy for Yale to open and some of the companies where we work in [this] room, but there's no guarantee that the construction workers are going to have access to it, right? And as long as there are people going to work and they're not being tested before they do things, there's still going to be community transmission, right? But it might be an answer for some of us to go back and do some of the things we want to do.

SIMON LEVIN:[18] Wonderful as always. In your modeling projections, how do you deal, if at all, with the differences between naturally acquired immunity and vaccine-acquired immunity, including the possibility of waning and more generally the heterogeneity that might exist within the population in terms of various attributes?

L. ANCEL MEYERS: So with vaccination, we are actually still in the process of building in our vaccine models, and we're very much trying to tie them to whatever clinical data come out that tell us anything—which we really have none yet, but there's some very early phase one and other experimental data that tells us something about antibody levels. And so we're trying to build whatever data we can into our models about specific vaccines, but we're waiting for that data to be able to do it.

S. LEVIN: Is it all antibody levels, or is it T cell levels too?

L. ANCEL MEYERS: I think the data we've looked at so far is just antibody levels, but those are models in progress. We'll take whatever insights there are, indirect or direct information, about what to assume. So there's a lot of uncertainty there. And those are really in development still. With respect to heterogeneity, are you talking about heterogeneity in susceptibility, those kinds of things?

S. LEVIN: Yeah, yeah.

L. ANCEL MEYERS: Okay. In the models that I talked about today, these are mostly garden-variety, elaborated, SEIR, age-structured models, where we have a number of different age groups, maybe five or ten in each age group, we have a low risk, high risk, and we have some age specific

---

18 Princeton University & SFI

assortativity that impact transmission rates based on what the age groups of the two people are. But in addition, we have also taken, and I don't know if everybody is familiar with this—there are a couple of preprints that have really gotten a lot of attention that claim that the herd immunity threshold is actually much lower than most of our models say it is, that it could be as low as like 7, 8, 9, 10% of people. And they've gotten a lot of traction, these studies. Although not totally explicit, they are the basis, for example, for the Great Barrington Declaration, which is written by a lot of well-known scientists, that asserts that we should open things up completely for young healthy individuals. And it's basically kind of a herd-immunity strategy.

. . . there probably is more heterogeneity in the world than we're capturing in our SEIR models, but not so much that we believe that the herd-immunity threshold is substantially lower than most of us are assuming at this point.

So we have also taken a closer look at those studies that claim these low herd immunity thresholds. In fact, we've reimplemented them and looked very closely at the assumptions. And what we find is that the results they get, where they get these very low estimates for the herd immunity threshold, they're based on fitting their model to data in four different European studies up through end of June or July, I forget. So before the new surge. And they make some assumptions about the magnitude of social distancing measures during that period of time, which are not crazy but probably not the most realistic assumptions. And it turns out that the numbers they get are very highly sensitive to those assumptions. So if you change even slightly what you assume to have been the impact of social-distancing measures during that time, you get much higher estimates for the herd immunity threshold.

So, in short, we believe that there probably is more heterogeneity in the world than we're capturing in our SEIR models, but not so much that we believe that the herd-immunity threshold is substantially lower than most of us are assuming at this point.

JOHN CHISHOLM: Lauren, thank you, especially for your service to the nation at the CDC. I have two questions. First, on the chart before "Texas on the Brink," I see some periodicity, perhaps weekly, in the data. Is that an artifact of people's behavior, when they tend to go to the hospital, how they act differently on weekends, an artifact of data collection, or something else? Second, I believe you said there's a broadly accepted level of herd immunity that you and others accept. What is that level, approximately? If people can be re-infected by the virus, might that level of herd immunity never be reached? Thank you.

L. ANCEL MEYERS: That's a great question, John. Okay. So first to answer the question of why there's this weekly signal in those reproduction number estimates: it's because there are two sources of data that are feeding our estimates. One is the hospitalization data and the other is the cell phone mobility data. And a lot of that is coming from the cell phone mobility data, just weekday, weekend patterns that are driving that. And part of that might be an artifact of the data and part of that might be real because we really are trying to estimate daily transmission rates. And so if people are out and about less on weekends or more, depending on where they're going, it could actually lead to upticks on the weekends when people are out doing things more recreationally. So that's the first question.

The second question about herd immunity: yes, absolutely. If immunity is super short-lived, then we don't get any herd immunity. If immunity lasts a year following infection and we're following vaccination, let's say, then we can build up herd immunity in a season. I mean, this is sort of what happens with flu, right? Like we can get flu multiple times during our lives. But we—actually, one of the reasons that flu tends to fade away in the spring and it's gone for the summer is we have sort of transient herd immunity where enough of the population is immunized for a short amount of time before there's enough susceptibility.

It does seem likely that this thing doesn't confer lifelong immunity and that our vaccines won't. And so we may be in one of those situations where we're not going to eradicate this virus through herd immunity, but we might be able to kind of control seasonal cycles or keep it low or keep it completely under wraps by vaccinating enough people on a regular basis.

And then what is the assumed? So the assumed is under the assumption that this is immunizing indefinitely or for a long enough period of time that... immunity is fairly complete. Under that scenario, depending on the model, I think that the estimates are somewhere between 50 and 80%, I would say, are what I hear from other modelers. It depends on the assumptions of the model. But there's a lot of uncertainty.

ANDY DOBSON:[19] Hi Lauren. That was a fabulous talk. I really enjoyed it. Good to see you. Gillian Tett, in another fabulous talk this morning, told us we had to stop having tunnel vision and look in different directions. Two thoughts on that:

One, everyone's hoping we get to herd immunity, but we won't start putting selection on the virus to change. We've seen lots of neutral selection, but we haven't seen any selective change. Although if mutation produces a strain with better transmission, it will quickly replace an earlier one with lower transmission. Selection will increase the rate it's going to evolve once we have herd immunity, and it will probably either evolve to be better at transmitting, or its virulence is going to change. How much does that worry you? And is this the iceberg that's hidden under herd immunity?

Two, then the second question is: The epidemiologists have now got fantastic models of age-dependent variation in transmission and virulence and how mixing between different age classes structures transmission. But we haven't got any models of how transmission between different sectors of the economy affects transmission. The host population is subdivided into people who are engineers, people who are doctors, people who are working in bars. Would that be a way to go with the next generation of models? You could then couple the epidemiological

19 Princeton University & SFI

models to economic models and look at what you're doing to control the disease impacts the economy (and vice versa) in the same framework.

L. ANCEL MEYERS: Wow. Those are great questions. I don't think I have answers to either one, but I want to follow up on both. As far as selection pressure that will change phenotypic characteristics of the virus, I don't know. I mean, I haven't looked at it. There are probably others who are trying to anticipate some of that. With flu, which I've spent a lot more time thinking about, we do see variation from one season to the next in terms of severity and the transmission rate. But some of that may come out of not intrinsic differences in a very quickly evolving virus, but it could also just come out from the changing landscape of immunity from one year to the next, right? Based on who's been infected in the past. So I'm kind of avoiding the answer. I don't really know. I think it's a really important question. And I think we don't have a lot of great precedents for what to expect. And I have heard from my friends who know more about coronaviruses than I do that they expect that it would be slightly less phenotypically pliable than influenza viruses, but I don't really know where that comes from.

With respect to thinking about structuring our population along occupational categories or industry domains, or I forget how you said it, that's really interesting. I certainly haven't thought about that. What we've seen in some communities is that there are clear hot spots that are around industries or specific things, right? We have the long-term care facilities, where a clearly delineated place that something special was going on, impacted in a certain way by the community and vice versa; our study on construction workers; we've seen these meat packing plants be these amplifiers in certain communities; schools, we're not seeing so much, but it certainly could be for a different kind of virus. We know that for influenza. And healthcare. I think it's really interesting. I'm still thinking from an epidemiological perspective, but coupling that with sort of economic modeling could be quite interesting. I know that Doyne Farmer has done some really interesting work looking at an interface between epidemic dynamics and industry economics.

GEOFFREY WEST: First of all, thank you, Lauren. Excellent talk as usual. I enjoyed it. But the part I liked best was the last five minutes, because you talked about something that Andy touched on, and that's what I'd like to bring up. And that is, you stopped talking just as an epidemiologist and you talked much more like an SFI person. And the big question is, How does this pandemic interface with the economy, with the social structures that we have, with the psychic well-being of citizens, and so on and so forth? Everything is interconnected. That's why there is an SFI. I'm wondering, first of all, have you actually given much thought to that? You have these huge collaborations—is it all "just" epidemiology or are there people thinking beyond the boundaries? And if not, then I want to echo what I think you said: Is this something that SFI ought to be doing, or, given what you know, is there any place doing any of this in a serious way, trying to see the thing much more in a systemic holistic way?

L. ANCEL MEYERS: I've been thinking about this a lot, and my motivation in creating this consortium was to try to bring in other perspectives. And we certainly have. We've talked to other people, and there are some people contributing to some of our models, but at the end of the day, our models are pretty much still in our lane of epidemiology. And I don't know of places that are really doing this from a holistic perspective. And that's why my last slide had Santa Fe Institute across the bottom. I think Santa Fe Institute is perfectly poised to do this. And I wish I could be in person at SFI right now, talking to all of you and really thinking about how we build those bridges and how we build those kinds of models. I don't think we're there yet, but I think it's desperately needed.

W. TRACY: Unfortunately we really are out of time. I want to thank Lauren again for an outstanding talk and discussion.

ACKNOWLEDGMENTS: COVID-19 MODELING CONSORTIUM CONTRIBUTORS

UNIVERSITY OF TEXAS

Spencer Fox, Associate Director

Becky Kester, Project Manager

Zhanwei Du, Research Associate

Kaitlyn Johnson, Postdoc

Spencer Woody, Postdoc

Emily Javan, Grad Student

Ciara Nugent, Grad Student

Remy Pasco, Grad Student

Mauricio Tec, Grad Student

Suzanna Wang, Grad Student

Cameron Matsui, Fellow

Tanvi Ingle, Undergrad

Riya Mahesh, Undergrad

Rogelio Mendoza, Undergrad

Maike Morrison, Undergrad

Michaela Petty, Undergrad

Lauren Ehrlich, Mol Bio

John Hasenbein, Engineering

Jungfeng Jiao, Architecture

Jason McClellan, Mol Bio

Varun Rai, LBJ

Lorenzo Sadun, Math

James Scott, Stats

Ravi Srinivasan, ARL

Peter Stone, CS

Gordon Wells, Center for Space Research

Emily Pease, Data Consultant

EXTERNAL MEMBERS

Michell Audirac, Inst Tec de Mexico

Daniel Duque, Northwestern

Michael Lachmann, Santa Fe Institute

David Morton, Northwestern

Simon Risanger, Norwegian U of Sci & Tech

Guni Sharon, Texas A&M

Bismark Singh, Nurnberg

Lin Wang, Institut Pasteur

Hoaxing Yang, Los Alamos National Labs

Jennifer Johnson-Leung, U Idaho

Benjamin Ridenhour, U Idaho

Holly Wichman, U Idaho

Simon Cauchemez, Institute Pasteur

Ben Cowling, U Hong Kong

Oscar Dowson, Northwestern University

Alison Galvani, Yale

Abhishek Pandey, Yale

Meagan Fitzpatrick, U Maryland

Neo Huang, Precima, LoyaltyOne

Alex White, Texas State

DECISION-MAKERS

Mayors Steve Adler & Sylvestor Turner

Matt Biggerstaff & Stephen Redd (CDC)

Mark Escott, Austin Health Authority

Mike Morath (TEA)

Stephen Pont, State Epidemiologist

Heads of Schools & School Boards

Art Markman & Cindy Froning (UT)

TEXAS ADVANCED COMPUTING CENTER

Joe Allen

Brian Beck

Joan-Yee Chuah

Maytal Dahan

Justin Drake

Victor Eijkhout

Erik Ferlanti

Kelly Gaither

Ethan Ho

Ari Kahn

Si Liu

Paul Navratil

Kelly Pierce

Pat Scherer

Dave Semeraro

Jason Song

Joe Stubbs

Matthew Vaughn

DELL MED

Dharlene Bhavnani

Parker Hudson

Clay Johnston

Elizabeth Matsui

Esther Melamed

Michael Pignone

Paul Rathouz

Victoria Valencia

Amy Young

Cory Zigler

SYMPOSIUM TALK NO. 004

INSTITUTIONAL TRANSITIONS & ECONOMIC OUTCOMES

DATE: *13 Nov. 2020* **FROM:** *Suresh Naidu, Columbia University; SFI*

WILLIAM TRACY: Our final speaker today is Suresh Naidu. Suresh is a member of SFI's external faculty, and also holds an appointment as a professor in economics and international and public affairs at Columbia University. Suresh has written extensively about institutional transitions resulting from exogenous shocks. In many ways, this is one of the most optimistic topics we will discuss during this year's symposium. The topics we discussed earlier today focused on the drivers of crisis, and processes through which crisis can spread to other domains. However, crisis can also unseat entrenched interests and rent-seekers and create an opportunity for social progress. On that note of hope, let me turn it over to you, Suresh.

SURESH NAIDU: Thank you. Thank you all for doing this. I wanted to present some evidence from my research, which I didn't really realize had this theme through it, until COVID hit. And I was like, "Oh, it turns out a lot of the things I had looked at involve some big economic shock happening and then looking at the subsequent changes." And so here we are. What I thought I'd talk about is some raw facts and how we, in economics, learn from historical data and can look at these big shocks that have happened in the past, and how they've changed things.

And what I'm going to do is present silver linings, if you will. . . . In retrospect, we think, "Man, that was probably a pretty good change that happened, that was really forced on a population or a country by some big negative shock." And it bounced that society out of, say, one

---

OPPOSITE: *Constantin Meunier, "The Peasant War—Assembling," 1875 (detail)*

equilibrium into another equilibrium, and generally, economic crises are precipitators of these kinds of big institutional changes, which are often pretty sudden. Institutions will go along for a long time, under their own inertia, and then very quickly in the space of less than a half a generation, they will be radically transformed. And that's what I'm after, and so I'm going to talk about three different levels at which we can look at these crisis-driven institutional transitions.

What I'm going to talk about first is a very high-level paper that's as cosmic as you can get, but it's basically on the relationship between democracy and economic development. . . . What we noticed that other people hadn't noticed is that unless you really handle the crisis that precipitates the transition to democracy, you're going to get the counterfactual all wrong.

And so, in order to form a correct counterfactual in which to assess the effects of democracy on economic growth and development, you really need to handle, empirically, the fact that countries don't randomly switch; they're precipitated by some economic crisis. And so what you want to do is form the control group, looking at other countries that went through a similar crisis but didn't undergo the same institutional change. That's what we do in this paper with Acemoglu and Robinson and Restrepo.[1] That's at the cross-country level. Then I want us to talk more about the cross-state level, in the US. And I'm talking about the peculiar transition to a partial social democracy that happened in the US in midcentury, and how the New Deal and World War II really pushed [the US] . . . into a different, more equal distribution of income.

Then, finally, what I want to talk about is historical racial inequality, and I'm going to draw on a paper I did a long time ago on the 1927 Mississippi flood.[2] And one of the interesting things about the 1927 Mississippi flood is that . . . the levees only break in some places and don't break in other places. And so you have this very interesting historical experiment where [in] the places that got flooded, it was basically a minor humanitarian disaster, but it also induced a huge Black

1 D. Acemoglu, S. Naidu, et al., 2019, "Democracy Does Cause Growth," *Journal of Political Economy* 127(1): 47-100, doi: 10.1086/700936

2 R. Hornbeck and S. Naidu, 2014, "When the Levee Breaks: Black Migration and Economic Development in the American South," *American Economic Review* 104(3): 963-90, doi: 10.1257/aer.104.3.963

out-migration. It's one of the push factors in the Great Migration; around the Mississippi Delta, the places that got hit with the Mississippi flood, that got flooded, experienced a large Black outmigration.

And what you can use that for is to figure out how much, for example, planters and farmers in that area valued Black labor, and look at what they did when that labor left. What I'll show you is how the agricultural systems in those counties that got flooded just completely transformed as a result of this Black outmigration. And so what we'll see is that the flooded counties here wind up modernizing their agriculture at a much earlier time than the counties right next door that didn't get flooded.

Let me talk about the first [study, with Acemoglu et al.,] in some more detail. It's one of the big questions, there have been endless papers written using both quantitative and qualitative data about the relationship between democracy and economic development. And basically, because the level of variation that democracy lives at is pretty much a country, the style of work in this area—and it's one of the oldest cross-country regressions you can imagine running—is, What's the regression of GDP per capita on whether or not a country is a democracy?

What you can do that's better than that is look at changes in democratization. If you look at all the countries that experience a transition from dictatorship to democracy, or from democracy to dictatorship, [you can ask], "Okay, when those countries change, does their rate of economic growth then change?" In that exercise, you're not trying to compare, like, the US to China, which is just confounded by hopeless numbers of things that you couldn't hope to control for. You're really trying to look at, while holding the other characteristics of the countries constant, when it changes from nondemocracy to democracy, does its level of economic development change? But even this experiment is pretty hard to get to behave properly; even when you run this very basic regression, you can get all kinds of answers depending on what you do . . . and I think what turns out to do the trick is that you have to account for the fact that a switch from dictatorship to democracy generally happens after an economic crisis. What you want to do is to form the experiment where the countries—here, let me show you the slides.

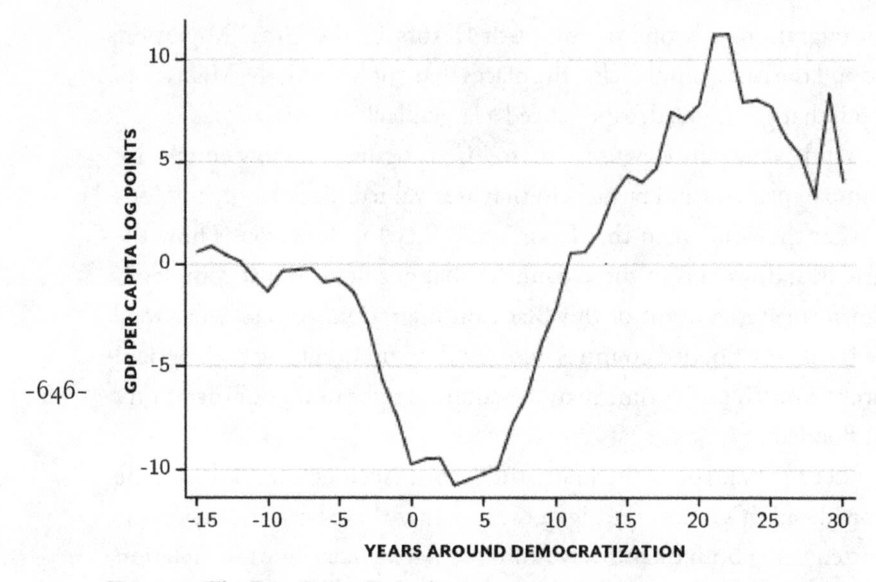

Figure 1: The Raw Data. *Figure plots GDP per capita in log points around a democratic transition. We normalize the average log GDP per capita in a country to zero. Time (in years) relative to the year of democratization runs on the horizontal axis.*

This is if you just look at the data of GDP per capita around a democratization (see fig. 1), you basically see that there's this big cliff that happens in the year around a transition to democracy: a country basically undergoes some economic crisis. And so, if you didn't handle this economic crisis that's leading you to switch into democracy, you would get a totally spurious effect on economic development, because that subsequent rebound would be confounded by just the rebound from the economic crisis.

What you want to do is form a counterfactual, where it's like you have two countries, both of which experienced the same kind of crisis, one of which switches into democracy and the other one doesn't. Once you do that exercise, you get a very clean and robust effect of democratization on economic development, which is roughly something like one, one and a half percentage points a year increase in GDP per capita. Over twenty-five years, you get something like a 20% increase in GDP, which is not too big, it's not too small—probably it's around the number that, in my mind, makes sense. Importantly, there's nothing prior to that; once you handle the four years of GDP before democratization, there's no preexisting structural trend; there are no other confounds that you really have to worry about. We document this in the paper that, basically, handling this economic crisis gives you a pretty good experiment, as far as you can go with this cross-country data. I'm sure that's going to be controversial, so I'm happy to talk about it.

The next thing I want to talk about is going into this labor market transition that I mentioned, and the Wagner Act and World War II. Also, the history is just interesting and relevant. The Wagner Act is this piece of law that basically legalized unions in the US. I mean, there's sporadic bits here and there, but really, unions, prior to 1935, were always under legal attack. And then, in 1935, the Wagner Act is passed in the context of a recovery from the Great Depression, and lots of people are still on unemployment. In the middle of all of this crisis, the Roosevelt administration passes the Wagner Act. It's also widely believed that it is not going to matter. Basically, everybody is convinced that the Supreme Court is going to strike it down within a year or so. The past rhyming with the present, Roosevelt threatens to pack the court in February of 1937 and then, lo and behold, the Supreme Court upholds the law in April of that year. One of the consequences of that is a sudden big increase in union density, one of the only moments that you've had a big increase in union density in the US. Then that just gets multiplied by the World War II War Labor Board basically mandating that if you are producing for the war, you have to be in compliance with the Wagner Act.

With these two things—the Wagner Act and then essentially taking 10% of the US economy and saying, "If you are producing for the war, you now have to respect the law protecting unions"—you get this really big increase in union density. And one of the interesting things about that is that, while lots of other things about the economy change in World War II, most of those things change back. Almost all of those things change back when the war ends. But unionization is one of those things that persists, and so you just think of the labor market in 1948, '49—it's not a war economy; it basically has turned back into a civilian economy.

But effectively one in every three households has a union member. And that's just a radically different labor market than what existed in 1934. It's a radically different labor market from what we have today, and it's a pretty drastic change. In this paper that I'm referring to, we basically produced some of the first estimates of union density back in this

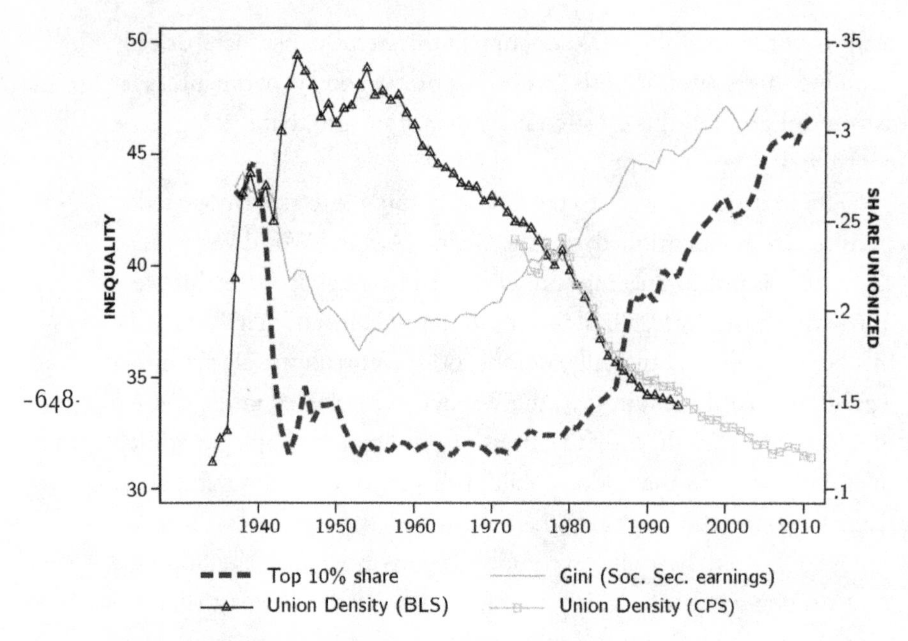

Figure 2: The Shape of Unions 1940–2010.

period, because it turns out no one asked in the census or the Bureau of Labor Statistics "Are you in a union?" until 1973, so most of what economists have known and thought about unions comes from 1973 on, which is a period of just monotonic union decline.

Let me show you this graph (fig. 2): this is the shape of unions. Union density is represented by the black triangles and the gray squares. The gray unbroken line is the Gini coefficient, and the thick dashed line is the top 10% share of income. Lots of people have shown graphs like this, but, importantly, before our work in this area, that line with squares was all that people really had with individual-level data. And what we noticed—what our contribution to this literature is—is that it turns out, you know who was always asking you about union status? Gallup polls. Basically pollsters, starting in 1936, always asked, "Are you a union member or not?" And we could go back and collect all of these old Gallup polls.

Once you start collecting them, you get roughly a million observations because they're done every month or a couple of months. Then you can form a new dataset that goes all the way back to the heyday of unions, and really look at who's in a union. In particular, you can look

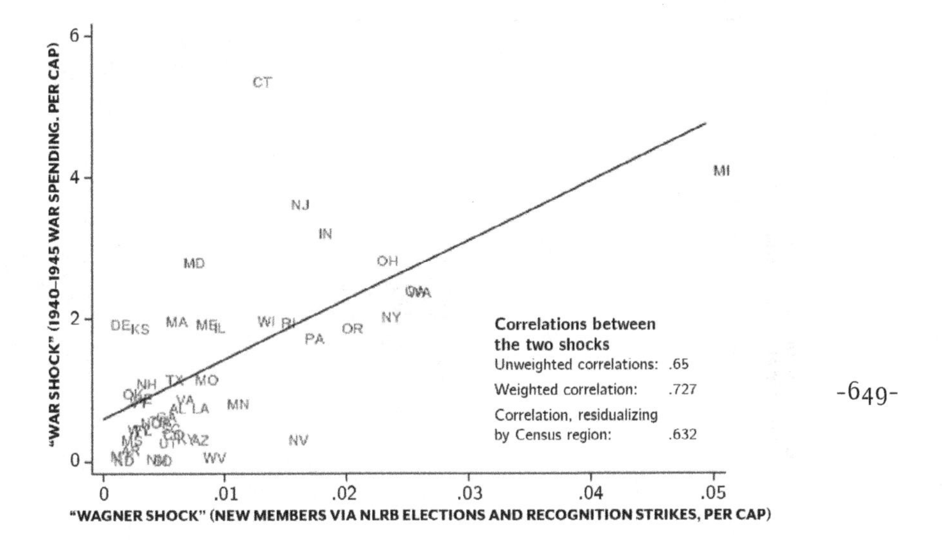

Figure 3: Correlation of the Two Policy Shocks. *On the x-axis is the (per capita) number of new union members by state, in the five years immediately following the passage of the National Labor Relations ("Wagner") Act. On the y-axis is the total value (in 1942 dollars) of military contracts given to firms, by state, from 1940 to 1945. The raw correlation reported is merely the fitted line depicted in the graph. The weighted correlation weights observations by 1930 population. And the residualized correlation is the unweighted correlation after controlling for four Census regions.*

at what states experienced the increase in union density, which is something you couldn't look at, really, before.

Let me talk about this graph on the right (see fig. 3). This is showing which states got a lot of new members via the Wagner Act on the *x*-axis, and then those same states then getting a big increase in war spending during World War II. I think Michigan is an obvious outlier here. It's like Michigan both was adding tons and tons of union members and was getting a lot of war money. Both of those things drove union density, and they drove them differentially in these particular states. Think of like Connecticut, New Jersey, Michigan, Ohio—these are states that are getting unions, and their unions are not going to Mississippi, Alabama, and Kansas in this period.

And you can compare these states that got unions in this period with these states that did not get unions in this period, and look at how inequality changed in this period, and that's what this next graph does. This is the difference between the states that got unions and the states that didn't get unions. The line with the small plus signs is the increase in

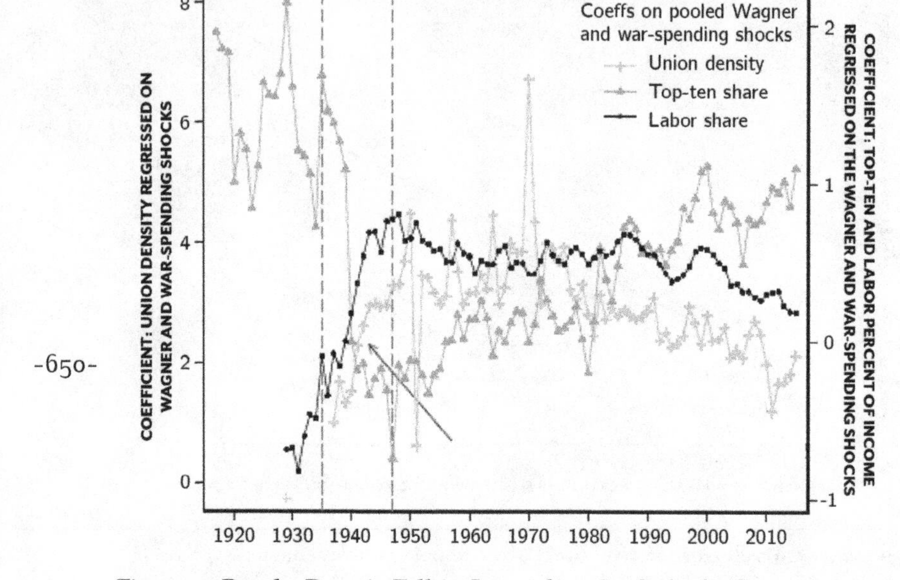

Figure 4: Result: Drastic Fall in Inequality—In Only the States Affected. *Regressing density and inequality outcomes on the pooled policy shock variable. The policy shocks are associated with significant increases (decreases) in union density and labor share (top-ten share) only during the treatment period.*

union density that's differential in these states that were unionizing in this period (fig. 4). Think Michigan, Connecticut, Ohio are along the plus-sign line, and they're different from all the other states. They experienced a big increase in union density, and at the same time those exact same states experienced a differential decline in the top 10% share of income and a differential increase in the labor share of income. The labor share of income is the total amount of wages times employment paid. That transition really happens only in that critical juncture, that 1935-to-1947 window of the New Deal and World War II. There's no action outside of that window; it's all pretty much flat. And so, it's this crisis, this period of American history, that really built social democracy in parts of the US. We had it for a while, and it's slowly been deteriorating, and again differentially deteriorating, in those same states since 1980.

Okay. My last little historical experiment here is looking at this Mississippi flood (fig. 5). For those of you that don't know, Jim Crow was this very repressive labor system in the US South, where Black Americans essentially worked in Southern agriculture, and it really was broken only

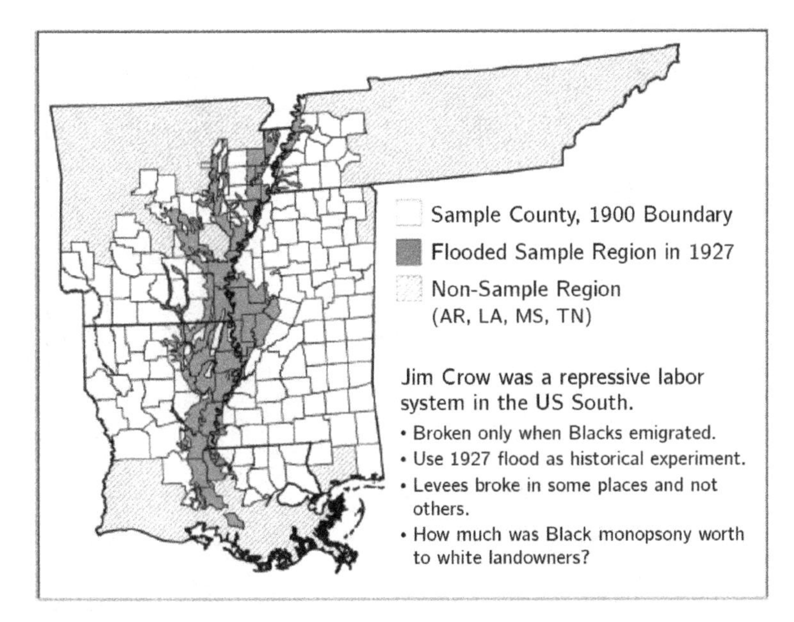

Figure 5: The Local Destruction of Jim Crow: 1927 Flooded Region and Sample Counties (1900 Boundaries). *The 163 sample counties' boundaries are based on county definitions in 1900. County-level data are adjusted to hold these boundaries fixed through 1970. The sample region flooded in 1927 is shaded gray, based on a map compiled and printed by the US Coast and Geodetic Survey. The non-sample region is cross-hatched. Excluded counties are missing outcome data in one of the analyzed years, have less than 15% of reported crop land in cotton in 1920, or have a Black population less than 10% of the total population in 1920.*

---

with Black families outmigrating, starting in 1910 and then accelerating over the twentieth century, to northern cities. Everyone thinks that immigration was really the cause of agricultural modernization in the South, but it's really hard to establish causality because you want something like an experiment, where you get some places that, for reasons unrelated to their agricultural system, had Black labor leave, and then look at the effects of that shock.

This is what the 1927 historical flood does for us, because the levees broke in some places and not others. And so, this map is showing you the places that got flooded, and then we are comparing the places that got flooded to the places just on the other side of the river that didn't get flooded in 1927. One of the things we can do with this is calculate—and one of the outcomes we look at, and I'll show you—is land values. You can use how much land values declined in the places where Blacks left as a measure for how valuable repressed Black labor was to planters,

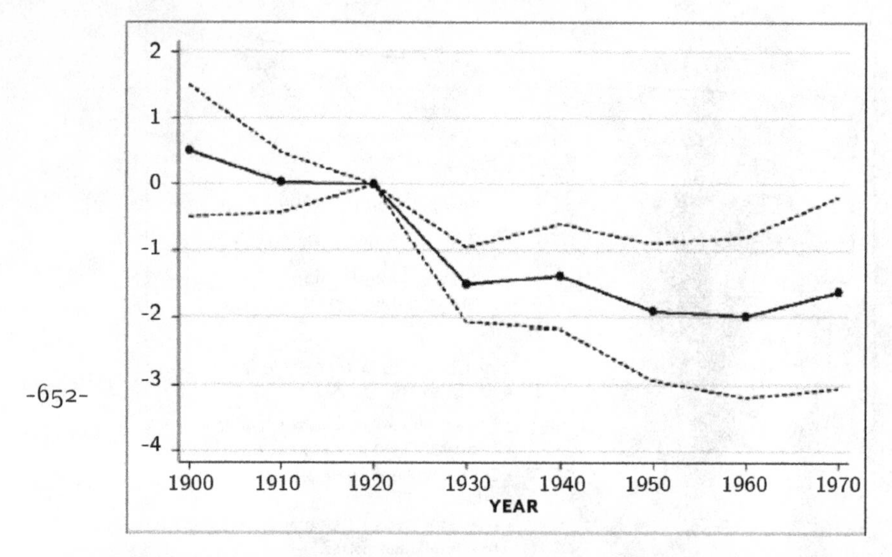

Figure 6: Flood Results in Big Change: Estimated Differences in Black Population in Flooded Counties. *This graph reports estimated differences in log Black population share between flooded counties and non-flooded counties, relative to differences in 1920. The outcome is regressed on the fraction of the county flooded, state-by-year fixed effects, and county fixed effects. The dashed lines indicate 95% confidence intervals, based on robust standard errors clustered by county.*

---

because that land value capitalizes the agricultural profits of those farmers. And so that's a way to convert this: by seeing the effects on land value, you can get a sense of the cost that this imposed on planters.

This is the difference between these flooded and nonflooded counties over time (fig. 6). You can see they're pretty similar leading up to the flood, and then after the flood the Black population falls and stays permanently lower. It's not that a new Black population comes in to replace the Black population that left; it's just like they leave and they're gone and no one comes back in to repopulate these counties. You first get this big outmigration due to the flood, and then you get this big increase in mechanization in those same flooded counties.

This is the pattern in farm capital (fig. 7). We also look at tractors and acreage being farmed. What you're seeing is that these places that lose Black labor wind up basically replacing Black labor with tractors; they pull a whole bunch more acreage, low-productivity acreage, into cultivation. Their output goes up enormously, so they're actually way more productive in agriculture, but their land values fall. It's a good example, right? Because the costs of production went up, even as the

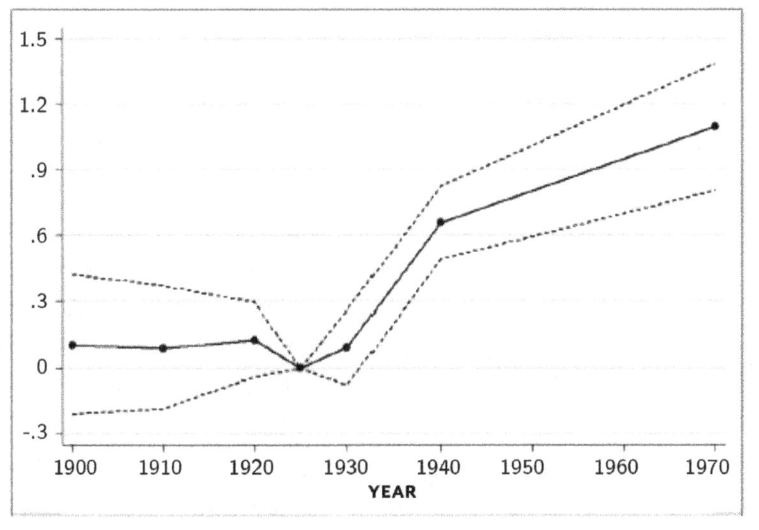

- Total population constant
- Output per acre increases
- But per-acre land values fall!
- Some is quality, but most probably reflects excess profits from Black labor

- Estimated elasticity of land value to Black labor share is 0.5.
- Total agricultural population is 50% Black+ $4.6 billion Southern farm value -> implies $250 billion today
  - Should be considered in any reparations calculation

Figure 7: Planters Respond by Mechanizing Early: Estimated Differences in Farm Capital in Flooded Counties, Relative to 1925. *This graph reports estimated differences in log value of farm equipment and machinery between flooded counties and non-flooded, relative to differences in 1925. The outcome is regressed on the fraction of the county flooded, state-by-year fixed effects and county fixed effects. The dashed lines indicate 95% confidence intervals, based on robust standard errors clustered by county.*

total scale of production went up, your profit can fall even though your output goes up a lot. You should think of it as like this Mississippi flood hit, and then it forced these planters to adopt this much more technologically advanced agricultural system, but they wouldn't have done it without that flood forcing them to do it, because the previous system was profitable for them. That's my last example of how a big crisis in these areas pushed these counties into a more technologically advanced agricultural system that was not necessarily profitable for them beforehand.

Okay. Let me just end here with some thoughts. I'm happy to take any kinds of questions. Why does crisis precipitate this kind of institutional change? Well, often it's the case that institutions persist just because it's too costly for a group to try something new, and a crisis can induce a particular group that was previously inertial or just trying to get by to do something new. Think of Blacks outmigrating after the flood—they could have outmigrated before, but it took the flood to push them to finally leave. Or think of the unions and the New Deal, when they had a really hard time organizing. Then the Wagner Act passes and they go out and are signing up members left, right, and center, and that forces everyone else to change how that whole little economy works.

The crisis works through a particular group deviating from the practices that were done before, and that changes how everything else is done. Or it could be that any particular subgroup was always pushing for a particular change, and then the crisis makes everyone else finally respond to this particular group that was pushing for change all along. You can have different accounts of this process, and probably both of these things are going on. In work I have with Sam Bowles and others at SFI, we have a bunch of formal models trying to capture these two different logics of institutional transitions. I'll leave it at that.

...a crisis can induce a particular group that was previously inertial or just trying to get by to do something new.

W. TRACY: Awesome. Thank you, Suresh. There is a big question that we've been discussing over the course of the day today, and will continue to think about tomorrow. We seem to be facing, as a global population, a number of correlated or interdependent crises right now. We have the pandemic; we have a great deal of social unrest, not just in the US but around the world; we have climate change. When we look at all of these crises coming at once, what types of institutional changes do you personally think we could see as a result of that? And, to the extent that you feel comfortable even speculating on it, what does that mean not just for short-term numbers like GDP growth but really the structure and long-term health of the economic system as it's currently in place globally?

S. NAIDU: I've found that the pandemic has forced me to narrow my lens—just keeping track of the policy environment in the United States has been hard enough. But I had been paying attention to the policy environment in the United States and . . . I know it is a catastrophe on many levels, but I will suggest that there are things inside the CARES Act, for example, that really point to a set of policies that we could do going forward in the future.

I think one of the things we learned is that a really generous unemployment insurance system that kicks in during a crisis like this is actually pretty good. And that the super UI that we did with the CARES Act was possibly one of the best policies that we've done in a long time. I'd like to think that, going forward, we'll think about ways to make the social insurance system in the US less vulnerable to the political process and much more on automatic stabilizers. So that when the economy hits a downturn, then . . . instead of having to go through Congress, we'll basically have a UI system that is pretty robust and flexible that will automatically kick in.

Here's something that you might not know. The unemployment system was topped up by the federal government with $600 a week over the course of the pandemic. So, if you were unemployed over the pandemic, you were getting $600 a week. I just want to point out [that] $600 a week is like $15 an hour for forty hours a week. That's more money for a lot of low-wage workers than they've ever seen in their lives. The reason that we did that $600 is because all the unemployment systems are run through the states, and they basically are still operating in Fortran. It was too complex to update the state software systems to do anything more complicated than just a flat payment of $600 a week.

I think one of the things that I would like—and this is just a small thing but it's in my wheelhouse—is that we've become much better at fixing the social safety net, so that when we're dealing with future recessions and things, we are much less vulnerable to the political idiosyncrasies and we can do it much better. That's a small thing that I hope we now have a lot of evidence from this pandemic about how we could do that.

AARON KING:[3] I thought that was a very interesting comment you made about the social insurance and its potential to act as a stabilizer in these situations. Along those same lines, you've thought a lot more

3 SFI External Professor

about this than I have. Have you thought about policies that we could put into place that would alleviate the need for major crisis as the instigator of social change? If your thesis is right that essentially you need a strong crisis on Earth to produce some of these social changes, we understand that, can we smooth the transitions?

S. NAIDU: Well, I mean, think . . . of the financial crisis of 2008–2009. It didn't really generate huge institutional changes, and that's partly because we managed it much better than we managed the Great Depression, even though it was not pretty by any means, but we didn't hit anything close to the Depression. Okay, that's why we didn't get institutional change. One would hope that what we've done is, for example, build a democratic and bureaucratic system that's much better at doing its own form of adaptive learning of figuring out like, "Okay, this thing happened, we'll build policies that will learn from that and will adapt going forward, so that it doesn't necessarily take the bottom falling out of the economy to rejigger the unemployment system," for example.

I'll put a pin in here for evidence-driven policymaking that it's a real thing in government to run experiments and to be continually adapting policies based on the evidence that's being generated by the government. That form of fine-tuning the regulatory environment would probably be better than having no change for a long period of time, and then all of a sudden rejiggering everything in a very short period of time. But I don't know, that's probably a ways off in the future.

A. KING: Certainly the most helpful thing I've heard in quite a while.

W. TRACY: I want to link that back to that first of the three vignettes that you shared with us. Do you see that process happening either in the US or in any of the other democracies that you're studying now? And I should say, I love democracy, I really do, but the only place where I've seen anything like what you've described would be China's economic transition under the Deng Xiaoping era, where they were running these experiments. And that was perhaps one of the least democratic systems that we've seen, doing large-scale economic experimentation in some time. Do you have examples of that from more democratic countries?

S. NAIDU: I do have thoughts about this because it's actually a really interesting thing that in order to run experiments on a population that you haven't completely subjected , you need to have some basic legitimacy, that people are going to accept that you're messing with them in their own interests. Any one of us that's had to fill out human subjects forms gets this. There needs to be some basic accountability between when you're running experiments with people; the people that are doing the experiments need to have some basic accountability to the people being experimented on.

When you're not in an authoritarian country, if you want to have this evidence-driven environment, you really need to have a broad mandate. And I'll actually suggest that Taiwan has been a democratic country that actually . . . just the response to COVID is amazing. But, also, the democratic ways in which they've induced citizen participation and contact tracing, and just had a really open and scientifically informed policymaking environment. We neglect Taiwan because it's a small country, but I think its handling of COVID was both on the outcomes really good and also procedurally pretty good.

W. TRACY: I would point out that Taiwan weren't doing the experiments that you were describing. They may have had scientifically driven policy but it was not any kind of experimentation, at least not to my knowledge.

S. NAIDU: Well, I mean, they do have small-scale experiments. But, that said, we actually have tons of experiments in the US all the time that are just done. Government funding experiments to be done happens, I think, both in Taiwan and in the US. The problem is that the outputs aren't integrated into the policymaking environment. We'll have experiments like Job Corps or the negative income tax experiments of the 1960s and '70s, and they happened; we got results back from them. One of the reasons we have earned income tax credit in the US is because of these outcomes, of these negative income tax experiments that were done in the '60s and '70s.

I don't think it's completely broken. I just think we have examples of them happening, and they need to be done much more systematically. I'll actually point out that the UK—actually, the conservative government in the UK has been pretty good about listening to evidence on things like the minimum wage, for example. They go out, they ask, they precommit to analysis of what will the effects of their minimum wage increases be on unemployment. If they found a negative effect, they would think about scaling it back. And so I think we do have, not a country that's doing it wholescale, but particular sectors of policy where you do have some of this adaptive, experimental feedback going on.

W. TRACY: I want to pivot a little bit. There's a conversation going on in the chat with Jenna Bednar, David Krakauer, and others about the intersection of COVID and inequality. And I wonder if you have any thoughts on what this crisis might do, particularly in the US context, for inequality?

S. NAIDU: I mean, I can talk about all the reasons it's a disaster for inequality. If we have another year of school closures, the intergenerational consequences of that are probably going to be big. But again, in my spirit of silver lining, let me give a silver lining. In particular, I think the movement of the super-high-value tradable services in the US to remote work could actually wind up moving a lot of economic activity out of the hubs of San Francisco, New York, LA, and into surrounding suburbs, into different towns. Really, you can imagine little clusters of software developers or telemedicine in South Dakota or in South Carolina, and there's a whole little economy built up around a set of remote workers that are doing the high-value services. That really deconcentrates a lot of the economy from these super-high-priced cities.

You can imagine that generally distributing economic activity more equally around the country might, in fact, have a larger egalitarian effect. There's my silver lining. Also, just another plug for CARES, I mean, that $600 a week for those three months, that radically compressed the income distribution, and there's very little evidence that it disincentivized work in any way. Maybe if we just funded it out of a tax on Amazon, that would be fine.

W. TRACY: I apologize for digging into the chat here, but Jenna, you had a slightly different opinion. Would you mind sharing that?

JENNA BEDNAR:[4] Absolutely. I'm wondering about your discussion about silver linings, which I like, of course, because I'm an eternal optimist as well. Do you have a theory behind it, in the sense of what causes a transition to be an opportunity rather than a failure? And then, is this inevitably a moment of greater inequality? For sure it seems to be, but

---

4 SFI External Professor

again here I'd love for you to share with everybody what you've been working on with equitable growth and shared prosperity. I've been doing some work on that as well, as have others—is there something in this moment that might cause people to wake up and reexamine the structure of our political economy? I'm thinking about the critique of neoliberalism—could that movement flip this moment into an opportunity for greater equity?

S. NAIDU: I think there is an opportunity. I worry, like I'm sure all of us worry, that there's just—and you know this better than me—too many veto points around the American political system that even if there is a big rethinking of the economy right now, that it's too hard to do it by the political system because of the structure of this rickety Constitution.

J. BEDNAR: No, I would say veto points, yes, but then there's a lot of complementarity. That is, we need to think a little bit differently about our political structure and that maybe when we have failure at one level, we could have successes and progress at another level, and learn from that. I'm sorry, I don't want to get in the middle—

S. NAIDU: No, no, that's interesting. Maybe it's just the past month that I'm just like, "Oh my God, this Constitution is a disaster," and that really the undemocratic components of the Constitution are really slapping you in the face these days. That's really interesting that there might actually be levels of consensus hiding, despite what looks like all of the obstacles to changing stuff. But I actually think you can win the war of ideas, but you just actually lose the actual war.

SAM BOWLES:[5] There are two things, quite different: one is about learning from social policy experiments. There are really two interesting cases to me. Around the turn of the last century, the Progressive movement in the upper Midwest, prior to the First World War—those upper Midwest states passed a lot of very progressive social legislation, workmen's compensation and a whole bunch of other things. They then became the basis of a lot of policies, which were pretty quickly passed after that, in the New Deal. That's a very short lag—that's a

5 SFI External Professor

twenty-five-year, thirty-year lag. And of course, I mean, it wasn't just copying them, but they were there and they seemed legitimate and so on.

But then contrast that with something that Suresh knows, because Suresh is from Canada . . . I mean, how can we possibly be living next to this country which we consider so similar to us, and there is so little learning about the Canadian healthcare system? I mean, it's an extraordinary fact that we just don't look at the data and say, "Oh, maybe we aren't the best healthcare system in the world," and it doesn't involve going and doing exotic calculations with other countries whose languages we don't understand. Those are two very different experiences in American social policy.

I wanted to come back, Suresh, to what you said at the end about the war of ideas. I think you're right. I don't think that it's silver lining. I think the COVID crisis and climate change together are really going to kill the idea that individualism and unregulated profit-seeking are a way to organize an economy. We have no idea what's going to replace it, and I think there's a lot of vacuums there, but it's really going to be very hard to sustain the neoliberal rhetoric or even the neoliberal models.

S. NAIDU: Yeah. I actually agree with you. I think that's dead, and my worry is more like this—that we're now splitting down on who's in the group and who's out of the group.

SAM BOWLES: Absolutely. I think it isn't at all clear. Wendy [Carlin] and I—and we've been talking with Suresh, also—have written that there are possibilities of changing to a way of understanding the economy and society which places much more importance on what we call community or civil society. It's not just a debate between governments and markets, but there are communities and other interpersonal relationships governed by values other than obedience and greed. Yes, that's fine, but among those motives, that they are very cooperative in civil society, is us versus them—call it tribalism. That's also a big possibility here, and, I mean, there's a lot at stake now. But I think, as Suresh said, it's a time that there's just a lot up for grabs, and it's worth spending some time working on it.

S. NAIDU: I can respond to the point about educating our inner-city children in the chat. I agree it's been ongoing for decades, but what's surprising is that we actually have had moments where we've made real dents in inequality. In racial inequality, for example, recent research has really shown that the 1966 expansion of the minimum wage to sectors that were exempted from the Fair Labor Standards Act (FLSA) really put a big dent in racial wage inequality, much more so than affirmative action and the Civil Rights Act.

It was actually just pushing up the wage in traditionally exempt sectors of the South, like textiles, and that really put a big dent. Education is important—no one will deny it—but it shouldn't be . . . I think there are lots of things you can do as well as educational reform that make a big dent in various measures of inequality.

W. TRACY: One of the other themes that we've seen earlier today, and I expect we'll see more of tomorrow, has to do with information and belief flows. Nancy, do you have a question related to this?

NANCY HAYDEN:[6] I was interested, going back to what you were talking about, the democratic transitions, and really appreciated what you were saying about looking back and finding use cases that had this same economic downturn preceding the economic upturn after a democratic transition. A different way of looking at democratic transitions that I found so fascinating was, in the literature and the democracy of transition scholarship, the ones that go back and look over history, we tend to think that democracy is a modern invention but in fact it is several thousand years old.

There's a fair amount of good historic data about the ancient Greek city-states, and some of the other experiments that preceded the modern one. And, as I recall, some of those comparative studies looking at why some city-states in ancient Greece evolve more democratically than others, or some isolated examples in Rome, that a lot had to do with the networks and road systems and communication systems. I was wondering if you saw anything like that in some of the more modern studies and comparisons you've done, and/or if you've ever gone back and tried to replicate these comparative studies in transitions that have been done historically and seen anything new with your approach?

S. NAIDU: Wow. The data gets a lot worse, the farther back you go. And it's really hard, or it takes these kinds of giant data-collection

6 Sandia National Laboratories

exercises, like the kind that Sam does, to produce these kinds of statistics for ancient societies. I actually suspect somebody has tried to do this for the city-states of ancient Greece, but I can't remember the reference. It's just really hard to get a measure of economic activity, and even measuring GDP in a classical economy is really hard. Most people don't work for a wage, so measuring prosperity is just a tough, tough thing to do, even for just one society in the ancient period.

N. HAYDEN: Do you have any thoughts—just conjecture even, maybe—would it be worth exploring the connection between the economic activity and the kinds of communication, information, proximity networks?

S. NAIDU: Oh, absolutely. I think it's one of our big challenges. We're just finally now getting the granularity in the data that would let us examine the things that we have always thought were really important for historical transitions.

For example, with the COVID crisis, the granularity of the data that's getting released in real time, that you can really see . . . I've been working with geolocation data where you can literally see people attending the Black Lives Matter protests, and you can see all the clumping of the device IDs in the streets. That's the kind of thing that we think really is a precipitator of these kinds of institutional changes, the kind I'm talking about, but we could really only measure these kinds of informal social gatherings, the networks, the patterns of communication . . . I think we've only just begun to start to be able to measure them in a systematic way.

N. HAYDEN: Just to keep digging, just one more little bit. Okay, I know that we find that interesting. What I was thinking about was the correlation—or not—with the economic trends that you were looking at, especially going back and saying, "Here was this downturn." It's putting those two together. We all know the story about the Arab Spring and the role that the networks played there, and that this was following, as you were saying, the crash in serial market. But you've got this larger set of economic data and these larger patterns of communications, even beyond social media. And for me, I feel like there's a deeper story there than just one about the crashing serial markets and social needs.

S. NAIDU: You mean something like the ideas in the air or what people are talking about?

N. HAYDEN: Yeah, and just the many layers of communication networks, and how it's not just communication networks, but it's the economic distribution networks. All of the information networks are much more than social media, and they underline the economic activity in the country. And the kind of network that underlies the economy in a transition to a democratic country is very different than the kind of network and information systems that you have that underlie finance and economics in an authoritarian country, for example. And it seems like there's a richer story than just, "Oh, we had GDP changes . . ."

S. NAIDU: Absolutely.

N. HAYDEN: Much more complex set of networks there.

S. NAIDU: Yeah. I'll just give you one anecdotal thing about what I've looked at and never turned into a paper of any form. But, with the value-added tax data, you can actually see the whole network of firm-to-firm interactions in a country, because every transaction between firms is logged for the tax authority. And one of the very interesting things is to look at this in an authoritarian country like Russia; because of that, you can look at the path.

You basically see all of the economic activity oriented toward the companies that are very close to the government—for example, the state-owned oil companies, the state-owned banks. Basically, the companies that are owned by people, by oligarchs, that are close to Putin, those are the kinds of companies that economic activity is directed toward. And you would think that once you remove the power of the government, you would probably get a total reconfiguration of that network, for example.

REFLECTION

CRISIS AND INSTITUTIONAL TRANSITIONS

DATE: *June 28, 2021*

FROM: *Suresh Naidu, Columbia University; Santa Fe Institute*

As I write this, I have Firefox tabs open about the economic changes engineered in the recovery from the pandemic. Low-wage workers are scarce, and wages are increasing at rates unheard of in recent years, particularly in traditionally low-wage sectors like retail and restaurants. Employers are not just raising pay, but altering working conditions. Whether this results in a sustained and sustainable improvement in Americans' working lives and a reduction in our staggering inequality remains to be seen, but the very possibility is indicative of a larger historical pattern: economic crisis often provides a moment for institutional changes, sometimes for good, sometimes for bad. I want to discuss evidence of the silver linings: when negative economic shocks propel institutional changes that result in something (arguably) good.

It is not controversial to say that societies are complex systems, which include people thinking about the complexity of those systems. What is less appreciated is how complex systems, by virtue of having feedback processes that are self-correcting locally (that is, the system is robust to small shocks), may also generate sudden transitions. The classic instance of this in economics is the business cycle, where the vast decentralized coordination enabled by the price system (e.g., profit-maximizing firms driving costs down by competing for customers) also gives rise to sudden downturns and crises.

One thing the pandemic has revealed is the importance of the economy as a *political* economy. In political economy, it is further acknowledged that the economy is linked inextricably to government policies, ideological changes, human community structure, and social norms of noneconomic interaction. Thus, the "general equilibrium" effects of economic shocks include endogenous policy responses, institutional changes, and even the evolution of ideas themselves. As John

Maynard Keynes wrote,[1] to do economics well, no part of nature or institutions must be entirely out of an economist's regard.

But further inputs into the political economy include our own knowledge of historical events and our abilities to make analogies. I spent some time thinking about parallels to the 1918 influenza pandemic, and while there are some , it is a qualitatively different experience today, partly because we now expect the state to manage disease much more effectively than we did in the early twentieth century. So looking at historical parallels wouldn't necessarily give the right answer to the policy questions today. To an economic historian, this is humbling!

It is clear that models must better reflect our understanding of the aforementioned role of endogenous policy, norms, and customs in general equilibrium theory, so that we can't think only in terms of economic feedbacks in societies where government allocates upward of 30% of GDP and intervenes massively in the economy. Further, any model of endogenous policy that worked would likely be self-negating as bureaucrats and politicians acted on it. I think we just need much more data and evidence on these complex interactions before we can start to make useful theories. I suspect the path here will look more like biology than anything else: a few high-level principles but mostly an endless concatenation of diverse pathways.

So why do institutions persist for long periods and then change in response to economic shocks? Institutions persist because it is too costly for most people to try something new, and a crisis can induce a particular group that was previously inertial to do something new (for example, workers go on strike, Black sharecroppers try to emigrate). This novel behavior of a subgroup can force everyone else to change, and this changes how the whole society works. Alternately, it could be that there was a small group always demanding an alternative set of institutions, and the crisis changed the economic and political pay-offs so that the whole society is now responding to these demands.

1 J. M. Keynes, 1924, "Alfred Marshall, 1842–1924," *The Economic Journal* 34(135): 311–372, doi: 10.2307/2222645

SYMPOSIUM PANEL DISCUSSION S-005

SFI FACULTY PANEL ON THE COMPLEXITY OF CRISIS

DATE: *14 Nov. 2020* **FROM:** *The Faculty of the Santa Fe Institute*

DAVID KRAKAUER: I just want to make a quick point that's obvious: we're living through a very dangerous time, and adherence to science, experiment, and evidence is more important than ever, and sometimes it feels we're moving in the wrong direction—hopefully, not for long.

In this morning's *New York Times*, there was an article about the discovery by Mordechai Feingold, who is Newton's biographer, that when the *Principia* was published in the late seventeenth century, it had a much larger readership than we realized. And I was kind of delighted by that idea that, in fact, this incredibly obscure, difficult, and important work was greeted with some enthusiasm, and it makes me think that perhaps the past is not quite as irrational as we thought and the present *might* be.

But we're here because of works like the *Principia*, works that are rigorous, and shocking, and beautiful. And today, hopefully, we'll hear rigorous, shocking, beautiful ideas. The way we're going to play this out is we have a series of questions. These questions come from friends of the Institute. Eight of those ten questions were videoed in, and you're going to see those, and then I'm going to turn to the panelists to answer them. If they're unable to answer those questions, I will turn it over to the audience, and perhaps they'll be able to answer them.

So, without further ado . . . I want to give you a sense of the kind of standards we're aspiring to in the question and answer. Let's run the first question, and the first question comes from Valerie Plame.

OPPOSITE: *Anonymous, "People Fleeing a Burning City," c. 16th century (detail)*

VALERIE PLAME:[1] This is Valerie Plame. My question for the "Complexity of Crisis" symposium panel revolves around the rise of deepfakes in AI and the consequences for our liberal democracy, where what we believe to be the truth can be so deeply subverted through AI, through deepfakes. ***What are the consequences for our society, broader democracies around the world, as well as the information ecosystem?*** Thanks so much.

D. KRAKAUER: I'm going to turn it over to our panelists to see which of you would be prepared to have the first stab at that question.

MAHZARIN BANAJI: Not that I'm the expert, but I have been attending meetings on deepfakes. And what I would say is that a country that cares so deeply about freedom of speech as this one has never quite come to terms with the fact that the evidence about how the brain works should make us at least think about what we mean when we say freedom of speech. And I, as a great believer in it for all the reasons David said—you have Newton saying crazy, amazing things, and we would want Newton to say crazy, amazing things—but we ought to know that when our brains hear a statement, whether it's true or not, the natural stance of the brain is to believe it. It is a believing organ. Something big has to happen for it to disbelieve something. Disbelief is not natural to us.

If I say it rains diamonds on Saturn, your brain will believe it. That one happens to be true, but even if it weren't, you would [believe it]. And now there are data showing that if I say it twice, "It rains diamonds on Saturn, it rains diamonds on Saturn," you see something called an illusory truth effect where you're more likely to believe it simply because it was said twice. Now, we see this even in young children. So I think those data have to just be put alongside everything else that we believe for us to adjudicate where we want to be.

D. KRAKAUER: Thank you, Mahzarin. Would someone else in our illustrious audience like to chime in on that question? If not, I'll go to the next question.

JESSICA FLACK: Before you jump to the second question, can I make a small remark? This wasn't a primary question I wanted to answer, but I want to point out and ask—because perhaps there's somebody in the

---

1 Author and Former CIA Agent

audience who knows better than I do—something I've been wondering about now for a while: do we really know that the quantity of mis- and disinformation that we are currently confronted with is greater than it has been in our past? My suspicion, if you think about ancient Rome and so forth, is that propaganda, we know, has always been with us, and if you do the corrections for time, and place, and mechanisms of getting information out there, it might not be the case that our situation is currently worse. So we've been, at least from a cultural context, evolving in the context of mis- and disinformation for an eternity.

. . . when our brains hear a statement, whether it's true or not, the natural stance of the brain is to believe it. It is a believing organ. Something big has to happen for it to disbelieve something.

RICARDO HAUSMANN: Can I jump in? I think that the problem is deeper in the sense that we scientists tend to believe that if the evidence says something, then you come to some conclusions, and that that's more or less how the brain works. But the brain has not evolved to help individual scientists find the truth. It has evolved to help us live in a social network. As a consequence, it is social to the core. Even reasoning is social, as argued by Dan Sperber and Hugo Mercier.[2] It evolved to justify our actions to others and to persuade others. People evolve beliefs about the nature of the world out there and their role in it. These beliefs are not just personal: they are shared, and sharing creates a sense of belonging. Beliefs are not just there to be *right*; they are there, in part, to belong, to be part of something bigger, to belong to a community.

So, if a fact contradicts your beliefs about the world, you disregard the facts. There's been a lot of experimental evidence on the idea that if you give people more information but you embed that information in a

2 H. Mercier and D. Sperber, 2019, *The Enigma of Reason*, Cambridge, MA: Harvard University Press, https://www.hup.harvard.edu/catalog.php?isbn=9780674237827

situation in which they have partisan beliefs about something, they will disregard that information. I don't think that it has to be deep AI. I don't think it has to do with the amount of information; I think it has to do with the underpinnings of the polarization that is at the core of social life.

D. KRAKAUER: Great. We don't need a video for the next section because Esther is asking the next question, and Esther is here.

ESTHER DYSON:[3] I was going to add that one of the things that's new is the tailoring of the information to the recipient. Different people see different things, and whether you're using what used to be called AI and is now just called advertising targeting, or something more clever—or something blunter—people are getting messages that are specifically designed for them.

I'd love to hear your thoughts about how human beings evolved to match a fitness landscape that was closer to the times of Newton or even before that. Now instead of scarcity and a small world filled with people you know, we've got a world where you now have to pay attention to what's going on in Africa and feel responsible for it (or maybe not); where there's too much food with too much sugar in it; where there are things that will attract your attention all the time... Rather than going out every day and seeing the same trees with a few leaves more each day, you get all these different videos or maybe some targeted information designed to entice you. ***Just speculate a little bit about what that means, how we should deal with the fact that we evolved for a completely different landscape.***

D. KRAKAUER: Thank you, Esther. Anyone want to jump in on that question—a question about the environment of evolutionary adaptiveness, I guess.

DAVID WOLPERT: I could say something that's sort of related. Putting on a bit of a big-history perspective, in a certain sense, kind of like what Jessica [Flack] did. Pandemics, historically, are actually amazingly inconsequential to the socioeconomic systems they're embedded in. Nobody would remember the Spanish flu if it weren't for the fact that we were going through COVID. Just going back a little bit farther,

3 Wellville

very few people remember that the last major plague pandemic, which killed many millions, was actually at the end of the nineteenth century.

Even when you go back to things like the Justinian plague, which potentially wiped out about a quarter of the eastern Roman Empire, as best as historians currently understood, it weakened the Byzantine Empire a little bit, and it helped to thwart Justinian's attempt to recover the western empire, but it wasn't the death knell of the eastern Roman Empire. In fact, people now think that the Justinian plague had something to do with, decades later, the serfs and so on had a little bit more bargaining power vis-à-vis the aristocracy. For that to be the sum total of the consequences of wiping out a quarter of the population is flabbergasting.

> Pandemics, historically, are actually amazingly inconsequential to the socioeconomic systems they're embedded in. Nobody would remember the Spanish flu if it weren't for the fact that we were going through COVID.

So, based upon that historical context, one might have predicted that COVID, with its mortality rates, would have been nothing, like Trump's flu plus plus. But there's something that's really extraordinarily different this time that touches on the question, which is that we now have communication mechanisms of baud rates that far exceed many, many orders of magnitude than was ever there before. We now have possibilities of things like a countrywide lockdown. Nobody in any pandemic before even had that as an option to do, never mind considering what the consequences might be. So, we're completely terra incognita in the social aspects of this.

If you were to look at a ratio of fatality rates to socioeconomic political consequences, COVID, I suspect, is going to be way off the charts for all these kinds of reasons that have to do with social amplification, which somewhat touches on Esther's question, because in a certain sense, humans are made for pandemics of the traditional type, where one hunter-gatherer tribe might put it on to the next hunter-gatherer tribe and so on, but this isn't even *social* media. This is something in the fact that you can have a government that can impose regulations that are more or less followed—in China, in Confucian countries, they actually *are* followed—that is completely new, independent of even things like social media, and the consequences for the economic systems and things like that are things that we have no experience with, that we weren't designed for, so to speak.

D. KRAKAUER: Thank you, David. I'll move to Doyne.

J. DOYNE FARMER: I guess I'll just maybe say that *we evolved*. The thing that makes human beings so powerful and unique is our ability to manipulate our environments. Many organisms can do that, but we can do it dramatically. Coming along with that is the fact that we're very plastic. We can adapt to remarkable situations, as we've already seen in the Industrial Revolution, when we put people in settings that were very different than jungles or savannas. I think that we're always being pushed to the edge of that envelope, because social evolution has become such a powerful force and we are selecting based on memes, and norms, and, as Ricardo said, we are social organisms, so we're being selected based on those things. We—being the means and ideas that propagate, the people that thrive or don't thrive—are the ones who participate in those, so we're being evolutionarily pushed not to be happy [but] to do what propagates. Now, it's always been that way, but as we get in increasingly artificial environments, as technology gets increasingly more powerful, then we're pushed to the envelope of that and things start to happen, like anxiety skyrocketing. Just go look at the numbers. It's quite amazing. I can't quote the statistics right off the bat, but I believe 80% of Americans had expressed feeling anxiety in the last year. And, ultimately, we might get pushed past where we can handle, and then the system has to go back, but it's always going to keep pushing us to that edge, and we're always going to see this interaction between our psyches, our social norms, and the technologies, and the technologies keep changing the game and driving us farther and farther to the edge of what can be done.

D. KRAKAUER: Thanks, Doyne. Ricardo, you're the last panelist.

R. HAUSMANN: Okay, sure. I think that in some sense, we're pulled farther away from the environment in which we evolve, but in other ways, technology might push us back to increase the salience of some things that were much more salient in our evolutionary past. For example, in our evolutionary past, we were a lot about "us and them," and the idea of the individual was really not part of that world view: it was really a creation of the Enlightenment. Before that, we thought of ourselves as part of a group, and we would die for the group, and so we would think about the group. In some sense, the polarization of the American society is sort of like the end of this weird period of individualism, and *weird*—I'm using it in the double sense of *weird*: Western, Educated, Industrialized, Rich, and Democratic—societies, which, as Joseph Henrich has argued, is very much based on the individual.

If you want to understand American behavior, including politics and elections, you might consider that people are behaving increasingly as members of a tribe. And with the tribe goes what you eat, where you buy, what you think about poverty in Africa, or about wearing masks, or about abortion and a thousand other things. If you do not share those beliefs, you will not be perceived as a true member of the tribe. So you align yourself with a pretty encompassing set of beliefs that signal not your careful reading of the facts, but your deep belonging to the group. Some of the new social media technologies might be pushing us to reduce our sense of individuality and reinforce more of the us and them mentality that we evolved to be programmed for.

D. KRAKAUER: That's interesting. Right.

J.D. FARMER: I'm going to channel Peter Turchin for a moment. It's worth taking a look at his book.[4] He has a measure of polarization and social dissonance that he gets from newspapers and media, and it's quite striking to look at how, before the Civil War, it skyrockets, it goes along, and then it just shoots up to an amazing level. After the Civil War, it simmers back down, and then in current times, it's back where it was just before the Civil War started. So, there may be some déjà vu.

D. KRAKAUER: Yes. Thank you, Doyne. Let's move on to the next question. And the next question is a virtual one.

---

4 P Turchin, 2016, *Ages of Discord*, Chaplin, CT: Beresta Books.

IAN MCKINNON:[5] All right. Well, it's always hard to pick just one question, but if I could only pick one question about the next year, it would be this: ***The pandemic has revealed that many complex systems have prioritized efficiency over redundancy, robustness, resilience. How much do you think the pandemic will alter that balance, and how lasting is that effect likely to be?***

D. KRAKAUER: Good one. Who would like to take the first stab at that? Jess, why don't you go first?

J. FLACK: Okay. So, will it change? I don't know. But I think perhaps one step in promoting change is to be clear about what design for robustness would entail. To do that, I want to introduce the concept that David and I and some others at SFI have been developing called emergent engineering (EE). This is very preliminary work, so take all of this with a grain of salt, but in a truly EE-capable system, there's no commitment to any particular outcome. The objective would be to let good solutions—be they policies, or social structures, or organizational designs—to challenges that the system encounters emerge through a process of collective intelligence. And this is a fundamentally different strategy—and philosophy, for that matter—than what we embrace now, and certainly than we've embraced in our history.

A somewhat weaker version of this would be to build an EE-capable system but constrain it so that it respects certain principles or values that we collectively deem important. I think I'm getting ahead of myself here to another question, but here's where ethics and science touch. As an example, rather than guaranteeing a particular form of government or governance, we might let an appropriate strategy emerge, given environmental challenges, but we guarantee something we agree on is important, like, for example—and maybe we don't all agree this is important—that there's a high minimum quality of life for all individuals through something like social insurance. Suresh Naidu was talking about that yesterday. The weakest form of an EE-capable system stresses process over outcome and robustness and adaptability, which is, of course, involved in all the other forms I just mentioned, and scenario planning over forecasting. And, I guess, most importantly, it proposes building tuning mechanisms into systems that control how

5 Sandia Holdings, LLC & SFI Trustee

REFLECTION

EMERGENT ENGINEERING FOR ROBUSTNESS & EFFICIENCY

Jessica C. Flack, Santa Fe Institute

A Darwinian view of life might seem to suggest nature should favor efficiency over robustness. Yet when one looks to nature, one finds robustness mechanisms are widespread and, given the complexity and intricacy of biological systems from cells on up to human societies, apparently also essential. After all, so much could go wrong but does not. In principle, robustness is hard to evolve, at least to rare perturbations, because in the absence of perturbation there is no incentive to pay the cost of maintaining robustness mechanisms, which then "drift" away from their robustness role. Whether in practice robustness is hard to evolve depends on many factors, including how common perturbations are, the timescale on which drift occurs (how fast do the mechanisms decay or get eliminated), and whether the mechanism contributing to robustness can also be used for other functions or the cost of maintaining it made negligible by distributing it over multiple components, each of which also plays other roles.

This basic logic applies broadly in biological systems. In the case of human social systems, there is increasingly an additional complexity. As better microscopic data on behavior become available, individuals and groups in human systems gain both the capacity to compute global information and greater power to intervene in outcomes. However, it is critical to realize that, in complex systems, greater capacity for intervention does not necessarily imply increased capacity to have a desired effect. Causality in complex systems, even with good microscopic data, is notoriously difficult. Reasons for this include nonstationarity, complex couplings, and concordant nonlinearity, which can lead to heavy-tailed distributions and the formation of spurious associations. Hence, even if substantial resources can be marshaled toward a given end, there is no guarantee that specific end will be achieved.

Many of us intuited during the COVID-19 pandemic that our systems are overdesigned for efficiency and underdesigned for robustness. But is this really the case? If these systems are indeed complex, in the absence of a long history of experimentation the probability is low that efficient solutions will be discovered. To answer this question rigorously we would need a series of empirical studies examining the following: (1) Are healthcare systems, financial markets, firms, and government agencies and institutions like the Senate designed to optimize particular functions (we know the answer to this in some cases—election design is one example)? (2) Does the design actually work—is optimization with respect to these functions achieved, and over what timescale? (3) If efficiency is achieved, is the system actually complex?

I think we will find that, although efficiency–robustness positions permeate policy debates around COVID-19 and other crises and consequently seem like implementable positions, in practice the consequences of human interventions with respect to achieving specific outcomes are murky at best, and our intuition that we have overdesigned for efficiency is overconfident.

Of course, being bad at efficiency does not imply we are good at robust design or even that we should design for robustness, but the reasons why efficient design is hard should give us pause and motivate us to rethink our design goals.

My colleagues and I have argued that one way forward is to develop a science of emergent engineering that emphasizes process over outcome and harnesses collective intelligence to identify problems on the horizon and temporary solutions to them as needed. This is an extremely preliminary idea, the crux of which is that we design our systems to have levers at multiple scales. Rather than controlling outcomes, the focus is on modulating fluidity by building systems with levers that facilitate or impede transitions. At the microscopic scale, levers include nudging and other forms of mechanism design to change behavioral strategies with the goal of obtaining "good-enough" collective output to solve the challenge at hand. At the mesoscale, we build levers that allow conditional changes in system states or regime shifts given crowd-sourced perception of environmental signals.

Under a modest program of emergent engineering we would, in principle, be moving social systems between regimes known to work in different environments much like a group of fish shifting between shoaling—a loosely aligned formation that allows foraging—and schooling when a predator appears. Under a bolder program of emergent engineering we might design levers that allow us to shift (temporarily, ideally) between a regime known to work in a relatively predictable environment and, when the environment becomes uncertain or demands innovation, a regime that incentivizes exploratory behavior.

FURTHER READING

J. Flack and M. Mitchell, "Uncertain Times," this volume on page 549

J. Flack and C. Massey, "All Stars," *Aeon*, November 27, 2020, https://aeon.co/essays/what-complexity-science-says-about-what-makes-a-winning-team

S. Strogatz, 2004, *Sync: How Order Emerges from Chaos in the Universe, Nature, and Daily Life*, New York, NY: Hyperion.

D.C. Krakauer and G.B. West, "The Damage We're Not Attending To," this volume on page 541

J. Pearl and D. Mackenzie, 2018, *The Book of Why: The New Science of Cause and Effect*, New York, NY: Basic Books.

A. Wagner, 2007, *Robustness and Evolvability in Living Systems*, Princeton, NJ: Princeton University Press.

N. Ay, 2020, "Ingredients for Robustness," *Theory in Biosciences* 139: 309–318, doi: 10.1007/s12064-020-00332-4

E. Maskin and A. Sen, "The Rules of the Game: A New Electoral System," *The New York Review of Books*, January 19, 2017, https://www.nybooks.com/articles/2017/01/19/rules-of-the-game-new-electoral-system/

fluid they are and how responsive they are by adding levers that allow for adjustment of timescales—in particular, control of timescale separation, adjustment of distance from tipping points, or critical points.

Now, if you're familiar with SFI's history, as I think everyone in this room is, you're probably thinking, "Well, we've moved into familiar territory here." These ideas have long been discussed at SFI, and some progress, of course, on levers has been made and implemented in various domains in the real world, in economic systems like defense and so forth. However, the idea that we can optimize for robustness and adaptability by tuning timescale separation between, for example, individual behavior and institutions is a relatively new idea and doesn't have many design precedents.

So, to make this a little bit more concrete, I want to give one example, and it comes from nature, from schooling fish. Most of the time, schooling fish—not all fish school, but schooling fish—they're in what's called a shoal; that means they're loosely aligned, foraging together. And then a predator, perhaps, appears, and one fish detects the predator, and they flip into a schooling state, when they're all closely aligned, and they can both confuse and/or evade the predator more effectively. So essentially what we've got here . . . it's a little bit under debate, but it's thought that this is made possible by the fact that the fish in the shoaling space sit near a critical point, so that information allows that observant fish, detecting the predator, to move across the system rapidly.

Now, of course, they don't know whether the predator is going to appear, they don't know when that's going to happen, but they know what to do when that predator does show up. So there is some uncertainty in this example. And what we get is, essentially, the fish have in their sort of social system repertoire, two states, schooling and shoaling, and they can move between these as the environment demands.

For some of the situations we've been dealing with, like COVID, where we haven't maybe experienced this kind of shock before—the fish have seen the predator for a long evolutionary history—we want to be able to move into a state that's exploratory, right? And we can probably guess by moving into this exploratory state that there are going to be some costs associated with that, but whatever we discover there is probably going to be better to the over-fixed state we were previously in, right? So that's an important point.

What I'm proposing here is that we—to come back to Ian's question—consider designing systems that can move between different aggregate states as the environment demands, and maybe we allow that exploration to be more sort of operational when we detect that the environment is becoming uncertain, and we control or keep the system from doing that exploration when we have a feeling that the environment is more stable. Finally, the sort of newer part, that, in addition to building mechanisms for shifting between states, we build in tuning mechanisms that allow us to control how sensitive our system is to noise. So that's sort of my long answer to Ian. I think there's a lot we could build on, based on work at SFI, to design systems for robustness and adaptability, and we definitely should be thinking hard about how to do it.

> ... in a complex system, you don't know what you need to be robust to because you cannot imagine all the things that can go wrong.

J.D. FARMER: This idea has been around a long time, the idea of resilience versus robustness and the trade-off that complex systems make. Joe Tainter, who has often appeared at SFI, had it in his book on collapse of civilizations[6] that civilizations collapse because they become more complex, and they become brittle, and they fall down. Around the beginning of the millennium, Jean Carlson and John Doyle had the theory of highly optimized tolerance, that complex systems evolved to be robust yet fragile.[7] So, they're robust to perturbations, to shocks, that they feel with some frequency, but they become fragile to perturbations that they never see because they're constantly tuning themselves to optimize, and optimization drives that.

6 J. Tainter, 1990, *The Collapse of Complex Societies*, Cambridge, UK: Cambridge University Press.

7 J.M. Carlson and J. Doyle, 1999, "Highly Optimized Tolerance: A Mechanism for Power Laws in Designed Systems," *Physical Review E* 60(2): 1412,. doi: 10.1103/PhysRevE.60.1412

Now, in the pandemic, I think what we're going see is, a year from now, two years from now, we're going to be much more robust to the next pandemic, but there are other things out there that we're going to be fragile toward. I think what Jessica is talking about is trying to look ahead and proactively think, "What are the perturbations, shocks that might hit us, that we're fragile to, and can we actually re-engineer things in society to fix that?" For example, we're working on robustness of supply chains. How can we re-engineer them to make them more robust? We have to begin by at least knowing what they are, which right now we don't even know. I think this is one of the principles of complex systems, so I think it's a very good question, that we as complex-systems scientists should be able to really do something useful on.

R. HAUSMANN: Yes, excellent question. Let me start by saying that you make a system robust to something you know that you have to be robust to, and in a complex system, you don't know what you need to be robust to because you cannot imagine all the things that can go wrong. For example, in the global financial crisis, nobody thought about subprime mortgages, and suddenly it was about subprime mortgages. And now nobody thought about COVID, and now it's COVID.

I split my time between managing a group of postdocs that do basic academic research and managing a group of people who help governments make economic decisions, and this year we've been working with twelve governments on helping them navigate through COVID. There was, in the global financial crisis, a difference of two orders of magnitude between what the costs of the shock say, in Belgium or Switzerland, vis-à-vis Latvia or Greece, and right now we're seeing enormous differences in what you might want to call resilience to the COVID shock in these twelve countries we are working on. And you might talk about resilience to COVID, but right now, not only did you just see Hurricane Eta, now there's Hurricane Iota, and it's going to go through the same path, through Honduras and Guatemala, that the previous one did, two countries that were already deeply hit by COVID. So there's just a limit to how much you can think usefully about resilience in the abstract.

I think that some systems achieve robustness by evolving solutions for the last crisis so as to prevent their reoccurrence. The last crisis captures all of the attention and creates a frame for what can go wrong. If there are enough crises over time, and if the system does not change too much, by this recursive method you may actually become resilient to many potential future crises. This is how we deal with airplane safety: we do a forensic analysis of accidents to figure out how to prevent them. But if the system is more complex and more significantly changing, say, like the financial system or the economy as a whole, this backward-looking approach may not make the system much more robust.

We still have a long way to go to understand why there's such a dramatic difference in a robustness to what appears like a common shock. COVID as a pathogen is more or less the same everywhere in the world. But the network of social contacts people have, the housing and transport infrastructure that they use, the organization of production, the ability of the governments to tax and borrow, the willingness of people to adopt prosocial behaviors, the effectiveness of their health systems, etc., are so different that the overall impact of COVID on outcomes differs by more than two orders of magnitude between different countries of the world.

J. FLACK: Two comments on that. First, I'm seeing a little bit of issue with the idea that robustness is about controlling things we've seen. I think that doesn't have to be the case, and there are lots of examples from nature where it isn't. A lot of robustness mechanisms like modularity, even things like relationship repair, are generic mechanisms that can be used in a lot of different situations, or which buffer damage that, say a perturbation takes out a component, downstream components feel. And so it's not necessarily about controlling tightly or maintaining a particular function tightly, sometimes it's about this buffering.

And the second point, where I totally agree with Ricardo, is that the assumption that continues to plague us is the assumption that we know what the problems are. COVID illustrates that we don't, so we design—and we overdesigned for problems we've just faced. The towers of the terrorism and so forth is a great example.

To come at this from a different angle, just to make clear where we're going, one change I would like to see in the collective-intelligence literature, particularly in the social sciences—this isn't so much of a problem in systems biology and so forth—is a move away from focusing on forecasting outcomes, like will Iran develop a nuclear arsenal, and prediction of ground truths, like what the weight of a steer is, and

REFLECTION

ON THE LIMITATIONS OF PREVENTATIVE ACTION

Ricardo Hausmann, Center for International Development, Harvard University; Santa Fe Institute

Should we wear masks to prevent contagion? Only indoors or also outdoors? Is it important to wash our hands and disinfect surfaces? Should we keep schools open? Should we vaccinate first the most vulnerable or the most likely to infect others? Should we approve use of a vaccine when we have good reason to believe that it is safe and effective, or should we wait for more evidence until we are super sure? Should we borrow massively to help people withstand the pandemic, or should we be more cautious to prevent a potential debt crisis? Should we require people to get vaccinated, or should we only encourage them to do so?

These are all questions that involve decisions. Answering them "correctly" involves knowing some positive features of the world (Is the virus transmitted through airborne means? Does it survive on surfaces?) and also how to assess some complex cost-benefit calculations such as the epidemiological benefits vs. the educational, social, and family costs of closing schools.

Reasonable people might disagree about the relative costs and benefits of these decisions, in part because of their own characteristics: people in their seventies with preexisting conditions will not necessarily value the risk of closing schools in the same way as parents in their thirties with kids in school. People with an ample cushion of savings and jobs that can be performed remotely may not value increased social transfers and fiscal deficits the same as a blue-collar worker who lives paycheck to paycheck.

Moreover, many of these decisions can be better evaluated with more information and knowledge, but this takes time, which can only be gained by making a decision today to postpone the decision, which in itself is a decision you may want to take some time to assess, which is a bit of a contradiction.

COVID-19 is an incredible natural experiment that has shocked the world in many different and intertwined dimensions.

All of this supposes that the implicit decision-maker behind all these choices is an emperor of the galaxy, sitting on some perch, observing the universe and pondering what is best. But in the real world, public decisions are made by individuals who have some delegated authority, and this authority can be revoked, in part depending on an assessment of the soundness of previous decisions they have made: a president might be impeached, a head of a public health agency fired, or a judicial decision overturned. So how should a person with delegated authority incorporate the impact of their decision not only on the pandemic but also on the likelihood of having their authority revoked? And what information would citizens need to have in order to make the decision to revoke that authority?

What if we agree on the inadequacy of the current decision-maker but are unable to agree on an alternative? What happens if there is a breakdown in the ability to agree on whom to delegate the authority to make decisions? And what if alternative solutions to the crisis involve imposing very different costs to different parts of society, and what if those costs follow cleavages that are already a source of social tension? What if this exacerbates political polarization in ways that prevent action?

COVID-19 is an incredible natural experiment that has shocked the world in many different and intertwined dimensions. Some countries may be months away from herd immunity and the return to a world that looks remarkably similar to 2019, except for the added degrees of freedom that Zoom provides and the growth prospects of new green industries. Some may even emerge with a renewed sense of confidence after having overcome a tough shared challenge. But for many others, it will be years before they can tame the pandemic and forgo social distancing with its enormous associated costs. Others are seeing the collapse of the political system that delegates authority and assures its stability and legitimacy. The recession triggered by the pandemic may morph into systemic banking, currency, and debt crises that may send societies down rabbit holes of political instability and economic collapse.

The social scientists of the future will be able to untangle the deep web of causality that made the COVID-19 shock percolate the way it did in different places. Unfortunately, we are left to act on the world today with inadequate knowledge and information. Let us hope that the costs of that inadequacy become a motivation for the next generations of scientists to advance our understanding of our very complex world.

toward thinking about individuals as sensors, components as sensors. They're information-processing sensors that can potentially detect unknown challenges in the environment, especially if the sensors are diverse, and then that information can get aggregated so the system can figure out actually what the problems are. So instead of assuming that the problems are known and asking how does the system know what the answer to that problem is, what is the problem in the first place?

A context for discussing this, a great one, is the Electoral College. If we just put aside for a moment Constitutional issues—of which there are many—the Electoral College, in my mind, is a misguided . . . a lot of people want to throw it out, especially among my liberal academic colleagues—we get rid of the Electoral College and just go with the popular vote. Well, [the Electoral College is] a misguided attempt, in some sense, to ensure geographic fairness, but I see its unrealized value as allowing information from different parts of the United States to inform aggregate decision-making.

I agree, it is not well designed to this end, but that's how we could think about it. The United States has many different needs; they're not just needs of people on the coasts, or the needs of geography and so forth, resources on the coast. How can we design an aggregation mechanism better than the Electoral College that thinks about individuals voting as sensors with valuable information?

D. KRAKAUER: Ian, what you unleashed, it's a big issue for us at SFI. It's a central one, I think. I don't know if you have a reply, Ian?

I. MCKINNON: No. I really appreciate the responses, and I guess for me, it's trying to find, with my economist hat on, that Pareto optimal curve where we can actually get some more robustness and perhaps not sacrifice some efficiency. I haven't seen enough on that, but I think there's a lot of fertile ground there, and this emergent engineering sounds right up that alley.

D. KRAKAUER: Fantastic. Thanks, everybody. Let's move on to the next question. This one is from Perry Chen.

PERRY CHEN:[8] Hi, everyone. I did a project on Y2K years ago where I stumbled upon the phenomenon I call politics of prevention. Imagine if a meteor was heading toward the Earth, and there was a 3% chance it would hit and kill tens of millions of people. Leaders understand the grim political calculus of preventing things. First, if you're preventing something that only has a 3% chance of occurring, the higher the chance that there will be evidence that you wouldn't have needed to take action, the larger political price you'll potentially pay. This is the snow-day problem for mayors canceling school.

Second, even if preventative actions were needed and effective, except in snow-day cases, where you have the snow on the ground the next day to judge, you've often reduced the evidence of the need for your costly preventative action, for example, like with something like COVID. So simply the best approach to preventative measures often results in outcomes that don't support or even appear to undermine those actions. Today, Y2K prevention seems like a joke, but many of the tens of thousands who worked on it for years would disagree. So, the question is, ***Does contextualizing this dynamic within a larger set of interdependent complex systems offer any useful insight for improving our species' ability to avoid future crises?*** Thank you.

D. KRAKAUER: All right. That's a good one, tough one. Who've we got? Who's going to be brave?

M. BANAJI: I'll take a shot at it only if nobody else who is better equipped will. Okay. Wonderful question, and it made me think a lot about all of the other successful ways in which we have dealt with things just like Y2K that we're not even counting today, and maybe ideas that come from complexity science have been somehow naturally a part of those solutions and we're not counting them. This is also like Jessica's questions: Is it really worse today than it was? And I don't think that those data are available. For example, we know that in the 1918 flu, some proportion of people actually did not want to wear masks. They said something very similar to what the residents of Florida or North Dakota are saying today. They said, "This is God's breathing system and we shouldn't interfere with that." Has there been any change? is

8 Kickstarter

the question. Does anybody in this audience even know if more people proportionally today believe in these crazy theories than did in 1918?

But I will give an example of a group I'm a part of that may say something about this particular question, and it goes on quietly, so we don't notice it. I'm on a committee right now that has been set up by the National Academy of Sciences to look at the planetary sciences, and every ten years, NASA asks a committee to tell it what it should do for the next ten years. And these are people who come from the public, because they're really interested in planets; these are experts in the field; this is a steering committee; these are institutions—it's a very interesting process. I'm just so impressed with both the transparency and the planning and telling NASA, "If we do *X*, *Y*, and *Z*, this is what should happen, and the Mars problem, the Moon problem." And I just think what you're raising with Y2K is a forgotten but very positive example that we should remember.

J.D. FARMER: I want to connect this question to another question, which is dealing with unknown unknowns, what's called Knightian uncertainty, fundamental uncertainty, where we don't know probabilities for things. Because these situations—Y2K is a decent example—what was the real probability that something really bad was going to happen from Y2K? I don't even think we know now how bad it would have been if we hadn't done all that stuff. These problems become particularly severe when we have something that's rare, with a really bad outcome if it occurs, and there's a lot of uncertainty about how bad that outcome would be and how likely that thing is to happen. As scientists, we're not very well equipped to deal with that because all the science is really built around, "Once you tell me the probabilities and the severity, then I can write down a plan to deal with it, and we can all agree what we should do." In these situations, it's typically not possible to agree, so we have to have adaptive mechanisms.

I think the way we do this is sort of groping our way around. Look at the evolution of our knowledge of climate change over the last century, first predicted in 1911 or something, by Arrhenius, and we slowly gain more information. We build science to reduce the unknown unknowns, but it's still a really hard problem, particularly when it involves heavy tails of serious outcomes. I think this is a place where science fails, and complex systems is desperately needed, because I don't think we really understand this question.

D. KRAKAUER: I actually think Arrhenius might have done that in the 1890s.[9]

J.D. FARMER: Yes, you might be right.

R. HAUSMANN: I don't know that the problem has a scientific solution. The problem, at its core, is a political problem in the sense that there are very few political benefits from preventing a problem because people never see the crisis you prevented. You do get political benefits from solving a crisis that has already erupted, but if you prevent a crisis that would have only happened with some probability, and is that solution comes at some cost, you will be blamed for the cost, for the waste of having done something that clearly was irrelevant because there was no problem, like Y2K.

D. WOLPERT: Very, very quick follow-up. Somebody mentioned also the successes that are not known. A striking example of this, which to a large degree was excuses but also had some truth, is in national-intelligence bureaucracies—the successes you never hear about. You hear about the failures, or the bad guys. Valerie's ex-bosses and so on, they would use that very often as an excuse: "Oh, we're not as bad as it appears, blah, blah, blah." Nonetheless, there is a big element of truth to that.

Now, what is the distinction between those examples of hidden successes, where people are doing the right thing, preventing stuff, versus these other ones? Quite honestly, actually, it's a matter of democracy. The ones who have the risk to their own political career, they are ones who are going to face the electorate; the ones who are doing the right thing, for whatever reasons, do not face the electorate. There's actually a big sobering example of that going on right now. There are parts of the world where COVID has been and gone, and everything's now hunky-dory, and we're back to galloping forward at greater than 5% growth rate. And that big fat, very, very dangerous, sobering, arguably evil, if you look at the concentration camps in Xinjiang, part of the world, they are gathering their success precisely by not being democratic.

---

9 S. Arrhenius, 1896, "On the Influence of Carbonic Acid in the Air upon the Temperature of the Ground," *Philosophical Magazine and Journal of Science* (41) 5 237-276, http://www.rsc.org/images/Arrhenius1896_tcm18-173546.pdf

D. KRAKAUER: Thank you. So, let's move to the next question, and I can see you're here, Ashton, but we've got you in virtual form. Let's play the next question.

You do get political benefits from solving a crisis that has already erupted, but if you prevent a crisis that would have only happened with some probability, and is that solution comes at some cost, you will be blamed for the cost . . .

ASHTON EATON:[10] Hello, everyone. This is Ashton Eaton, two-time Olympic gold medalist in the decathlon, retired, currently a father and a product development engineer at Intel, but, more importantly, friend of SFI.

My question is, ***What type of content, text, image, video, amount of content, one or more than one instance, and source of that content, social media, news outlet, friend or family, is required to change the belief of at least 50% of a population from one side to the other of a fairly bilateral issue that has a lot of evidence for one side and limited evidence for another?*** An example of a bilateral issue, in my opinion—and these are just examples—would be "wearing a mask helps reduce the spread of virus." Another example would be "the government uses vaccines to control its citizens." ***So, if you had an entire population who believed an issue like this was either true or not true, what type, amount, and source of content, if any, would it take to convince at least half that population to the other side?***

I think the fundamental question is, ***Do we, as a society, have a critical-thinking problem, and can it be identified?*** What constitutes something being a problem, I don't really know, but basically, current situations around the world from the pandemic, to the former administration, to climate change, things like that, have, at least to me, seemed to kind of shed light on the idea that perhaps a larger percentage of society is

10 Olympic gold medalist

susceptible to believing things that have limited evidence than originally thought. That's my question. Thank you for the time and looking forward to seeing what you guys do.

J. FLACK: I have a brief comment. I was going to suggest Mirta [Galesic] to answer that part two of Ashton's question.[11] But I'd like to point out something I think we all sort of implicitly know—a nice study showed it in 2019 in *PNAS*—and that is, How does legislation passed by governments influence our attitude?[12] This relates to Ashton's question in the sense of, What are the mechanisms that lead to switches and the expression of beliefs, right?

The authors of this study investigated this question in the context of views on gay marriage. They measured it over a twelve-year window, and they found that, while anti-gay bias had been decreasing over time following local same-sex marriage legalization, anti-gay bias decreased at roughly double the rate when there was federal legislation. The takeaway from the study is that the legislation obviously influenced people's willingness to express their beliefs, that's very important, and it seems to be the case that many individuals were already fine with gay marriage but were unwilling to state it because they thought the consensus was against them.

Legislation can potentially play this important role, a sort of top-down effect of providing social safety nets, and then this whole thing gets amplified because as more people express the fact that they don't care at all, more people are willing to say that.

MATT JACKSON:[13] It's a fascinating question. I can mention one study that I was involved in with Roland Fryer and Philipp Harms.[14] What we did was give people abstracts of articles about climate change and about the effectiveness of the death penalty. First of all, we asked

---

11 See Mirta Galesic's reflection on page 689

12 E.K. Ofosu, M.K. Chambers, et al., 2019, "Same-Sex Marriage Legalization Associated with Reduced Implicit and Explicit Antigay Bias," *PNAS* 116(18): 8846–8851, doi: 10.1073/pnas.1806000116

13 Stanford University & SFI

14 R.G. Fryer, Jr., P. Harms, et al., 2019, "Updating Beliefs when Evidence is Open to Interpretation: Implications for Bias and Polarization," *Journal of the European Economic Association* 17(5): 1470–1501, doi: 10.1093/jeea/jvy025

REFLECTION

BRINGING FACTS OUT OF ISOLATION

Mirta Galesic, Santa Fe Institute

It is possible to change people's beliefs about many issues. For example, most people were not convinced when Trump first showed up on the political scene. But very soon a substantial portion of the electorate gave him their vote, including people from the seemingly opposite end of the political spectrum. Similarly, in the US beliefs about the danger of COVID-19 and the usefulness of vaccines were changing quite substantially as the pandemic progressed and as vaccines were linked to different political factions. More generally, fashions in clothing and art, investments, and even science change all the time.

And yet we scientists often find it difficult to affect public beliefs about scientific issues from the value of vaccination to the danger of climate change. Why is that? I would argue that for too long scientists have relied on inadequate norms of rationality to describe how people make up their minds. The traditional ideal of a scientific thinker who is unaffected by emotion, irrelevant beliefs, or others' opinions, represents neither how we scientists really think nor how the general public forms their beliefs.

Decades of research in psychology have shown that we all often rely on strategies that do not conform to ideal notions of individual rationality but help us to navigate our complex and uncertain worlds. One set of such strategies is social: we rely on opinions and behavioral examples of others around us, especially those who seem confident and have a high status in our social circles. We also take into account the social costs of believing something that could hurt the relationships that we rely on daily for financial, informational, and emotional support. This reliance on our social networks is not surprising: our ability to learn from each other and cooperate is, after all, the very cornerstone of our species' success.

Another set of strategies relies on our core values: we decide whether to accept novel information based on how well it fits with other beliefs we hold dear. This helps us to make choices that are in accord with our personal and social context. Our core values often summarize years of experience with specific problems in our local environment and with social norms in our society. Choosing whether to accept novel information based on its correspondence to our core values often helps us choose what is right for us, even if it does not appear rational.

For example, I admit that much of my belief in anthropogenic climate change comes not from careful reading of hundreds of relevant articles but from following the opinions of other scientists who I trust have done good research about it, at least in part because everyone around me trusts them as well. I am also aware that my colleagues might object if I were to loudly pronounce climate-change skepticism. And climate-change skepticism would be in contrast with my core values of respect for nature and care for others' well-being. Similarly, a climate-change skeptic might hold that belief not because of careful research but because she believes certain sources of information that are highly esteemed in her social circle, and because certain policies related to halting climate change strike her as being in contrast with her core values of patriotism and independence.

We at the Santa Fe Institute model these social and moral influences as networks of beliefs that are continuously adjusting to reach and maintain a state of low dissonance or relative consistency between different beliefs. Seeing our reasoning this way helps us understand how we form our beliefs and when and why they persist or change. Many phenomena that look biased when viewed through the static lens of an individual rational thinker suddenly appear as reasonable solutions to the multitude of social and ecological problems we need to juggle.

Modeling belief change this way also helps us understand what kind of educational programs can change people's beliefs. Specifically, most helpful programs highlight rather than hide the dissonance between scientific facts and people's core values, but at the same time provide information that helps to resolve that dissonance. Successful communicators are sensitive to other people's life experiences and values, and make effort to explain how new information can be helpful within people's specific personal and social context. For example, it might be futile to provide facts about climate change without also seriously addressing concerns about how climate-change policies affect our nation's global competitiveness and our individual freedoms.

The way other people understand the world is often difficult to imagine, so scientists need to make an effort to research and understand their audiences and adjust the way they communicate without being dismissive. For example, many people have difficulty understanding ratios, percentages, and conditional probabilities, but only because people often do not have sufficient recent experience with these numerical formats. Once these concepts are presented in more intuitive ways, most people can understand them. For example, people from all walks of life can compare ratios easily when these ratios use common denominators, and people who have little experience with percentages can understand them when percentages are presented visually using icon arrays. And, when conditional probabilities are presented in terms of natural frequencies, even children can solve Bayesian reasoning problems.

In sum, if critical thinking means impassioned consideration of facts in isolation from everything else we know and need, then this was never the way most people were thinking anyway. Instead, we as scientists might have been projecting our specific way of thinking onto others and are only now learning how to communicate better.

REFERENCES

J. Dalege and T. van der Does, 2021, "Changing Beliefs About Scientific Issues: The Role of Moral and Social Belief Networks," arXiv:2102.10751 [cs.SI], https://arxiv.org/abs/2102.10751

M. Galesic, H. Olsson, et al., 2021, "Integrating Social and Cognitive Aspects of Belief Dynamics: Towards a Unifying Framework." *Journal of the Royal Society Interface* 18, doi: 10.1098/rsif.2020.0857

T. van der Does, D.L. Stein, et al., 2021, "Moral and Social Foundations of Beliefs About Scientific Issues: Predicting and Understanding Belief Change," OSF preprint, doi: 10.31219/osf.io/zs7dq

people's views beforehand on what they thought about these things, and then we showed them the abstracts. Two things, I think, stuck in my mind about this. One is that there was a heterogeneity in how people reacted. Some people you could think of as being very rational and really paying attention to what they were seeing and moving their beliefs based on what was reported in the abstracts; and then there were other people who would interpret the findings of the article based on their own beliefs . . . so you could see the same article being interpreted completely differently in terms of whether it was evidence for it or against. So, depending on their opinion, some people would look at exactly the same evidence and have a very different interpretation of that piece of evidence, and in fact became more polarized by looking at exactly the same information.

It's not completely obvious how you can get people to come to agree just by showing them the same evidence; I think that there's also social aspects of this, but there are a lot of confirmation biases that we know well from psychology, and different people exhibit different extents of this. I think it's a fairly complex question in terms of exactly how we convince people of something that we think is true from scientific evidence, because even when you show them that evidence, it seems to be interpreted very differently based on their perspectives.

D. KRAKAUER: Thanks, Matt. Anyone else like to jump in? Mahzarin?

M. BANAJI: This will be an answer to this particular question but also to a later question, the mayor's question about poverty and what levers need to be moved in order to change how we think, and we'll come to that in a second. But I thought I would just say—and, interestingly, it's the same example that Jessica used that will be relevant here as well—we've been looking at implicit forms of bias and how quickly they're changing, and we expected, based on everything we knew, that they would not be changing, but they are changing.

What is interesting is, of course, they're all over the place: some are changing, some are not, some are getting worse. But the remarkable one is the anti-gay bias, which was pretty high in 2005, and then, starting in 2006, something happens and it's dropped off by 50% on survey measures and 33% even on measures where people can't control what they're thinking. So, this tells us that it's possible, and we need to understand why. Jessica gave us one reason about legislation coming in while individuals are prepared. And so, I would say to the mayor, and also in response to this question, that maybe we really . . . This is where my having spent time at SFI has changed the way I do my work now, that I feel like I cannot be the person looking at these little pieces separately, that I've got to spend the time to put the system together and look at it.

We're learning some remarkable things. For example, it's when all three are lined up that we think this kind of change can happen. Individual change didn't just happen; it happened because grandmothers and parents and friends and coworkers found out that their friends and children were gay, so that produced individual-level reckoning, at least. But then institutions—like the military or SFI—came in and said, "Look, the country is not there yet, but we're going to recognize same-sex couples, we're going to do something locally to act as if you are straight in the support that we might give you." And then, slowly, it's Hollywood and the Supreme Court that come in and move a major social lever. But I'm going to argue that if you don't see changes on race, anti-elderly bias, disability, and so on, it's because these three have not been tuned in to operate together.

D. KRAKAUER: David, I'm a bit concerned about time, unless it was really an unbelievably profound remark you were going to make.

D. WOLPERT: Very much so. I'll be very quick with it. I'm very interested in what you just said, Mahzarin, and what other people have said, but there's a chicken-and-egg problem. What causes all of those in the first place? This is actually a huge mystery which many people have written about, [like] Steven Pinker, but many before him. There are many different kinds of great divergences between the West and the East that happened around the eighteenth and nineteenth century. One of the big ones, the most striking ones, is that the inclusiveness of society, the expansion of the set of entities that are considered "us versus them," has been galloping by many measures—Simon DeDeo has also got these things in his Old Bailey records[15]—in the West. Didn't occur anywhere else, didn't occur historically to get up to these last levels. I mean, you can make the glib assessment as well, because the West became rich, and

15 S. Klingenstein, T. Hitchcock, and S. DeDeo, 2014, "The Civilizing Process in London's Old Bailey," *PNAS*, doi: 10.1073/pnas.1405984111

you only ascribe rights to others once you're rich enough. Could be, but nobody that I know of has actually done any proper study of that underlying cost of all these phenomena.

D. KRAKAUER: Ashton, since you're Superman, I feel I should allow you to make a remark since you asked the question.

A. EATON: Oh, wow. I appreciate that. First of all, thank you guys for using your thought power to kind of answer it, and second of all, I think it's complex, there's a lot of things that go into it, and I was trying to break it down to something simple. The reason I asked it was because it seems actually easy for a lot of folks to go one way but harder to go another. And I think it's easy for folks to go in a way of something that's a little bit more obscure and that . . . I don't know, like a Roswell or something might be a good or bad example, and I think once people go down that road, it's actually harder to get them out of it. And I was just wondering if we could quantify how to do that.

D. KRAKAUER: Thanks, Ashton. Let's go to our next question, and that would be Toby.

TOBY SHANNAN:[16] As preamble, I feel I must say that asking questions into my camera is a poor substitute to sitting in the hotel bar of the Inn at Loretto sipping preprandials with Doctors Hausmann, Miller, and Tracy, and with any luck, we'll find ourselves same place, same time next year.

My question. There seems to be a relationship, an obvious one, between knowledge and prediction. We just went through an event [the 2020 US presidential election] where we had a class of professional predictors, pollsters, and pundits do an exceptionally bad job at prediction, maybe not a catastrophic job but it was much more of a near-run thing than we had anticipated. ***Meditating on this relationship between knowing and predicting, what is it that you think this class of professional predictors don't know that they need to know?*** Thanks.

R. HAUSMANN: I think they just use a very lousy instrument. They use surveys, and they have enormous difficulty in having a random sample of people answer the question truthfully. And so, we would need more unconscious, indirect measures of people's sentiment there. It would be a different technology. I think what we're seeing is the end of a method that doesn't work.

16 Shopify

J.D. FARMER: So, I want to make a sort of irritating geek answer to this question by saying that I think knowledge and the ability to predict are almost the same thing. This is essentially proved by Jorma Rissanen basically showing that data compression and forecasting power are equivalent, and that what models allow you to do is compress information. In fact, the statement revolves around the idea that, to compress information with a model, you have to look at how many bits are in the model and how many bits of information it takes to write down the deviations between the predictions and what actually happened. And so in fact it shows they didn't really have knowledge. This is in a way what Ricardo was saying. I mean, they had flawed models. They didn't really understand what the system was about and so they couldn't predict it.

So, I'm in a sense picking at a semantical thing, but this is actually really fundamental when you start doing prediction, realizing this and realizing the difference between how well you're fitting data, and how well you're able to predict. They're very different things in general. We sort of fit the last data very well; we didn't fit this data.

D. KRAKAUER: Thank you, Doyne. Let's jump to the next question, and that would be the Seamus Blackley question.

SEAMUS BLACKLEY:[17] Hi, Seamus Blackley here. I have a question. We have a seeming global problem with the proliferation of conspiracy theories and other disinformation, and specifically the problem of QAnon, anti-vax, etc. It's a complex process, the way that this information appears to be disseminated, consumed by individuals that are susceptible potentially to it, also the relative susceptibility of a population to this information, so that's very interesting. ***How does complexity theory inform our ability to address these issues in public policy?***

D. KRAKAUER: Who would like to take that one first? Ricardo.

R. HAUSMANN: Let me popularize the work of Simon DeDeo here.[18] What's nice about conspiracy theories is that they explain the world; they explain many things about the world. The problem of us living in the world is that we can get a lot of information, but what we need is

17 Pacific Light & Hologram

18 Z. Wojtowicz and S. DeDeo, 2020, "From Probability to Consilience: How Explanatory Values Implement Bayesian Reasoning," *Trends in Cognitive Sciences* 24(12): 981-993, doi: 10.1016/j.tics.2020.09.013

not information; what we need are theories, that is, stories or narratives that explain why things happen. Now, how do we change our theories on the basis of the information we get? That's an evolutionary process of how we evolve our beliefs.

The essence of the story is that there's a lot of things that we cannot explain, and we want theories that are somewhat consilient in the sense that they can explain many things. David started with Newton, saying Newton's beauty is that with a simple framework he explained many things, and QAnon with a simple framework explains many things, and these conspiracies make sense of many things about the world. It may be a fluke in our reasoning, but it is at the core of this Bayesian problem we have of trying to make sense of the world and searching for frameworks that allow us to make sense of the world. If you just want to go to the simplest one, it's God. God is the ultimate conspiracy theory. Why does it rain now? God wanted it to. So we're in some sense wired for these things.

The problem of us living in the world is that we can get a lot of information, but what we need is not information; what we need are theories, that is, stories or narratives that explain why things happen.

D. KRAKAUER: Thank you, Ricardo. Would anyone else in this audience who has expertise in this area like to chime in? If not, let's move on to the next question, and that's from Santa Fe Mayor Alan Webber.

ALAN WEBBER: Thanks for the opportunity to ask a question. The general topic of my question is poverty. As a result of COVID-19, a lot of things that were evident but had escaped our attention for quite a while have now become inescapable, and I think the most significant one of all is the issue of poverty. We are a country that has one of the greatest Gini indices in the world; the gap between the rich and the poor, of any developed country, in the United States has grown larger. Unique among Western democracies in the United States, when we have a policy intervention around income and wealth creation, our policy interventions at the national level, historically and in recent times, have made things more unequal. Poverty continues to be the underlying story for people in Santa Fe, in New Mexico, and of course around the country.

So, the question is, ***Where are the levers that could most directly affect, improve, change the lives of people who are living in poverty? What are the systems that keep them in poverty? And where can that systems approach begin to change things for the better as we look at a new administration in Washington, and perhaps post-COVID efforts to change the path and improve the lives of people who are trapped in poverty?*** I hope you'll come back with some good answers. Thank you.

D. KRAKAUER: Thank you, Alan. I know that Mahzarin has an answer to this question.

M. BANAJI: Mayor, in general, in America, we speak about poverty but not about the poor. Poverty is an affectively cold concept. The poor, on the other hand, have faces; they evoke emotion. There is a sense that something must be done. And so, I think partly we have to begin with the linguistic representations that we use to help with this deception and to such a sufficient magnitude that even the poor today believe themselves to be middle-class. So what are the levers to move? I think earlier I mentioned that if we could do something that simultaneously included individual-level change, institutional-level change, and societal-level change, that we may have a better chance of moving the needle, and there's a whole conference's worth to think and talk about that.

But there is one other concept that I want to give to you to show you what the resistance to this is going to be, and it comes from Susan Fiske's idea that when we think about other humans, or even human groups, we seem to do something peculiar.[19] We either give them warmth and love, or we give them respect, but not both, and rarely neither. So it's always the off diagonal.

So, mothers are warm but not competent; fathers are competent but not warm; CEOs are competent but not warm, as are feminists and

19 S.T. Fiske, A.J.C. Cuddy, and P. Glick, 2007, "Universal Dimensions of Social Cognition: Warmth and Competence," *Trends in Cognitive Sciences* 11(2):77-83, doi: 10.1016/j.tics.2006.11.005; and S.T. Fiske, A.J.C. Cuddy, et al., 2002, "A Model of (Often Mixed) Stereotype Content: Competence and Warmth Respectively Follow from Perceived Status and Competition," *Journal of Personality and Social Psychology* 82(6): 878-902, doi: 10.1037/0022-3514.82.6.878

Jews and Asians; but there are other groups that are nice and that we might like, but they're not very competent. And the only exceptions to this, the only exceptions to this trade-off, are drug addicts and the poor, and they are seen as neither competent nor good. So, they evoke in us feelings of pity and contempt. And if we begin by acknowledging that that is the case, then our solutions are going to have to be very different than the ones that we might imagine for any of these other groups.

D. KRAKAUER: Thank you, Mahzarin. Anyone else like to jump in?

R. HAUSMANN: These issues are kind of like my day job, so I could go on, but if you want to put it in the language of Santa Fe, you have to ask yourself the question, Is poverty something or is it the absence of something? In the same way as physicists were discussing whether heat is something or whether cold is something, or whether cold is the absence of heat.

I want to think that poverty is the absence of something, and that something is something that humans are able to produce when they connect in constructive ways. So, poverty is really, in some sense, disconnectedness. It's exclusion. It's people who are not sufficiently networked into the rest of society to benefit from the things that are produced in human networks. I would say a strategy to reduce poverty is a strategy to include those that are not being sufficiently included. Let me stop there.

D. KRAKAUER: Thanks, Ricardo. Alan, do you have a response?

A. WEBBER: Thanks, everybody, for an amazing gathering. I'm glad to be able to lurk in the off-camera part and just listen. If you were to go back to some of the earlier questions about why pollsters have been wrong or what do we do to change the trajectory of people's need to believe in things that are demonstrably false, I think a lot of it does in fact track back to whether you call it poverty, or the lives of poor people, or a system of keeping people excluded. And this is the challenge of potential already coming apart at the seams because of exacerbated inequalities and perceptions of political and economic favoritism.

I love the idea that there are three things to get aligned. One of the things that I remember from reading recently—not recently [published] but rereading recently—was an essay on inequality in America that pointed out that we are just about the only Western democracy that holds poor people at fault for being poor. Most other countries look at the lives of poor people and say, "It's not their fault. We need to step up and do something to alleviate the problem."

There's a significant number of Americans, historically, who have looked at poor people and said, "It is their fault, and if they can't pull themselves up by their bootstraps, then it's their own hard luck." And that overarching sensibility, or lack of sensibility, is something I don't know how you change at the top level, but I believe it really is important to get. It's just as much of a bias as being biased against gay people or anyone else who we look down on and exclude and then keep excluding

in a systematic way. So, thanks for taking up my question, and I will keep listening and I will go back to being off-camera so I can listen without being seen. Thank you for having me.

You can think about entropy as *disorder*, but you can also think of it as a number of options.

D. KRAKAUER: Thanks, Alan. We have the last question from Adam Messinger.

ADAM MESSINGER:[20] Hi, everyone. I've been a friend of SFI since I worked for Melanie Mitchell in 1994 and '95 as a research intern. My question for the panel is, ***Given SFI's interdisciplinary nature, can you imagine developing a system of ethics that sort of spans philosophy and science in a useful way?*** A concrete example of what I'm thinking of here is you could imagine maybe defining evil as the same as entropy, and you'd have some consequences of such an ethical system where wasting energy would be seen as morally wrong; breaking complexity, breaking structures would be morally wrong; life would be celebrated, because it's the best way we know to turn energy into complexity and to structure. Higher forms of life with more complexity would be preferred morally over lower forms of life; we should all be

20 Twitter

vegetarian or something. So that's the basic idea, and I hope it makes sense. Look forward to hearing your answer.

D. KRAKAUER: Doyne wrote to me and said, "I want a go at that one."

J.D. FARMER: Well, I love this question. It's in a sense what all the other questions are about. This encompasses everything. Mark Bedau, a philosopher at Reed College, and I have been working on this exact thing for a while, although it keeps getting pushed to the background by our day jobs. But I think that what you're saying is really the fundamental principle that drives my metaphysics, and that I think potentially explains the world. Purposeful behavior follows from propagation and life, and it's always in support, ultimately, of propagating patterns.

When you think about ethics, you're always going back to those processes that propagate patterns. Now, I'll pick something in this, that entropy . . . unfortunately, it's not the same as entropy. That's what I originally hoped for when I was an undergraduate in physics, but you eventually realize that it's not just entropy, because entropy . . . well, a crystal has low entropy; any orderly structure has low entropy. You really have to get at functional complexity, and that's a very slippery thing to define, which is why this is hard to do in a rigorous way.

But let's say at the heart of my own moral belief system, you have to believe that life is good, and if you believe that life is good, then all the other stuff that you're talking about here follows.

J. FLACK: I like this question too because I like the concept of entropy a lot, and I think about it often. I want to add to some remarks Doyne made there about it, which highlight, in some sense, the trickiness of this proposal. The first point I'll make is that you can think about entropy as disorder, but you can also think of it as a number of options. And when you think of it as a number of options, it's not so clear that it's inherently negative, and I would even argue that disorder in itself is not necessarily negative, because it can bring about change, and in the way I sort of alluded to when I was discussing critical points earlier. So, I'm not sure that we gain much in going this direction. We risk versions of the naturalistic and normative fallacies, so we need to worry a lot about that, maybe not in the conventional sense but in some other sense.

I do think that there is a possibility to bridge the sort of ethics–science divide, if we come back to the emergent engineering concept I mentioned, in that more intermediate form in which we design systems for emergent engineering, but we impose constraints that come from our ethical systems. So the system is free to explore its state space under these constraints, right? And, of course, deciding what those constraints will be is really difficult, but that's a potentially productive conversation we could have where in our engineering we're incorporating both ethical principles and empirically grounded design principles for robustness and adaptability.

R. HAUSMANN: I don't know that we can design moral systems. I think we have to understand why we are moral beings. And I think that morality, as a sentiment, evolved in us to solve two cooperation problems. One is, How do you make the individual cooperate with a group, and how do we make the group fight its external enemies? And when you feel that you were supposed to cooperate and you didn't, you feel guilty. When other people find out, you feel ashamed; when others did not cooperate with you, you feel outraged. So, our moral sentiments are there to assure cooperation.

The way I would frame the issue is, What is the cooperation problem that you want to address, and what are the moral sentiments that would solve that cooperation problem?

D. KRAKAUER: Thank you, Ricardo. I'm going to thank all of the panel for their fantastic thoughtful answers. I want to thank all of the questioners for their fantastic thoughtful questions, and all of you for participating. And let's have our beautiful canned round of applause, if we have such a thing.

[CANNED APPLAUSE]

BONUS QUESTION

Jonah was unable to attend the symposium due to his filming schedule, but we are delighted to include his question and David's subsequent reflection here. —Eds.

JONAH NOLAN:[21] My question is related to disinformation, freedom of speech, and "error correction" in our new forms of communication (building on David's theory of the "singularity" beginning with the advent of spoken language—and social networks being the latest advance/

21 Screenwriter, producer, and director

wrinkle in knitting together our collective cognitive processes). The early internet seemed like such a robust and revolutionary form of communication, yet the "algorithm" (or what passes for one, as most of these systems are, in fact, mechanical Turks puppeted by engineers, shareholders, and code) has been captured by commerce and driven by a reward system that capitalizes on ancient bugs of the human mind.

Can we fix social media? And how? That's certainly one of the biggest problems in my mind, as it has become one of our collective sense organs, and we can't fix the other grave problems we're facing if we can't agree on what those problems are in the first place.

D. KRAKAUER: Behind this question is a very challenging set of issues related to the philosophy of tools. None of us would ask whether we should "fix" pencils, or slide rules, or even the abacus. Certainly, we might improve the design of a calculator but only toward more efficient and reliable calculation. There are some tools, let's call them weapons, where the moral dimension is unavoidable. Here in Santa Fe, we live intimately with the legacy of the atomic bomb. In his book, *Adventures of a Mathematician*, Stan Ulam, who is partly responsible for the Teller–Ulam design for staged fusion in the hydrogen bomb, wrote that he had few objections to doing the theoretical work, but that the strategic use of the weapon was a sociopolitical problem of the gravest kind and beyond his expertise. The same relation of science to society motivated the 1975 Asilomar[22] conference on limiting the use of recombinant DNA.

If social media is a tool that can be weaponized—which, to your point, it most certainly is—then we have precedents in both atomic energy and genetics to which we might better attend. My own view is that it makes sense to separate tools from application and to focus on the ethics of use, delineating very carefully the military and market forces that are guiding their adoption.

---

22 P. Berg, 2008, "Asilomar 1975: DNA Modification Secured" *Nature* 455: 290–291 (2008), https://doi.org/10.1038/455290a

— Appendix —

GLOSSARY

ACE [angiotensin-converting enzyme] receptor a protein spanning the cell membranes of some cells that can transport materials into the cells and which also enables the SARS-CoV-2 virus to enter human cells and cause the disease COVID-19

aerosol a fine spray of tiny particles, ranging from about 0.1 μm to 10 μm that can remain suspended in the air for up to three hours; the smallest respiratory particle that can contain SARS-CoV-2 is thought to be about 4.7 μm

antibody a protein in the blood that recognizes and binds to specific viruses, bacteria or other threats, marking them for destruction so that the immune system can identify and neutralize them

apex/peak in epidemiology, the maximum number of cases, deaths, or recovered individuals, usually referring to the top of a graphed curve within a specific time frame

armillary sphere an astronomical instrument; a model usually consisting of metal rings illustrating the routes of various objects such as the stars and planets as they move through the sky

astrolabe an ancient instrument used to observe and calculate the positions of celestial bodies such as stars and planets

asymmetrical immunity immunity in which one strain of a virus induces host immunity against itself and all other strains, while other strains of the same virus induce host immunity only against themselves

asymptomatic in the context of COVID-19, the period of time in which an individual is infected with the SARS-CoV-2 virus and is able to infect others, yet feels either no symptoms or symptoms so mild that they are not recognized as COVID-19

OPPOSITE: *Gustave Doré, "The Tower of Babel," 1866 (detail)*

B-cell (or B-lymphocyte) a type of immune cell that displays antibody on its surface that can bind to viruses and other pathogens with unique specificity. Activated B-cells differentiate into plasma cells that can secrete hundreds to a thousand antibody molecules per second. B-cell antibodies are so specific they can distinguish different strains of coronavirus.

backcasting essentially the opposite of forecasting, a planning method in which a desirable future is identified and the present state is adjusted (ex. through policies and programs) to achieve that specified future

Bayes' theorem in probability and statistics, a mathematical formula that describes the likelihood of an outcome occurring based on previous outcomes

Bayesian particle filter in probability and statistics, samples gathered through noisy measurements at each time step; a Monte Carlo sampling method that permits inferences about the state of a state-space model (characterized by the state of a system evolving over time)

Bayesian updating the process of going from a prior probability $P(H)$ to the posterior $P(H|D)$

bet-hedging in biology, a generalist strategy that lowers mean fitness (selective advantage from generation to generation) in order to reduce variation in fitness over many generations

beta (β) in epidemiology, the average number of susceptible individuals that an infected individual infects per unit of time; a highly infectious disease has a high β

bifurcate to split into two branches; in mathematics, a sudden change in system behavior

CARES Act the Coronavirus Aid, Relief, and Economic Security Act, the largest aid package in US history, which provided $2.2 trillion in economic assistance for American workers, families, small businesses, and industries during the COVID-19 pandemic

C. elegans a small, harmless nematode, or roundworm, that lives in soil or rotting fruit and feeds on microbes such as bacteria; extensively used as a model species in laboratory experiments

cell surface receptors molecular receptors embedded in cell membranes (which enclose all cells) and to which other molecules bind, often altering the behavior of the cell

chaperone proteins proteins that help other proteins fold into their proper shape

clade a group of organisms that share a common ancestor; for example, humans, orangutans, gorillas, and chimpanzees all share a common ancestor and constitute a single clade

community spread the spread of a contagious disease within a community

coronavirus a group of about forty-five species of RNA viruses that cause disease in mammals and birds, including MERS, SARS, and COVID-19; all coronaviruses are members of the family *Coronaviridae*; they are large, single-stranded RNA viruses with a lipid envelope studded with a "crown" of distinctive "spike proteins"

counterfactual a hypothetical alternative; a statement that is contrary to fact

COVID-19 a potentially severe respiratory illness first identified in Wuhan, China, in December 2019; COVID-19 is caused by the coronavirus SARS-CoV-2 and mainly contracted through virus-laden respiratory droplets and aerosols; the illness is characterized by fever, cough, and shortness of breath and may progress to pneumonia and respiratory failure; other symptoms include fatigue, chills, body aches, headache, loss of taste or smell, sore throat, runny nose, nausea, vomiting, and diarrhea

CRISPR (clustered regularly interspaced short palindromic repeats) a bacterial derived genetic system used as a tool for editing the genomes of humans and other organisms

cyclicals in investing, stocks that rise and fall with the economy; e.g., automobile or appliance manufacturers

deepfake an image or recording manipulated to convincingly misrepresent someone as having said or done something they did not

defective interfering particles (DIPs) virus particles with intact structural proteins but altered or truncated genomes that can only replicate in the presence of a helper virus, and interfere with the replication of a non-defective virus; a term coined by Alice Huang and David Baltimore

defective viral genomes (DVG) viral genomes with mild to severe changes that cannot replicate in the absence of a co-infecting standard virus

density-dependent transmission in which rates of transmission of a disease depend on population density, increasing with a higher density and decreasing with a lower density

deterministic in population biology, models described by classical dynamics with no noise, this implies that the time evolution of a system depends only on the initial conditions and the dynamics

DNA deoxyribonucleic acid, a double helix of paired "bases" that encode information; the molecular basis of heredity in nearly all organisms

Draconian harsh, severe, or cruel

droplet in the context of COVID-19, drops of mucus or saliva coming from the mouth or nose following a cough, sneeze, singing, or loud talking that are about 5–100 μm in diameter; larger and heavier than aerosolized mucus or saliva and more likely than the tinier "aerosols" to quickly come to rest on the nearest surface

dynamical models in the field of dynamical systems, models that consider time dependence; synonymous with dynamics

endogenous caused by factors internal to an organism or system

endonuclease an enzyme that breaks a nucleotide chain into two or more shorter chains

epidemic a widespread occurrence of an infectious disease or other undesirable phenomenon within a population, community, or region at a particular time

epigenetic cellular strategies for controlling gene expression that do not change the DNA itself; more broadly, various kinds of developmental plasticity that are sometimes heritable over a limited number of generations

epigenetic landscape in developmental biology, a mathematical metaphor for the integration of gene regulation, induction, and other factors that determine the phenotype

epigenetic modification cellular strategies for controlling gene expression, especially in response to the environment, that do not change the DNA itself but that result in a stable, heritable phenotype, for one to several generations

epistemic relating to knowledge or the degree of its validity

epithelial cells cells that form sheets of tissue and line the outer surfaces and cavities of organs, including the skin, blood vessels and urinary tract

ex ante expected, forecast, or intended

exogenous caused by factors or agents external to an organism or system; generated from outside

Fermi calculation (also Fermi estimate) a "guesstimate" or back-of-the-envelope calculation; named after the physicist Enrico Fermi who was famous for his ability to make estimations with surprising accuracy

fitness landscape (or adaptive landscape) in evolutionary biology, a visual metaphor introduced by Sewell Wright in 1932, to portray the distribution of fitness values in space; now also in reference to an experimental, systematic construction that allows for analysis of combinations of predefined sets of genetic mutations

flattening the curve in epidemiology, the goal of reducing the infection rate to avoid overwhelming health care and other systems

Fortran a computer programming language widely used for scientific applications with notation resembling that of algebra

frequency-dependent transmission in which rates of transmission of a disease depend on the frequency of strategies in a population; not on population density

generalist viruses viruses able to infect viral hosts from different species and even from higher taxonomical units; generalist viruses are more common in plants

Gini coefficient in demography, a measure of income (or wealth) inequality based on a measure of statistical dispersion; the Gini coefficient ranges from zero to one; zero indicates perfect equality, where everyone receives an equal share of income, and one indicates perfect inequality, where only one individual or group receives all of the income

heavy tail in probability, a distribution that tends toward zero more slowly than a distribution with exponential tails

Hegelian sometimes summarized as "the rational alone is real"; characteristic of Hegel, his philosophy, or dialectic method, belonging to the German period of idealism following Kant

herd immunity when most of a population is immune to an infectious agent; herd immunity reduces the risk that a susceptible individual will encounter infected individuals

high fidelity a faithful replication of an effect such as a sound or image

homeostatic the tendency toward a relatively stable state

host cell a living cell taken over by or capable of being taken over by a pathogen, such as a bacterium or virus

hyperobject things that are real and which we can study but which we cannot see directly, such as the biosphere or global warming

hysteresis a nonlinear phenomenon in which movement between equilibria depends on history; hysteresis emerges in systems with multiple equlibria with the addition of feedback

incubation period the period of time between when an individual is first infected by a pathogen and when signs and symptoms of illness first manifest

kilobase one thousand base pairs, a unit of measure for the length of a nucleic-acid chain

Lagrange multiplier a technique for finding the maximum or minimum of a multivariable function when the input values are subject to constraints

long tail a subset of heavy-tailed distributions whose tails taper off gradually instead of dropping off sharply; *see also heavy tail*

low-P/E in investing, the price-to-earnings ratio; an investor might use a company's current share price relative to its per-share earnings, P/E, as a measure of value; a low P/E indicates that the current stock price is low relative to company earnings

mass spectrometry an instrumental method that identifies the chemical makeup of a substance by separating gaseous ions according to their mass to charge ratios

meiotic spindle a protein structure that pulls the genetic material towards the two poles of a dividing germ cell, ultimately resulting in a haploid cell

memetic the transmission of ideas, whose study has been analogized to evolutionary biology; generally, the cultural spread of ideas, behavior, styles, or usage; also amusing or interesting pictures or videos that spread widely online, especially through social media

memetic transmission the passing of cultural traits from person to person, which allows cultural evolution to be understood as analogous to biological evolution, subject to the same basic mechanisms of reproduction, variation (mutation), dispersal, and selection

messenger RNA (mRNA) a single-stranded RNA transcribed from the DNA that serves as a template for protein formation by ribosomes; information in the DNA is transcribed into messenger RNA, which is read by the ribosome as it builds the corresponding polypeptide

mimetic relating to, characterized by, or displaying mimicry

monoculture in agriculture, the cultivation of a single crop in a given area

Moore's Law an axiom in the field of microprocessor development that processing power doubles every two years

morphology in biology, the overall form of an organism, organ, or other body part

mutable capable of or prone to mutation or change

N95 in the context of COVID-19, N95 respirators and surgical masks are personal protective equipment (PPE) used to protect the wearer from virus-laden airborne particles; N95 masks are intended to form a seal around the nose and mouth and block 95% of particles that are 0.3 μm or larger

Networked Society a society whose practices and structures constituting every aspect of social life are (re-) shaped by interwoven information and communication structures, with an implied approach of a more prosperous, stimulating, and benevolent society, otherwise known as the digital good life

noise the introduction of randomness or stochasticity into a signal where the color of noise is determined by the distribution of the stochastic process; noise can either obscure a signal or be used to amplify it

null model a statistical or mechanical model with the absolute minimum number of assumptions that operates as a baseline for expectations and comparison

ontology the theory of categorizing things or ideas into hierarchical systems or taxonomies

overfit an error in modeling in which a model is too closely aligned to a limited set of data, resulting in a useless model that performs well with the initial data set but poorly with any new data

pandemic a worldwide or multi-continent epidemic

Pareto optimal an economic state in which no further changes will increase an individual's economic standing without jointly decreasing that of another individual

path-dependencies in mathematics, path dependence indicates that the current state (of a system, for example) is the result of past actions leading up to that time

personal protective equipment (PPE) in the context of COVID-19, protective equipment such as suits, gloves, face masks, and face shields worn to minimize an individual's exposure to the virus

perturbation a change in the usual state of things in a system, whether due to an outside (exogenous) influence or as a result of the nature of the system itself (endogenous)

Pollyanna an unfailingly optimistic person with a tendency to find the good in everything

polymerase chain reaction (PCR) machines PCR is an automated, cycling process used to copy small segments of DNA, directed by a machine called a thermocycler, which alters the temperature of the reaction every few minutes to allow for DNA denaturing and new synthesis

posterior the updated probability of an event occurring based on new information, in other words, the probability that event A occurs given that event B occurs

prior the probability of an event occurring according to current knowledge at some point (before running an experiment to gather data to further inform said probability), also known as the "prior probability distribution"

protoplasm the colorless material inside a cell's plasma membrane, including the cytoplasm, nucleus and organelles

Q.E.D. indicating the ending of a mathematical proof, an abbreviation for the Latin "*quod erat demonstrandum*," meaning "which was to be demonstrated"

quantitative polymerase chain reaction (qPCR) a technology used to measure DNA with PCR (which turns a small amount of DNA into a larger amount of DNA), necessary for gel electrophoresis and most DNA sequencing and useful for studying gene expression

quasi-species a population structure wherein collections of closely related genomes undergo a continuous process of genetic variation, competition, and selection; the distribution of descendent genomes clustered around the average master sequence

RAG system a system that immune cells use to create antibodies that are customized to recognize millions of different viruses, bacteria and other pathogens

recombination-activating genes (RAGs) antigen-generating genes that are unique to vertebrates and critical to generating millions of custom antibodies

relative immunity incomplete immunity to a virus, resulting from a fluctuating equilibrium between the pathogen and the host's defense mechanisms

renin-angiotensin pathway a multi-organ hormonal system that regulates blood pressure and fluid and electrolyte balance

river system in the context of computers, a way of cooling server farms (a set of many interconnected servers kept in the same location) that uses water to prevent overheating and breakdown of the system

reproduction number (R_0, R sub zero) in epidemiology, a measure of infectiousness; R_0 is the average number of people to which a single infected person transmits the disease at the start of a pandemic; a value of R_0 greater than 1 means transmission will increase exponentially, while a value of R_0 less than 1 means the infection will decrease and die out

RNA ribonucleic acid, a nucleic acid polymer used as the basis for viral genomes such as the coronavirus; also a critical part of the regulation and synthesis of proteins in all of life

RNA vaccine a vaccine which encodes an antigentically active part of a virus, using a short sequence of mRNA (messenger RNA) delivered to cells in a lipid particle; in contrast to traditional vaccines that contain typically attenuated virus or parts of a virus

R(t) in epidemiology, a measure of infectiousness at time *t* during a disease outbreak; the effective reproduction number

SEIR a compartmental mathematical model used to simulate disease transmission and recovery, where individuals move from the categories of susceptible (S) to exposed (E) to infected (I) to recovered/removed (R)

server farm a set of interconnected computer servers located in the same physical facility

SIR a compartmental mathematical model used to simulate disease transmission and recovery, where individuals move from the categories of susceptible (S) to infected (I) to recovered/removed (R)

SIRD a compartmental mathematical model used to simulate disease transmission and recovery, where individuals move from the categories of susceptible (S) to infected (I) to recovered/removed (R) or dead (D)

smoothing in graphing, a smoothing line is a line fitted to data to help visualize potential relationships between two variables without fitting a specific model (for example, a regression line)

social distancing the practice of maintaining a distance of six feet or more from other individuals and avoiding direct contact with people or objects, especially in public places; social distancing helps minimize exposure of individuals to pathogens and therefore transmission of infection during outbreaks of contagious diseases such as COVID-19

social graph illustrates connections among individuals and organizations for mathematical analysis

social optimum the point on the utility-possibility frontier (used in welfare economics to illustrate the maximum amount of societal utility of individuals) that maximizes social welfare

specialist viruses infect just one or very few viral host species

spike protein a glycoprotein protruding from the surface of some viruses (such as coronaviruses) that binds to a receptor on the surface of a host cell in order to fuse the membranes of the viral and host cells, enabling the viral genome to enter the host cell

stasis in evolution, a period of stability in a lineage, during which little or no evolutionary change occurs

stay-at-home order a government order that limits individuals' trips outside the home in order to limit new infections during a pandemic

super spreader a event where a person or gathering infects an unusually high number of others (tens or hundreds of people)

symptomatic in the context of COVID-19, an individual who exhibits signs and symptoms of infection

T-cell receptors a protein complex on the surfaces of T cells that binds to antigens on abnormal cells, causing the T cells to attack and fight infection or other diseases

technocratic suggestive of a technical expert or a society managed by technical experts; typically used pejoratively

telos an ultimate end goal or aim (Greek: τέλοζ)

trophic networks in food webs, idealized levels of a food chain; the first level consists of plants, which store energy and nutrients through photosynthesis using solar energy; the next trophic level consists of organisms that consume plants; and so on

trophic niche in ecological communities, the emptying of a trophic niche through extinction provides opportunities for new organisms to evolve to exploit the empty niche

variant in virology, a new strain of a virus characterized by one or more mutations and new capabilities

zoonotic referring to an infection or disease that is transmissible between animals and humans

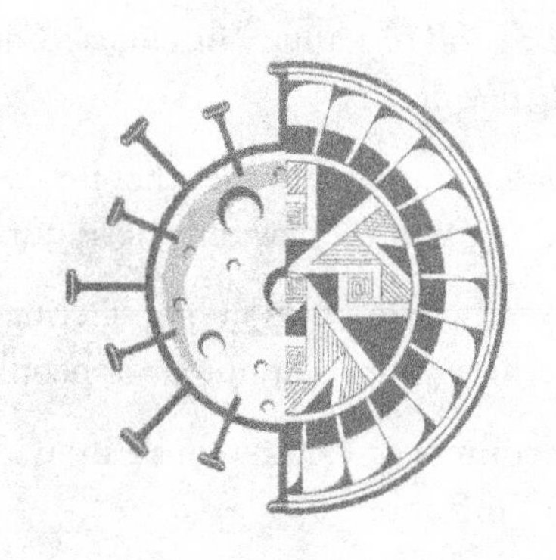

EDITORS

DAVID KRAKAUER is the president and William H. Miller Professor of Complex Systems at the Santa Fe Institute. His research explores the evolution of intelligence on Earth. This includes studying the evolution of genetic, neural, linguistic, social, and cultural mechanisms supporting memory and information processing, and exploring their shared properties. He served as the founding director of the Wisconsin Institutes for Discovery, the co-director of the Center for Complexity and Collective Computation, and professor of mathematical genetics, all at the University of Wisconsin, Madison. He has been a visiting fellow at the Genomics Frontiers Institute at the University of Pennsylvania, a Sage Fellow at the Sage Center for the Study of the Mind at the University of California, Santa Barbara, a long-term fellow of the Institute for Advanced Study, and visiting professor of evolution at Princeton University. In 2012, he was included in the *Wired Magazine* Smart List: Fifty People Who Will Change the World. In 2016, he was included in *Entrepreneur Magazine*'s list of visionary leaders advancing global research and business.

GEOFFREY WEST is the Shannan Distinguished Professor and former president of the Santa Fe Institute and associate senior fellow of University of Oxford's Green Templeton College. He has a BA from Cambridge and a PhD from Stanford. A theoretical physicist whose primary interests have been in fundamental questions ranging from the elementary particles and their cosmological implications to universal scaling laws in biology and a quantitative science of cities, companies, and global sustainability, his work is motivated by the search for simplicity underlying complexity. His research includes metabolism, growth, aging and lifespan, sleep, cancer and ecosystems, the dynamics of cities and companies, rates of growth and innovation, and the accelerating pace of life. His work has been featured in numerous publications, podcasts and TV productions worldwide. His work was selected in 2006 as a breakthrough idea by the *Harvard Business Review* and he was named to *Time* magazine's list of "100 Most Influential People in the World" in 2007. He is the author of the best-selling book *Scale*.

THE SANTA FE INSTITUTE PRESS

The SFI Press endeavors to communicate the best of complexity science and to capture a sense of the diversity, range, breadth, excitement, and ambition of research at the Santa Fe Institute. To provide a distillation of discussions, debates, and meetings across a range of influential and nascent topics.

To change the way we think.

SEMINAR SERIES
New findings emerging from the Institute's ongoing working groups and research projects, for an audience of interdisciplinary scholars and practitioners.

ARCHIVE SERIES
Fresh editions of classic texts from the complexity canon, spanning the Institute's thirty years advancing the field.

COMPASS SERIES
Provoking, exploratory volumes aiming to build complexity literacy in the humanities, industry, and the curious public.

SCHOLARS SERIES
Affordable and accessible textbooks disseminating the latest findings in the complex systems science world.

ABOUT THE SANTA FE INSTITUTE

The Santa Fe Institute is the world headquarters for complexity science, operated as an independent, nonprofit research and education center located in Santa Fe, New Mexico. Our researchers endeavor to understand and unify the underlying, shared patterns in complex physical, biological, social, cultural, technological, and even possible astrobiological worlds. Our global research network of scholars spans borders, departments, and disciplines, bringing together curious minds steeped in rigorous logical, mathematical, and computational reasoning. As we reveal the unseen mechanisms and processes that shape these evolving worlds, we seek to use this understanding to promote the well-being of humankind and of life on Earth.

COLOPHON

The body copy for this book was set in EB Garamond, a typeface designed by Georg Duffner after the Ebenolff-Berner type specimen of 1592. Mrs Eaves Italic, designed by Zuzana Licko in 1996 and styled after Baskerville (the transitional serif typeface designed in 1757 by John Baskerville) and named after Baskerville's housekeeper, Sarah Eaves, whom he later married, is used as an accent type.

Headings are in Kurier, created by Janusz M. Nowacki, based on typefaces by the Polish typographer Małgorzata Budyta. For footnotes and captions, we have used CMU Bright, a sans serif variant of Computer Modern, created by Donald Knuth for use in TeX, the typesetting program he developed in 1978.

The SFI Press complexity glyphs used throughout this book were designed by Brian Crandall Williams.

SANTA FE INSTITUTE

COMPLEXITY GLYPHS

ZERO

ONE

TWO

THREE

FOUR

FIVE

SIX

SEVEN

EIGHT

NINE

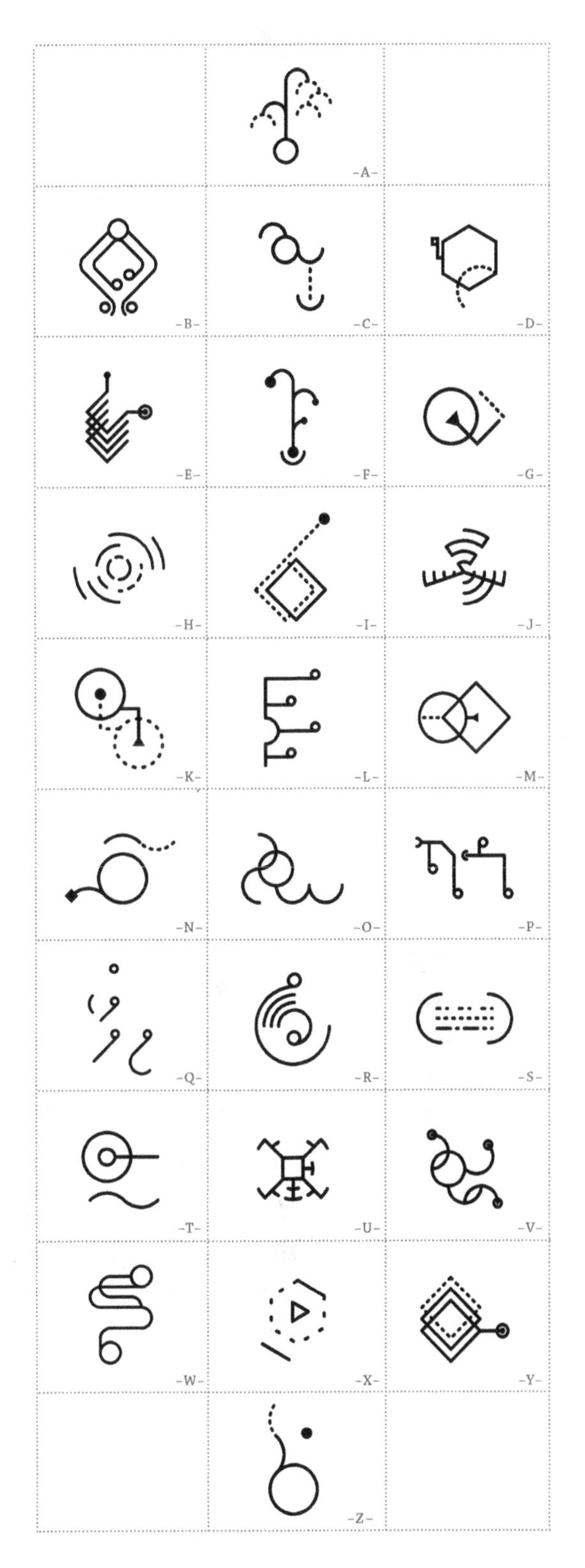

SFI PRESS
COMPASS SERIES

CPSIA information can be obtained
at www.ICGtesting.com
Printed in the USA
BVHW081246211221
624585BV00016B/1155/J

9 781947 864405